Heritage Studies

The idea to publish this scientific series emerged as a result of the transformation process of heritage from a cultural and natural asset that provides history and identity to a commodity with economic interests. Its contextual framework is provided by the UNESCO Convention Concerning the Protection of the World Cultural and Natural Heritage (1972), the UNESCO Convention for the Safeguarding of the Intangible Cultural Heritage (2003) and the UNESCO Memory of the World Programme. The research focus of the series is the wide range of applications and constructions of heritage associated with the above-named standard-setting instruments and their corresponding perceptions and paradigms. The reason for this is the fact that despite – or perhaps because of – these standard-setting instruments on the protection of heritage, there is an enormous variety in the understandings of what heritage is, could be or should be.

Different interpretations of heritage are evident in diverse structures and perceptions, from material to immaterial, from static to dynamic or even from individual to social or cultural. These interpretations were expressed in paradigms formulated in very different ways, e.g. saying that heritage has an inherent cultural value or ascribing importance for sustainable human development to heritage. Diverse perceptions of heritage are associated with conservation and use concepts as well as with their underlying disciplines, including inter- and transdisciplinary networks. Regionally and internationally, theoretically and practically, individually and institutionally, the epistemological process of understanding heritage still finds itself in its infancy. Insofar the new series Heritage Studies is overdue.

The series aims to motivate experienced and young scholars to conduct research systematically in the broad field of Heritage Studies and to make the results of research available to the national and international, theoretically- and practically-oriented, disciplinarily and interdisciplinarily established heritage community.

The series is structured according to the key UNESCO conventions and programmes for heritage into three sections focusing on: World Heritage, Intangible Cultural Heritage and Memory of the World. Although the conventions and programmes for heritage provide a framework, the series distinguishes itself through its attempt to depart from the UNESCO-related political and institutional context, which dominates the heritage discourse today, and to place the theme of heritage in a scientific context so as to give it a sound and rigorous scientific base. To this end, each of the three main sections addresses four dimensions of the heritage discourse broadly framed as Theory and Methods, Paradigms, History and Documents, and Case Studies.

Marie-Theres Albert • Roland Bernecker
Claire Cave • Anca Claudia Prodan
Matthias Ripp
Editors

50 Years World Heritage Convention: Shared Responsibility – Conflict & Reconciliation

Springer

Editors
Marie-Theres Albert
Institute Heritage Studies at the
Internationale Akademie Berlin
(INA) gGmbH
Berlin, Germany

Roland Bernecker
Chair of Cultural Management
Brandenburg University of Technology
Cottbus-Senftenberg
Cottbus, Germany

Claire Cave
School of Archaeology
University College Dublin
Dublin, Ireland

Anca Claudia Prodan
Institute Heritage Studies at the
Internationale Akademie Berlin
(INA) gGmbH
Berlin, Germany

Matthias Ripp
Organisation of World Heritage Cities
Regensburg, Germany

Federal Foreign Office, Germany

Heritage Studies
ISBN 978-3-031-05659-8 ISBN 978-3-031-05660-4 (eBook)
https://doi.org/10.1007/978-3-031-05660-4

This Springer imprint is published by the registered company Springer Nature Switzerland AG
The registered company address is: Gewerbestrasse 11, 6330 Cham, Switzerland

Disclaimer

The authors are responsible for the choice and presentation of information contained in this volume and for the opinions expressed therein, which are not necessarily those of the editors or the institutions they are affiliated with or mentioned in this volume, and they do not commit either organization.

Acknowledgements

The eighth publication in the Heritage Studies series was conceptualized in 2021 by the Institute Heritage Studies and developed through a series of think tanks and an international conference, organized around six conflict areas, Global Governance, Urban Transformation, War and Terrorism, Climate Change, Technological Change and Commodification. Thematically, these conflict areas were coordinated by a group of experts, without whose support this book would not have been possible. Therefore, we wish to extend special thanks to Roland Bernecker, Claire Cave, Nicole Franceschini, Matthias Ripp, Friedrich Schipper, Thomas Schmitt and Alexander Siegmund. The preparation of this publication was further made possible through the support of the Editorial Advisory Board of the Heritage Studies series, credited previously in the title pages, and of a group of independent peer reviewers, who have evaluated the contributions submitted for this volume. Whereas the Editorial Advisory Board helped align the publication with the paradigmatic direction and multidisciplinary approach of the Series, the independent experts ensured that it reflects thematic diversity. Their constructive feedback has greatly contributed to the quality and clarity of the contributions and we would like to express our deepest appreciation to Verena Aebischer, Renate Bornberg, Christoph Brumann, Christina Cameron, Claire Cave, Annalisa Christie, Magdalena Droste, Aitziber Egusquiza Ortega, Stephen Espiner, Karen Foley, Sieglinde Gauer-Lietz, Placido Gonzalez Martinez, Jennifer Harris, Markus Hilgert, Caroline Jaeger-Klein, Adam Kane, Ping Kong, Johanna Leissner, Tobias Matusch, Irmengard Mayer, Gerd Michelsen, Lutz Moeller, Falk Schmidt, Jutta Ströter-Bender, Michael Turner, Carsten Wergin and Solmaz Yadollahi.

The publication of a book on the occasion of the 50th anniversary of the World Heritage Convention would not be possible without Birgitta Ringbeck, who has closely accompanied the implementation of the World Heritage programme over the past 20 years at the national and international level. She is to be thanked because, despite all the national euphoria about the inscription of World Heritage Sites, she has repeatedly pointed out the general obligation to protect monuments and nature and to cooperate internationally, as well as the need for a geographically balanced World Heritage List. She has promoted substantially the implementation of management plans and heritage impact assessments as tools to preserve Outstanding Universal Value even in the face of dynamic developments, both in theory and practice. Birgitta Ringbeck is one of the few people worldwide who has always emphasized the socio-political value and peace mandate of the Convention and – as in her contribution “World Heritage and Reconciliation” – placed it in the context of conservation concepts. As editors, we thank her for continuously sharing information on current discourses and developments. We and many authors in this book have known Birgitta Ringbeck for a long time and have benefited from her competence, critical guidance and empathy.

Similarly, we want to thank all of our contributors for their efforts and unfailing dedication throughout the publication process, and Mary Cain and Shane Cullen for supporting the preparation of the book through copy editing. Last but not least, we would like to acknowledge the financial support of the Federal Foreign Office, which funded the project 50 Years World Heritage Convention: Shared Responsibility – Conflict & Reconciliation, the series of think tanks and the conference, making the open access publication of this book possible.

Federal Foreign Office

Contents

Climate Change

Technological Change

List of Figures

List of Tables

Editors and Contributors

About the Editors

Marie-Theres Albert (Germany) is a retired professor and former Chairholder of the Chair Intercultural Studies and the UNESCO Chair in Heritage Studies at the Brandenburg University of Technology Cottbus-Senftenberg, where she co-initiated and presided over the World Heritage Studies Master Program (1999–2010) and founded the PhD program Heritage Studies (2010). In 2014, she founded the Institute Heritage Studies (IHS) at the Internationale Akademie Berlin (INA) gGmbH. In the context of the IHS, she developed a new paradigmatic approach to Heritage Studies together with her former students. This approach interprets our heritage in the interest of human development. Since then, she has successfully implemented it in many international projects.

Roland Bernecker (Germany) is Visiting Professor for the Chair of Cultural Management at Brandenburg University of Technology. He was Secretary-General of the German Commission for UNESCO (2004–2019). He took part in the negotiations of the legal instruments of UNESCO for the Intangible Cultural Heritage (2003) and the Diversity of Cultural Expressions (2005) and was Member of the Panel for the European Heritage Label (2013–2017). From 2002 to 2004, he worked in the German Federal Foreign Office in Berlin. He has also been a Director of the French-German Cultural Institute in Nantes, France, and a lecturer at the University of Nantes.

Claire Cave (Ireland) is a lecturer and coordinator of the World Heritage Management programme at University College Dublin (UCD). Claire holds a PhD in Zoology from UCD with a focus on conservation biology. Her work centres on protected area management in view of globalization and sustainable development. Claire founded the innovative UCD distance learning course in World Heritage Conservation and leads the UCD World Heritage PhD programme which is engaged in diverse research in different countries. Claire is currently a member of the expert advisory group supporting the revision of Ireland's World Heritage Tentative List.

Anca Claudia Prodan (Romania/Germany), PhD, is a research consultant at the Institute Heritage Studies, Berlin, and guest lecturer at the Latvian Academy of Culture, Riga, where she helped establish a Master's programme in Cultural Heritage Governance and Communication. She holds degrees in Philosophy-Anthropology and (World) Heritage Studies, and she initiated academic research on the UNESCO Memory of the World Programme. She is a corresponding member of the Sub-Committee on Education and Research of the MoW Programme, and a member of the Editorial Advisory Board of the Heritage Studies Series, the Association of Critical Heritage Studies, and the International Association of World Heritage Professionals e.V.

Matthias Ripp (Germany), PhD, historical geographer, is coordinating the Old Town of Regensburg with Stadtamhof site. He specializes in integrated heritage management, heritage communication, governance and resilience of urban heritage, and chairs the German Association of World Heritage Cities. He is also an appointed member of the European Heritage Panel and the Regional Coordinator for Northwest Europe and North America in the Organization of Heritage Cities (OWHC). He very much enjoys teaching at the University of Regensburg and the Ostbayerische Technische Hochschule Regensburg.

Contributors

Esteban Avigliano (Argentina) holds a PhD in Biology and a postdoc degree in Biological Sciences from the University of Buenos Aires. He is a graduate teacher in Water Quality and Fisheries. He is a researcher at the National Science and Technology Council (CONICET) and is Director of the CIAR biological centre in Misiones Province, Argentina. He has published more than 60 scientific research articles and books. He is leading on field projects for the NGO Argentinean Native Forest for Biodiversity.

Juan Carlos Barrientos García (Honduras) was born in Tegucigalpa, Honduras. He specialised in International Law at the Central American Technological University, with a Master in World Heritage Studies from the Brandenburg University of Technology (BTU) in Cottbus, Germany, and did a postgraduate course on International Nature Conservation at the International Academy for Nature Conservation. He was a legal intern at IUCN Environmental Law Centre in Bonn, Germany. He has contributed to the publication "Heritage Conservation and Ideologies – a reader" (2017 chair of Monument Conservation at BTU). He worked with the International Cultural Youth Exchange Programme in Honduras, Germany and Austria. He is currently part of European Heritage Volunteers' staff as Coordinator of the Educational Volunteering Programmes in Weimar, Germany.

Kristen Barrett-Casey (United Kingdom) is an MSc Built Environment: Sustainable Heritage student at the Institute for Sustainable Heritage, University College London. She is currently conducting research for the Council of Europe's European Routes of Industrial Heritage on the future of management at industrial heritage sites across Europe, and is interested in heritage management, heritage values and public engagement generally. Her previous and ongoing research focuses on post-conflict heritage management and sustainability in the MENA region, the politics of the past, and the looting and trafficking of cultural property. Her most recent publication is "Is World Heritage Politically Sustainable?" in the forthcoming *Handbook for Sustainable Heritage*.

Lia Bassa (Hungary) graduated as a English/French teacher for linguistics and literature in 1978, and received a PhD in English literature in 1982. She is an assistant professor at Technische Universität Berlin (TUB), established a postgraduate course for technical interpretation, and has been a professional consecutive and simultaneous interpreter throughout her career. She was a senior consultant on the Hungarian National Committee of UNESCO World Heritage (2000–2004). She was an assistant professor at TUB, Department of Information and Knowledge Management (2004–2007). She is Managing Director of the Foundation for Information Society, and researcher at the UNESCO World Heritage Information Management Research Centre (2005–). She is the author of numerous heritage-related articles and is an invited lecturer on culture, heritage and communication, responsible for World Heritage MA education at the Budapest Metropolitan University (2010–).

Akeem O. Bello (Nigeria) holds a Master in Peace and Security Studies from Ekiti State University, Nigeria. He lectures in the Department of Political Science, Bamidele University of Education, Science and Technology, Ikere-Ekiti, Nigeria. He is presently a research associate with the International Center for IT and Development (ICITD) at Southern University, Baton Rouge, and a visiting research scholar with the ICITD, East Carolina University, Greenville, North Carolina. He researches on terrorism, urban violence, Security Governance and the Security of Cultural Heritages. He has published in journals and contributed to book chapters. His doctoral research is on the security of cultural heritage during violent conflicts.

Antonio Benavides Castillo (Mexico), Dr, is a graduate of the National School of Anthropology and History and has been with the National Institute of Anthropology and History, doing research, excavations and architectural restoration in various Mayan sites of the Yucatan Peninsula. He has wide experience working on Mayan cities in Quintana Roo, Yucatan and Campeche. In recent years, he has directed research and architectural conservation work at *Jaina* and *Edzna*. He supervises a maintenance programme for the conservation of a large number of heritage buildings in sites not open to the public, in the State of Campeche.

Sushma Bhatta (Nepal) is a Nepal-based lead researcher and photographer for Dignity Without Danger, a project led by Liverpool John Moore University, and a research coordinator for the MyGHTi social impact surveys. She is also a blogger focusing on the day-to-day lives of people throughout the country. Fluent in five languages, she has a background as a translator for the Carter Centre during Nepal's 2013 Constitutional Elections, and remote area social work and agricultural development.

Robin Boustead (UK) is a consultant on sustainable agriculture and tourism development. He currently works on a range of projects in Nepal and Bhutan for The Netherlands Government and the European Centre of Eco and Agro Tourism. He is also a certified sustainability auditor for Travelife and the founder of the Great Himalaya Trail. His recent publications include "Assessment of the Social Impacts of Community Based Tourism" (2021), "Reframing the Sustainable Tourism Paradox" (2020), Nepal Trekking & the Great Himalaya Trail (Trailblazer Publications, 2020, 3rd edition), a topographic map series of Nepal Himalaya (2019, 3rd edition) and Pictorial Guide to the Great Himalaya Trail (Himalayan Map House, 2019, 2nd edition).

Christina Cameron (Canada) held the Canada Research Chair in Built Heritage at the University of Montreal from 2005 to 2019 where she directed a research program on heritage conservation. She previously served as a heritage executive with Parks Canada for more than 35 years. She has worked with the World Heritage Convention since 1987, chairing the Committee in 1990 and 2008 and co-authoring *Many Voices, One Vision: The Early Years of the World Heritage Convention* (Routledge, 2013). She is a founding member of OurWorldHeritage.

Caroline Capdepon de Bigu-Poirrier (France) is a Master's student in World Heritage Studies and Cultural Heritage at Brandenburg University of Technology Cottbus-Senftenberg in Germany. She received her undergraduate degree in Art History and Archaeology at the School of the Louvre, in France. She recently participated in research for a BTU study project on civil society, global governance and the making of heritage policy. She is currently writing her Master's thesis on a critical approach to French gastronomy and the Intangible Cultural Heritage nomination of the French gastronomic meal. Her main interests include intangible heritage, communities support and cultural diversity.

Boyoon Choi (Korea) is a Master's student in World Heritage Studies at Brandenburg University of Technology Cottbus-Senftenberg. She obtained a Bachelor's degree in Tourism Management and is currently leading the Korean culture community Sopung in Germany, which was founded as part of the Hallyu Com-on project funded by KOFICE. Boyoon has research experience for a BTU study project on tourism marketing for World Heritage Sites, the World Monuments Watch programme and the making of cultural policy. She is working on her Master's thesis regarding the impact of normative instruments on the safeguarding of traditional ecological knowledge at the national level.

Cathy Daly (UK) is a senior lecturer in Conservation at the University of Lincoln, UK. Cathy is an archaeological conservator and holds an MA in World Heritage Studies from BTU Cottbus. Her research focuses on climate change and cultural heritage. She is a bureau member of the international Council of Monuments and Sites (ICOMOS) International Working Group on Climate Change and co-lead of working group 4 of the Climate Heritage Network. Cathy was amongst the expert group that advised the World Heritage Centre on revising the 2007 policy (Vilm workshop 2017) and member of the ICOMOS WG that provided comments on the 2021 policy draft. She can be reached at www.drcathydaly.com.

Philippe De Maeyer (Belgium), Dr, is emeritus senior full professor in cartography and GIS. From 2008 to 2020 he was the chair of the Department of Geography. Currently he is full member of the Royal Academy of Overseas Sciences (RAOS) and since January 2021 Permanent Secretary of the RAOS. He is Vice-President of the International Cartographic Association. He is involved in research on the history of cartography and map making and the use of GIS in historical (map) studies, among other areas, as well as the use of GIS and remote sensing in archaeology (Silk Road in Xinjiang, [China], Campeche [Mexico]).

Zeina Elcheikh (French-born German-Syrian) holds an MSc in Integrated Urbanism and Sustainable Design (IUSD) from the University of Stuttgart. She worked with the German International Cooperation (GIZ) in Syria. Her research interests include community development, heritage interpretation and museum education, fields in which she has published and worked in Germany, Syria and Egypt. Zeina is the author of the book *Die Moschee: Architektur zwischen Geistlichem und Weltlichem* (Neulingen, J.S. Klotz Verlaghaus 2020).

Yonca Erkan (Turkey) is the UNESCO Chair on the Management and Promotion of World Heritage Sites: New Media and Community Involvement at Kadır Has University, Istanbul (since 2015) and Professor of Heritage Studies at the University of Antwerp. In 2018, she worked as senior consultant at the UNESCO World Heritage Centre, as the HUL/World Heritage Cities Programme Coordinator. Currently, she is coordinating the research project Horizon 2020 Marie-Sklodowska Curie – RISE Project titled "Sustainable Management of Industrial Heritage as a Resource for Urban Development" (2021–2025). She has been a jury member of the European Union Cultural Heritage – EUROPA NOSTRA Awards since 2015.

Irene Fogarty (Ireland) is a PhD candidate at University College Dublin. Previously, she worked for The Irish Times DAC as Research Executive and chaired the International Newsmedia Association's media research expert group. Her doctoral work focuses on co-management of Canada's current and tentative World Heritage Sites by Indigenous peoples and State/provincial authorities. She hopes her work will impact heritage conservation, highlighting the urgency of Indigenous-led, equitable, rights-based approaches.

Kalliopi Fouseki (Greece/United Kingdom) is a Professor of Sustainable Heritage Management at the UCL Institute for Sustainable Heritage. She has previously worked in museums in positions focusing on participation and social inclusion. Her research interests lie in sustainable transformations of historic urban areas, with particular emphasis on understanding the complex dynamics of historic cities and the ways in which heritage can be a catalyst for sustainable development goals. Currently, she is involved in the JPI-JPHE research project, CURBATHERI-DEEP CITIES, which focuses on curating urban transformation through heritage (www.deepcities.eu).

Nicole Franceschini (Italy) is an Italian heritage practitioner with a background in archaeology and World Heritage Studies. Between 2016 and 2021, she worked at the Heritage Management Unit of the Chair of Cultural Management at BTU Cottbus-Senftenberg. She is currently working as online activities and networking coordinator for the ICCROM-IUCN World Heritage Leadership programme and is pursuing her doctoral degree.

Constanze Fuhrmann (Germany), MA, MSc, heads the Unit Environment and Cultural Heritage at the German Federal Environmental Foundation (DBU) where she promotes innovative projects on topics such as climate change and sustainability in the field of cultural heritage. Prior to this she worked in different heritage-related roles, for example at the Fraunhofer Institute for Computer Graphics Research IGD (Competence Center for Cultural Heritage Digitization) and the Fraunhofer Office in Brussels. She is trained as a conservator and has an MA in Art History, Cultural Studies and History from Technical University and Humboldt-University Berlin, as well as an MSc in Sustainable Heritage from University College London.

Claudia Grünberg (Germany) holds a Bachelor's in Cultural Studies (Europa Universität Viadrina, Frankfurt Oder) and a Master's in World Heritage Studies (BTU Cottbus-Senftenberg). In her Master's thesis, she dealt with the potential of UNESCO World Heritage for education and implemented various World Heritage projects at UNESCO Associated Project Schools. Since 2017 she has been a research associate at the Institute Heritage Studies and developed and implemented World Heritage education projects such as Our World Heritage: The mining cultural landscape Erzgebirge/Krušnohoří. Her focus is curriculum development and implementation of World Heritage in school education.

Christer Gustafsson (Sweden), PhD, is full-time Professor in Conservation at Uppsala University, Sweden. He has broad international experience, for example in connection with EU, UNESCO, OECD, EIT, as well as national, regional and local authorities. He is a member of the ICOMOS Scientific Council and Advisory Committee, as well as Secretary-General for the International Scientific Committee on the Economics of Conservation. His trans-disciplinary research focuses on boundary-spanning challenges and opportunities for cultural heritage as a driver for heritage-led social/sustainable/innovation-driven regional/urban development.

Friederike Hansell (Germany) completed the international Master's programme World Heritage Studies at BTU Cottbus. Since 2010, she has been a research assistant responsible for the transboundary World Heritage nomination Erzgebirge/Krušnohoří Mining Region at the Institute of Industrial Archaeology, History of Science and Technology at the Technical University Bergakademie Freiberg. In addition, **Friederike** has been the focal point for World Heritage in Saxony since 2015. Since 2020, she has been the World Heritage Desk Officer at the Saxon State Office for the Conservation of Monuments. She is a member of the ICOMOS Germany and the German TICCIH National Committee.

Mario Hernández (Mexico), Dr, has had a long professional career working in satellite remote sensing. At UNESCO he implemented the European Space Agency and UNESCO Open Initiative on the use of space technologies to support World Heritage Sites. His current activities are related to assisting heritage managers on the use of applied geomatics and mentoring Earth Observation projects. He is Regional Representative for Latin America at the International Society for Photogrammetry and Remote Sensing (ISPRS), and co-chair of the European Association of Remote Sensing Laboratories (EARSeL) group on EO for Natural and Cultural Heritage.

Lorika Hisari (Republic of Kosovo) is a PhD researcher at the UCL Institute for Sustainable Heritage. She has a background in architecture and is experienced in post-war reconstruction and built heritage conservation. Her research interests lie in urban heritage and area-based conservation with a focus on socio-political dynamics in historic urban environments in conflict and post-conflict contexts. Her recent publication is "Post-War Cultural Heritage Preservation in Kosovo: Rethinking the Implementation of Ahtisaari Plan Annex V" in the Special Issue of *heritage*, "Urban Heritage Management in Conflict and Post-Conflict Contexts for Inclusive, Resilient, and Sustainable Recovery".

Valasia Iakovoglou (Greece) is Director of the Ecotourism Sector of the UNESCO chair Conservation and Ecotourism in riparian and deltaic ecosystems, Con-E-Ect and has more than 25 years of international research/teaching experience as an expert in restoration/conservation of ecosystems with emphasis on biodiversity under climate change challenges. She has received numerous scholarships, awards and recognitions. She is an editor/reviewer of more than 15 international journals with more than 100 publications in books/book chapters and peer-reviewed papers. She is also a Board Member of the Association of Inter-Balkan Woman's Cooperation Societies (AIWCS) and active in many scientific societies such as the MedECC.

Mariko Ikeda (Japan), DSc, is a Japanese human geographer and an assistant professor at the faculty of Art and Design at the University of Tsukuba, Japan. Her research focuses on the cultural and social dimensions within multi-dimensional urban transformation processes and is based on long-term fixed-place observations and theoretical and empirical research on Berlin's inner-city districts, exemplifying the role of culture and/or artists in urban development. In recent years, with an awareness of the potential role of geographies, she has been leading the discussions on the intangible role of tangible in order to construct global urban heritage studies.

Zachary M. Jones (USA) is a research fellow at the Politecnico di Milano, where he completed his PhD in urban planning, design and planning. His teaching and research activities span architectural and urban design, planning, heritage studies, mega-event planning and cultural policy. His recent book *Cultural Mega-Events: Opportunities and Risks for Heritage Cities* (2020) is part of the Routledge Research in Planning and Urban Design Series.

Michael Kloos (Germany), Prof Dr, studied architecture and urban planning at Karlsruhe University, Germany. He focuses his work on the interface between preservation and sustainable development of UNESCO World Heritage Sites. Since 2003 he had been a scientific assistant at RWTH Aachen University, Germany, where he co-established the UNESCO Chair in World Cultural and Urban Landscapes in 2012 and obtained his doctoral degree in Engineering Sciences with distinction in 2014. Since 2016 Michael has held the professorship for Preservation and Sustainable Development of Cultural and Historic Urban Landscapes at RheinMain University of Applied Sciences, Wiesbaden, Germany, where he is Chairholder of the UNESCO Chair of Historic Urban Landscapes and Heritage Impact Assessments since 2022.

Maritta Koch-Weser (Germany) is anthropologist, environmentalist and social entrepreneur. She has spent her professional life in international development, including in managerial positions at the World Bank in Washington, D.C., for some 20 years, and subsequently as Director-General of IUCN – the World Conservation Union. She is Founder and President of Earth3000, a non-profit organization that supports strategic innovations in governance for environment and development, and she coordinates the Amazonia em Transformação programme at the Institute for Advanced Studies at Sao Paulo University, Brazil. In 2015 she co-founded and became President of World Heritage Watch – an international civil society network for the protection of UNESCO World Heritage Sites.

Elisabeth Korinth (Germany) is Vice-President of Blue Shield Germany and curator of the Interactive Heritage Map of Syria, a virtual community project affiliated to the Museum for Islamic Art and the Syrian Heritage Archive in Berlin, which focuses on the development of participatory approaches and digital tools for the documentation of intangible heritage under threat. Elisabeth studied Near and Middle Eastern Studies with a focus on Assyriology and Arabic Studies (BA) and holds a Master's degree in World Heritage Studies.

Jan Küver (Germany/Tanzania) is an anthropologist and heritage practitioner holding an MA in Sociology and Ethnology from the University of Göttingen and a PhD in Heritage Studies from the BTU Cottbus-Senftenberg in Germany. Since 2007, he has served as a lecturer and administrator at the University of Iringa, Tanzania. He is also the director of fahari yetu Tanzania, an NGO implementing applied culture and heritage conservation projects in the country. Jan's project work served as case study for his PhD research, which investigates the link between cultural heritage and community development and seeks to refine ethnographic approaches to heritage studies.

Claudia Lozano (Argentina), is a graduate in Sociology from the University of Buenos Aires (1989). She holds a PhD in Philosophy specializing in Latin American Studies, from the University of Berlin (1999), a postdoctoral Fellow of CONICET (2000–2003), a research and teaching fellow of the Equal Chances for Women Program, Humboldt University of Berlin (2004–2007); a lecturer at the Latin American Institute of the Free University of Berlin (2004–2015); and an independent researcher at the Social and Cultural Theory Psychoanalysis Program, German Psychoanalytic Association, Berlin Psychoanalytic Institute, Karl-Abraham-Institute (2014–2018). Claudia is also a scientific consultant for the Violence Prevention Network (since 2018). Her areas of interest include sociocultural theories, research methods applied to women's history and Andean ethnography.

Kurt Luger (Austria) is a professor of Transcultural Communication and holder of the UNESCO Chair Cultural Heritage and Tourism in the Department of Communication Studies at the University of Salzburg, Austria. He is also Chairman of the Institute for Interdisciplinary Tourism Research and of EcoHimal, Society for Cooperation Alps-Himalaya. His recent publications include *World Heritage, Place Making and Sustainable Tourism* (StudienVerlag, 2020, ed. with M. Ripp); MediaCultureTourism (2018); *Alpine Travels* (2017, ed. with F. Rest); *Tourism and Mobile Leisure* (2015, ed. with R. Egger); *Culture, Tourism and Development in the Himalayas* (2014); *Cultural Heritage and Tourism* (2014, ed. with K. Wohler).

Fergus T. Maclaren (Canada) is a Canadian destination planning and management professional with 25 years of international experience, currently focusing on World Heritage Sites and tourism-related facets of the 2030 Sustainable Development Goals. He has taught sustainable tourism at McGill University and lectured on cultural heritage management at post-secondary institutions internationally. He serves as President of the ICOMOS International Cultural Tourism Committee and as Senior Advisor to Augtraveler. He also works in an expert capacity for UNESCO, UNWTO, the Organization of World Heritage Cities, the World Monuments Fund, and the Heritage and Cultural Society of Africa.

Pankaj Manchanda (India) is the Founder of Augtraveler, an inter-disciplinary start-up working in the heritage and culture travel space. The project builds on AR and geolocation to interpret and enhance visitor experiences at UNESCO World Heritage Sites and museum spaces. He has over 24 years of proven experience in the Education Technology domain and has built global businesses and partner ecosystems across the US, Europe and Asia. He is also a member of the ICOMOS SDG working group and an expert member of the ICOMOS International Cultural Tourism Committee.

William P. Megarry (Ireland), Dr, is a senior lecturer in Archaeology at the School of Natural and Built Environment, Queen's University Belfast. His research explores the intersections between cultural heritage and climate change and he is a bureau member of the ICOMOS Working Group on Climate Change and Heritage. He was a lead author on the Future of our Pasts report. He is Principal Investigator on the CVI Africa Project and was the ICOMOS Project Lead on the Heritage on the Edge project, exploring the impacts and resilience of climate change at five World Heritage properties.

Lynn Meskell (Australia) is Penn Integrates Knowledge Professor at the University of Pennsylvania. She is Richard D. Green Professor of Anthropology at the School of Arts and Sciences, Professor of Historic Preservation at the Weitzman School of Design, and Curator in the Middle East and Asia sections at Penn Museum. She serves as AD White Professor-at-Large at Cornell University. She holds Honorary Professorships in the UK, India and South Africa. Her most recent award-winning book, *A Future in Ruins: UNESCO, World Heritage, and the Dream of Peace* (Oxford University Press, 2018), reveals UNESCO's early forays into a one-world archaeology and its later commitments to global heritage.

Webber Ndoro (South Africa) is Director-General of ICCROM. Before joining ICCROM in January 2018, he was Director of the African World Heritage Fund, based in Johannesburg, South Africa. He is also Associate Professor at the University of Cape Town. He was Project Manager at ICCROM in Rome where he worked on the Africa 2009 programme from 2000 to 2007. He worked at Great Zimbabwe as the site manager. His recent books and edited collections include *Great Zimbabwe: Your Monument our Shrine* (2000), *Cultural Heritage and the Law: Protecting Immovable Heritage in sub-Saharan Africa* (2009), *The Archaeological Heritage of Africa* (2014) *and Managing Heritage in Africa: Who Cares?* (2017).

Roger Negredo Fernández (Spain) is an emerging (World) Heritage Practitioner with a background in International Relations, Geopolitics and Diplomacy, and his two main areas of work are heritage management and cultural policy. After working at UNESCO-Beirut on the Emergency Safeguarding of the Syrian Cultural Heritage Project, he focused his research on heritage destruction by non-state armed groups.

Now at BTU, he is focusing on politicization of the World Heritage processes, mainly during the Committee listing procedures, while working for the Institute for Heritage Management on several nomination dossiers worldwide. He is actively involved in several civil society platforms, such as ESACH and ICOMOS-Spain.

Azeez Olaniyan (Nigeria) is a professor of political science at the Federal University, Oye Ekiti. He is also the director of Advancement of the university. His research interests revolve around issues related to peace and conflict, security, and governance. He was a postdoctoral fellow at the University of KwaZulu-Natal, Durban, South Africa. He is also a fellow of the Rachel Carson Center for Environment and Society, Munich Germany, African Peacebuilding Network and the African Humanities Program. He has attended several conferences and his articles have appeared in a number of reputable local and international outlets.

Birgitta Ringbeck (Germany) graduated in History of Art, Archaeology and Ethnology from Münster, Rom und Bonn. She is Head of the World Heritage coordinating body at the Federal Foreign Office of Germany. She is also the cultural expert in the German Delegation to UNESCO's World Heritage Committee, and a member of the World Heritage Committee; the Council of ICCROM; the German Commission for UNESCO; the German World Heritage Foundation, ICOMOS; and ICOM. She lectures on World Heritage and World Heritage Management and has published on architecture history, monument conservation and the UNESCO World Heritage Convention.

Dennis Rodwell (Scotland) is a consultant architect-planner and independent researcher, based in Scotland, UK, and works internationally in the field of cultural heritage and sustainable urban development, focusing on the promotion and achievement of best practice in the management of the broadly defined historic environment. Previously a principal in private architectural practice, he has also served in local government posts as architect, conservation officer, urban designer, principal planner and project manager. He writes and publishes widely on the theme of conservation and sustainability in historic cities. For further information see http://www.dennisrodwell.co.uk.

Michael Rohde (Germany), Prof Dr, has been Gardens Director of the Prussian Palaces and Gardens Foundation Berlin-Brandenburg (SPSG) since 2004 and honorary professor for garden history and garden monument preservation at the TU Berlin (ILaUP). He is a member of specialist societies/ advisory boards, including the German National Committee of ICOMOS (since 2006), and an advisory member of the ICOMOS-IFLA International Scientific Committee on Cultural Landscapes (ISCCL), and of the Dessau Wörlitz Cultural Foundation (since 2018 as founding member and Chairman). He has also held teaching and research positions and he is specialized in landscape architecture. His most recent publication is *Historic Gardens and Society* (Schnell & Steiner, 2020) (eds.).

Isabelle Rupp (Austria/USA) is a dual degree Master's student in World Heritage Studies and Cultural Heritage at Brandenburg University of Technology Cottbus-Senftenberg in Germany and Deakin University in Australia. She is originally Austrian-American but grew up abroad in various countries and holds an undergraduate degree in Anthropology and Classics. Isabelle recently completed research for a BTU study project on civil society, global governance and the making of heritage policy. She is currently working on her Master's thesis dealing with museum exhibitionary practices and decolonization. Her other research interests include heritage interpretation, the history of heritage-making, and contested heritage.

Nahuel Schenone (Argentina), Dr, is the executive project director of the NGO Argentinean Native Forest for Biodiversity. He has a PhD in Biology and a postdoctoral degree in Environmental Science from the University of Buenos Aires. His research field is social environmental issues and sustainable development, and he has published more than 20 research articles. He is the editor-in-chief of the national magazine *BIODIVERSITY*. His work includes the coordination of UNDP-GEF projects in Argentina, and he is a national and sub-national government senior consultant and water management consultant at national level.

Friedrich Schipper (Austria) is an archaeologist and deals with the cultural history of the Near East, cultural property protection, illicit trafficking in cultural property and civil–military cooperation, among other areas. He is the chairperson of the Competence Centre for Cultural Property Protection at the University of Vienna. He is serving as cultural property protection expert officer at the Austrian Peacekeeper Association as well as at the Theresian Military Academy of the Austrian Armed Forces. Schipper is CEO of the Austrian National Committee of the Blue Shield, the UNESCO-affiliated NGO network for the protection of cultural property in the event of armed conflict.

Eike Tobias Schmedt (Germany) received his Bachelor's in Cultural Anthropology and Philosophy from the University of Kiel. He joined the World Heritage Studies programme at BTU Cottbus-Senftenberg and graduated with a Master's degree in 2014. After recognizing the increasing politicization of World Heritage during his studies, he joined the Global Governance and Human Security program at the University of Massachusetts Boston to develop his research on the underlying structures of the convention. He graduated with a PhD in 2021 after creating the World Heritage Site Index as a new tool of comparative heritage assessment.

Thomas M. Schmitt (Germany) is Human geographer and has been researching, among others, on the governance of the World Heritage system, on the transformation of urban and cultural Heritage landscapes, on cultural policies, on the UNESCO concept of intangible heritage, on energy transitions, and on interreligious conflicts. Since 2020 he has been professor for Cultural Heritage at Heidelberg University/Germany.

Alexander Siegmund (Germany) studied geography, business administration and education; he received his PhD from the University of Mannheim in 1997. He held the position of academic assistant at the Chair of Physical Geography and Regional Studies, University of Mannheim and was Professor of Geography and its Didactics, Karlsruhe University of Education. Since 2004 he has been Professor of Physical Geography and its Didactics, Heidelberg University of Education. He is Honorary Professor at the University of Heidelberg and holder of the UNESCO Chair on Observation and Education of World Heritage and Biosphere Reserve. He is Managing Director of the Heidelberg Center for Education for Sustainable Development and is a member of the European Science and Technology Advisory Group (E-STAG) by the UNDRR.

Giulia Tomasi (Italy) completed an M.A. in Arts and Cultural Management at the University of Melbourne. Having worked in collection and exhibition management within international organizations, she held a position as a gallery manager in Berlin. Passionate about interdisciplinary and collaborative projects for cultural enterprises, she currently holds a position as manager within the arts and creative industry and is Secretary General of the Youth Association for UNESCO, based in Italy – the biggest volunteering Association for UNESCO worldwide, recognized by UNESCO Paris and the Italian National Commission for UNESCO as a best practice.

Sabine von Schorlemer (Germany), Prof Dr Dr, holds the Chair of International Law, Law of the EU and International Relations and the UNESCO Chair in International Relations, at Technische Universität Dresden. She is an elected member of the German Commission of UNESCO and served, inter alia, as State Minister for Higher Education, Research and the Fine Arts and as an Independent Expert for the UNESCO Director-General. She has published widely on various UN topics, including cultural heritage, terrorism and security issues.

Fabienne Wallenwein (Germany) is a postdoctoral researcher at the Heidelberg Centre for Transcultural Studies at Heidelberg University, Germany. In her work, she combines theories and methods from Chinese Studies with social sciences, focusing on the areas of economic development, urbanization, regional disparities and cultural heritage. Her book *Tackling Urban Monotony: Cultural Heritage Conservation in China's Historically and Culturally Famous Cities* traces China's urban conservation development, from the first emergence of conservation principles to the implementation of a nationwide conservation system. Currently, she is exploring the potential of cultural landscapes for social innovation as part of Heidelberg University's Flagship Initiative Transforming Cultural Heritage.

Tanja Willhalm (Germany) is a Double-Degree Master's student in World Heritage Studies and Cultural Heritage. She is completing her degree at Brandenburg University of Technology in Cottbus-Senftenberg, Germany, and Deakin University in Australia. She was born in Munich, where she also obtained her undergraduate degree in Tourism Management. Her research interests range from museology, contested heritage and art history to feminist perspectives on heritage. Recently, she engaged in a study project on civil society, global governance and the making of heritage policy. She is currently working on her Master's thesis dealing with the display of sexuality in Western art museums.

George N. Zaimes (Greece/USA) is an associate professor in the Department of Forest and Natural Environment Sciences, International Hellenic University, Greece. He is also the vice-chair of the UNESCO Con-E-Ect Chair. He received his PhD from the Department of Natural Resources Ecology and Management, Iowa State University. His research interest focuses on the sustainable management of water resources, restoration of riparian areas, soil protection conservation practices, climate change, flood and drought management, ecohydrology, protected areas, ecotourism and World Heritage Sites.

Klaus-Christian Zehbe (Germany) graduated from the World Heritage Studies Programme at Brandenburg University of Technology (BTU) Cottbus and worked subsequently as an academic assistant at the UNESCO Chair in Heritage Studies at BTU Cottbus. For his doctorate, he analysed educational transformation processes of performers in the traditional Japanese theatre form of Kyōgen in Japan. He currently works as a postdoctoral researcher at TU-Dortmund University, specializing in educational theory, effects of digitalization on human transformation processes, transformative education and Global Citizenship Education.

Luc Zwartjes (Belgium), MSc, is a researcher and pre-service teacher trainer at UGent and pedagogic advisor for geography, civic society and tourism. Since 2011 he has been working in the Department of Geography, as a teaching assistant and researcher. His research focuses on the use of geo-ICT tools and education and he is involved in projects dealing with geospatial thinking via international projects. He is a Board Member of EUROGEO and Chairman of the VLA – Flemish Geography Teachers Association.

Abbreviations and Acronyms

ALIPH	International Alliance for the Protection of Heritage in Conflict Areas
ANSBS	Association of National Committees of the Blue Shield
APN	National Parks Administration
APSA	Association for the Protection of Syrian Archaeology
AQIM	Al-Qaida in the Islamic Maghreb
AR	Augmented Reality
ARF	Atlantic Rainforest
ASEAN	Association of Southeast Asian Nations
AWHF	African World Heritage Fund
BBC	British Broadcasting Company
BELSPO	Belgian Science Policy Office
BMBF	German Federal Ministry of Education and Research
CC	Climate Change
CIAR	Centro de Investigaciones Antonia Ramos
CIVVIH	ICOMOS International Committee on Historic Cities, Towns and Villages
CLCS	Cultural Landscape Compatibility Study
CoC	Capital/ City of Culture
CoE	Council of Europe
CONAE	Argentina National Space Activities Commission
COVID-19	Coronavirus disease 2019
CPPI	Culture, Place and Policy Institute
CVI	Climate Vulnerability Index
DBU	German Federal Environmental Foundation
DGAM	Syrian Directorate General of Antiquities and Museums
DNA	Deoxyribonucleic acid
DPBR	Delta del Paraná Biosphere Reserve
DRMC	Disaster Risk Management Committee
EBC	Everest Base Camp
ECoC	European Capital of Culture
EEAS	European Union External Action Service

EIA	Environmental Impact Assessment
ESA	European Space Agency
ESD	Education for Sustainable Development
EU	European Union
FBNA	Fundacion Bosques Nativos Argentinos para la Biodiversidad
FhG	Fraunhofer-Gesellschaft
FooP	Future of our Pasts
GCEd	Global Citizenship Education
GCP	Ground control points
GDP	Gross domestic product
GERICS	Climate Service Center Germany
GHGs	Greenhouse gases
GIS	Geographic Information Systems
GLOFs	Glacial lake outburst floods
HCCH	Heidelberg Center for Cultural Heritage
HCE	Heidelberg Centre for the Environment
HEPPs	Hydroelectric Power Plants
HIA	Heritage Impact Assessment
HUL	Historic Urban Landscape
iBOL	International Barcode of Life Program
ICA	International Council of Archives
ICBS	International Committee of the Blue Shield
ICC	International Criminal Court
ICCROM	International Centre for the Study of the Preservation and Restoration of Cultural Property
ICMM	International Council on Mining and Metals
ICOMOS	International Council of Monuments and Sites
ICORP	International Scientific Committee on Risk Preparedness
ICRC	International Committee of the Red Cross
IDA	Institute for Digital Archaeology
IFLA	International Federation of Library Associations
IHL	International Humanitarian Law
IIPFWH	International Indigenous Peoples' Forum on World Heritage
IMAP	Input, Management, Analysis and Presentation
INAH	Instituto Nacional de Antropología e Historia
INEGI	Mexican Institute for Geography and Statistics
InSAR	Interferometric synthetic aperture radar
INTERPOL	International Criminal Police Organization
IPBES	Intergovernmental Science-Policy Platform on Biodiversity and Ecosystem Services
IPCC	Intergovernmental Panel on Climate Change
ISDE	International Society for Digital Earth
ISIL	Islamic State of Iraq and the Levant
ISIS	Islamic State in Iraq and Syria
ISTAT	Italian National Institute of Statistics

ISWAP	Islamic State in West Africa Province
IUCN	International Union for Conservation of Nature and Natural Resources
IWGIA	International Work Group for Indigenous Affairs
JPICH	Joint Programming Initiative on Cultural Heritage and Global Change
LAC	Latin America and the Caribbean
MAB	Man and the Biosphere Programme
MDB	Monitoring Advisory Body
MENA	Middle East and North Africa
MINUSMA	United Nations Multidimensional Integrated Stabilization Mission in Mali
MoTCA	Ministry of Culture, Tourism and Civil Aviation, Government of Nepal, Kathmandu
NbS	Nature-based solutions
NGO	Nongovernmental organization
NPA	National Protected Area
OAS	Organization of American States
OSCE	Organization for Security and Co-operation in Europe
OUV	Outstanding Universal Value
PA	Protected areas
PIK	Potsdam Institute for Climate Impact Research
PNLA	Los Alerces National Park
POI	Points of interest
RGB-D	Red-green-blue-depth
RSOUV	Retrospective Statements of Outstanding Universal Value
SC	Security Council
SDGs	Sustainability Development Goals
SEAs	Strategic Impact Assessments
SfM	Structure from motion
SNP	Sagarmatha (Mt Everest) National Park
SNPBZ	Sagarmatha (Mt Everest) National Park Buffer Zone
SoC/SOC	State of conservation
SOC	Serbian Orthodox Church
SOPHIA	Social Platform for Holistic Heritage Impact Assessment
SOUV	Statement of Outstanding Universal Value
SP	State Party
SPAB	Society for the Protection of Ancient Buildings
SPCC	Sagarmatha Pollution Control Committee
SPSG	Prussian Gardens
STEM	Science, Technology, Engineering and Maths
TUA	Temporary Use Agency
UAS	unmanned aircraft systems
UKCoC	UK City of Culture
UN	United Nations
UNDP	United Nations Development Programme

UNDRIP	United Nations Declaration on the Rights of Indigenous Peoples
UNESCO	United Nations Educational, Scientific and Cultural Organization
UNFCCC	United Nations Framework Convention on Climate Change
UNIDROIT	International Institute for the Unification of Private Law
UNODC	United Nations Office on Drugs and Crime
UN PKO	United Nations Peacekeeping Operations
UNWTO	United Nations World Tourism Organization
VR	Virtual Reality
WH	World Heritage
WHC	World Heritage Committee
WHE	World Heritage Education
WHEP	World Heritage Education Programme
WHIPCOE	World Heritage Indigenous Peoples' Council of Experts
WHITRAP	World Heritage Institute of Training and Research for the Asia and the Pacific Region
WHL	World Heritage List
WHS	World Heritage Site
WHSI	World Heritage Site Index
WHVI	World Heritage Vulnerability Index
WNF	World Nature Forum
YBR	Yabotí Biosphere Reserve

Part I
Introduction

Chapter 1
Introduction into the Overall Message of the Book: Destruction of Heritage Is Destroying Identity – Shared Responsibility Is Therefore Our Common Task for the Future

Marie-Theres Albert, Roland Bernecker, Claire Cave, Anca Claudia Prodan, and Matthias Ripp

Abstract The Introduction to the book "50 Years World Heritage Convention: Shared Responsibility – Conflict & Reconciliation" sets the stage by presenting the key message, background and content of the book. The identity-building function of heritage and its sustainable protection assume a central role. Accordingly, the chapter provides an overview of policy tools and academic debates engaging with this matter, while emphasizing the critical issues undertaken with the volume at hand. These are to reflect on whether the goals and content of the World Heritage Convention have been implemented accordingly; on the conflicts that have been affecting it over time and the need for sustainable strategies; and on perspectives for the future. This chapter further emphasises the requirements for diversity, arising from the World Heritage Convention and the variety of heritage properties, and it is reflected in the thematic, geographic and disciplinary diversity of the contributions

M.-T. Albert (✉) · A. C. Prodan
Institute Heritage Studies at the Internationale Akademie Berlin (INA) gGmbH, Berlin, Germany
e-mail: albert@inaberlin.org; prodan@inaberlin.org

R. Bernecker
Chair of Cultural Management, Brandenburg University of Technology Cottbus Senftenberg, Cottbus, Germany
e-mail: bernecker@outlook.de

C. Cave
School of Archaeology, University College Dublin, Dublin, Ireland
e-mail: claire.cave@ucd.ie

M. Ripp
Organisation of World Heritage Cities, Regensburg, Germany
e-mail: ripp.matthias@regensburg.de

M.-T. Albert et al. (eds.), *50 Years World Heritage Convention: Shared Responsibility – Conflict & Reconciliation*, Heritage Studies,
https://doi.org/10.1007/978-3-031-05660-4_1

in this book. For illustration, the chapter provides brief descriptions of conflicts affecting heritage, categorized into six areas, as well as summaries of the chapters which address them.

Keywords Heritage · Sustainability · Identity · Responsibility · Diversity · Conflict

1.1 Introduction

Heritage creates identity. This is the message of eminent scientists behind UNESCO's founding ideas including the concept of peace, such as structural ethnologist Claude Lévi-Strauss and Norbert Elias, co-founder of critical sociology. Consequently, based on this message, individuals and societies are responsible for the sustainable safeguarding of their heritage. Therefore, the sustainable protection of heritage is also the focus of significant international policy documents such as charters, declarations or conventions.

The UN General Assembly in New York had already adopted the most vital commitment in 1948 – the "*Universal Declaration of Human Rights*" (United Nations, 1948). Considering the diversity of the world regions, further political statements must be mentioned. The Organization of American States (OAS) adopted the "*Social Charter of the Americas*" on 20 September 2012 in Cochabamba, Bolivia (Organization of American States, 2012); the Association of Southeast Asian Nations (ASEAN) adopted the "*2000 Asean Declaration on Cultural Heritage*" on 25 July 2000 in Bangkok, Thailand (ASEAN, 2000); the African Union adopted the "*Charter for African Cultural Renaissance*" on 24 January 2006 in Khartoum, Sudan (African Union, 2006); the Council of Europe adopted the "*Faro Convention*" on 27 October 2005 in Faro, Portugal (Council of Europe, 2005); the General Conference of UNESCO adopted the "*Convention for the Safeguarding of the Intangible Cultural Heritage*" in 2003, which reflects on the function of intangible heritage for sustainability (UNESCO, 2003).

The sustainable protection of heritage has also been the focus of several academic publications. Examples include "*UNESCO World Heritage and the SDGs – Interdisciplinary Perspectives*" (von Schorlemer et al., 2020), "*World Heritage and Sustainable Development: New Directions in World Heritage Management*" (Larsen & Logan, 2018), "*World Heritage Conservation: The World Heritage Convention, Linking Culture and Nature for Sustainable Development*" (Cave & Negussie, 2017), "*Theory and Practice in Heritage and Sustainability: Between Past and Future*" (Auclair & Fairclough, 2015) and "*40 Years World Heritage Convention: Popularizing the Protection of Cultural and Natural Heritage*" (Albert & Ringbeck, 2015). However, it is the message of the most successful convention for the protection of the heritage of humankind, the "World Heritage Convention", which is the particular focus of this publication. It indicates that the destruction of the heritage of humankind is multidimensional and, as the preamble of the Convention says, "the deterioration or disappearance of any item of the cultural and natural heritage constitutes a harmful impoverishment of the heritage of all the nations of the world" (UNESCO, 1972). Following this view, as already said, individuals and societies are consequently responsible for the sustainable safeguarding of their heritage.

On the 16th of November, 2022, the World Heritage Convention is celebrating its 50th anniversary. The celebration of this birthday is the right time and the appropriate setting to reflect on whether the goals and content of the Convention have been implemented accordingly, and this is the critical reflection that we have undertaken with our book *50 Years World Heritage Convention: Shared Responsibility – Conflict & Reconciliation*. Among other important ideas, we reflect on the identity-building function of heritage, on the multidimensional conflicts and destruction of heritage and discuss conflict-solving strategies. These have been addressed at the various World Heritage Committee meetings over the years, and they have resulted in standard-setting instruments[1] and other policies focused on the diversity of problems according to the different types of sites and the damage they suffer in the changing world. An important political statement on these developments and needed responses were expressed in the "*Policy Document for the Integration of a Sustainable Development Perspective into the Processes of the World Heritage Convention*" (UNESCO, 2015).

Concerning the different types of conflicts and corresponding solving strategies presented in the book, we refer on the one hand to political "recommendations"[2] or "declarations"[3] or other standard-setting instruments, which have been adopted over time. On the other hand, we encouraged the authors to develop theoretical visions and practical proposals. We identified some well-known conflicts that have been considered in the above-mentioned documents, for example, the conflicts and challenges arising from the transformation of cities and their historic environment. Such conflicts have been addressed, for example, in the *"Recommendation on the Historic Urban Landscape"* (HUL), adopted by UNESCO in 2011. This document is also relevant to several questions concerning technological change. However, technological change has not yet become the core of a specific policy.

There is also a lack of a specific policy dedicated to the increasing devaluation of World Heritage through commodification. Several processes are associated with

[1] "Standard-setting instrument" is a general designation used by UNESCO to refer to different forms of policies, with different degrees of authority, which it adopts formally. These are Conventions, Recommendations, Declarations and Charters (UNESCO, n.d.-a). For details about each type and the formal procedure of adoption, please see: http://portal.unesco.org/en/ev.php-URL_ID=23772&URL_DO=DO_TOPIC&URL_SECTION=201.html

[2] Recommendations are instruments "in which the General Conference formulates principles and norms for the international regulation of any particular question and invites Member States to take whatever legislative or other steps may be required – in conformity with the constitutional practice of each state and the nature of the question under consideration – to apply the principles and norms aforesaid within their respective territories" (UNESCO, 2012, Article 1b)

[3] Declarations set forth universal principles to which the community of States wished to attribute the greatest possible authority and to afford the broadest possible support. http://portal.unesco.org/en/ev.php-URL_ID=23772&URL_DO=DO_TOPIC&URL_SECTION=201.html

this devaluation, such as those resulting from tourism development with damaging effects, for example, on infrastructure and other facilities of historic urban landscapes. The recovery from this threat is also recommended through HUL. Nevertheless, until today, commodification as a complex mechanism in the World Heritage system has not been recognized as a threat. If at all, the dangers have been considered in formal reports on the state of conservation of World Heritage properties. Interesting developments in the political treatment of damage caused by armed conflicts or natural hazards have been addressed recently in the "*Warsaw Recommendation on Recovery and Reconstruction of Cultural Heritage*", which also proposes strategies that may lead to reconciliation (UNESCO World Heritage Committee, 2017).

In addition to standard-setting instruments, we also reflect on future perspectives in terms of implementing Agenda 2030, where the responsibility of humans, in general, has been defined in 17 Sustainable Development Goals. These are more or less mirrored in all of the chapters of this book. Last but not least, we reflect on and discuss theoretical and practical concepts of responsibility, reconciliation, sustainability and education.

The protection of World Heritage requires knowledge of potential conflicts and their avoidance and appropriate implementation strategies. As explained above, such strategies hardly exist so far. Furthermore, the conflicts have changed over time, and suitable mediation and conflict-avoiding strategies have not been transmitted. Knowledge of heritage protection and how it should be accomplished must be developed and implemented in a participatory and sustainable manner. This was the aim of our project in producing this book. The project was developed through think tanks, in which potential conflicts threatening our World Heritage and possible solutions were presented and discussed. The book was developed further through a conference where we selected and highlighted the topics presented in the publication. In the book, based on these steps, we first identify the various forms of heritage destruction and analyse their causes. Only through knowledge of the reasons, backgrounds and intentions of heritage destruction can short-, medium- and long-term responsibilities be defined and sustainable protection strategies be developed and implemented. Thus, the book also explores the development of conflict avoiding and solving strategies based on integrating heritage into an overall human development strategy, namely the Agenda 2030.

Today, 1154 sites in 167 countries[4] are inscribed on the World Heritage list. Of these, nearly 50% are in Europe. The rest of the world shares the other 50%. An unbalanced distribution is also evident in the inscription of cultural and natural heritage; this imbalance is why the inscription procedure defined in the Operational Guidelines and, primarily, the international community's consciousness must be changed. Over time, some improvements – like the Global Strategy and the 5 C's, specifically the initiative to involve communities – have been implemented;

[4] Number of properties as of December 2021.

however, the conflicts have not been avoided or even resolved.[5] Looking at the sites inscribed on the World Heritage List in Danger, the distribution of inscriptions is the opposite. Of the 52 sites in danger worldwide, 21 cultural (UNESCO, n.d.-b) sites are in Arab States and 15 (4 cultural and 11 natural) in Africa. Of these cultural sites, some in Mali, Iraq or the Syrian Arab Republic have been damaged by attacks during war, while most have been damaged due to poorly managed disasters or illegal activities (UNESCO, n.d.-b).

In addition to the war-related destruction, which also affects the Outstanding Universal Value (OUV) of natural sites, African natural World Heritage sites are in particular danger due to various development projects, mining and illegal activities. On the one hand, these activities consist of poaching important and partly endangered mammals or illegal fishing. On the other hand, there is forest clearing and the mining of ores, gold, oil etc. Most of these activities require infrastructural developments that also affect the OUV. For many people in Africa, these are long overdue developments. In the context of UNESCO World Heritage, there are hardly any solutions.[6] This means that the future perspective of World Heritage must include not only the analyses of the causes of heritage in danger but also strategies to overcome this unbalanced distribution of sites inscribed on the World Heritage List and the World Heritage List in Danger.

However, our interest is not focused on the properties currently on the World Heritage List in Danger. As our perspective is the future, we need to reflect upon the overall and worldwide social, cultural, economic and ecological developments that the heritage of humanity is currently facing. Based on the concept of identifying conflicts, developing resolving strategies and perceiving the future through the integration of heritage into the Agenda 2030, we reflect upon the overall conflicts endangering heritage. However, determining responsibility, reconciliation and sustainability is as important as analysing the causes of conflicts and is discussed in detail in the section "The Day After Tomorrow".

1.2 Diversity

The adoption of the World Heritage Convention on the 16th of November 1972 was based on the internationally communicated concern to sustainably preserve, for future generations, cultural and natural heritage with an exceptional and universal

[5]The problems mentioned here are not new. Causes and solution strategies are constantly being re-discussed. For an example see *40 Years World Heritage Convention: Popularizing the Protection of Cultural and Natural Heritage* by Marie-Theres Albert and Birgitta Ringbeck.

[6]A very precise and personal view of this conflict was given in the article "The Dilemma of Zambia's Barotse Plains Cultural Landscape Nomination: Implications for Sustainable Development" by Kagosi Mwamulowe published in the book *Going Beyond: Perceptions of Sustainability in Heritage Studies No. 2* edited by Marie-Theres Albert, Francesco Bandarin and Ana Pereira Roders in 2017.

value. Its creation goes back to the 1920s in the cultural field, "continued after the Second World War with an added focus on natural resources protection" (Cameron & Rössler, 2013, 1) and was finally triggered by the destruction of the temples of Abu Simbel due to the construction of the Aswan High Dam.

As already mentioned, 1154 sites (2021) in 167 countries have been inscribed on the World Heritage list, which is not only a great success of the Convention but a testament to the world's extraordinary cultural and natural diversity and the creative power of the world's cultures. The diversity of World Heritage is therefore already given quantitatively. This also applies to the number of states in which World Heritage is designated. After all, 194 countries[7] have ratified the Convention. There is also diversity in the justifications for inscriptions formulated by the states and the interpretations of the "properties" defined in Articles 1 and 2 of the Convention (UNESCO, 1972). Diversity is related to the respective types of World Heritage properties and the justifications for their structures or chosen materials, their architectural or technical design, their immaterial contexts, etc., which accompany the inscriptions; or, in the case of natural heritage, the unique qualities of their regional and national biological or geological characteristics. The characteristics of properties are influenced by environmental, political, social, cultural or economic developments and arise from the diversity of the sites themselves; for example, from how they came to be and how they need to be preserved in a correspondingly sustainable way. Diversity is, therefore, a central category with which we must assess the implementation of the World Heritage Convention over the past 50 years.

Consequently, diversity is also required in our book *50 Years World Heritage Convention: Shared Responsibility – Conflict & Reconciliation*, in the presentation and interpretation of conflicts and in the presentation and discussion of solutions. This also applies to interpretations of the Convention's mandate, as set out in Article 8 (UNESCO, 1972).[8] For example, there are distinct differences between advisory bodies that see their mandate as the conservation of World Heritage properties, such as the International Council of Monuments and Sites (ICOMOS) and the International Union for Conservation of Nature and Natural Resources (IUCN), and communities around the world that hope World Heritage will enhance economic development processes. These, in turn, differ from the education and science target groups addressed by Article 27,[9] which prioritize the implementation of the Convention in the sense of international, transcultural and sustainable development for society as a whole, as intended, for example, with the 1994 Global Strategy (UNESCO,

---

[7] As of December 2021.

[8] This article stipulates the establishment of the World Heritage Committee, composed of 21 States Parties to the World Heritage Convention, and the participation at its meetings of representatives of the International Centre for the Study of the Preservation and Restoration of Cultural Property (ICCROM), a representative of the International Council of Monuments and Sites (ICOMOS) and a representative of the International Union for Conservation of Nature and Natural Resources (IUCN), as well as other intergovernmental or non-governmental organizations.

[9] This article stipulates the development of educational and informational programmes aimed to strengthen appreciation and inform about the threats.

1972).[10] In other words, we would like to emphasize this publication only refers to a selection from the broad spectrum of World Heritage interpretations and inscriptions. This is also the case for conflicts. At no time was our intention to name the totality of conflicts that developed for World Heritage due to the changing world over 50 years.

Knowing that conflicts and approaches to resolving them can also be overarching, we made an exemplary selection of six areas of conflict, which experts then interpreted and reflected upon based on conflict resolution or avoidance strategies. The selection was based on development processes of sustainable relevance across society, internationally and transculturally, which, accordingly, affect World Heritage. We are very aware that this is only a selection and that the discourse initiated here must be driven forward actively. We want to create an incentive to continue working on these topics through our book. We would also like to emphasize the cultural and scientific diversity – especially in the World Heritage context – intrinsic to the Convention itself. Therefore, we have not imposed uniform standards on the epistemological approaches to the interpretation or treatment of topics, which are culturally and contextually specific. The same applies to the use of scientific methods. In this respect, the book is diverse, both thematically and methodologically, precisely because of its international and scientific ambitions; we hope thereby to mobilize the broad spectrum of target groups that move in the context of World Heritage to protect their heritage and World Heritage through its sustainable use.

1.3 Conflict Areas

Heritage destruction has many facets and dimensions. Though heritage has been continuously destroyed over time through war and terrorism, climate change, technological change, modernization, commodification, international policies and urban transformation processes etc., the effects of those processes on peoples and societies have always been the same. The destruction of heritage destroys identity and impoverishes the heritage of all nations. Consequently, the book systematically identifies various forms of heritage destruction and analyses their causes within the following six conflict areas.

Global Governance

In a globalizing world, the system of the United Nations is the most important institutional arena for developing global standards that provide universal guidelines for national cultural policies, including World Heritage protection. Today, this form of international cooperation is challenged by a reinvigorated sense of national

[10] The Global Strategy for a Representative, Balanced and Credible World Heritage List was launched by the World Heritage Committee in 1994 in order to ensure that the List reflects the world's cultural and natural diversity.

unilateralism and increasingly antagonistic geopolitical powerplays. Consequently, the ability of UN organizations such as UNESCO to manage conflicts and promote reconciliation in heritage issues is likely to weaken considerably in the long term.

Urban Transformation
Urban transformation and change in urban systems have occurred increasingly since the nineteenth century, when much of the rural population in many parts of the world began to migrate to cities, resulting in significant changes to their habitats and living conditions. Today, due to modernization, urbanization is one of the phenomena that encompasses urban and rural areas in many positive and negative ways. This chapter examines some of the impacts of transformation and change within the system of urban heritage and develops strategies to help address these impacts.

War and Terrorism
Heritage destruction resulting from acts of war and terrorism has become one of the key problems of the twenty-first century. By destroying monuments and other tangible heritage, terrorists and warring countries aim to destroy cultural heritage and identity and recruit followers to their revisionist ideology. However, the destruction of heritage for purposes of destroying identity is not a new phenomenon; on the contrary, it can be seen throughout history. Therefore, this chapter reflects upon the different aspects of heritage destruction in war and with terrorist intentions.

Climate Change
The phenomenon of climate change is as overarching as the previously presented phenomena. This means that it would be presumptuous to assume that all the effects of climate change on people's tangible, intangible and natural heritage and their societies are already known fully. The sustainability of world heritage requires a dynamic approach to resilience and adaptation policies. This chapter includes reflections on the endangerment of cultural and natural heritage as a consequence of changing climate and extreme weather events and the associated disruption to society, economy and culture.

Technological Change
The urge for technological progress is an integral part of the human being and thus closely linked to the history of humankind. Many World Heritage Sites bear witness to the impressive technological achievements of different eras. On the one hand, there is a danger of abandoning structures and relics that have grown over centuries and therefore losing important parts of World Heritage. On the other hand, technological change always creates new possibilities and perspectives. Therefore, the chapter is dedicated to understanding technological change both as a threat to World Heritage and as an opportunity for its preservation.

Commodification of Heritage
Over time, the value of tangible and intangible heritage in peoples' minds has changed fundamentally. This can be seen mainly in the change from heritage being valued as a cultural good to a product, or, in other words, the commodification of cultural heritage values in contrast to the goals defined in the cultural heritage

conventions. Regrettably, neither the benefits nor the disadvantages of the commodification processes are known; they have not been reflected upon at all. Therefore, the variety of impacts of commodification processes on people and on the heritage of humankind itself have to be investigated.

1.4 The Contributions in Brief

Global Governance

Roland Bernecker and Nicole Franceschini focus on the World Heritage Convention in relation to global governance. The World Heritage Convention is increasingly exposed to criticism due to its "infection by politics". The political dynamics of the World Heritage system reflect broader transformations in global governance. As an international organization, UNESCO has not escaped the continuous weakening of multilateralism. States parties to the 1972 Convention are becoming used to treating it mainly as a proxy for power and international conflict. The global narrative of World Heritage is slowly being corrupted. The authors argue that to understand developments in the World Heritage system, we need to develop a broader perception of the transformations in international relations and make the best use of the still-emerging concept of global governance.

The lifeways of indigenous peoples in their cultural and natural landscapes are a recurring theme in the World Heritage context, exacerbated by climate change and the human rights discourse. Against this backdrop, **Irene Fogarty** critically examines theoretical interpretations of indigenous peoples' life expressions and their consideration in World Heritage documents. However, she also interprets practice. Notably, she demonstrates interpretations of "oral traditions" as "legacies" of colonial and Eurocentric history, explicitly commenting on the colonial legacy that still informs our scholarly thinking. The concluding case study on "Pimachiowin Aki" exemplifies her findings, which should be considered in future discourses on World Heritage.

In his paper, **Eike Schmedt** tackles the complex governance system of the World Heritage Convention, which includes many actors on various levels. He developed the World Heritage Site Index, a database that allows comparative assessments of properties, regardless of their typology; the index can show the governance structures that influence protection and offers perspectives on potential steps to ensure that the governance structures work in favour of conserving the sites. The World Heritage Site Index reveals the need for comprehensive legal frameworks, adequate resources and improved community involvement and stakeholder communication, and it validates the relevance of existing measures, such as the 5Cs.

The article compiled by **Roland Bernecker and Nicole Franceschini** presents several experts and young professionals' personal reflections on global governance and its influence on the World Heritage Convention over 50 years. Webber Ndoro draws on the African experience to reflect upon the distinction between local and global forms of governance. Christina Cameron speaks about opportunities for more inclusive governance and the broader participation of varied stakeholders. Maritta Koch-Weser continues to reflect on opportunities, offering five recommendations for adjusting the system. Lynn Meskell builds her perspective around 5Cs developed by her to present the challenges to the workings of World Heritage. These views are complemented by a group of master's students discussing the impact of civil society initiatives on the governance of the World Heritage system.

Urban Transformation

Urban transformation goes hand in hand with the development of human communities as a continuous process; this is the summary of the introductory paper by **Christer Gustaffson and Matthias Ripp**. In this respect, the conflicts that arise from urban transformation processes are also multidimensional. Some exemplary conflicts directly related to urban transformation and the World Heritage Convention are discussed. These include "over-tourism and gentrification" in particular, while other challenges such as climate change, war or terror are attributable to overall social developments. This chapter discusses how these challenges can be addressed in the context of urban transformation based on a "systemic approach". Although the facets of this area of conflict can only be hinted at in this introduction, they shed light on the complex problem of urban transformation.

In her contribution, **Mariko Ikeda** deals with the vacancy of buildings in urban centres and the exposure of the heritage of historic old towns to various dangers. Her proposed solution is a "temporary use" of endangered buildings to enable preservation measures without substantial interventions in the material substance. In her theoretical justifications, she refers to a publication which analysed ten European countries. It focuses on Berlin, a city whose particular history shows ambivalent effects on the urban structure and on very diversely positioned actors such as the squatter movement and its political ambitions, alternative artist groups or people in need of housing in the former eastern part of the city after the fall of the Wall. The paper draws on case studies to highlight the spectra of short-term uses.

Zachary M. Jones deals with how festivalization in World Heritage cities can contribute to the sustainable development of such cities. The author is aware of the adverse effects of such events, which are explicitly mentioned but, for the professional discussion, it is relevant that they are related to the Historic Urban Landscape Approach (HUL). The author also formulates positive developments beyond tourism. He focuses on the intensification and improvement of the images of such cities and thus refers to ideal values. These are interpreted multidimensionally via images, making new attractions possible. This development has a new meaning for World Heritage, which enriches the material values and directs the perspective towards the future.

Dennis Rodwell engages with several objectives of heritage protection, the concept of sustainability, broadly defined by him and the representation and interpretation of urban heritage in the context of the World Heritage Convention. He further discusses how heritage is transformed in current development processes, for example, through relevant societal changes resulting from the adoption of the World Heritage Convention. The justifications behind the adoption of the Convention no longer grasp the current reality, as it can and must be illustrated by some of the central categories such as authenticity or integrity. The author deals with such developments and especially with processes of commodification of heritage. The discussion on the sustainable development of urban heritage versus an "orthodoxy of heritage" provides an intriguing new reflection.

The article by **Jan Küver** deals with the city and region of Iringa in Tanzania. This contribution is fundamental for the publication. With the "shared heritage" approach, which he introduces here for the sustainable development of urban landscapes in post-colonial societies, he anticipates a long overdue discourse in the context of World Heritage. This contribution is about the place and the region and especially the people and their development in Iringa, colonized by Germany in the late nineteenth century, whose diverse historical and contemporary experience the author addresses with the "shared heritage" approach. Jan Küver's perspective on urban development is holistic and, as World Heritage discourse often has a colonial history, constitutes a forward-looking contribution to this vital period of history.

War and Terrorism

In his paper, **Friedrich Schipper** is concerned with the impacts of armed conflicts on the protection of cultural properties, including World Heritage. The author discusses the reasons behind acts of destruction and consequences for the direct communities. Often the reasons are typical of cultural cleansing, but the recent destruction of cultural property by Islamic terrorists shows a different pattern related to the interpretation and definition of Islam. Existing normative instruments, such as the UNESCO 1954 Hague Convention and its Second Protocol, have not been very successful, and the question of why these instruments fail to meet expectations remains unanswered. Nevertheless, research shows that the protection of World Heritage properties in areas of armed conflict cannot rely exclusively on expert opinion, and it requires the integration of military-compatible skills.

The destruction of heritage is the destruction of identity and the past to achieve new power structures. This introduction to our book is implicitly also the context of **Zeina Elcheikh**'s contribution, with which she explores ISIS's motivation for destroying Palmyra. Another reason for the destruction by ISIS could be that the inscription of Palmyra as a World Heritage Site does not reference its Arab and Muslim history. At the same time, the reconstruction of the World Heritage Site means reconciliation for the local people. In her article, Zeina Elcheikh presents facets of the history of Palmyra also in the context of the Convention, interpreting potential destruction interests of ISIS and formulating visions for the future.

Azeez Olaniyan and Akeem O. Bello deal with the functionalization of religions for killing people and destroying their heritage. In their interpretation, religions motivated rulers to erect monuments that could present faith and power as one. Many such monuments are inscribed today as World Heritage Sites. Starting with a brief introduction of diverse religions, they narrow their discussion to Islam. Narratively rather than theoretically, they focus on how diverse groups interpret religious texts differently to advance their own political and power interests. Their statements are supported using excerpts from the Quran and several heritage preservation measures. According to their study, religion is not only misused as a means to an end but also as a legitimization for murder and crime.

The paper of **Lorika Hisari, Kristen Barrett-Casey and Kalliopi Fouseki** addresses a topic that has hardly been reflected upon so far, namely the extent to which heritage in former war zones can be mobilized for sustainable peace. They analyse the framework necessary for this and the role of the categories of "reconciliation" and "sustainable development". Relevant to this book is a critical examination of the possibilities and limitations of UNESCO. They show that the peacebuilding activities of the local population are significantly hampered due to technocratic regulations. The authors focus on Kosovo and Iraq as representative examples of post-war regions and substantiate their critical views on UNESCO's political practice. At the same time, they formulate visions of how UNESCO could or should open up to future perspectives and how reconciliation can lead to peacebuilding.

The issues of war and terrorism and the economically motivated destruction of heritage are only partially addressed in the World Heritage Convention. If at all, attention can be drawn to these problems by putting destroyed World Heritage on the List in Danger, allowing the initiation of international assistance. In this context, **Sabine von Schorlemer** demonstrates that national and international laws, instruments and jurisdictions must be established with which such forms of destruction can be legally punished and, if not prevented, at least practically minimized. With her "integrated approach", she explains the mercilessness of the destroyers and the relative powerlessness of the international community. She opens up new perspectives by reflecting on the Nicosia Convention of the Council of Europe.

Climate Change

Claire Cave's chapter introduces the impacts of climate change on World Heritage sites, considering both direct and indirect impacts and how these are being addressed. It also presents conflicts between climate change mitigation and heritage conservation measures. Climate change has been identified as the most significant potential threat to natural World Heritage and substantially affects cultural heritage. Yet, the author suggests that we need to view heritage not simply as a passive victim of climate change but as a tool that can be used proactively to mitigate its threats. Thus, emphasis should be placed on local management interventions appropriate for the diversity of World Heritage and the role World Heritage can play in generating sustainable and climate-resilient changes in human behaviour.

William P. Megarry engages with conflicts arising from "static" provisions of the World Heritage Convention and the inevitable transformations resulting from climate change. His contribution argues that the Convention's central concepts may need to adapt to the climate crisis and its impacts on World Heritage properties and provides suggestions for how this could be achieved. In this context, the use of a proactive values-based vulnerability assessment tool is promising, as illustrated by a case study from Tanzania. However, it is necessary to consider the impacts of climate change on World Heritage properties more consistently by providing more precise and explicit guidance within the existing proactive mechanisms of the

Operational Guidelines and through an inclusive approach to values, which reflects the spirit of the Convention.

Cathy Daly focuses on potential conflicts related to the World Heritage system and the processes of Climate Action. These refer to different policy instruments, the World Heritage Convention and the 2015 Paris Agreement, respectively, whose reconciliation appears necessary. To explore the intersections, Cathy Daly focuses on issues arising from the Policy Document on the Impacts of Climate Change on World Heritage Properties adopted by the World Heritage Committee in 2007, providing an exploration of how policy approaches, including the policy reframe of 2021, changed over time. Four keywords guide the exploration, i.e., mitigation, change, loss and responsibility, and it reveals the areas where improvements are needed to integrate Climate Action with the World Heritage Convention.

Esteban Avigliano and Nahuel Schenone argue that holistic management is necessary to counteract the worldwide decline in biodiversity and climate change. To this end, they developed a research approach that is holistically oriented and empirically implemented within a case study. They focused on non-protected areas to better understand the diversity of land use in contrast to protected areas such as World Heritage sites. They argue that there is a proven need to expand bioscience research with respect to such areas. Using a multi-institutional analysis, they identified the participants required for the case study and captured the diversity of behaviours and needs. The results of this paper reveal the relevance of such a holistic research approach and the findings obtained through its use.

Michael Rohde addresses the challenges associated with climate change for those managing historic gardens and examines the relevant philosophical works, history and policy. The link between heritage, climate change and historic gardens has not been significantly addressed in international literature so far. Thus, this chapter represents an innovative subject, exploring new areas and providing a basis for further research on the subject. Its core argument is that while gardens are affected by climate change, they can also represent scientific laboratories to help understand conservation. The chapter further promotes a view of gardens as cultural monuments, and it explores how they could help change people's perception of nature, prompting a more responsible and humane way of life.

Sushma Bhatta, Robin Boustead and Kurt Luger discuss the impacts of climate change on natural heritage through the case of Sagarmatha National Park in Nepal. As tourism is one of the few sources of income in the country, it is the focus of both positive and negative attention. While it reduces poverty in the country, it also exacerbates the effects of climate change. The research focuses on discussing with local target groups how to positively engage in tourism while avoiding further exacerbating climate change impacts. They address a complex set of questions to very different target groups in the region (community leaders, business owners, individuals etc.), demonstrating that World Heritage sustainability is multidimensional and must be considered in tourism management.

Technological Change

Alexander Siegmund and Anca Claudia Prodan reflect upon the consequences of technological change on World Heritage through examples of the negative impacts, e.g., from mining or urban sprawl, and positive developments caused by digital technology. The examples indicate that the pace and scale of technological change are critical impacting factors, but there is currently no methodology for assessing the consequences for World Heritage. Insights can be drawn from the experience of the Historic Urban Landscape approach, the UNESCO Man and the Biosphere Programme and the application of heritage and environmental impact assessment methods. While these are useful in several regards, the authors conclude by emphasizing the need for a methodology tailored to the impacts of technological change on World Heritage properties against the background of the Sustainable Development Goals.

Yonca Erkan's contribution focuses on water technologies and their use, impacts and implications for people and local communities. While water presents many uses, this contribution engages with the conflicts and problems that result from overusing water as a source of energy at the expense of its use for daily activities. At the same time, the paper argues that infrastructural remnants represent an invaluable part of cultural heritage and opportunities for the traditional knowledge they embody to mitigate the impacts of climate change. Such views are supported with examples of changes in water use and related technologies, some ancient, in Anatolia and the implications for World Heritage Sites in the region. The lessons learnt in this case, which the author summarizes in conclusion, offer a basis for a more sustainable way of treating water.

Friederike Hansell describes the challenges concerning mining activities within World Heritage cultural landscapes, arguing that there is a need for adapted conservation measures and appropriate management. The chapter narrates the experience of the Erzgebirge/Krušnohoří Mining Region, a transboundary historic mining landscape shared by Saxony and the Czech Republic. The description illustrates the influence of mining on the landscape, people and their traditions and its contribution to the OUV. This contribution further discusses how these could be affected by the resumption of mining activities and introduces appropriate solutions and approaches. The literature that the author provides with the description includes a range of sources relevant to understanding the discussions and approaches to mining and World Heritage, in particular within the most relevant international organizations.

Against the backdrop of changing challenges to World Heritage and the associated inscription policies, **Michael Kloos** presents technical needs for protection and conservation in response to changing management requirements. He refers explicitly to cultural landscapes, which suffer more complex changes than other types of World Heritage properties, focusing on the case of the Upper Middle Rhine Valley with its ambivalent challenges of development and heritage protection. He identifies a set of necessary tools and explains how they can be successfully implemented through practical management with technological support. He also notes that while the iconic significance of World Heritage properties was once safeguarded for the future, it is now dependent on "efficient systematic tools" that enable the integration of OUV conservation and sustainable development.

The contribution by **Mario Hernández, Philippe De Maeyer, Luc Zwartjes and Antonio Benavides Castillo** sums up the aim of this conflict area with the statement that "modern technologies have significantly changed the way our societies behave and operate". Positive and negative developments must be considered in implementing the World Heritage Convention. The central message of this paper is related to the use of technological developments for the support of experts in the heritage context. The authors present this through a case study, primarily conducted at the Maya archaeological site of "Edzná" in Mexico. As an example, they describe how digital scientific and technological methodologies, which they call "Geoheritage", can support the management of sites and the local communities from the surroundings.

While acknowledging that tourism can have negative impacts, **George N. Zaimes, Valasia Iakovoglou, Fergus T. Maclaren and Pankaj Manchanda** present the potential of digital technology for sustainable tourism at World Heritage Sites. The authors' opinion is based on their own experience with new-age technologies for tourism in Greece and India. The technologies they present stand out through their sensitive approach to nature and people, helping develop, diversify and extend eco-tourism to less frequently visited or unknown sites; engaging the local population and promoting local traditions and crafts; and integrating heritage conservation and education. The authors have professional relationships with UNESCO and ICOMOS, and their activities have been recognized as best practices by these organizations.

Commodification of Heritage

The article of **Thomas M. Schmitt** focuses on the commodification of World Heritage, which occurs through different "markets" that include tourism and media and the inscription processes by the World Heritage Committee. He follows Marxian categories and insights from philosophical anthropology to highlight commodification areas, focusing on aspects of "exploitation", "alienation" and "fetishism". This is accompanied by examples of how these commodification processes are reflected in, for example, the use of the World Heritage label or in the relationships between residents, visitors and the World Heritage site. Thomas Schmitt uses these notions as conceptual lenses capable of revealing problem areas, and his discussion shows that possible solution strategies have to apply different levers, usually with a combination of structural changes, changed awareness and individual practices.

"World Heritage must be preserved, maintained and... made public for the transmission of cultural knowledge" are the words of **Lia Bassa** regarding the tourism concept of the Hungarian World Heritage Site "Millenary Benedictine Abbey of Pannonhalma". In her short case study, she explains how World Heritage can be financed with tourism, as the site's management succeeded by implementing the market approach of commodification while integrating sustainable development. A selection of tourists was targeted and comprehensively informed about the World Heritage site and involved in conservation processes, which was complemented by local community participation. This case represents an example of consciousness-raising and a counter-model to the market regime of the consumer society.

Claudia Lozano addresses commodification conflicts and their causes in using natural sites such as national parks. She analyses them theoretically and describes them practically before demonstrating how they might be prevented. She focuses on Los Alerces National Park in Argentina and its buffer zones, inscribed as a World Heritage Site in 2017. Management is aligned according to the site's framework, and this is where her conflict analysis comes in. It extends the common understanding of commodification as commercialization through uses of natural landscapes that no longer correspond to traditional and biological forms. This concerns the use of such areas through grain export, agro-industrial production or mineral mining. At the same time, the uses of natural landscapes contribute to social development processes. Such ambivalences are elaborated in this contribution.

Fabienne Wallenwein discusses commodification of heritage through tourism in the Chinese rice terraces "Honghe Hani Rice Terraces", inscribed as World Heritage Cultural Landscape in 2013. She discusses the ambivalent impacts of tourism on the cultural landscape and the region, on people and nature, and their evaluation by local communities and the Chinese government. The ecosystem and its use by local people with their local and indigenous knowledge were threatened long before tourism was stylized as an economic mechanism. Globalization, climate change, industrialization and urbanization were the triggers. Although tourism promotes these developments, it also contributes to economic and social development. By applying the concept of "politics of scale" in the Rice Terraces, such developments should be regulated in the context of the criteria for World Heritage protection.

The Day After Tomorrow

The theme of responsibility presented by **Marie-Theres Albert** not only runs through the World Heritage Convention but is also a constituent element of all conventions and declarations adopted by UNESCO. In this respect, the analysis of the various justifications for the assumption of responsibility by societies and individuals and their attributions based on norms, values and laws is an essential basis for the peaceful coexistence of people. This also applies to the assumption of responsibility for the protection of World Heritage. In her contribution, Marie-Theres Albert discusses three relevant theoretical approaches to the concept of responsibility in the context of World Heritage: Hans Jonas's concept of people's overall social responsibility for their heritage, Max Weber's notion of political responsibility, and Hannah Arendt's notion of personal responsibility.

Birgitta Ringbeck focuses on post-war reconstruction interpreted as a way to reconciliation in the context of World Heritage. Beyond protecting unique properties, World Heritage has the power to create a sense of a global community, in line with UNESCO's aim to achieve the intellectual and moral solidarity of mankind. While this remains a challenge, post-war recovery provides opportunities, as shown by how approaches to reconstructions have changed over time. Birgitta Ringbeck shows this with examples of policy tools and cases of destruction and reconstruction from countries such as Mali, Syria, Afghanistan and Bosnia Herzegovina. These cases show that the reconstruction of cultural properties after conflicts is not simply the restoration of physical materials, but a process by which social cohesion can be restored and cultural identity strengthened.

The World Heritage Convention is presented by **Constanze Fuhrmann** against the deteriorating living conditions of people worldwide, which she contrasts with the significance of the Convention for its identity-forming function and its sustainable development mandate. She points to real problems of implementation related to the SDGs themselves and their potential contradictory goals. A particular focus of this contribution is the management of World Heritage, which has to solve the contradiction of preserving sites based on the provisions of the Convention and in the context of the climate crisis, leading to what she calls dilemma situations. In addition, World Heritage is increasingly at the mercy of economic interests, which often conflict with sustainable development. In this respect, various conflict resolution models are required, which she presents and discusses in this paper.

Claudia Grünberg and Klaus-Christian Zehbe take a critical look at Article 27 of the World Heritage Convention, which deals with education. Based on a comprehensive study concerning its realisation, they identify significant pitfalls and possibilities of World Heritage Education (WHE). By analysing theoretical paradoxes of World Heritage sites, they identify a gap between the claims of WHE and its current implementation. To overcome these paradoxes, they suggest an approach grounded in local cultures but oriented by a global perspective of minimal morality, which includes sustainability. For the future, they recommend that WHE needs to be underpinned by local educational theories and needs to be practically expanded to tie in with current UNESCO educational concepts, such as Education for Sustainable Development (ESD).

The positions of young professionals provide reflections on the state of the art of World Heritage and visions for the future of the Convention. Such views are presented in this chapter compiled by **Roland Bernecker**, which is the result of a roundtable discussion with young professionals representing different organizations. As a result, this chapter is not designed according to the traditional framework of the other chapters. Instead, diverse positions are presented and discussed based on introductory questions. The young professionals assessed the implementation of the Convention and underlined the need for critical views on Eurocentrism, unbalanced listing and lack of funding, especially in the countries of the global south. They further emphasized a need for comprehensive educational measures for diverse target groups – a demand mentioned several times in the context of this publication.

References

African Union. (2006). *Charter for African cultural renaissance*. African Union.

Albert, M. T., & Ringbeck, B. (2015). *40 years World Heritage Convention: Popularizing the protection of cultural and natural heritage*. Springer.

Albert, M. T., Bandarin, F., & Pereira Roders, A. (Eds.). (2017). *Going beyond: Perceptions of sustainability in Heritage Studies No. 2*. Springer.

Association of Southeast Asian Nations. (2000). *2000 Asean Declaration on Cultural Heritage*. ASEAN.

Auclair, E., & Fairclough, G. (Eds.). (2015). *Theory and practice in heritage and sustainability: Between past and future*. Routledge.
Cameron, C., & Rössler, M. (2013). *Many voices, one vision: The early years of the World Heritage Convention*. Ashgate Publishing.
Cave, C., & Negussie, E. (2017). *World Heritage conservation: The World Heritage Convention: Linking culture and nature for sustainable development*. Routledge.
Council of Europe. (2005). *Convention on the value of cultural heritage for society*. Council of Europe.
Larsen, P. B., & Logan, W. (Eds.). (2018). *World Heritage and sustainable development: New directions in World Heritage management*. Routledge.
Mwamulowe, K. (2017). The Dilemma of Zambia's Barotse Plains cultural landscape nomination: Implications for sustainable development. In M.-T. Albert, F. Bandarin, & A. Pereira Roders (Eds.), *Going beyond: Perceptions of sustainability in Heritage Studies No. 2* (pp. 189–200). Springer.
Organization of American States. (2012). *Social Charter of the Americas*. OAS.
United Nations. (1948). *Universal Declaration of Human Rights*. UN General Assembly.
UNESCO. (1954). *Convention on the protection of cultural property in the event of armed conflict*. UNESCO.
UNESCO. (1972). *Convention concerning the protection of the world cultural and natural heritage*. UNESCO.
UNESCO. (2003). *Convention for the safeguarding of the intangible cultural heritage*. UNESCO.
UNESCO. (2011). *Recommendation on the historic urban landscape, including a glossary of definitions*. UNESCO.
UNESCO. (2012). *Rules of procedure concerning recommendations to Member States and international conventions covered by the terms of Article IV, paragraph 4, of the Constitution*. UNESCO.
UNESCO. (2015). *Policy document for the integration of a sustainable development perspective into the processes of the World Heritage Convention*. .
UNESCO. (n.d.-a). *General introduction to the standard-setting instruments of UNESCO*. UNESCO. http://portal.unesco.org/en/ev.php-URL_ID=23772&URL_DO=DO_TOPIC&URL_SECTION=201.html
UNESCO. (n.d.-b). *List of World Heritage in Danger*. World Heritage Centre UNESCO. http://whc.unesco.org/en/danger
UNESCO World Heritage Committee. (2017). *Warsaw recommendation on recovery and reconstruction of cultural heritage*. UNESCO.
von Schorlemer, S., Maus, S., & Schmermer, F. (Eds.). (2020). *UNESCO World Heritage and the SDGs – Interdisciplinary perspectives*. UNESCO Chair in International Relations at the University of Dresden.

Chapter 2
50 Years World Heritage Convention: Founding Idea(s) and Implementation – Reflections on Important Developments Over Time

Birgitta Ringbeck

Abstract The World Heritage Convention has the highest ratification rate in the world. Not only quantitatively, but also conceptually and politically, the World Heritage programme can be regarded as a great success. Based on the principle of equality of all cultures and societies, it combines the protection of the world's cultural and natural heritage; regardless of state borders, the preservation of these unique properties should be secured by international cooperation and assistance. This programme is not static, but rather evolves with what we continually redefine as heritage from different technical and political perspectives. Even if the members of the World Heritage Committee do not always advocate for the conservation principles of the Convention, the annual Committee meeting has become the heritage forum for the global community and has proven to be a viable platform for the safeguarding of heritage.

Keywords World Heritage · Commitment · Political attention · Preconditions · Conflicts and challenges

2.1 Introduction

UNESCO's flagship program is the World Heritage Convention. International attention and recognition, as well as solidarity and international cooperation, for the protection of monuments and nature and endangered heritage were the guiding principles leading to the ratification of the World Heritage Convention by UNESCO's General Conference in 1972. The World Heritage list became its figurehead, and the World Heritage Committee is its key forum. The development and composition of

B. Ringbeck (✉)
Federal Foreign Office, World Heritage Coordinating Body (retired November 2022), Berlin, Germany
e-mail: birgitta.ringbeck@freenet.de

M.-T. Albert et al. (eds.), *50 Years World Heritage Convention: Shared Responsibility – Conflict & Reconciliation*, Heritage Studies,
https://doi.org/10.1007/978-3-031-05660-4_2

both of them are yardsticks for the credibility of the convention based on the universal claim to represent the global community as well as to protect and conserve the natural and cultural heritage worldwide for future generations. In a time of disregard for cultural diversity and the deliberate destruction of cultural properties in armed conflicts, the vision of the unifying power of common heritage underlying the convention's idea is needed more than ever.

2.2 Committee and Political Attention

The World Heritage list as well as the number of States Parties to the Convention, which reached the universal ratification rate of 194 in 2020, have been constantly on the increase over the last five decades. While the World Heritage list is still growing and subject to constant observation, the number and the distribution of mandates on the Committee were not in focus for a long time. This changed radically in 2013; the General Assembly of the World Heritage Convention ended in a great uproar. One year after the 40th anniversary of the Convention, none of the African States Parties running for the Committee had been elected. Senegal was the only remaining Committee member representing the African continent for the following two years. By contrast, Asia/Pacific went out of this election as the strongest regional group with four newly elected States Parties and a total of seven Committee members. The African States Parties were deeply indignant about the result, and an immediate change in the electoral mode was demanded.

Without delay, a working group was set up whose proposals formed the basis for the adoption of amended rules of procedure at the extraordinary General Assembly in the following year (UNESCO, 2014). The election system was completely converted. Prior to this, only two places on the Committee were reserved, one for a State Party without a World Heritage site and one for a State Party that has never served as a member of the Committee; today, only 4 seats are not reserved. In order to achieve a balance in the geographical distribution of seats or mandates, 17 seats are assigned to the electoral groups as follows: 2 seats for group I (Western European and North American States), 2 seats for group II (Eastern European States), 2 seats for group III (Latin-American and Caribbean States), 3 seats for group IV (Asian and Pacific States), 4 seats for group V(a) (African States), 2 seats for group V(b) (Arab States). An additional seat has to be allocated for Group III and Group IV on a rotational basis. Moreover, it is recommended to consider the election of at least one State Party which has never served as a member of the World Heritage Committee.

The changes take effect only gradually. Seven years after the change of the electoral mode, group I still has the most mandates and years on the World Heritage Committee. This is also due to the fact that a mandate lasted 6 years during the first three decades of the convention and was not limited to 4 years as it is today. It was only decided in 2014 that there must be a seven-year gap between two mandates of a state party.

The service on the Committee within the electoral groups is as unbalanced as the distribution of mandates between the regional groups. In all electoral groups, there are States Parties running regularly. In group I, it is France with a total of 25 years on the Committee, the Russian Federation with 12 years in group II, Brazil with 27 years in group III, Australia also with 27 years in group IV, Senegal with 19 years in group Va and Egypt with 23 years in group Vb. Among the top twelve States Parties serving on the Committee between 20 and 27 years, 3 are from group I, none of group II, 3 from group III, 3 from group IV, none from group 5a and 3 from group 5b (UNESCO, 2021).

Membership in the Committee obviously depends crucially on political attention and implementation of the Convention, stipulating in Article 5 that each State Party shall "...adopt a general policy which aims to give the cultural and natural heritage a function in the life of the community and to integrate the protection of that heritage into comprehensive planning programmes" (UNESCO, 1972a).

2.3 Commitment and Preconditions

The *Recommendation concerning the Protection, at National Level, of the Cultural and Natural Heritage*, which was adopted by the General Assembly of UNESCO on the same day as the World Heritage Convention on 16 November 1972, provides a comprehensive framework and operational methodology for implementing the Convention as an integral part of the national heritage protection system (UNESCO, 1972b). In seven chapters, it provides definitions of cultural and natural heritage and the basics for a national policy, general principles, the organization of services, protective measures, educational and cultural action and international cooperation. It has never been evaluated whether and, if so, how this recommendation has been applied to strengthen the implementation of the World Heritage Convention as an international treaty relying on the commitment of its States Parties.

However, it can be assumed that those States Parties that shaped the way in which the Convention worked in the first two decades after ratification had such a national heritage protection system and enough human and financial resources at their disposal. This applies in particular to those European States Parties, which were very active right from the beginning. In the 1970s, the political awareness of heritage protection was promoted not only by the ratification process of the 1972 Convention but also through the launch of the European Heritage Year of the Council of Europe in 1975. The result was the amendment of many monument protection acts, the strengthening of governing structures and the development of new policies; much of this has been included in the *Recommendation concerning the Protection, at National Level, of the Cultural and Natural Heritage*.

The States Parties that are well represented on the Committee established their governing structures in the nineteenth century at a time when the legacy of the local communities in colonized nations was being robbed. How profoundly the deprivation of their cultural assets has affected the development of heritage policies and protection structures in the regions concerned, to this day, can only be imagined. It has not been investigated how colonial looting of heritage interrupted established protection and preservation mechanisms as well as traditional management and skills.

A significant light has been shed on this topic by the debate on the Luf boat from New Guinea, which came to Germany under dubious circumstances (Götz, 2021) in the colonial era and is one of the most prominent exhibits of the ethnological collections in the recently opened Humboldt Forum. According to a video message from Roselyn Stanley and Stanley Inum, descendants of one of the boat builders, a New Guinean delegation wants to come to Berlin to study the traditional shipbuilding of the almost exterminated indigenous people of Luf Island and to learn how to reconstruct such a boat from the original preserved cultural asset (Staatliche Museen zu Berlin, 2021). The Humboldt Forum very much welcomed the mission and promised to support the project; at the same time, it was very pleased that restitution claims were not being made (Staatliche Museen zu Berlin, n.d.). What is completely disregarded in the discussion is the fact that colonial looting means not only the deprivation of cultural properties but also the destruction of intangible heritage and the historical context. The traditional men's house on Luf Island, also used as a boathouse, would have had the potential to convey outstanding universal value and could have been inscribed on the World Heritage list. As an exhibit in Berlin, it has lost its integrity and essential attributes of authenticity.

The losses and breaks caused by colonization still have an impact on the awareness of and the access to heritage, as well as on the possibility to build on conservation policies that have evolved over time, on political attention and, last but not least, on active participation in the implementation of the Convention. In Africa, for instance, the weakness of governance structures is still one of the main challenges. As of October 2021, 95 of the 194 States Parties have never served on the World Heritage Committee, including 26 (55.32%) of the African States Parties. Although all African states have ratified the World Heritage Convention, the prestigious World Heritage Committee met only once in Africa, in Durban in 2005.

2.4 Concepts and Strategies

The prerequisites for the implementation of the Convention and the commitment of the States Parties could not have been more different. In particular, the European States Parties benefit from the advantages afforded by an established heritage protection system and human and financial resources, as well as long periods of peace and economic prosperity. From the outset, they made extensive use of the right to nominate sites enshrined in the Convention. This has led to an unmistakable Eurocentrism of the World Heritage list with 545 World Heritage sites in Europe and North America; that is a share of 47.23% of the 1154 World Heritage sites. Africa, as the continent with the largest number of states, has 98 World Heritage Sites, a share of only 8.49%. In addition to this geographical imbalance and the rapidly increasing number of World Heritage sites, the accumulation of sites in comparable categories is still a problem. Natural heritage continues to be completely underrepresented, with only 218 properties.

In 1994, the World Heritage Committee adopted the Global Strategy to counteract this development that was already evident at the time. If the quantitative trends and the geographical balance of the World Heritage list are applied as the only benchmark, the Global Strategy has failed. But the approach was much broader, even if the starting point was to achieve a representative, balanced and credible World Heritage list. As strategic objectives, the 5 Cs, namely credibility, conservation, capacity-building, communication and communities, stand for the provision of a comprehensive framework and operational methodology for implementing the Convention and a better reflection of the founding ideas of the World Heritage program.

Measures to strengthen the credibility (C 1) of the World Heritage list became apparent even before 1994. The conceptual approach of the World Heritage list started to shift from "best of the best" to "representative of the best" (Cameron, 2005) already in the 1980s. Instead of identifying icons, the categorization of registered and potential World Heritage sites and thematic studies came to the fore in order to achieve a universal and representative list. In addition, the introduction of "cultural landscape" as a new category emphasizes that the World Heritage Convention is a dynamic concept and reacts to the fact that the understanding of "heritage" is changing. Finally, in 2004 the analysis of the World Heritage list and Tentative Lists of ICOMOS and IUCN and follow-up action plans were presented to strengthen the credibility of the list. As a result, gaps in the World Heritage were closed, unfortunately, primarily in Europe.

As mentioned at the beginning, effective conservation (C 2) of World Heritage properties, as well as cultural and natural heritage at a national level, is one of the founding ideas. The comprehensive framework and operational methodology developed for this strategic objective include the mandatory introduction of the statement of Outstanding Universal Value and management plans (2005), as integrated planning and action concepts that set goals and measures for the protection, conservation, use and development of World Heritage sites, as well as policy papers, for example, on Climate Change (2005) and Sustainable Development (2015). Last but not least, UNESCO's General Assembly adopted *the Recommendation on the Historic Urban Landscape* (2011) to address the need to better integrate and frame urban heritage conservation strategies within the larger goals of overall sustainable development to facilitate the implementation of the Convention and supporting policies.

Due to the different prerequisites for the implementation of the Convention, especially concerning human resources and education and training opportunities, the development of effective Capacity-building (C 3) in States Parties is essential. In 2006, the African World Heritage Fund (AWHF), one of the most-targeted initiatives, was launched by the African Union member states that signed the World Heritage Convention. As the first regional coordinating and funding body under the auspices of UNESCO, it pursues the training of heritage experts and site managers to strengthen the identification of potential World Heritage sites in Africa and the preservation and conservation of listed sites, as well as effective and sustainable management. The work of the trust is emphatically accompanied and supported by ICCROM, ICOMOS and IUCN as advisory bodies of the Convention.

The AWHF demonstrates the importance of public awareness and political attention, in particular, for the involvement and support for the World Heritage Convention through Communication (C 4) on all levels. Without any doubt, its founding was a follow up of the action plans presented by ICOMOS and IUCN concerning the gaps on the World Heritage list two years earlier in 2004. Even more than the Global Strategy itself, the sober communication of the figures obviously alerted the African Union member states of the World Heritage Convention to the need for action and governing structures. However, no thought was given to adding a 6th C to the Global Strategy to clearly address the unbalanced commitment on the political level, which is essential for giving "cultural and natural heritage a function in the life of the community" according to article 5 of the Convention. That might have earlier led to the insight that the election of the members of the World Heritage Committee on the basis of the composition of the electoral groups of UNESCO is crucial for a credible World Heritage program.

In 2007, at the initiative of New Zealand, the fifth and probably last C was added to the Global Strategy to enhance the role of Communities (C 5) in the implementation of the World Heritage Convention. In particular, New Zealand wanted to emphasize the rights and role as well as involvement of indigenous people in protection, conservation and management. The urgent need for political adherence to C 5 became clear during the 44th Committee meeting in 2021: the decision-making in favour of the inscription of the Kaeng Krachan Forest Complex went surprisingly smoothly with only one veto, despite human rights violations against the Karen indigenous people brought forward by the UN High Commissioner for Human Rights and IUCN's recommendation to defer the dossier as in the previous year. Such decisions jeopardize the credibility of the World Heritage Convention far more than an unbalanced World Heritage list. This case again makes clear how important a sixth C would be, namely the non-negotiable commitment of all Committee members and States Parties to the World Heritage Convention as a vision for peace, respect and reconciliation.

2.5 Conflicts and Challenges

If only the quantitative development of the World Heritage list is used as a yardstick, then the Global Strategy has failed. This would also reduce the importance of the World Heritage programme to the World Heritage list. The *Operational Guidelines for the Implementation of the World Heritage Convention* are indeed very much focused on the nomination process, but the concept is much broader. Due to the holistic approach related to nature and culture protection, it was realized at an early stage that the impacts of climate change and the need for sustainable development are not only a matter of environmental and nature protection but also essential for the preservation of cultural heritage. The *Policy Document on the Impacts of Climate Change on World Heritage Sites* adopted by the General Assembly in 2007 (UNESCO, 2007) and the *Policy on the Integration of a Sustainable Development Perspective into the Processes of the World Heritage Convention* adopted by the General Assembly in 2015 (UNESCO, 2015) became standard-setting instruments. The initiatives and projects addressing the rise of violent extremism and the destruction of cultural heritage were similarly groundbreaking and led to the further development of monument conservation principles. These and many other initiatives within the framework of the implementation of the World Heritage Convention mean that culture is now an integral part of international politics, as confirmed by the Rome Declaration of the G20 Ministers of Culture in July 2021 (G20 Research Group, 2021), addressing all challenges which have been tackled by the World Heritage Committee since the beginning of the 2000s.

194 States Parties and 1154 World Heritage sites identify the World Heritage Convention as one of UNESCO's most successful programs. No other Convention has such a high rate of ratification and such a high level of awareness. The World Heritage program can be regarded as a great success, not only quantitatively but also conceptually and politically. Based on the principle of equality of all cultures and societies, it combines the protection of the world's cultural and natural heritage; regardless of state borders, the preservation of these unique properties should be secured by international cooperation and assistance. Even if the members of the Committee do not always advocate for the conservation principles of the Convention, the annual World Heritage Committee meeting has become the heritage forum for the global community and has proven to be a viable platform for the safeguarding of heritage. For the future viability of the World Heritage programme, the involvement of civil society is more important than ever.

References

Cameron, C. (2005). *Evolution of the Application of "Outstanding Universal Value" for Cultural and Natural Heritage*. Paper presented at the Special Expert Meeting of the World Heritage Convention: The Concept of Outstanding Universal Value; Kazan, Russian Federation, WHC-05/29.COM/INF.9B. https://whc.unesco.org/archive/2005/whc05-29com-inf09Be.pdf

G20 Research Group. (2021). *Rome declaration of the G20 Ministers of Culture*. University of Toronto. http://www.g20.utoronto.ca/2021/210730-culture.html

Götz, A. (2021). *Das Prachtboot. Wie Deutsche die Kunstschätze der Südsee raubten*. S. Fischer.

Staatliche Museen zu Berlin. (2021, September 24). *Luf-Boot: Heute und morgen*. [Video]. Youtube. https://www.youtube/4tSB1zdo_JI

Staatliche Museen zu Berlin. (n.d.). *Ethnologisches Museum*. Staatliche Museen zu Berlin Preußicher Kulturbesitz. https://www.smb.museum/museen-einrichtungen/ethnologisches-museum/sammeln-forschen/auslegerboot-von-der-insel-luf/

UNESCO. (1972a). *Convention concerning the protection of the world cultural and natural heritage*. UNESCO.

UNESCO. (1972b). *Recommendation concerning the protection, at National Level, of cultural and natural heritage*. UNESCO.

UNESCO. (2007). *Sixteenth session of the General Assembly of the States Parties to the Convention concerning the protection of the world cultural and natural heritage: Item 10 of the Provisional Agenda: Policy document on the impact of Climate Change on World Heritage properties* (WHC-07/16.GA/10). UNESCO. https://whc.unesco.org/en/sessions/16GA/documents/WHC-07/16.GA/10

UNESCO. (2014). *First extraordinary session of the General Assembly of the States Parties to the Convention concerning the protection of the world cultural and natural heritage: Item 4 of the Agenda: Closure of the session: Resolutions adopted by the General Assembly of States Parties to the World Heritage Convention at its 1st extraordinary session* (WHC-14/1 EXT. GA/4). UNESCO.

UNESCO. (2015). *Twentieth session of the General Assembly of the States Parties to the Convention concerning the protection of the world cultural and natural heritage: Item 13 of the Provisional Agenda: World Heritage and Sustainable Development. INF.13: Draft Policy Document for the integration of a sustainable development perspective into the processes of the World Heritage Convention* (WHC-15/20.GA/INF.13). UNESCO. https://whc.unesco.org/en/sessions/20GA/documents/WHC-15/20.GA/INF.13

UNESCO. (2021). *Twenty-third session of the General Assembly of the States Parties to the Convention concerning the protection of the world cultural and natural heritage: Item 5 of the Provisional Agenda: Election to the World Heritage Committee* (WHC/21/23.GA/INF.5B). UNESCO. https://whc.unesco.org/archive/2021/whc21-23ga-inf5B-en.pdf

Part II
The Destruction of Heritage Is Multidimensional – Theoretical Reflections, Case Studies and Narratives

Chapter 3
UNESCO's World Heritage Convention and Global Governance

Roland Bernecker and Nicole Franceschini

> *While power produces rationality and rationality produces power, their relationship is asymmetrical. Power has a clear tendency to dominate rationality.*
>
> *Bent Flyvbjerg (2002, 361)*

Abstract The World Heritage Convention is increasingly exposed to criticism mainly due to its "infection by politics". The transforming dynamics of the World Heritage system reflect broader transformations in global governance. As an international organization, UNESCO does not escape the continuous weakening of multilateralism. States parties to the 1972 convention are getting used to dealing with it mainly as a proxy for power and international conflict (Meskell). The global narrative of World Heritage is slowly being corrupted. The authors argue that in order to understand developments in the World Heritage system we need to acquire a broader perception of the transformations in international relations, and to make the best use of the still emerging concept of global governance.

Keywords Governance · Multilateralism · Civil society · International relations

3.1 A Successful Convention

The self-enforcing effectiveness of international law, its authority and legitimacy are established through the "cumulative actions of the system's stakeholders" (Ku, 2018, 37). There are 1154 sites inscribed on the emblematic World Heritage list (as

R. Bernecker (✉)
Chair of Cultural Management, Brandenburg University of Technology Cottbus-Senftenberg, Cottbus, Germany
e-mail: bernecker@outlook.de

N. Franceschini
Independent Heritage Practitioner, Rome, Italy
e-mail: nicole.franceschinin.work@gmail.com

M.-T. Albert et al. (eds.), *50 Years World Heritage Convention: Shared Responsibility – Conflict & Reconciliation*, Heritage Studies,
https://doi.org/10.1007/978-3-031-05660-4_3

of November 2021). The effectiveness and legitimacy of the convention and its authority as an international legal instrument are technically substantiated by the annual gathering of the World Heritage Committee during which the state of conservation of inscribed properties and newly proposed inscriptions are assessed.

The legal framework of World Heritage has not only achieved global acceptance by states, it has gained significant visibility for a broad public and the renown of a highly valued label for heritage sites on a global scale, relevant not least for a continuously growing tourism industry.

Even if counterfactual proof cannot obviously be provided, we claim that the World Heritage Convention has significantly contributed to raising global awareness about the importance of protecting cultural and natural heritage. It has created global cooperation dedicated to the worldwide establishment of policies and measures for enhanced protection of heritage sites. This is precisely one of the ambitions set out in the convention as a legal provision, often overlooked due to the usual focus on the much more conspicuous selection mechanism for the inscription of sites in the global list.

Article 5 of the convention stipulates that States Parties have the duty to establish a comprehensive policy for the protection, conservation and presentation of the cultural and natural heritage situated on their territory, including its consideration in planning, the establishment of public services with appropriate staff and means, sufficiently resourced processes for the identification of heritage, and the development of technical and scientific research, as well as capacity building and training infrastructures. It does not come as a surprise that this far-reaching provision was softened, in the course of the negotiations for the convention, with the insertion, in the introductory *châpeau* of article 5, that each State Party to the Convention shall only "endeavor, in so far as possible, and as appropriate for each country", to put into practice all of the above. Those having been involved in this kind of negotiation know that "shall" is a good start to give a provision some authority. Which is, however, immediately lost with "endeavour", and finally loses any appearance of rigor when followed by "in so far as possible, and as appropriate for each country". Government delegations, when negotiating legal obligations that will, through the possible adoption and ratification of a convention, become binding for their countries, try to push for the minimum possible level of commitment. The noble moral underpinning of such an obligation is, in the perspective of political realism, no reason to consent to a legal commitment.

3.2 A Global Narrative

It is not the technical provisions for the protection of cultural and natural heritage that have resulted in the success of the World Heritage Convention. Its popularity and visibility are fundamentally linked to a powerful narrative the convention has contributed to create and to establish. The deconstruction of nationhood as "imagined communities" (Anderson, 1983) reflects a growing sense of the legitimacy of a concept which transcends the contingencies of a "nation": the concept of humankind, a convergence of universal values in the notion of a world, creating a political awareness of its common past as a heritage of humanity, or World Heritage. At the core of the fascination with and the success of World Heritage lies precisely the idea of a global and united humankind, and of the wealth and the deeper, intrinsic connectedness of its highly diversified cultural and natural sites.

The label "UNESCO World Heritage" still has an enthralling sound. It has become an internationally valued brand, reflecting the noble ambitions of a humanity exposed to persistent political fragmentation and intensifying global pressures of environmental and social crises.

3.3 A Crisis

In the World Heritage community – all actors engaged in the implementation of the convention and involved in the "cumulative actions of the system's stakeholders" (Ku, 2018) – there is a remarkable enthusiasm for the endeavours sustained by this narrative, and for the exploratory links created by the World Heritage list across a global map of highly diversified cultural and natural places. But what is most striking in the approach of the 50th anniversary of the adoption of this legal text by UNESCO's Member States in 1972, is a widespread unease, intensifying frustrations and criticisms towards the implementation of the convention. These are all the more serious as they are voiced by those communities, institutions and professionals who are more closely related to World Heritage.

Half a century into its existence, the World Heritage programme is facing its most challenging crisis yet. A crisis not visible to everyone, not to most of the tourists travelling to World Heritage-designated sites, nor to a public willing to give credit to this wonderfully encouraging silver lining in an otherwise depressingly mundane business of international affairs. But the crisis is there, and it is strikingly evident to all those involved in the implementation of the convention. As for every crisis, the question is: does it mean that a situation is changing, or that it is confirmed?

The miscontent found its way into an article of *The Economist* (2010), which stated with regard to decisions taken at the World Heritage Committee meeting in Brazil in 2010, that "in its care for precious places, the UN cultural agency is torn between its own principles and its members' wishes; the principles are losing ground". As a reason, the article diagnosed that decisions regarding "danger listings" or concerning inscriptions into the World Heritage List "are getting infected by politics" (The Economist, 2010). In 2011, an independent evaluation by the external UNESCO auditor presented a detailed analysis of the 1994 *Global Strategy for a credible, representative and balanced World Heritage List* that came to similar conclusions. Among the many issues raised in relation to the implementation of the Global Strategy (Labadi, 2005; Gfeller, 2015; Franceschini, 2016), the 2011 evaluation highlighted that:

> Credibility was defined as enforcing a rigorous application of the criteria established by the Committee […]. But many States Parties do not adhere to these definitions, which is a source of dissatisfaction and misunderstanding. Some States Parties consider the nomination of a site to the List as a right. This claim diverges from the spirit and the letter of the 1972 Convention that provides that only properties having an outstanding universal value can be considered for inscription. (UNESCO, 2011b, 7)

In another document for the same session, the Auditor concluded that "due to the prestige of the World Heritage List, nomination becomes increasingly a geopolitical stake and not a heritage one for the benefit of all humanity and future generations" (UNESCO, 2011a, 6). Developments that contributed, in the eyes of the Independent External Auditor, "to a drift towards a more political rather than heritage approach to the Convention" (UNESCO, 2011a, 3).

This critical evaluation does not only document the confrontation of a political with a technical agenda. Meskell has pointed out that "World Heritage is being mobilized as a proxy for international conflict" (Meskell, 2018, 155). The World Heritage Convention is an excellent instance of multilateral cooperation, inspiring as well multifaceted bilateral and agency support. It's a microcosm of all sorts of intersecting cultural, economic, local and global issues. It mirrors the ongoing deep transformations that are taking place in global governance in all areas. The prestigious multilateral arena of World Heritage has become an increasingly contentious pressure point for a new vigour in the articulation – and successful assertion – of the geopolitical ambitions of this twenty-first century. This is why we think that a critical analysis of this instrument at the occasion of its 50th anniversary needs to be put in a broader context.

3.4 Decline of Multilateralism

The processes, decision-making and institutional set up directing all activities within the World Heritage framework are part of the wider architecture of multilateral cooperation. We cannot reflect on the increasingly evident flaws and deficiencies of World Heritage without considering them in the light of a broader crisis of multilateral cooperation. The foundational text of this framework is an international treaty, adopted by sovereign states after a process of unflinching negotiations, some of which were collected in the excellent research done by Christina Cameron and Mechtild Rössler (2013). During these negotiations, every single issue is put in the perspective of the specific spin of political interests which – in a highly fragmented international arena – are as divergent as they are hard-wired. Conventions are usually carefully constructed in such a way that no relevant curtailment of state sovereignty is likely to occur in the wake of a possible adoption and ratification by the relevant national authorities. But the main challenge is how states "behave" in putting into practice such an international treaty. Much depends on how processes and procedures in the dynamic of its implementation are understood and handled by the actors, in relation to the foundational spirit of the convention. To what extent can an organization like UNESCO, as "trustee" of this treaty, impact the implementation with its secretariat?

Whereas from the outside, it might seem that the UNESCO secretariat, with its stock of regional units and international officers in its headquarters in Paris, is the body in charge of handling the World Heritage convention, a closer look at the convention's text and procedures reveals that decision making is in the hands of member states, through decisions of the 21 States Parties elected to the World Heritage Committee and in a more limited way by the General Assembly of States Parties to the convention. Multilateralism is transforming with a change in how states and their governments are willing to use international organizations for their own agendas. In international relations, an anniversary of 50 years is a very long period. Even longer are the 77 years (in 2022) since the foundation of UNESCO in 1945.

In April 2021, the renowned think-tank Chatham House came to the conclusion that

> the multilateral system is outdated. A refit is long overdue. Multilateralism is facing a crisis of confidence. The system designed in the 20th century has successfully curbed great-power conflict and advanced humanitarian and development aims. But it has significant deficits in terms of legitimacy, transparency, accountability and equitable representation (Chatham House, 2021).

In one of the roundtables of the think-tank, a participant noted "that the system may be working effectively at the 'thin', interest-based level [among states], but that the deeper values of multilateralism are under threat" (Chatham House, 2021).

The multilateral architecture of international cooperation, of which UNESCO is only a small element, is massively challenged by reinvigorating strategies of national unilateralism and increasingly antagonistic struggles for global and regional hegemonies from which new geopolitical situations are emerging. These developments are accompanied by an increasing renunciation of diplomatic subtlety. We all are witnesses of a significant decrease in the ambition of governments to hide, justify or rationalize controversial political manoeuvres.

The uneasiness with a continuous weakening of the multilateral architecture is more felt by middle powers who most benefit from its functioning. Reacting to these developments, during the joint French and German Presidencies of the UN Security Council in March and April 2019, both countries presented their plan for an *Alliance for Multilateralism* in New York, on 2 April 2019, to a group of 14 countries. This informal network "aims to renew the global commitment to stabilize the rules-based international order, uphold its principles and adapt it, where necessary" (Alliance for Multilateralism). A commitment that needs to be renewed is a commitment that is failing.

UNESCO must be regarded in this broader context as a rather weak organization, a paradoxical perception against its still immense global renown, in particular owed to the prestige of the World Heritage programme. The funding UNESCO has received from its member states over decades must be described as negligible when held against the organization's aims and objectives. The situation became even more challenging with the second withdrawal of the United States from the organization in the wake of the decision taken by UNESCO's General Conference on 31st October 2011 to admit Palestine as a full member of UNESCO. As a consequence of the immediate stop of US contributions, "as from 2011, the regular budget had to be reduced by almost 25 per cent, which implied a reduction of activities by 30 per cent and more" (Hüfner, 2017).

3.5 Global Governance

In the introduction to their handbook *International Organization and Global Governance*, Weiss and Wilkinson (2018) argue that the time has come to make a decisive shift in the research for international organizations and international relations towards the broader concept of global governance. The reason they see is that this concept captures the pluralization of the international political arena of today, with its large variety of players, including civil society organisation:

> [...] global governance refers to the totality of the ways, formal and informal, the world is governed. The emergence and widespread recognition of transnational issues that circumscribe state capacity along with the proliferation of non-state actors responding to perceived shortfalls in national capabilities and a willingness to address them in the context of a perceived crisis of multilateralism combined to stimulate new thinking (Weiss & Wilkinson, 2018, 9).

The concept of global governance appeared, in its current usage, with the creation of an independent commission, "supported by the UN secretary-general and chaired by then Swedish prime minister, Ingvar Carlsson, and former Commonwealth secretary-general Shridath Ramphal" (Murphy, 2018, 33), that gave itself the name of "Global Governance Commission" when issuing its report in 1995. The commission addressed four global problems, concerning the environment, a more equitable global economic order for developing countries, an increase in productivity with commitments to sharing the benefits of growth, and finally the strengthening of the UN system for peacekeeping and humanitarian interventions.

This terminology seemed to respond to a conceptual need for the analysis of international relations. Murphy relates its immediate impact. In the same year 1995, a new journal *Global Governance: A Review of Multilateralism and International Organizations* was launched by the Academic Council on the UN system. "In 1999, the fifth year that the phrase was used, Google Scholar reports that there were over 1000 articles and books published that used it. Twelve years later, there were about ten times that number [...]. The use of 'global governance' continues to grow faster than that of either of these other terms [i.e. 'international security' and 'international political economy']" (Murphy, 2018, 33; for World Heritage cfr. Schmitt, 2011).

As stated by Weiss and Wilkinson, it was the concept of global governance "that really captured the post-Cold War Zeitgeist and that has enabled IR (international relations) scholars to begin to grapple more fully with how the world is organized in all of its complexity" (Weiss & Wilkinson, 2018, 9). This conceptual shift is largely driven by the recognition of the increasingly important role of non-state actors. It is a scientific concern for greater inclusivity.

This concern is echoed by the informal network Alliance for Multilateralism. One of the three objectives the alliance proposes for the advancement of multilateral organizations through reforms, is to make "multilateral institutions and the global political and economic order more inclusive and effective in delivering tangible results to citizens around the world" (Alliance for Multilateralism, 2021). Aiming at greater inclusivity and more effective links to the everyday life of "citizens around the world" is a key plea not only for these politicians. The Chatham House think-tank, in its reflection on global governance, arrives at the same conclusion: "Greater inclusion is an important element of re-engineering global governance for today's world. The influence of new agents of change is showing that states do not control the governance equation as they once did" (Chatham House, 2021, 2, Roundtable Perspectives).

It is important to note that this conceptual shift to a more inclusive and broader understanding of "the totality of the ways, formal and informal, the world is governed" (Weiss and Wilkinson) is connected to another conceptual shift: the perspective of strategic intervention. Surprisingly for an analytical paper of academic research, the authors see the real value of the concept of global governance in its use to provide political direction: "Yet global governance's primary utility lies not only in working out the theoretical and empirical parameters but rather in reorienting the way we ask questions about the world around us" (Weiss & Wilkinson, 2018, 11). They repeat this claim a page further with an even greater emphasis on the potential of this new thinking to mobilize citizens to get involved with how the world is governed: "Global governance thus should help us understand where we came from and why we have got to where we are, as well as *a way to develop strategies for where we should be going*" [our italics] (Weiss & Wilkinson, 2018, 12). Interestingly, in his essay on the emergence of global governance, Murphy puts the same emphasis on this point: "It may be helpful to close this opening chapter to the subject by suggesting that the most fruitful use of the term [global governance] has been contributing to our understanding of how the world works and what we might do to change that" (Murphy, 2018, 33).

The crisis of the World Heritage system, its politicization, is directly linked to a lack of inclusivity, a reduced influence of non-state actors, and to an insufficient concern for effectively "delivering tangible results to citizens around the world" (Alliance for Multilateralism, 2021). The weakness of multilateralism in its current form to deliver on these promises is increasingly reflected in UNESCO's World Heritage programme – understood as all activities initiated and carried out in the framework of the implementation of the 1972 convention.

3.6 A Global Narrative

The promise of inclusivity, however, was given by founding member states of UNESCO in the solemn adoption of its constitution. In the fifth paragraph of the *consideranda* of the constitution,

> the Governments of the States Parties to this Constitution on behalf of their peoples declare: […] That a peace based exclusively upon the political and economic arrangements of governments would not be a peace which could secure the unanimous, lasting and sincere support of the peoples of the world, and that the peace must therefore be founded, if it is not to fail, upon the intellectual and moral solidarity of mankind (UNESCO, 2020, 6).

The concept of global governance, in its ambition to embrace and mobilize citizens, to stimulate their sense of global solidarity, and to create, in this strategic perspective, a peaceful global cooperation, is part of UNESCO's DNA. It is at the heart of its foundational narrative.

In its unique and fascinating way, UNESCO's *Convention Concerning the Protection of the World Cultural and Natural Heritage* takes up the ambition to respond to a newly emerging, and increasingly urgent sense of global responsibility, through the creation of a heritage protection system based on international cooperation and shared responsibilities. In its preamble, the convention highlights that challenges of decay and destruction of heritage are a concern for "all the nations of the world" (preamble, para 3) and "all the peoples of the world" (para 6), and, concluding in a pathetic crescendo, "of mankind as a whole" (para 7). The concept of "world heritage" mirrors the awareness of global threats that call for different, globally agreed and shared approaches to governance, to be established by the "international community as a whole" (para 8), in the form of an "effective system of collective protection" (para 9).

This sense of globality had become particularly conspicuous in 1972. The Cold War had petrified a political confrontation of planetary dimensions. The stabilizing threat was that of mutual extinction. There were calculations of how many times accumulated nuclear weapons could annihilate the entire world population. A peripheral element helps to capture the atmosphere: in 1972, the construction of a huge government bunker in Ahrweiler, Germany, was completed. Today a heritage site, the bunker was built to provide shelter to some 3.000 members of government, military and administration in case of a nuclear attack. A quite worrying detail: the speech that West-Germany's President would have delivered to the people in the case of a nuclear war, was already drafted and is today on display in the museum.

In this context, the historic landing on the moon of Apollo 11 on July 20th 1969, one of the first globally televised events in history, meant to win the Cold War space race, projected the presence of humanity beyond the colonized and increasingly threatened sphere of our planet. This move into space can be seen as the practical completion of our post-modern concept of globality. Only a few weeks after the adoption of the World Heritage Convention, the inspirational iconography of the Blue Marble, "one of the most iconic images – not just of our time, but of all time" (Petsko, 2011) was produced by Apollo 17 in December 1972. The imagery created by the Apollo missions contributed to a concrete perception of the globality as well as of the fragility of human endeavours in the cosmic dimension. The picture sent in February 1990 by Voyager 1 from over 6 billion kilometres distance, in which earth is a mere "pale blue dot" with less than a pixel of the photograph, should contribute to further calming our planetary antagonisms and to promoting the reasonable logic of a "intellectual and moral solidarity of mankind", idealized in UNESCO's constitution. The exploration of space has led to the paradoxical insight that the emergence of globality in the political sphere is less an effect of accumulating space for human agency, than a result of the inevitable perception of the narrowing limits of growth.

This was the main message when, again in 1972 – a pivotal year in a transformational twentieth century – the Club of Rome published the *Limits to Growth*, a report that challenged the concept of continuous economic growth and found an astonishing global resonance. When revisiting the report in 2000, Matthew Simmons commented: "Its conclusions were stunning. It was ultimately published in 30 languages and sold over 30 million copies. According to a sophisticated MIT computer model, the world would ultimately run out of many key resources. These limits would become the 'ultimate' predicament to mankind" (Simmons, 2000).

3.7 New Agendas

In the 1970s, the safeguarding of heritage was rising higher on the international agenda, encouraged by campaigns like the rescue and safeguarding of monumental Nubian monuments such as Abu Simbel. In the long term, however, environmental concerns became a more pressing force in driving political agendas. In these years in which mankind learned to imagine itself inhabiting an increasingly limited planetary space – a heritage and resource in its own right – the first UN-Conference on the Human Environment was held in Stockholm in June 1972. At that early stage of the rising environmental agenda, UNESCO halted plans of IUCN to propose a convention on the safeguarding of natural sites for adoption in the Stockholm conference. UNESCO was, in parallel, preparing a convention on the protection of cultural sites for adoption in November 1972 in Paris. When UNESCO officials became aware of the plans of IUCN, they feared that a separate international legal instrument for natural sites, coordinated by IUCN, would lead to a significant reduction in UNESCO's own mandate for nature conservation (Batisse & Bolla, 2005). This would have been a setback for UNESCO, which only two years prior had launched its intergovernmental programme for the establishment of Biosphere-reserves.

As conceptually fertile as the subsequent wrapping of the cultural and natural heritage in one legal instrument may have been for the future, in the decades after the adoption of the World Heritage Convention culture and cultural heritage increasingly lost connection with the continuously rising agenda of sustainability and sustainable development. Attempts to establish a politically relevant cultural dimension in the sustainability agenda were not successful in the Millennium Development

Goals in 2000, and only partially for the Sustainable Development Goals in 2015. A new attempt was made to create a more substantial linkage with the *Policy on the integration of a sustainable development perspective into the processes of the World Heritage Convention,* adopted by the 20th General Assembly of the States Parties in November 2015 (UNESCO, 2015).

The World Heritage programme, notwithstanding its mandate for both the protection and conservation of cultural and natural heritage has, as yet, not developed concrete statutory or substantive programmatic links to the UN Sustainable Development Agenda. After the introduction of the category of cultural landscapes, historic urban landscapes, and the inclusion of provisions for the involvement of local communities in heritage identification and management, there is still the need to further modernize the convention's perspectives on cultural and natural heritage, also by continuing the exploration of the conceptual potential of their convergences. This is today being advanced through the work of the Advisory Bodies and structured capacity building efforts established within the World Heritage system: from the Capacity Building Strategy launched in 2011, to the ICOMOS-IUCN Connecting Practice project, to the joint ICCROM-IUCN World Heritage Leadership programme looking at improving conservation and management practices for culture and nature through people- centred and place-based approaches.

An entire dimension of that heritage has been cut out of its remit, with the creation of a new convention for intangible heritage. The 2003 Convention emerged as a result of the surge of interest in protecting cultural diversity and traditional knowledge in the face of globalizing forces as well as the inability of the 1972 Convention to adequately address these issues. Admittedly, the personal agenda of the then Director-General of UNESCO was a crucial element in this dynamic. The creation of the 2003 Convention can however be read as a result of the perceived shortcomings of the World Heritage Convention, including its bias towards European heritage and a failure to reflect properly the intangible values associated with natural and cultural sites in many regions of the world.

What the astonishing prestige of the World Heritage programme could have achieved in terms of conceptual innovation, international cooperation, mobilization of expert networks, involvement of communities and transformation of narratives, has been significantly limited and overshadowed by the heavy-handedness of States Parties, increasingly determined to cash in the prestige given by the World Heritage brand in the currency of unilateral political profit and economic development. This is why States Parties prefer to fully exert what they claim as their right: to outweigh a global ambition of humanity, set out in UNESCO's constitution and given concrete shape in the World Heritage Convention, with plain political power.

The formula coined by the think-tank Chatham House fails to capture this reality:

> Greater inclusion is an important element of re-engineering global governance for today's world. The influence of new agents of change is showing that states do not control the governance equation as they once did.

This may be true for rating agencies and global software giants which have accumulated sufficient economic power to escape, at least partially (and probably only temporarily), the control of national governments. In the World Heritage programme, the influence of new agents of change does not seem to be a reality. This is precisely the reason why the crisis of the World Heritage programme is worsening, why its deeper narrative is being corrupted.

The challenge we are facing has been captured poignantly by the philosopher Peter Sloterdijk in his formula of the synchronized time of the world:

> The many cultures must understand that they look back at primarily distinct pasts and forward to primarily shared futures. [...] Local narratives are increasingly compelled to coordinate the *idiochronic* horizons of their constructs of history with the virtual synchronic horizon of a common world time (Sloterdijk, 2018).

We come from different histories, but we are heading towards the same future. This perspective is intended to reverse the antagonistic fragmentation of humanity. It seems that unleashed power politics have little interest in sustaining narratives of inclusivity and human solidarity, which are inscribed in UNESCO's constitution and are part of its institutional DNA. We might need the emerging concept of global governance not only, as Murphy (2018, 33) has stated, to contribute "to our understanding of how the world works", but also to get a better grip of "what we might do to change that".

References

Alliance for Multilateralism. https://multilateralism.org/the-alliance/. Accessed 30 May 2021.

Anderson, B. (1983). *Imagined communities. Reflections on the origin and spread of nationalism*. Verso.

Batisse, M., & Bolla, G. (2005). *The Invention of "World Heritage"* (History Papers, Paper 2). Associations des Anciens Fonctionnaires de l'UNESCO.

Cameron, C., & Rössler, M. (2013). *Many voices, one vision: The early years of the World Heritage Convention*. Routledge.

Chatham House. (2021, April 15). Reflections on building more inclusive global governance. Ten insights into emerging practice. *Other resource*. https://www.chathamhouse.org/2021/04/reflections-building-more-inclusive-global-governance/02-roundtable-perspectives. Accessed 30 May 2021.

Flyvbjerg, B. (2002). Bringing power to planning research: One researcher's praxis story. *Journal of Planning Education and Research, 21*, 353–366.

Franceschini, N. (2016). *A representative, balanced and credible World Heritage List? A critical analysis of the 1994 UNESCO Global Strategy*. Master's thesis. BTU Cottbus-Senftenberg.

Gfeller, A. E. (2015). Anthropologizing and indigenizing heritage: The origins of the UNESCO Global Strategy for a Representative, Balanced and Credible World Heritage List. *Journal of Social Archaeology*, 1–21.

Hüfner, K. (2017). *The financial crisis of UNESCO after 2011: Political reactions and organizational consequences*. (First published: 24 August 2017). https://onlinelibrary.wiley.com/doi/full/10.1111/1758-5899.12459

Ku, C. (2018). The evolution of international law. In T. G. Weiss & R. Wilkinson (Eds.), *International organization and global governance* (2nd ed., pp. 37–50). Routledge.

Labadi, S. (2005). A review of the global strategy for a balanced, representative and credible World Heritage List 1994–2004. *Conservation and Management of Archaeological Sites, 7*(2), 89–102.

Meskell, L. (2018). *A future in ruins*. Oxford University Press. Kindle Edition.
Murphy, C. N. (2018). The emergence of global governance. In T. G. Weiss & R. Wilkinson (Eds.), *International Organization and Global Governance* (2nd ed., pp. 25–36). Routledge.
Petsko, G. A. (2011). The Blue Marble. *Genome Biology, 12*(4), 112. Published online 2011 Apr 28. https://doi.org/10.1186/gb-2011-12-4-112
Schmitt, T. M. (2011). *Cultural governance. Zur Kulturgeographie des UNESCO-Welterberegimes*. Franz Steiner Verlag.
Simmons, M. R. (2000, October). *Revisiting the Limits to Growth: Could the Club of Rome have been correct, after all? An energy White Paper*. https://www.mudcitypress.com/PDF/clubofrome.pdf. Accessed 27 May 2021.
Sloterdijk, P. (2018). *Kulturen leiden am Gegeneinander von Tradition und Zukunft* (Interview for the Global Thought Leader Index, March 7, 2018). https://www.gdi.ch/de/publikationen/trend-updates/peter-sloterdijk-kulturen-leiden-am-gegeneinander-von-tradition-und
The Economist. (2010, August 26). *UNESCO's World Heritage sites. A danger list in danger*. International Edition.
UNESCO. (1972). *Convention concerning the protection of the world cultural and natural heritage*. UNESCO.
UNESCO. (2003). *Convention for the safeguarding of the intangible cultural heritage*. UNESCO.
UNESCO. (2011a). *Independent evaluation by the UNESCO External Auditor on the implementation of the global strategy for a credible, representative and balanced World Heritage List and the Partnership for Conservation Initiative (PACT)*. WHC-11/35.COM/9A. Paris, May 27, 2011.
UNESCO. (2011b). *Implementation of the global strategy for the credible, balanced and representative World Heritage List. Final report of the Audit of the Global Strategy and the PACT initiative*. WHC-11/35.COM/INF.9A. Paris, May 27, 2011.
UNESCO (2015). *Policy document for the integration of a sustainable development perspective into the processes of the World Heritage Convention.* .
UNESCO. (2020). *Basic texts*. UNESCO.
Weiss, T. G., & Wilkinson, R. (Eds.). (2018). *International organization and global governance* (2nd ed.). Routledge.

Chapter 4
Coloniality, Natural World Heritage and Indigenous Peoples: A Critical Analysis of World Heritage Cultural Governance

Irene Fogarty

Abstract This essay analyses synergies and antagonisms of World Heritage cultural governance in respect of Indigenous peoples' participation and rights. In tandem with recognition of nature-culture interlinkages, the World Heritage Committee has demonstrated a growing concern with rights-based approaches, moving Indigenous peoples' rights to a more normative position in the Convention's implementation. However, the Convention follows a Statist approach and adheres to a Eurocentric conceptualisation of nature, reproduced through World Heritage cultural governance. These issues can result in power asymmetries, coloniality of knowledge and the relegation of Indigenous peoples' worldviews and rights.

Keywords Coloniality · Cultural governance · Natural heritage

4.1 Introduction

The special relationships of Indigenous peoples with their landscapes and maintenance of their knowledge systems are vital for sustainable landscape custodianship and climate change adaptation (IPCC, 2014, cited in UNESCO, 2017, p. 28). However, the World Heritage Convention predates some of the most significant steps taken in international law to recognise and protect Indigenous peoples' rights (Tauli-Corpuz, 2014, xii). Examples abound of States Parties exacerbating human rights violations under the auspices of implementing the Convention, particularly where sites are inscribed under natural heritage criteria alone (see Amougou-Amougou & Woodburne, 2014; IWGIA and Forest Peoples' Programme, 2015; Muchuba Buhereko, 2014). Resultantly, the World Heritage Committee has

I. Fogarty (✉)
University College Dublin, School of Archaeology (World Heritage Major), Dublin, Ireland
e-mail: irene.fogarty@ucdconnect.ie

M.-T. Albert et al. (eds.), *50 Years World Heritage Convention: Shared Responsibility – Conflict & Reconciliation*, Heritage Studies,
https://doi.org/10.1007/978-3-031-05660-4_4

demonstrated a growing concern with human rights-based approaches, including integrating principles of the United Nations *Declaration on the Rights of Indigenous Peoples* (UNDRIP) through all aspects of the Convention's implementation (see UNESCO World Heritage Centre, 2019, Paras. 12, 39–40, 64, 119, 123).

In this essay, research from primary and secondary sources is used to discuss synergies and antagonisms of World Heritage cultural governance in respect of Indigenous peoples' participation and rights. While the structural dimension of governance refers to the institutional framework of World Heritage, a wide conceptualisation of cultural governance speaks to the "[s]ocial negotiation and control of the production of social sense and meaning, cultural orientation systems and their symbols, and cultural and artistic forms of expression" (Schmitt, 2009, p. 105). Rather than examining cultural governance at State level, the essay examines its exercise by the World Heritage Committee and its Advisory Bodies. Grounded in Aníbal Quijano's seminal work on coloniality (Quijano, 2007), the essay begins by outlining the historical and ideological contingency of the nature/culture schism, a dualism represented within the World Heritage Convention. With reference to case study sites, the essay examines how the locus of decision-making power and the coloniality of knowledge that inform World Heritage cultural governance remain problematic, particularly in respect of "natural heritage". However, the essay also discusses how synergies between World Heritage Cultural Landscapes and a growing concern for rights-based approaches can move Indigenous peoples' rights and worldviews to a more normative position, broadening global social meanings and social values imparted through World Heritage cultural governance.

4.2 Coloniality, Modernity and the Nature/Culture Schism

While caution must be taken against essentialising Indigeneity and Indigenous knowledges, numerous Indigenous scholars describe a universality of shared principles and recognise commonalities of protocols, laws and ontologies (Chilisa, 2011; Little Bear, 2000; Watson, 2014; Whyte, 2018). Indigenous knowledges are relational and rooted in environment, family, animals and the spirit world (Little Bear, 2000; Watson, 2014; Whyte, 2018). Knowledge transmission in Indigenous communities is multifaceted and, according to Leanne Simpson, "might come to us from relationships, experiences, story-telling, dreaming, participating in ceremonies, from the Elders, the oral tradition, experimentation, observation, from our children, or from teachers in the plant and animal world" (2001, p. 142). These relational worldviews stand in stark contrast to the positioning of nature and culture as discrete entities, a dualism linked to the emergence of intellectual secular thought in Europe during the period between the Renaissance and the Eighteenth Century. Parallel to the European industrial revolution, the scientific and intellectual revolutions of this period saw a shift in academic scholarship where nature, the self and society were treated as separate knowledge projects (Love, 1989). However, the European mode of rationality, based on a subject/object relation, reified an externalised view of nature that paralleled the global colonisation of Indigenous peoples (Quijano, 2007). As Sylvia Wynter asserts, rationalisation and domination of culture and nature were intellectually cemented within Western science disciplines that reflected and reproduced the "colonizer/colonized relation that the West was to discursively constitute and empirically institutionalize [...] on the mainlands of the Americas" (2003, p. 264).

While colonialism describes the political and economic relations that subsume the sovereignty of other nations or peoples within an empire, coloniality refers to the long-standing patterns of power that emerge from colonialism (Maldonado-Torres, 2007; Mignolo, 2007; Quijano, 2007). Fundamentally, coloniality is the invisible and constitutive side of modernity: there is no modernity without coloniality (Mignolo, 2007, p. 451). Using race as the most basic classification for the domination and exploitation of peoples, coloniality goes beyond the limits of colonial administrations, arising in the global division of labour; concentration of economic resources in the global North; racism and sexism; intersubjective relations; and forms of cultural and knowledge production (Balaton-Chrimes & Stead, 2017; Maldonado-Torres, 2007). Coloniality of knowledge is part of a power matrix that underpins the modern capitalist patriarchal world-system and privileges Eurocentrism in knowledge production (Grosfugel, 2011). Hegemonic Eurocentric paradigms have informed western philosophy and sciences, asserting "neutrality" and "objectivity" as universal and desirable ways of knowing (Grosfugel, 2011), while simultaneously relegating place-based knowledges such as those of Indigenous peoples, which are informed by lived experience and social relations (Poloma & Szelényi, 2019, p. 637). Eurocentric knowledge is based on hierarchical dualisms such as masculine/feminine, reason/body and nature/culture. A relationship of externality to nature has been normalised within centres of knowledge production including Western academia (Lander, 2009; Polona & Szelényi, 2019). The externalisation of nature appears in conceptualisations of "natural" World Heritage which, combined with the Convention's Statist approach, can reproduce power asymmetries and coloniality of knowledge that constrain Indigenous peoples' equitable participation and rights.

4.3 Coloniality and World Heritage Cultural Governance

4.3.1 The Locus of Power

Through discursive practices tied to the institutional setting of UNESCO, World Heritage cultural governance regulates, controls and imparts heritage meanings and values to uphold UNESCO's universalistic assertion that the "disappearance of any item of the cultural or natural heritage constitutes a harmful impoverishment of the heritage of all the nations of the world" (UNESCO, 1972). However, the locus of the Convention's decision-making power rests upon a "three-legged stool" of governance comprised of States Parties, the World Heritage Committee (consisting of States Parties representatives) and the Committee's Advisory Bodies, with limited governance opportunities for non-state actors such as NGOs, communities and civil society (Larsen & Buckley, 2018, p. 90). The Convention is grounded in a Statist approach that reifies State sovereignty and property rights covered by national legislation, with States Parties maintaining responsibility for nominating properties for World Heritage inscription (UNESCO, 1972). Evaluations of site nominations are carried out on behalf of the World Heritage Committee by two professionalised Advisory Bodies. The International Council on Monuments and Sites (ICOMOS) and the International Union for Conservation of Nature (IUCN) evaluate cultural and natural heritage site nominations respectively, but have adopted a more cohesive approach in recent years through joint rather than separate evaluations of mixed properties. Nevertheless, while the Advisory Bodies offer recommendations, inscription decisions rest with the World Heritage Committee.

As heritage scholars and practitioners, policy makers and non-Western governments have undermined Eurocentrism in heritage discourses, the World Heritage programme has moved to a more inclusive and participatory approach (Coombe, 2012, p. 376). However, while the Committee now encourages States Parties to adopt a human-rights based approach in their implementation of the Convention (UNESCO World Heritage Centre, 2019, Para. 12), it cannot compel them to do so. Furthermore, as an international treaty, the Convention operates in the absence of a third-party effect unless sanctioned by member states, and an initiative to create a World Heritage Indigenous Peoples Council of Experts as an additional Advisory Body was ultimately rejected by the World Heritage Committee (Meskell, 2013). Committee members hold sway over the Convention's implementation standards and may create allegiances based upon, inter alia, historical and potentially colonial linkages (Meskell, 2013, p. 168).

4.3.2 Outstanding Universal Value and the Nature/Culture Divide

The conceptual tool of Outstanding Universal Value (OUV) is the threshold by which a site is deemed worthy of World Heritage status. While Articles 1 and 2 of the World Heritage Convention establish general types of cultural and natural properties to be considered of OUV respectively, the definition and application of OUV hinge on several pillars including a list of ten inscription threshold criteria, six for assessing cultural heritage OUV and four for natural heritage (UNESCO World Heritage Centre, 2019, Para. 77). The OUV criteria are periodically revised by the Committee and described in the *Operational Guidelines for the Implementation of the World Heritage Convention* (henceforth Operational Guidelines). The concept of OUV is reflexive and linked to changing heritage values and discourses. The Committee's recognition of subjectivity in cultural heritage values (UNESCO

World Heritage Centre, 2019, Para. 81) is paralleled by ICOMOS' constructivist approach in evaluating nominated cultural properties (Schmitt, 2009, p. 111), enabling a broad conceptualisation of cultural heritage values that now include nature-culture interlinkages. In 1992, the Committee introduced the concept of Cultural Landscapes, based on Article 1 of the Convention, which asserts how the "combined works of nature and man" can demonstrate Outstanding Universal Value (UNESCO, 1972). Cultural Landscapes are particularly relevant to Indigenous peoples' landscapes, where material culture can be absent and it is the associative values of a site that are significant (Mitchell, 2008, p. 26).

However, the conceptualisation of natural World Heritage contrasts with Cultural Landscapes and the more holistic perspectives of other conservation instruments such as Australia ICOMOS' Burra Charter and IUCN's Category V Protected Seascapes/Landscapes. Despite other heritage-related instruments accepting amendments or protocols in line with evolving heritage discourses, the Convention's text remains static (Forrest, 2009, p. 40). Article 2 presents an externalised view of nature linked to a Eurocentric form of rationality that is reproduced in the natural heritage OUV criteria. In fact, the Committee's changes to these criteria have proven regressive. Prior to alterations in the 1994 Operational Guidelines, natural OUV criterion (iii) (now criterion (ix)) recognised the OUV of sites demonstrating exceptional combinations of natural and cultural elements (UNESCO, 1992, p. 8). Sagarmatha National Park's inscription included this criterion, with IUCN describing how the area "is of major religious and cultural significance in Nepal since it abounds in holy places like the Thyangboche and also is the homeland of the Sherpas whose way of life is unique, compared to other high altitude dwellers" (IUCN, 1979). Natural heritage OUV criteria now exclude nature–culture interlinkages except for values related to "exceptional natural beauty and aesthetic importance" (UNESCO World Heritage Centre, 2019, Para. 77).

In a submission to the Expert Mechanism on the Rights of Indigenous Peoples, the International Work Group for Indigenous Affairs (IWGIA) and Forest Peoples Programme assert that inscribing World Heritage sites solely under natural OUV criteria "often comes at the expense of indigenous peoples, their livelihoods, and the protection, exercise and development of their cultural heritage and expressions. It is obvious that this can have far-reaching human rights implications" (2015, p. 7). In 2014, IUCN called upon the World Heritage Convention to "fully and consistently recognise Indigenous Peoples' cultural values as universal, and develop methods for recognition and support for the interconnectedness of natural, cultural, social, and spiritual significance of World Heritage sites" (IUCN, 2015, p. 8). Thus, in addition to power asymmetries driven by the Convention's Statist approach, a coloniality of knowledge underpins meanings and values attached to natural World Heritage and reproduced through cultural governance. The case of Pimachiowin Aki World Heritage Site in Canada is emblematic of these issues. Here, a culturally insensitive Advisory Body evaluation converged with the constraints of natural OUV threshold criteria, exposing the weaknesses of World Heritage cultural governance in upholding Indigenous peoples' worldviews and human rights.

4.4 The Case of Pimachiowin Aki

Pimachiowin Aki is the traditional territory of the Anishinaabeg people and is Canada's only mixed World Heritage site. Aside from its rich archaeological heritage of Anishinaabeg petroglyphs, the site is a Cultural Landscape of the Anishinaabeg who have occupied this vast territory of approx. 30,000 km^2 for more than 6000 years (Jones, 2014, p. 441). The site is also inscribed for its natural values as an outstanding example of the North American Boreal Shield (IUCN, 2018, p. 71). The name Pimachiowin Aki translates from Ojibwe as "the Land that Gives Life" and is used by the First Nations who are its custodians. While Pimachiowin Aki achieved World Heritage status in 2018, its inscription campaign began in 2002 when four First Nations set forth a shared stewardship accord to protect their ancestral lands and resources from incursions by extractive industries being encouraged by Canada's provincial and federal authorities (Poplar River First Nation et al., 2002). The accord described how initiatives undertaken by the First Nations to protect their lands "represent a unique and internationally significant opportunity to demonstrate the value of First Nation traditional knowledge in protecting and taking care of the land in the spirit of cooperation and harmony with other First Nations, other governments and the larger society". World Heritage inscription was cited in the accord as a means of recognising these goals (Poplar River First Nation et al., 2002).

Profoundly, considering the colonisation, cultural genocide and attempted assimilation wrought by the British colonial and federal Canadian governments upon Indigenous peoples,[1] the First Nations described how the bid for World Heritage inscription would involve cooperation with Canada and the international community. This perspective speaks to the First Nations' relational worldview, standing in sharp relief against the entanglements of colonialism and capitalism perpetuated by Canada, which prompted the World Heritage bid in the first instance. In 2004, Canada's Minister of the Environment endorsed the First Nations' World Heritage bid and the site's nomination dossier was forwarded to the World Heritage Committee in 2012 (Rabliauskas, 2020, pp. 10–11). A key aspect of the site's nomination under cultural OUV criterion (v) was the relationship between the First Nations and their landscape. Careful custodianship of the site is maintained through the Anishinaabeg stewardship of Keeping the Land, where oral traditions, Indigenous knowledge systems, beliefs, customary governance and cosmology are integral to the continuity of traditional land-use practices (Pimachiowin Aki Corporation, 2012, xii, p. 202). The underlying expression of interconnectedness in the ethic of Keeping the Land (Pimachiowin Aki Corporation, 2012, p. 117) illustrates the inseparability of the First Nations from their territories, a worldview that contrasts with the acultural and objectified constitution of nature in natural World Heritage.

[1] See reports by the National Inquiry into Missing and Murdered Indigenous Women and Girls (2019) and the Truth and Reconciliation Commission of Canada (2015).

Aware of requirements for a comparative analysis between Pimachiowin Aki and similar sites, the First Nations expressed discomfort about demonstrating the site's exceptionality. In additional information provided to ICOMOS, they explained how they "did not want to make judgments about the relationships of other First Nations with their lands and thus make comparisons (ICOMOS, 2013, p. 39). This viewpoint is described by Sophia Rabliauskas, a member of the Poplar River First Nation who acted as spokesperson for Pimachiowin Aki. From the time she was a child, her father, grandfather and community Elders "talked about the Circle of Life, where all forms of life, the animals, fish, birds, insects, plants, everything, including human beings fit on a circle side by side, no one more important than the next" (Rabliauskas, 2020, p. 10). Despite notification of this cultural norm, ICOMOS' evaluation found that the Anishinaabeg relationship with the land is "not unique and persists in many places associated with indigenous peoples in North America and other parts of the world […] What has not been demonstrated is how this strong association between the Anishinaabeg and the land in the area nominated can be seen to be exceptional – in other words of wider importance than to the Anishinaabeg themselves" (ICOMOS, 2013, p. 39). Thus, value judgments were based on a Eurocentric conceptualisation of exceptionality that involved the positioning of Indigenous peoples against one another. Ironically, the Operational Guidelines have long stated that "[j]udgments about value attributed to cultural heritage...may differ from culture to culture, and even within the same culture. The respect due to all cultures requires that cultural heritage must be considered and judged primarily within the cultural contexts to which it belongs" (UNESCO World Heritage Centre, 2005, Para. 81).

As the site was a mixed nomination, IUCN separately evaluated the site's natural heritage value. In light of the Anishinaabeg relationship with Pimachiowin Aki being a crucial element of the site's significance, IUCN was concerned about an Indigenous-led nomination being inscribed under natural heritage criteria alone, "given the community-led nature of this nomination, and the central premise that traditional use would be recognized as intrinsic to the values of the property, if inscribed" (IUCN, 2013, pp. 141–142). IUCN evidently recognised the danger of misrepresenting the site's significance, drawing attention to the importance of recognising the interconnectedness between Indigenous peoples and their landscapes. ICOMOS also noted the limitations of the OUV criteria in inscribing Pimachiowin Aki under natural criteria alone, stating that there is "no way for properties to demonstrate within the current wording of the criteria, either that cultural systems are necessary to sustain the outstanding value of nature in a property, or that nature is imbued with cultural value in a property to a degree that is exceptional" (ICOMOS, 2013, p. 45).

These issues shed light upon the limitations of World Heritage cultural governance in supporting Indigenous peoples' rights to maintain, control, protect and develop their cultural heritage: a vital principle asserted by Article 31 of UNDRIP (UN, 2007, p. 22). The case prompted the UN Special Rapporteur on the Rights of Indigenous Peoples to write to the Director of the World Heritage Centre citing concerns over inconsistencies in UNESCO's approach to the natural and cultural world heritage of Indigenous peoples and the separate evaluation processes (Anaya, 2013). The Special Rapporteur stated how these issues were not new—UNESCO had been previously urged to modify its approach so that "indigenous peoples' rights and worldviews are fully valued and respected in all current and future World Heritage site designations as well as in the overall implementation of the World Heritage Convention" (Anaya, 2013). Pimachiowin Aki was resubmitted for evaluation in 2017 under cultural OUV criteria (iii) and (vi) and natural OUV criterion (ix), with the Advisory Bodies undertaking a joint evaluation. In 2018, ICOMOS found that the site met the OUV thresholds, recognising the strength of the First Nations' tradition of Keeping the Land as an exceptional example of a belief of universal significance (ICOMOS, 2018, p. 28). IUCN also supported the site's inscription and acknowledged the essential role of the Anishinaabeg in maintaining ecosystem health and sustainability, describing how "[t]raditional use by Anishinaabeg, including sustainable fishing, hunting and trapping, is also an integral part of the boreal ecosystems in Pimachiowin Aki" (IUCN, 2018, p. 70). IUCN also asserted that Pimachiowin Aki's nomination could serve as a compelling model for future nominations that wish to demonstrate the indissoluble links between nature and culture and the relationship between cultural and ecological integrity (IUCN, 2018).

4.5 Cultural Governance and Indigenous Peoples' Rights: A More Synergistic Approach?

To shift the production of knowledge from its confines in European rationality/modernity, Anibal Quijano advocates for the decolonisation of knowledge through interchange of experiences and meanings, in order to realise an alternative form of rationality that may "legitimately pretend to some universality" (2007, p. 177). Recognition of worldviews outside the Western canon is pivotal for broadening understandings of World Heritage beyond Eurocentric confines. Numerous current and tentative World Heritage sites demonstrate links between relational worldviews, Indigenous custodianship and site sustainability. At Pimachiowin Aki, the Anishinaabeg and all other beings, the animals, the trees and plants, the fish, the waters, are understood and safeguarded as one living entity (Pimachiowin Aki Corporation, 2012). In Australia, Uluru–Kata Tjuta National Park is managed using the traditional methods of the Anangu people and governed by *Tjukurpa*—a sacred Law that encompasses "the relationship between people, plants, animals and the physical features of the land" (Uluru-Kata Tjuta Board of Management, 2010, pp. 2–3). The tentative World Heritage site of Gwaii Haanas, Canada, lies within the unceded territory of the Haida Nation and is an exceptional landscape of unique biodiversity and living Haida culture. Gwaii Haanas is managed as an "interconnected ecosystem of land, sea and people under the Haida principle of *gina 'waadluxan gud ad kwaagid*, which means that everything is connected to everything else" (Council of the Haida Nation & Government of Canada, 2018, p. 6, p. 29).

In 2017, the World Heritage Committee formally recognised the International Indigenous Peoples' Forum on World Heritage (IIPFWH), considering it "an important reflection platform on the involvement of Indigenous Peoples in the identification, conservation and management of World Heritage properties" (UNESCO World Heritage Committee, 2017, pp. 12–13). However, IIPFWH holds consultative status only, and the Convention's Statist approach means that decisions to further empower IIPFWH or similar forums are contingent on the Committee's composition in a given cycle. Nevertheless, recent Committee decisions evidence a growing human rights concern. Furthermore, while Cultural Landscapes fail to resolve issues of natural World Heritage, they offer an entry point in understanding synergies between World Heritage recognition of nature–culture interlinkages and international human rights law. The values under which a Cultural Landscape is inscribed can assert cultural rights and cultural continuity linked to land custodianship as normative components of World Heritage significance, with the recent inscription of Budj Bim, Australia being an instructive case.

Budj Bim is the traditional territory of the Gunditjmara peoples, who manage the site through traditional practices (UNESCO, 2019). ICOMOS' site evaluation describes how "continuity of associated Gunditjmara practices, traditions and knowledge [...] is essential to the conservation of the nominated property. This memory resides with elders of the Gunditjmara and the young people are being mentored to continue traditions and practices. ICOMOS considers that this is indeed of critical importance..." (ICOMOS, 2019, p. 107) In recognising the importance of Gunditjmara cultural continuity, ICOMOS' exercise of cultural governance aligned with international legal principles on the collective right to culture. UNESCO's (1966) *Declaration of the Principles of International Cultural Cooperation* calls upon "governments, authorities, organizations, associations and institutions responsible for cultural activities" to be guided by the principle that "[e]very people has the right and the duty to develop its culture" (UNESCO, 1966). Furthermore, the Committee's decision to inscribe Budj Bim supports Article 31 of UNDRIP, which states that Indigenous peoples have the right to maintain, control, protect and develop their cultural heritage (UN, 2007, p. 22). As with Pimachiowin Aki, the site's inscription broadens the global social meanings and social values imparted through World Heritage cultural governance, normalising interlinkages between recognition and protection of Indigenous peoples' worldviews, human rights and heritage conservation.

4.6 Conclusion

This essay critically examined synergies and antagonisms of World Heritage cultural governance in respect of Indigenous peoples' rights and worldviews. A reflexive approach to cultural heritage value and the advent of World Heritage Cultural Landscapes demonstrate a broadening of heritage discourses imparted through World Heritage cultural governance. Resultantly, a more synergistic approach is developing between World Heritage values and the upholding of Indigenous peoples' rights. However, the historical and ideological constitution of "nature" as an external entity is intertwined with colonial relations and reproduced through Eurocentric knowledge production. The conceptualisation of natural World Heritage and the associated natural OUV criteria adhere to a Eurocentric form of rationality, belying the relational worldviews of Indigenous peoples who recognise the interconnectedness of nature, culture and all beings. In tandem, the Convention's state-centric approach and its relegation of Indigenous authorities such as IIPFWH to consultative status within World Heritage governance remains problematic, constraining the equitable participation of Indigenous peoples.

Acknowledgements The author gratefully acknowledges support from the Irish Research Council/Government of Ireland Postgraduate Scholarship and the National University of Ireland's Travelling Doctoral Studentship.

References

Amougou-Amougou, V., & Woodburne, O. (2014). The Sangha trinational world heritage site: The experiences of Indigenous Peoples. In S. Disko & H. Tugendhat (Eds.), *World Heritage sites and indigenous peoples' rights* (pp. 103–118). Transaction Publishers.

Anaya, J. (2013). *Reference: OL Indigenous (2001–8) OTH 10/2013, Mandate of the Special Rapporteur on the rights of Indigenous peoples*. https://spcommreports.ohchr.org/TMResultsBase/DownLoadPublicCommunicationFile?gId=18157

Balaton-Chrimes, S., & Stead, V. (2017). Recognition, power and coloniality. *Postcolonial Studies, 20*(1), 1–17. https://doi.org/10.1080/13688790.2017.1355875

Chilisa, B. (2011). *Indigenous research methodologies*. SAGE Publications.

Coombe, R. (2012). Managing cultural heritage as neoliberal governmentality. In R. Bendix, A. Eggert, A. Peselmann, & S. Mißling (Eds.), *Heritage regimes and the State* (pp. 375–389). Göttingen University Press.

Council of the Haida Nation & Government of Canada. (2018). *Gwaii Haanas Gina "Waadluxan KilGuhlGa" Land-Sea-People Management Plan 2018*. Government of Canada. https://pcacdn.azureedge.net/-/media/pn-np/bc/gwaiihaanas/WET4/info/plan/PDFs/Gwaii-Haanas-LSP-Plan-2018_EN_lowres.pdf

Forrest, C. (2009). *International law and the protection of cultural heritage*. Routledge. https://doi.org/10.4324/9780203865194

Grosfugel, R. (2011). Decolonizing post-colonial studies and paradigms of political economy: Transmodernity, decolonial thinking, and global coloniality. *TRANSMODERNITY: Journal of Peripheral Cultural Production of the Luso-Hispanic World, 1*(1), 31.

ICOMOS. (2013). *Evaluations of nominations of cultural and mixed properties. ICOMOS report for the World Heritage Committee, 37th ordinary session, Phnom Penh, June 2013* (WHC-13/37.COM/INF.8B1) (p. 344). ICOMOS. https://whc.unesco.org/archive/2013/whc13-37com-8B1inf-en.pdf

ICOMOS. (2018). *Evaluations of nominations of cultural and mixed properties. ICOMOS report for the World Heritage Committee, 42nd ordinary session, Manama, 24 June—4 July 2018* (WHC-18/42.COM/INF.8B1) (p. 276). ICOMOS. https://www.icomos.org/images/DOCUMENTS/World_Heritage/Volumes_Evaluation/ICOMOS_Evalauation_Volume_WHList_2018_EN.pdf

ICOMOS. (2019). *Budj Bim Cultural Landscape: Advisory Body evaluation*. ICOMOS. https://whc.unesco.org/document/176137

IUCN. (1979). *IUCN Review. World Heritage Nomination, Sagarmatha National Park*. IUCN. https://whc.unesco.org/document/152019

IUCN. (2013). *IUCN Evaluations of Nominations of Natural and Mixed Properties to the World Heritage List. IUCN*. https://whc.unesco.org/archive/2013/whc13-37com-8B2inf-en.pdf

IUCN. (2015). *World Heritage at the IUCN Parks Congress 2014. The Promise of Sydney* (p. 8). IUCN. https://whc.unesco.org/document/139839

IUCN. (2018). *IUCN World Heritage Evaluations 2018. IUCN Evaluations of nominations of natural and mixed properties to the World Heritage List* (WHC/18/42.COM/INF.8B2) (p. 106). IUCN. https://whc.unesco.org/archive/2018/whc18-42com-inf8B2-en.pdf

IWGIA & Forest Peoples Programme. (2015). *Promotion and protection of the rights of Indigenous peoples with respect to their cultural heritage in the context of the implementation of UNESCO's World Heritage Convention* (p. 29). International Work Group for Indigenous Affairs (IWGIA) and Forest Peoples Programme. https://www.ohchr.org/Documents/Issues/IPeoples/EMRIP/CulturalHeritage/IWGIA.pdf

Jones, G. (2014). The Pimachiowin Aki World Heritage Project: A collaborate effort of Anishinaabe First Nations and two Canadian provinces to nominate a World Heritage Site. In S. Disko & H. Tugendhat (Eds.), *World Heritage Sites and Indigenous Peoples' Rights* (pp. 441–458). Transaction Publishers.

Lander, E. (2009). Eurocentrism, modern knowledges and the 'natural' order of capital. *KULT 6: Special Issue–Latin America*. http://postkolonial.dk/kult-6-special-issue-latin-america/

Larsen, P. B., & Buckley, K. (2018). Approaching human rights at the World Heritage Committee: Capturing situated conversations, complexity, and dynamism in global heritage processes. *International Journal of Cultural Property, 25*(1), 85–110. https://doi.org/10.1017/S0940739118000048

Little Bear, L. (2000). Jagged worldviews colliding. In M. Battiste (Ed.), *Reclaiming indigenous voice and vision* (1st ed., pp. 77–85). UBC Press.

Love, N. S. (1989). Foucault & Habermas on discourse & democracy. *Polity, 22*(2), 269–293. https://doi.org/10.2307/3234835

Maldonado-Torres, N. (2007). On the coloniality of being. *Cultural Studies, 21*(2/3), 240–270. https://doi.org/10.1080/09502380601162548

Meskell, L. (2013). UNESCO and the fate of the World Heritage Indigenous Peoples Council of Experts (WHIPCOE). *International Journal of Cultural Property, 20*(2), 155–174. https://doi.org/10.1017/S0940739113000039

Mignolo, W. D. (2007). DELINKING: The rhetoric of modernity, the logic of coloniality and the grammar of de-coloniality. *Cultural Studies, 21*(2–3), 449–514. https://doi.org/10.1080/09502380601162647

Mitchell, N. J. (2008). Considering the authenticity of cultural landscapes. *APT Bulletin: The Journal of Preservation Technology, 39*(2/3), 25–31.

Muchuba Buhereko, R. (2014). Kahuzi-Biega National Park: World Heritage site versus the Indigenous Twa. In S. Disko & H. Tugendhat (Eds.), *World Heritage sites and indigenous peoples' rights* (pp. 131–144). Transaction Publishers.

National Inquiry into Missing and Murdered Indigenous Women and Girls. (2019). *A legal analysis of genocide: Supplementary report of the National Inquiry Into Missing and Murdered Indigenous Women and Girls* (p. 43). https://www.mmiwg-ffada.ca/wp-content/uploads/2019/06/Supplementary-Report_Genocide.pdf

Pimachiowin Aki Corporation. (2012). *Pimachiowin Aki World Heritage Project: Nomination for inscription on the World Heritage List* (p. 288). Pimachiowin Aki Corporation.

Poloma, A., & Szelényi, K. (2019). Coloniality of knowledge, hybridisation, and indigenous survival: Exploring transnational higher education development in Africa from the 1920s to the 1960s. *Compare: A Journal of Comparative and International Education, 49*(4), 635–653.

Poplar River First Nation, Pauingassi First Nation, Little Grand Rapids First Nation, & Pikangikum First Nation. (2002). *Protected areas and first nation resource Stewardship: A cooperative relationship* (p. 15). https://pimaki.ca/wp-content/uploads/Accord.pdf
Quijano, A. (2007). Coloniality and modernity/rationality. *Cultural Studies, 21*(2–3), 168–178. https://doi.org/10.1080/09502380601164353
Rabliauskas, S. (2020). An indigenous perspective: The case of Pimachiowin Aki Mixed cultural and natural heritage, Canada. *Journal of World Heritage Studies*. Special Issue 2020: Proceedings of the Fourth Capacity Building Workshop on Nature-Culture Linkages in Heritage Conservation in Asia and the Pacific. September 24 – October 4, 2019, Tsukuba, Japan, ISSN 2189-4728. https://tsukuba.repo.nii.ac.jp/?action=repository_uri&item_id=56132&file_id=17&file_no=1
Schmitt, T. M. (2009). Global cultural governance. Decision-making concerning World Heritage between politics and science. *Erdkunde, 63*(2), 103–121.
Simpson, L. (2001). Aboriginal peoples and knowledge: Decolonizing our process. *Canadian Journal of Native Studies, 21*(1), 137–148.
Tauli-Corpuz, V. (2014). Foreward. In S. Disko & H. Tugendhat (Eds.), *World heritage sites and indigenous peoples' rights* (pp. xii–xiv). Transaction Publishers.
Truth and Reconciliation Commission of Canada. (2015). *Honouring the truth, reconciling for the future: Summary of the final report of the Truth and Reconciliation Commission of Canada* (p. 536) http://www.trc.ca/assets/pdf/Honouring_the_Truth_Reconciling_for_the_Future_July_23_2015.pdf
Uluru-Kata Tjuta Board of Management. (2010). *Uluru-Kata Tjuta National Park, Management Plan 2010–2020*. Australian Government: Director of National Parks. https://www.environment.gov.au/system/files/resources/f7d3c167-8bd1-470a-a502-ba222067e1ac/files/management-plan.pdf
United Nations. (2007). *United Nations declaration on the rights of indigenous peoples*. United Nations. https://www.un.org/development/desa/indigenouspeoples/wp-content/uploads/sites/19/2018/11/UNDRIP_E_web.pdf
UNESCO. (1966). *Declaration of principles of international cultural co-operation*. UNESCO. http://portal.unesco.org/en/ev.php-URL_ID=13147&URL_DO=DO_TOPIC&URL_SECTION=201.html
UNESCO. (1972). *Convention concerning the protection of the world cultural and natural heritage*. UNESCO.
UNESCO. (1992). *Operational guidelines for the implementation of the World Heritage Convention*. UNESCO. https://whc.unesco.org/archive/opguide92.pdf
UNESCO. (2017). *Local knowledge, global goals*. UNESCO.
UNESCO. (2019). *Budj Bim cultural landscape*. UNESCO. https://whc.unesco.org/en/list/1577/
UNESCO World Heritage Centre. (2005). *Operational guidelines for the implementation of the World Heritage Convention*. UNESCO. https://whc.unesco.org/archive/opguide05-en.pdf
UNESCO World Heritage Centre. (2019). *Operational guidelines for the implementation of the World Heritage Convention*. UNESCO. https://whc.unesco.org/document/178167
UNESCO World Heritage Committee. (2017). *Decisions adopted during the 41st session of the World Heritage Committee (Krakow, 2017)* (WHC/17/41.COM/18) (p. 290). UNESCO. https://whc.unesco.org/archive/2017/whc17-41com-18-en.pdf
Watson, I. (2014). Re-centring first nations knowledge and places in a Terra Nullius space. *AlterNative: An International Journal of Indigenous Peoples, 10*(5), 508–520. https://doi.org/10.1177/117718011401000506
Whyte, K. (2018). Settler colonialism, ecology, and environmental injustice. *Environment and Society, 9*, 125–144.
Wynter, S. (2003). Unsettling the coloniality of being/power/truth/freedom: Towards the human, after man, its overrepresentation—An argument. *CR: The New Centennial Review, 3*(3), 257–337.

Chapter 5
Governing World Heritage – Taking Stock of the Structures that Determine the Protection and Conservation of World Heritage Sites

Eike Tobias Schmedt

Abstract The World Heritage Convention combines efforts of heritage protection and conservation on the global, national and local level. It has been adopted by almost 200 countries and has a complex governance system with actors on every level. While these actors are critical to the protection and conservation of World Heritage Sites, very limited research is available that assesses their role and the importance of governmental and managerial structures on a holistic level. This study assesses different governance structures and illustrates how they influence protective efforts. The World Heritage Site Index, which is a comprehensive database of information from almost 900 World Heritage Sites, creates a unique perspective that allows for the comparative assessment of sites regardless of their designation or typology. Utilizing a mixed-methods approach, the study reveals clear governance structures that influence the protection of World Heritage Sites and offers a perspective on potential steps toward ensuring that these structures work for and not against the protection and conservation of these sites.

Keywords Heritage governance · Mixed methods · Comparative assessment · Community involvement · Governance communication

5.1 Governing the World's Most Outstanding Sites

Governing and managing sites is one of the central elements of World Heritage. Every site has some form of governance structure or bodies that are responsible for its management and protection. Yet, very little attention has been focused on this subject on a larger scale. Many studies have evaluated the structures of individual

E. T. Schmedt (✉)
Center for Governance and Sustainability, University of Massachusetts Boston, Boston, MA, USA
e-mail: Eike.schmedt@gmail.com

M.-T. Albert et al. (eds.), *50 Years World Heritage Convention: Shared Responsibility – Conflict & Reconciliation*, Heritage Studies,
https://doi.org/10.1007/978-3-031-05660-4_5

cases and sites, but larger comparative assessments or the study of global institutions within the World Heritage framework such as the World Heritage Centre, ICOMOS, IUCN or UNESCO are severely lacking. Even the state parties and complete assessments of national systems of World Heritage have received little to no attention from researchers until now. There has been some research on aspects such as financial support for culture within a country (Trupiano, 2005), rethinking resource distribution on a larger scale (Bertacchini et al., 2011), how sites play into the identity of a nation (Labadi, 2007) or how national conservation systems transition between political models (Bonini Baraldi & Ferri, 2019). However, this research has been limited in scale and does not hone in on the institutions that are critical for World Heritage Sites. On an even smaller scale, numerous case studies focus on specific sites or aspects of national heritage management and their underlying management and governance structures (Lindholm & Ekblom, 2019; Maksić et al., 2018; Yakusheva, 2019; Zan & Bonini Baraldi, 2013). Many of these resources are not even directly related to UNESCO World Heritage and speak more about heritage and protected areas in general. Works by researchers such as William Logan (2001), Lynn Meskell (2013), Thomas M. Schmitt (2009, 2011) and Luke James and Tim Winter (2017) are some of the few notable exceptions in this underdeveloped area. These authors have gone beyond sole case studies and focused on the underlying systematic structures that are of significance in a governance context. Though these researchers have provided important insights into the World Heritage System and its governing bodies, there are very few of them focusing on this area. Questions around the functioning of the underlying governing structures of UNESCO World Heritage and how they impact the individual sites are incredibly underdeveloped. After 50 years, it is appropriate to take stock of global and national systems, how they operate, what is working well and what may need some improvement. In this context, the guiding question in this chapter is whether different governance structures are important and if there are certain stakeholders that have a primarily positive or negative influence on the protection of World Heritage Sites.

Several issues arise when trying to compare governance structures. While it is possible to draw out and compare individual governance components from two countries, there is no tangible way to say whether one is performing better than the other. As such, the World Heritage Site Index (WHSI) (Schmedt, 2021) was created as an entirely new way of assessing World Heritage Sites and answering these underlying questions. It is the first large scale database that can compare sites with one another on a variety of topics, including governance-related issues. The underlying data of the WHSI is based on the assessment of Periodic Reports from the second reporting cycle. These reports are standardized questionnaires created by the World Heritage Centre and answered in regular intervals by each country (Section I) and site (Section II) represented on the World Heritage List. The index takes the individual reports of every World Heritage Site in Section II that participated in the second reporting cycle and creates a matrix for all codable questions. In total, 184 individual questions in the reports of 882 individual sites create the backbone of this index. The resulting database is the largest of its kind outside of UNESCO itself and is the only comprehensive set of data that enables comparison between any two or more sites, regardless of their typology, designation, country or region. Within the 184 questions, a subset of 74 questions is directly related to governance structures and issues. These questions address aspects such as legal frameworks, management structures, cooperation amongst stakeholders and finances and are the basis for the following assessment.

To create an overview of the performance of the individual sites, questions are coded on a Likert-Scale of 0–5. The resulting overall score represents a unique indicator that is comparable between all cultural, natural and mixed sites and allows a perspective of whether a site appears to perform well or not. It must be noted that there are limitations to the value of this indicator. These limitations mainly arise from the fact that the periodic reports are self-assessments and are prepared by site managers, governmental officials and a variety of other groups and individuals within the respective countries and sites. Therefore, the WHSI does not represent an independent mode of evaluation through expert assessments but rather represents the perception of the protective efforts from the point of view of the individuals preparing the reports.

5.2 Assessing the WHSI Data and Its Implications

To assess such a large set of data and create an analytical overview of the underlying structures, specific statistical methods are applied to answer the underlying research question. Furthermore, a set of factor analyses is carried out to identify and evaluate the governance structures of World Heritage sites. These analyses take the 74 governance-related questions within the WHSI and correlate them with one another. The results form groups of questions that are highly correlated with one another and are referred to as components. Each of these components essentially suggests a specific underlying structure that is responsible for how the questions were answered. In other words, as all of the questions are governance-related, each component indicates a specific structure that is responsible for the governance of World Heritage Sites. As all 882 sites are included in this analysis, the results may not reflect each site but are rather a general representation of the most common structures.

Most of the components created through the factor analyses follow very intuitive patterns that are found across cultural and natural sites alike. These components and their underlying governance structures focus on community involvement, training opportunities, availability of professionals, management, maintenance and monitoring involvement of different stakeholders. Yet, there are certain differences and separate components that are notable and suggest more complex underlying structures at many sites. For example, monitoring involvement and training opportunities for conservation and research are separate at natural sites, while they are addressed together at cultural sites. This indicates that the treatment of these aspects differs significantly from a structural perspective. Another even more prominent difference is evident in questions and components related to the legal framework and the sites' boundaries and buffer zones. While natural sites combine these two groups of questions into one component, cultural sites separate them. This suggests that the boundaries and buffer zones of natural sites are likely to be designated by the same entity or specific governing structure that also determines the legal framework. Compared to this, cultural sites have one structure that is responsible for the creation of the legal framework while an entirely different structure is responsible for the designation of boundaries and buffer zones. A more extensive and detailed evaluation of this assessment can be found in previously published work (Schmedt, 2021).

While some of these differences are interesting and demonstrate how sites are structured, they do not clarify whether such structures are beneficial for the individual sites or if they are relevant to a site's protective efforts. The WHSI helps address this issue by calculating the average score of sites within a country. While not fully comparable due to the vastly different numbers of sites in each country, this mean average of a country creates a compelling perspective that can be correlated with specific governance structures. As Fig. 5.1 illustrates, there is a significant variance in scores between countries and no one region is performing below average. While there are no low-performing countries in the Europe and North America Region, every other region also has high-performing representatives. For example, the overall highest performing country is Malaysia in the Asia and Pacific Region with an average score of 4.56. Within the top 10 countries, five are from Europe and North America (Azerbaijan (2) 4.32, Germany (3) 4.42, San Marino (4) 4.24, Israel (7) 4.11 and Greece (9) 3.98), three are from Asia and Pacific (Malaysia (1) 4.56, Turkmenistan (8) 4.03 and Japan (10) 3.98 one from Africa (Botswana (6) 4.13), and one from Latin America and Caribbean (Saint Kitts and Nevis (5) 4.21). Only the Arab States did not have a country within the top 10. Iran is the highest performing country from this region at rank 23 with an average score of 3.82.

Even though the regions of Europe and North America and Asia and Pacific overshadow the higher ranks, the remaining regions are also well-represented and there

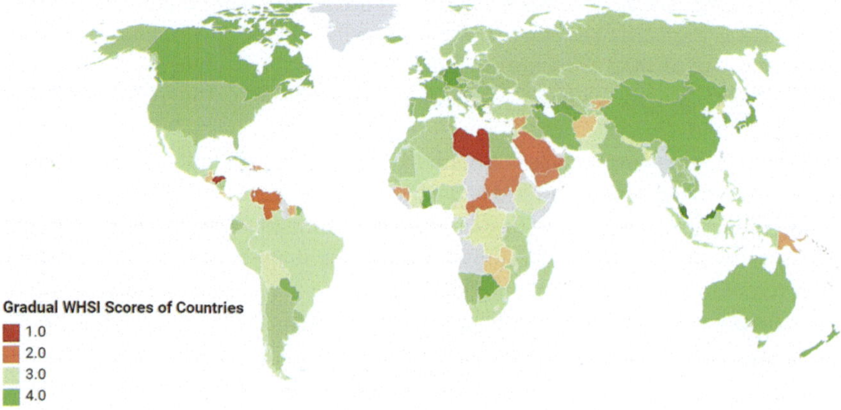

Fig. 5.1 Average WHSI Scores of Sites within a Country. (Note. Prepared by Eike Schmedt, 2021)

is no direct connection between specific regions and overall performance. This pattern is further confirmed by a second statistical assessment that compared the overall scores of the WHSI with established indices such as gross domestic product (GDP) or the Global Competitiveness Index. Through regressions models, these comparisons aimed to assess the potential connection between economic aspects, protective efforts outside the World Heritage Convention and other similar elements of individual World Heritage Sites. While most of these models were statistically significant, they were not strong enough to represent a viable argumentative basis. For example, a high GDP is positively correlated with the performance of a country's sites in the WHSI, but not significantly enough to argue that a higher GDP can determine the performance of sites on the World Heritage List. As this is the case throughout all regression models, it is clear that no single variable can be utilized to explain why certain countries perform higher or lower on the WHSI. Yet, as most of them are statistically significant, it could be argued that these established indices might not be determinants but single aspects of the protection of World Heritage Sites that must be considered.

5.3 Governance Structures and Their Impact

The findings concerning internal structures and components as well as the comparison with established indices create valuable insights into the performance of World Heritage Sites and aspects that might have an impact on their protection. However, they do not clearly answer the underlying research question. This is achieved by assessing some of the countries within the index and comparing their respective structures to the factor analyses as well as the overall scores. To maintain the scale of the WHSI, the structures of 10 countries (2 from each region) are assessed and compared to the findings of the statistical assessment. Connecting the statistical assessments with real-world examples and structures allowed for concrete evaluations of different approaches to managing and governing World Heritage Sites. It included countries with highly complex governance structures such as Germany and Mexico as well as countries with flat governance structures such as Tunisia.[1]

Comparing the existing structures with the WHSI scores and analytics revealed a large number of governance aspects that have positive as well as negative influences on the protection of World Heritage Sites. This comparison also directly answers the research question as it demonstrates that governance and management structures are important and that they have a significant influence on how well World Heritage Sites perform on the World Heritage Site Index. To be more precise, the evaluation revealed that the national level of governance, which includes national ministries, offices and organizations, is the most important component of almost

[1] The 10 countries included in the assessment are: Bulgaria, Egypt, Germany, India, Japan, Mexico, Peru, South Africa, Tanzania and Tunisia. Case selection followed criteria independent from the WHSI and was based on site distribution and representation of the World Heritage List.

every World Heritage governance structure. This hardly comes as a surprise as the World Heritage Convention itself asserts that the individual countries and their respective bodies have absolute authority over their respective sites. The only exception to this focus on the national level is Germany, which has a highly decentralized structure, especially for its cultural sites. This is primarily due to the federalized structure and organization of the German governance system but goes beyond the simple decentralization of authority.

However, decentralization by itself is not indicative of performance. While the German system highlights the positive elements of decentralization, some other countries indicate that it could have a negative impact. Countries such as Mexico, which have a mixed structure of centralized and decentralized authority perform much lower than Germany for example. On the flipside, flat hierarchical systems such as Japan perform very well. A deeper assessment revealed that the number of entities and levels involved in the World Heritage governance system of a country do not predict performance, but that the coordination, communication and clear structures of authority are critical (Schmedt, 2021). This held true for cultural and natural sites with very few exceptions. The assessment further indicates that high performing countries, in particular, have very strong structures to include communities on the local and provincial/state level. Such structures comprise the involvement of communities in general management, monitoring, interpretation and preservation efforts. All countries that had such structures in place and had an active network of entities involved in the protection and conservation of World Heritage Sites performed very well.

Additional elements that are crucial for the protection of sites within a country include adequate training of professionals and staff members, sufficient funding and strong legal frameworks that reach beyond the scope of World Heritage. Adequate and secure funding in particular appear as essential elements and are directly related to the national level of governance in most countries. Flexibility is also an important aspect of World Heritage governance systems, and each site requires its own specific structure to properly function. While some of these aspects seem to contradict one another, they do paint a very clear picture of which authorities, responsibilities and different entities are needed in the governance of World Heritage Sites for them to perform well (Schmedt, 2021).

The best example to explain all of these aspects is the case of Japan. Japan ranks at number 10 of 152 countries included in the WHSI and has a very centralized system with the Agency for Cultural Affairs at its core. This agency, which is a body of the Ministry of Education, Culture, Sports, Science and Technology, is responsible for almost all sites in the country, including two of the three natural sites. While it does work together and share responsibilities with the Ministry of Finance, Ministry of Agriculture, Forestry and Fisheries, and the Ministry of the Environment at some sites, it is the central body within the country that governs matters of World Heritage. The critical difference to other centralized countries is that, while the agency has the final authority for projects and measures taken at each site, the day-to-day management is in the hands of local stakeholders such as cities, religious authorities and management offices. As such, the local communities and stakeholders are much more involved in the management, protection and conservation of the sites and only large-scale projects require the involvement and approval of the Agency for Cultural Affairs. The agency further offers continuous support for the local stakeholders and provides expertise and information if requested. This governance structure takes the best parts of centralized and decentralized systems and combines them in a unique but very efficient manner. Having a centralized entity allows resources and expertise to be combined and easily shared amongst all sites. Yet, this entity is only involved if necessary or requested, and the day-to-day aspects of managing and preserving the World Heritage Sites are in the hands of local stakeholders and communities. This shared responsibility and adequate communication are a highly effective way of governing World Heritage Sites and create a clear structure that every involved party can easily understand, while still allowing local and national entities to provide their specific expertise for the benefit of the site.

In comparison, Mexico, which ranks at 95 in the WHSI, showcases how the variation of centralized and decentralized structures can lead to lower performances if not implemented fully. On the surface, the Japanese and Mexican governance models appear very similar. Both have centralized agencies with authority and several other ministries and national offices involved in the governance of World Heritage Sites. The differences start with the consistent involvement of local communities and stakeholders. While Japan has local structures for every individual site, a large number of the sites in Mexico do not have any associated local organizations or stakeholders. As such, the centralized agency in Mexico, the Institute of History and Anthropology, has more authority over the individual sites but also increased responsibility and pressure to handle the day-to-day management in addition to larger strategic planning efforts. The individual reports further mention the large disconnect between governing and managing entities within the Mexican World Heritage system and showcase the critical importance of communication and coordination within each individual system.

To recap, the comparison of the performance of countries in the WHSI with their individual governance structures has revealed several findings that indicate a positive influence on the performance of World Heritage Sites. First, governance at the national level is critical and in almost all cases is responsible for creating and implementing an adequate legal framework and organizational structure that includes all the necessary stakeholders and entities. Second, funding structures must be clarified and dependable for adequate planning. While most countries and their respective sites primarily rely on national-level support for financial resources, it is not a requirement and various different models of funding perform very well as long as the resources are sufficient and secure. Third, neither simple nor complex governance structures are indicative of the performance of a country. Yet, decentralized structures perform better if the communication between all involved stakeholders is adequately organized and the responsibilities are clear. Fourth, the more local communities are actively involved in the management and governance process and engage in communication, protection and conservation efforts, the better the overall performance of a site and country.

5.4 The Way Forward

After almost 50 years and 1121 sites in 167 countries, it is safe to say that the World Heritage Convention has been interpreted and implemented in a vast number of different ways. Correlating the underlying governance structures with the individual performance of sites and countries allows us to identify how these interpretations and implementations affect the sites and what improvements can be made in the future. The analysis above demonstrates what aspects of governance structures are important for the protection of World Heritage Sites. Many of these aspects are directly related to the national level governance in each country and how the convention is implemented and supported. It is clear that a strong and comprehensive legal framework is necessary and that adequate resources, both personnel and financial, are crucial for the performance of individual sites. As such, these two aspects are key indicators for high-performing countries on the World Heritage Site Index and can help guide other countries in their implementation and structuring.

The findings that certain structures and the involvement of communities are significant indicators of high performing countries and sites are even more important than the recognition of significant elements on the national level. These aspects of communication between different stakeholders and the involvement of communities are crucial for a site and country to perform well overall. The importance of these aspects has been well-established as part of the strategic objectives of the World Heritage Convention and the 5 Cs (Albert, 2012). Established in the Budapest Declaration on World Heritage in 2002 and expanded upon in 2007, the 5 Cs aim to foster a more balanced World Heritage List and further promote heritage as a tool of sustainable development through mutual understanding between societies. While this research did not intend to include any existing strategies in the evaluation, it is telling that the results of a comparative assessment of the entire World Heritage List highlight some of the core components of an existing strategy. The presented research confirms, through statistically significant results, that Communication, Community Involvement and in conjunction, Capacity Building in local communities are crucial elements for the performance of World Heritage Sites on a larger scale (Schmedt, 2021).

Even though the results support existing strategies in their scope and aim, they also indicate that the implementation of these strategies could be improved in many countries. As the example of Mexico demonstrated, there is immense potential for improvement in community involvement, communication and capacity building. The reports themselves identify these structural issues and should be taken into account when addressing specific issues or evaluating individual sites. In other words, the World Heritage community knows what actions are required to protect World Heritage Sites and has created a strategic framework for this purpose but has fallen behind in its execution. This study creates a clear argument for committing more resources and attention to the implementation of the 5 Cs, as it demonstrates their direct benefits for the performance of World Heritage Sites. These issues are at the heart of the Convention, and identifying them as aspects that can directly influence the performance of sites is the first step in addressing concerns and encouraging necessary changes. To address these issues, global and national institutions in the World Heritage community must come together and develop a concrete plan of implementation with the necessary resources. However, this brings up debates that have been ongoing in the field of World Heritage for years.

The World Heritage Convention was born out of the recognition that certain places in the world must be preserved and protected for future generations and that this requires the combined efforts of the international community. While the establishment of the 5 Cs indicates that specific needs are recognized among the World Heritage community, more resources must be dedicated to their implementation. A site is not guaranteed adequate protection and support just because it receives the status of UNESCO World Heritage. The World Heritage community must address the issue of an increasing number of inscriptions on the World Heritage List, with fewer resources for their protection and insufficient support from the Advisory Bodies, World Heritage Centre and national institutions, which are already operating at their capacity. There must be a renewed focus on existing sites to take stock of what is working and what might need to be changed to protect these sites for future generations. In essence, it comes down to whether the World Heritage System can return to its roots as an international community to protect the world's most outstanding sites. Moving forward into the next 50 years, the World Heritage community, and the involved states parties, in particular, will need to decide on their priorities. Do they want to follow the fundamental idea of the Convention and protect World Heritage Sites for future generations, or do they want to have more and more tourist destinations and commodified places? It is the most important decision to make, and it will decide if the Convention continues to be hailed as a success or if it succumbs to politicization which could lead to a degradation of its original intentions and the protection and conservation of these outstanding sites for future generations.

References

Albert, M.-T. (2012). Perspectives of World Heritage: Towards future-oriented strategies with the five 'Cs'. In M.-T. Albert, M. Richon, M. J. Viñals, & A. Witcomb (Eds.), *Community development through World Heritage* (pp. 32–38). World Heritage Centre, UNESCO.

Bertacchini, E., Saccone, D., & Santagata, W. (2011). Embracing diversity, correcting inequalities: Towards a new global governance for the UNESCO World Heritage. *International Journal of Cultural Policy, 17*(3), 278–288.

Bonini Baraldi, S., & Ferri, P. (2019). From communism to market: Business models and governance in heritage conservation in Poland. *Journal of Management and Governance, 23*(3), 787–812.

James, L., & Winter, T. (2017). Expertise and the making of World Heritage policy. *International Journal of Cultural Policy, 23*(1), 36–51.

Labadi, S. (2007). Representations of the nation and cultural diversity in discourses on World Heritage. *Journal of Social Archaeology, 7*(2), 147–170.

Lindholm, K.-J., & Ekblom, A. (2019). A framework for exploring and managing biocultural heritage. *Anthropocene, 25*, 100195.
Logan, W. (2001). Globalizing heritage: World Heritage as a manifestation of modernism and the challenge from the periphery. In *20th century heritage: Our recent cultural legacy: Proceedings of the Australia ICOMOS National Conference 2001, 28 November–1 December 2001, Adelaide, the University of Adelaide, Australia*, pp. 51–57.
Maksić, M., Dobričić, M., & Trkulja, S. (2018). Institutional limitations in the management of UNESCO cultural heritage in Serbia: The case of Gamzigrad-Romuliana archaeological site. *Land Use Policy, 78*, 195–206.
Meskell, L. (2013). UNESCO's World Heritage Convention at 40: Challenging the Economic and Political Order of International Heritage Conservation. *Current Anthropology, 54*(4), 483–494.
Schmedt, E. (2021). *World Heritage Governance* (28415337). Doctoral Dissertation, University of Massachusetts Boston. ProQuest, https://www.proquest.com/docview/2545588337?pq-origsite=primo&accountid=28932
Schmitt, T. M. (2009). Global cultural governance. Decision-making concerning world heritage between politics and science. *ERDKUNDE, 63*(2), 103–121. https://doi.org/10.3112/erdkunde.2009.02.01
Schmitt, T. M. (2011). *Cultural Governance: Zur Kulturgeographie des UNESCO-Welterberegimes*. Franz Steiner Verlag.
Trupiano, G. (2005). Financing the culture in Italy. *Journal of Cultural Heritage, 6*(4), 337–343.
Yakusheva, N. (2019). Managing protected areas in Central Eastern Europe: Between path-dependence and Europeanisation. *Land Use Policy, 87*, 104036.
Zan, L., & Bonini Baraldi, S. (2013). The heritage chain management. General issues and a case study, China. *Journal of Cultural Heritage, 14*(3), 211–218.

Chapter 6
World Heritage and Global Governance: Thematic Reflections

Roland Bernecker, Nicole Franceschini, Webber Ndoro, Christina Cameron, Maritta Koch-Weser, Lynn Meskell, Caroline Capdepon de Bigu-Poirrier, Boyoon Choi, Roger Negredo Fernández, Isabelle Rupp, and Tanja Willhalm

Abstract The article compiled by Roland Bernecker and Nicole Franceschini presents the personal reflections of several experts and young professionals on global governance and on how its evolution is affecting the implementation of the World Heritage Convention. Webber Ndoro reflects upon the distinction between local and global forms of governance, considering an African perspective on the colonial bias of World Heritage. Christina Cameron emphasizes the need and opportunities for a

R. Bernecker (✉)
Chair of Cultural Management, Brandenburg University of Technology Cottbus-Senftenberg, Cottbus, Germany
e-mail: bernecker@outlook.de

N. Franceschini
Independent Heritage Practitioner, Rome, Italy
e-mail: nicole.franceschini.work@gmail.com

W. Ndoro
International Centre for the Study of the Preservation and Restoration of Cultural Property, Rome, Italy
e-mail: wn@iccrom.org

C. Cameron
Université de Montréal, Montreal, Canada
e-mail: christina.cameron@umontreal.ca

M. Koch-Weser
World Heritage Watch, Berlin, Germany
e-mail: maritta.koch-weser@world-heritage-watch.org; mkochweser@earth3000.org

L. Meskell
Department of Anthropology, School of Arts & Sciences, and Weitzman School of Design, Penn Museum, University of Pennsylvania, Philadelphia, PA, USA
e-mail: lmeskell@upenn.edu

C. Capdepon de Bigu-Poirrier · B. Choi · R. Negredo Fernández · I. Rupp · T. Willhalm
Brandenburg University of Technology Cottbus-Senftenberg, World Heritage Studies, Cottbus, Germany
e-mail: capdecar@b-tu.de; choiboyo@b-tu.de; negrerog@b-tu.de; ruppisab@b-tu.de; Tanja.Willhalm@b-tu.de

M.-T. Albert et al. (eds.), *50 Years World Heritage Convention: Shared Responsibility – Conflict & Reconciliation*, Heritage Studies,
https://doi.org/10.1007/978-3-031-05660-4_6

more inclusive governance based on broader participation of stakeholders not foreseen in the Convention. Based on her experience in the institutional limits of international cooperation, Maritta Koch-Weser develops five recommendations for adjusting the system. Lynn Meskell builds her analysis on 5Cs, in response to those elaborated in the context of the Convention, and addresses the power of politics in the implementation of the Convention. These contributions are complemented with that of a group of master's students, who discuss the impact of civil society initiatives on the governance of the World Heritage system.

Keywords Global governance · Local communities · International cooperation · Civil society

6.1 Introduction

R. Bernecker (✉)
Brandenburg University of Technology, Chair of Cultural Management, Cottbus, Germany
e-mail: bernecker@outlook.de

N. Franceschini
Independent Heritage Practitioner, Rome, Italy
e-mail: nicole.franceschini.work@gmail.com

For a meeting organized in March 2020, we invited eminent experts to contribute personal reflections on global governance, a thematic issue progressively emerging as one of the main topics in the debates addressing the achievements and failures of the World Heritage Convention in the context of its 50th anniversary. Focusing on the political core of what we refer to as the World Heritage system, these reflections are personal stances distilled from decades of experience and insights into procedures, institutional rituals and political logic, weighed against the World Heritage Convention's ambitions for conservation, cooperation and transcultural opportunities. How does the evolution of global governance impact the dynamics of the Convention's implementation 50 years after its adoption? The reflections collected here offer broadened perspectives on trends that have become evident in the last decades.

The future of global heritage governance will inescapably shape the future of heritage. The question of governance is therefore of particular relevance to young and emerging experts in the heritage profession. Therefore, the voices of Master students from World Heritage Studies at Brandenburg University of Technology are also included, with their analysis of the impact of civil society initiatives on the governance of the World Heritage system.

6.2 International Versus Local Governance

W. Ndoro
International Centre for the Study of the Preservation and Restoration of Cultural Property, Rome, Italy
e-mail: wn@iccrom.org

The World Heritage Convention is an instrument mostly used for celebratory appreciation of diversity and heritage. To most governments it is an instrument of recognition, rather than of conservation and management. This clearly is the reason for the success of the Convention itself. Experts and international organizations like ICCROM are more concerned about the actual practice of conservation and management, which is precisely the core intention of the Convention.

We have to recognise upfront that it is an intergovernmental Convention where States Parties take the decisions. In my view, therein lies the fundamental issue of the Convention. For nation states, governance is about identity and self-preservation. The current pandemic with its vaccine nationalism and origins blame game (e.g., referring to COVID-19 as the Chinese virus or the South African deadly variant) clearly demonstrates the confrontational dynamics of states' interests and requisites of international governance.

We have been discussing in the past linking global and local in the implementation of the Convention. Perhaps we are missing a crucial point in this discussion – what is happening at the national level? How are national governments working to bridge the gaps between international and local practices and realities, other than by nominating new sites for inscription? National governments use heritage as an instrument of politics and power to further national priorities and identitarian claims. Yet, conservation, in whatever form, is practised at the local level. Global and local issues are also linked to the international order of governance; there are democratic and non-democratic states. How local issues are dealt with mainly depends on the governance of nation states, which impacts the practical processes of accepting, presenting and interpreting multiple narratives of heritage places. How can these local processes be accommodated in a World Heritage governance system that increasingly promotes culturally loaded approaches at the expense of locally driven systems and thought processes? Experts come from particular countries and regions.

From my non-expert experience work in Africa, it is evident that World Heritage does have some success stories to tell. Particularly where it has generated political interest at the national level. In turn, this can lead to greater attention to heritage organizations and better management, benefitting some cultural sites (for example, in Burkina Faso with the nomination of Loropéni). But we have to ask fundamental questions on the whole process. In places like Africa, what has been the advantage for local communities? Are they even considered in the governance of these sites, which are in protected zones and managed by conservation experts? Has the label of World Heritage caused any improvements in their livelihoods? Perhaps that is not the concern of the Convention, considering the process ends for most governments with the nomination of a site.

There are also important issues to consider in light of the predominance of cultural sites in Africa focused on archaeological or European heritage. Even the so-called "outstanding universal value" of a site like Great Zimbabwe refers to the Queen of Sheba and the role of the site as a medieval capital – despite archaeological evidence to the contrary. World Heritage concepts and practices could learn from recent discussions on heritage and museum developments arising from the *Black Lives Matter* movement. Heritage sites in Africa celebrate colonial history; rarely do we find places dedicated to African achievements and liberation struggles. The *Rhodes Must Fall* movement has clearly demonstrated the need to balance colonial history and African heritage. World Heritage experts who dismiss these movements as political machinations far removed from the realities of the hard conservation science and architectural analysis, give credence to the idea that African heritage begins and ends with colonialism.

6.3 Harnessing the Global Village to Conserve World Heritage Sites

C. Cameron
Université de Montréal, Montreal, Canada
e-mail: christina.cameron@umontreal.ca

The 1972 UNESCO World Heritage Convention is a treaty among nation states. From a governance perspective, States Parties consider that they control the process and the outcomes. From a legal viewpoint, they are correct. The Convention does not mention civil society in its text and has been correctly criticized for its concentration of power at the level of national governments. Initially uninterested in the participation of civil society, the World Heritage Committee modified its position in 2005 by explicitly encouraging the participation of a wide variety of communities, stakeholders and non-governmental organizations in the work of World Heritage, albeit not as decision makers.

Much has changed in fifty years, especially in the field of communications. Information technologies have connected people throughout the world, fostering an informed and engaged citizenry. Reconsideration of global governance of World Heritage is therefore timely, especially in light of the many pressures on World Heritage sites that nation states alone have not been able to respond to. To fulfil the promise of the World Heritage Convention, broad intersectoral engagement in conservation is needed to confront the threats of our time.

In the 1960s, when television was in its infancy, a visionary Canadian university professor, Marshall McLuhan, predicted that "the new electronic interdependence would recreate the world in the image of a global village" (1962, 1964). McLuhan invented this catchy term "global village" to suggest that the world had grown closer through this new technology. While the World Heritage Convention's vision to conserve significant cultural sites and natural areas remains valid, the phenomenon of an interconnected world has affected the implementation of the Convention and exposed some fault lines in the system.

From the outset – and still today – States Parties consider that they "own" the Convention. Governance is in the hands of national governments who engage in intergovernmental decision-making and provide direction to implement the Convention within their countries. This is essentially a top-down exercise, a closed system that limits its effectiveness. Within countries, links between central authorities and individual site managers are often not strong, given the myriad levels of government that interrupt the flow of information and stifle dialogue. Even weaker – or non-existent – are the connections between States Parties and non-state actors. The top-down process is exclusive. As a result, there are no access points for the excluded.

Among those with an interest in World Heritage but not mentioned in the Convention, are organizations, including, among others, governmental and non-governmental organizations, universities, research institutions, civil society, communities and individuals. Many are well aware of World Heritage activities through traditional and social media. Many also play a significant role in protecting and conserving World Heritage sites. Yet, the Convention does not assign any official role to them. States Parties, the advisory bodies (ICOMOS, IUCN and ICCROM) and the UNESCO secretariat all enjoy a statutory legitimacy that outsiders do not. In fact, voices from external groups have no official place in World Heritage processes. Moreover, if representatives from these groups are admitted as observers to committee sessions, their voices are only heard at the discretion of the chairperson – and only after the committee has taken its decision on their issue.

Beyond the exclusion of voices, the governance model is also weakened by increasing politicization, well documented by Lynn Meskell (2018) in *A Future in Ruins: UNESCO, World Heritage and the Dream of Peace*. The live streaming of the annual sessions of the World Heritage Committee in the past decade has revealed this situation to a global audience. Observers from around the world can now better understand how this politicization occurs, particularly in the processes for new inscriptions and additions to the List of World Heritage in Danger. It is arguable that many sites are in reality national priorities – presented by state actors – and *not* sites of global significance as required by the Convention. Despite that, committee delegations collaborate among themselves to inscribe sites on the World Heritage List and keep imperilled sites off the In Danger List, often against the recommendations of the advisory bodies. In his recent book, Christoph Brumann comments that "the practice of agreeing to the wishes of member states has simply become too well-established.... The idea of solidarity and the multilateral protection of heritage valuable to all of humanity has been pushed to the background" (2021). The politically charged atmosphere at committee sessions leads to the conclusion that the governance model is no longer effective.

Shortcomings can also present opportunities. Weakened governance can make space for new actors and diverse voices. The exponential growth in virtual meetings during the COVID-19 pandemic has connected a remarkable number of groups and people interested in World Heritage matters, ranging from academics, civil society associations, community groups, the private sector and so forth. So far, the World Heritage Committee has not granted official standing to any of them.

The 2017 establishment of the International Indigenous Peoples' Forum on World Heritage (IIPFWH) provides a useful example of how governance is expanding beyond the States Parties and UNESCO. It is acknowledged that Indigenous peoples have long and deep connections to their lands and waters. They bring traditional knowledge and a holistic understanding of the unbroken bonds between culture and nature, a perspective that is important for enriching an understanding of World Heritage.

Yet in 2001, the World Heritage Committee flatly rejected a proposal to create a World Heritage Indigenous Peoples' Council of Experts (WHIPCOE) whose purpose was to mobilize Indigenous voices and complementary competencies to protect and manage World Heritage sites. While some committee members acknowledged the special role that Indigenous peoples could play, others questioned the definition of Indigenous peoples and the relevance of such a distinction in different regions of the world. One committee member made the bold – and erroneous – statement that there are no Indigenous peoples in Asia! As a result, the proposal was not approved (UNESCO, 2001). Fast forward to 2017, when a group of Indigenous peoples took charge and created an independent organization called the International Indigenous Peoples' Forum on World Heritage (IIPFWH). The purpose of this standing global body is to engage with the World Heritage Committee during its meetings and represent the voices of Indigenous peoples with regard to the World Heritage Convention. Since the committee has no power over the creation of the IIPFWH, at the 2017 session in Krakow it simply took note of the establishment of the forum as an important reflection platform (UNESCO, 2017).

Other organizations have adopted the same model, including the independent World Heritage Watch, which mobilizes citizens and local groups to document problems at specific World Heritage sites and attempts to influence decisions through two-minute interventions at committee sessions. Among the many other organizations that attend committee meetings as observers are Greenpeace, Wildlife Conservation Society, Rivers without Boundaries, The Wilderness Society, the World Monuments Fund and the World Wide Fund for Nature.

A new NGO is on the horizon, a fledgling organization led by an informal coalition of concerned individuals. The newly minted OurWorldHeritage Foundation seeks to involve citizens, site managers, and civil society groups, as well as professionals, scholars, the private sector and emerging professionals, in the protection and conservation of cultural and natural sites. Throughout 2021, the initiative hosted twelve monthly virtual debates on critical issues facing World Heritage today, such as threats from climate change, overtourism, disasters and conflicts, as well as the listing process and new approaches to heritage conservation. The initiative hopes to mobilize a robust global network to renew the original spirit of the Convention and reinforce heritage protection for the next 50 years.

The fiftieth anniversary of the World Heritage Convention creates an opportunity to consider more inclusive governance and broader participation in the conservation of sites of importance to us all. Consider the familiar proverb observing that it takes a village to raise a child. Since each World Heritage site is precious, like a child, let us harness the energy of our global village – to use McLuhan's phrase – to look after our children, these special places on this earth.

References

Brumann, C. (2021). *The best we share: Nation, culture and world-making in the UNESCO World Heritage arena*. Berghahn.

IIPFWH. (n.d.). International Indigenous Peoples Forum on World Heritage. Available at https://iipfwh.org/

McLuhan, M. (1962). *The Gutenberg galaxy: The making of typographic man*. University of Toronto Press.

McLuhan, M. (1964). *Understanding media: The extensions of man*. McGraw Hill.

Meskell, L. (2018). *A future in ruins: UNESCO, World Heritage and the dream of peace*. Oxford University Press.

UNESCO. (2001). *Report of the 25th session of the World Heritage Committee (WHC-01/CONF.208/24, XV)*. UNESCO. https://whc.unesco.org/archive/2001/whc-01-conf208-24e.pdf

UNESCO. (2017). *Decisions adopted during the 41st session of the World Heritage Committee (WHC/17/41.COM/18, 41)*. https://whc.unesco.org/archive/2017/whc17-41com-18-en.pdf

6.4 Governance Challenges

M. Koch-Weser
World Heritage Watch, Berlin, Germany
e-mail: maritta.koch-weser@world-heritage-watch.org; mkochweser@earth3000.org

I want to pick up on a word that Christina Cameron used: The 50th anniversary is an opportunity – an opportunity for correcting the current system. Opportunities are not just about correcting shortcomings. In this text, I present five recommended adjustments that we want to see over the coming years.

The first recommendation is a 180-degree shift toward the conservation and maintenance of all sites that are already inscribed, rather than emphasizing this politicized process of continuously nominating new sites. With that shift comes a re-emphasis of the importance of civil society. World Heritage Watch works on the premise that long-term monitoring is very effectively done by an engaged local and regional civil society. They can truly make a difference and contribute to the shift toward conversation and maintenance.

Our governance discussion is not just about adding to the rulebook. Rather, I was also wondering about a word that has been missing from much of our discussions – accountability. Having access, a voice, to be heard, to have a role and a narrative are all important, but not enough. We must consider accountabilities. One example of this lack of clear accountabilities is the case of Canaima in Venezuela, which was highlighted in the World Heritage Watch report this year. Canaima is a fantastic region becoming overrun with wildcat miners and where it is likely that the government is a perpetrator just as much as a State Party. We need to dedicate more thought to these conflicts in all those celebratory meetings that we will have over the coming year. Let me emphasize that enabling civil society to participate in a more structured way will not be enough. There also need to be more clearly defined accountabilities for responding to the issues raised by civil society actors.

With the challenge to make a 180-degree switch to conservation and maintenance also comes the second question, whether we are going to see an inflation in the nominations of new sites? If we do not want the nominations process to become inflationary over time, what could be done? I think it would be worthwhile over the coming months to think about an annual cap on new nominations using a sliding scale: instead of twenty a year, at first there should be a maximum of eighteen a year, followed by a maximum of sixteen the following year, and so on until there is a small number that can be maintained. Some very pragmatic and operationally poignant thought needs to be given to the total number of sites to be nominated – sites we really want to conserve and maintain in perpetuity.

The third recommendation is that the opportunities offered by modern communication must be exploited. These communication capabilities could be used to integrate World Heritage with national education systems in states that are party to the Convention, and provide them with routine access to modules on World Heritage. Such an education programme would contribute to the maintenance of sites "in perpetuity". We must catch up with one generation of young people after another. A top-down declaration of love for culture is not enough; it has to permeate the education system. Today's communication tools can facilitate this.

The fourth recommendation is related to a point that has been raised frequently: there is a need to reinforce connections between World Heritage and other UN goals, such as the Sustainable Development Goals, the Climate Agenda, the Biodiversity Agenda and Indigenous Peoples' Human Rights. For example, the World Heritage Committee could dedicate a regular segment of their sessions to improving these connections and the opportunities for sites that arise from them, instead of focusing on the nomination of new sites. These other programmes are often well-funded and could be of benefit for World Heritage, which leads to my final recommendation.

My fifth and final point is about money, which is always in short supply. The World Heritage Fund for 2021 has a total of $5.6 million. This level of funding is insufficient to build climate resilience in so many areas, let alone achieve any other goals. Either existing funds in the international system, such as the World Heritage Fund, need to be supplemented, or new funding sources need to be identified or created, for example, a World Heritage Global Environment Facility or a standalone World Heritage Resilience Facility. My point is simply: something needs to happen. If we do not put the money where the mouth is when objectives are declared, we will forever remain in financial trouble and fail to achieve a self-reliant World Heritage system.

These are my five recommendations:

1. A 180-degree shift to conservation and maintenance;
2. A gradual cap on the nomination numbers;
3. Developing educational systems that integrate World Heritage modules;
4. Greater integration of World Heritage with other UN programmes and goals (and their respective often well-funded programmes);
5. And finally, the need for a more suitably structured and adequately financed World Heritage Fund or the creation of a World Heritage Resilience Facility of some sort in one of the larger United Nations system institutions.

Addressing younger, emerging experts, I would like to make a final overall remark on the UN, the World Bank, and the regional development bank systems. If you are a student, study the whole system, study all the sources of finance, all the programmatic lines and chart them. Seek out ways the guidelines in each of these institutions and funding lines could be adapted to provide support for World Heritage. We did this at the World Bank in the 1990s. World Heritage often gets lost between funds and institutions that care for culture on the one hand, and institutions that care for the environment on the other. And when you come with the integrated World Heritage mandate – an intelligent and wonderful package – they say "we would love to do it, but we are not structured to do it". So please look at these broader structures and make them work better for World Heritage in the next fifty years.

6.5 UNESCO and Global Governance

L. Meskell
Department of Anthropology, School of Arts & Sciences, and Weitzman School of Design, Penn Museum, University of Pennsylvania, Philadelphia, PA, USA
e-mail: lmeskell@upenn.edu

This brief contribution is a reflection on the contemporary challenges of global governance in World Heritage. The themes that emerge are deeply interconnected as a result of the history of UNESCO as an institution and the particular political evolution of the 1972 Convention. Briefly, I would like to touch upon issues comprising a set of five Cs: communities and rights, conservation, colonialism, conflict and civil society. Of course, more than a decade ago, UNESCO devised its own five Cs, which were designated strategic objectives for the World Heritage Convention: credibility, conservation, capacity-building, communication and communities. UNESCO judged those criteria to be central in sustaining the operations of the Convention, whereas my own selection is directed at the challenges to the workings of World Heritage out there in the world, to some crucial issues that can no longer be side-lined. I hope it can also provide a platform to begin a fuller reflection on the 50th anniversary of the 1972 World Heritage Convention.

6.5.1 Communities and Rights

One major issue to be confronted in terms of governance is the disconnect between the World Heritage Convention and other long-standing United Nations conventions and declarations. This is particularly pertinent given the lack of harmonization between the 1972 Convention in regard to human rights, specifically in terms of Indigenous groups and their rights to heritage and involvement in the processes that guide inscription and conservation. After all, UNESCO was established to contribute to peace and security globally by promoting collaboration among nations through education, science, culture and communication so as to further universal respect for justice, the rule of law, human rights and fundamental freedoms set out in the Charter of the United Nations. However, today's World Heritage regime has yet to fully incorporate the living aspects of heritage that would further necessitate rights of inclusion, access, use, and benefits. Many complain that even at the administrative level there are constraints. Indigenous representatives report that simple attempts to contact the World Heritage Centre go unanswered and that their concerns are not addressed. Since UNESCO is tasked with culture in the UN, this situation remains unsatisfactory. Other UN bodies have moved ahead; for example, there is a UN council that deals with Indigenous Peoples' rights and business ethics. This suggests that there is an urgent need to strategize across agencies since a World Heritage inscription is often part of a broader system that can further marginalize communities and, in some cases, lead to human rights violations.

Conservation can easily be deployed to marginalize or force out local people, as recently happened at the World Heritage site of Hampi (India). This checkered conservation past escalated in 2010 when the Archaeological Survey of India took control of the Virupaksha Temple and Hampi Bazaar and declared its residents to be squatters and stripped them of any legal rights. The stalls, shops, restaurants and dwellings around the main temple were deemed illegal encroachments despite having their own living histories. Then in July 2011, bulldozers levelled shops, stalls and hotels, damaging the original medieval fabric of the bazaar in their wake. This is why many Indigenous communities are sceptical of UNESCO processes and the listing of sites. In dealing with their own governments, and with national bodies, the protectors and the perpetrators can be one and the same. This is one serious challenge and there is need to create a legal mechanism to hold states accountable in respect to rights. Several of us are working with international lawyers in the preparation of a policy document to be presented at the General Assembly of States Parties, focused on enforcing and respecting the legal obligations provided for by the 1972 Convention.

6.5.2 Conservation

It is hard to imagine what a world without World Heritage conservation might look like today. Almost certainly we would be witnessing even greater degrees of development and exploitation than are being challenged by multilateral agencies, NGOs, and civil society. From this perspective, UNESCO has proven highly successful in mobilizing external commitment and intellectual resources for its projects, World Heritage perhaps being the ultimate example. Indeed, it is hard to critique UNESCO's nobility of purpose or desire to forge intellectual ties around the globe. However, its norms remain expert-driven and Eurocentric, while its treaties lack incentives for compliance and implementation amongst its Member States.

Today we are witnessing the never-ending pushback from States Parties on matters related to conservation and danger listing. States Parties privilege their socio-economic self-interest and will resist any decision that impinges upon those directives. States Parties are entirely entrepreneurial in their strategies to avoid danger listing, as seen with Venice, Vicenza, Vienna, London, and Liverpool. States perceive World Heritage as an instrument to pursue political objectives. Those with more political leverage and stronger lobbying capacities can exert greater influence on decisions in the World Heritage system. The question remains: If European nations are exerting their power through expansive lobbying, why should other states not do the same? There is a need to re-focus on conservation and communities in the governance strategies of World Heritage.

The World Heritage List ensures that a global public is more aware of conservation issues. Yet, the List creates a competitive arena for nationalist aspirations and rivalries where the potential economic developments that accompany listing increasingly outweigh preservation priorities. In the past decade, the List has been manipulated to secure and maintain the UNESCO brand in tandem with unhindered development, exploitation and commercialism. Countries such as Italy, France, Germany and the United Kingdom have relentlessly marketed the UNESCO brand, taking full advantage of their long experience within the World Heritage system, at the expense of the conservation principles it represents. Poorer, less powerful nations are more readily criticized and have less bargaining power to fend off censure. Conservation, once so central to UNESCO's mission and the raison d'être of the 1972 Convention, is now itself in danger.

6.5.3 Colonialism

The issues outlined above reveal much about the Eurocentric nature of World Heritage and how specific nations operate within UNESCO. In previous work with my colleagues, we have demonstrated that former colonial ties are still the strongest link in the support that States provide to each other in the World Heritage system. These ties play a key role in alliance building. It has to be acknowledged that the UN was built in the mid-20th century mainly to find ways for the victors of the war to keep their colonies. These colonial histories recall the particular European genealogy of international organizations from the League of Nations to UNESCO and their limitations to inculcate a culture of peace and to reconcile fundamentally different worldviews.

Successful World Heritage nominations are linked to the support not just of one or two neighbours but also of a wide array of countries spanning the globe. Moreover, these interdependencies are no longer confined to a single group of countries (e.g., the West or industrialized democracies) but have expanded to encompass a diverse range of economic regime types, religions, and cultures. Indeed, support for listing is more closely tied to former colonial relationships and trade partnerships than to regional, religious, or linguistic affiliations. In our analysis over more than a decade, we found that support for inscribing a site is more likely to occur if a member of the World Heritage Committee was a former colonizer of the country proposing a World Heritage nomination. This can be explained considering that several former colonies often nominate sites of their colonial past that are also historically linked with their former colonizer. This is part of a wider process of support that begins during the initial stages of the nomination process, with technical and/or financial assistance funnelled to former colonies through Fund in Trust agreements with UNESCO.

If we look at just one region that continues to experience the negative effects of colonial power relations, representatives from Africa increasingly complain that World Heritage processes are stacked against them. They have repeatedly denounced programmes like the Global Strategy, which failed to boost African regional representation but instead created a proliferation of new heritage categories, opportunistically used by overrepresented nations to inscribe more and more of their properties. They have also been vocal in demanding that African sites should not be targeted for the Danger List while those of more powerful countries avoid discussion and are subject to much less censure. African delegates have also deployed the rhetoric of UNESCO's 'Priority Africa' in challenging the Convention guidelines, specifically advocating that a balance be struck between conservation and sustainable development in the African context. They have argued for increased use of African expertise in all aspects of the implementation of the Convention.

6.5.4 Conflict

Forged in the twilight of empire and led by the victors of the war and major colonizing powers, UNESCO's founders sought to expand their influence through the last gasps of the civilizing mission. Beginning as a programme of reconstruction for a war-ravaged Europe, indeed, UNESCO was born of war and established in the wake of conflict with the dream of overcoming future conflict. But its establishment was bound up with the actions of nation states rather than the variety of actors and armed groups present today. What we are witnessing now is not only the difficulty of dealing with devastating conflicts that also impact material heritage but those proxy wars that play out in the World Heritage arena.

There are also political conflicts within the workings of the Convention. While outspoken about heritage destruction in Syria and Iraq by non-state actors, UNESCO has been relatively silent about the devastation in nations like Yemen. Neither the organization nor the majority of the Member States wants to criticize powerful nations at the forefront of such conflicts because some of these countries currently make large financial contributions, including Saudi Arabia, the Russian Federation, and the Gulf States. It is also important to keep in mind that we are not simply talking about long-term preservation of heritage sites, but we are talking about the fate of people. UNESCO's response to the crisis in Mali has often been criticized for its singular approach to material destruction and reconstruction, since many observed that World Heritage seemed primarily concerned with stones.

I would further argue that the system is now faced with an increasing number of proxy wars being played out in the World Heritage forum. Moreover, the Convention can be used to gradually chip away at state sovereignty and borderlines or to reignite historic conflicts, some recent examples involve China and Japan, Thailand and Cambodia, as well as Turkey and Armenia. Additionally, these proxy conflicts are mobilized across the Conventions, and what happens in the context of the 1972 Convention can be later paid back within the scope of another convention, for example, the 2003 Convention for the Safeguarding of the Intangible Cultural Heritage. We might see these as the consequences of predatory states, and such aggressive steps are overtaking previous commitments to intellectual and cultural cooperation. In this regard, the politicization of UNESCO is no different from other UN agencies, and heritage is no different from other political issues. Cultural and natural heritage might seem to operate only within the remit of soft power, but instead are often used in the political realm as a hard power.

6.5.5 Civil Society

The role of civil society has become a subject of discussion at UNESCO, particularly in their committee meetings and public-facing activities. Yet, as we have seen with the very limited Indigenous and local community participation, civil society actors have limited to no power within the system. NGOs and civil society organizations can bring their perspectives to participants in World Heritage Committee sessions, for example, but only after legally binding decisions have been taken. This has resulted in a type of courtesy performance that has no bearing on decision-making or the implementation of the Convention. It is important to keep in mind that there are also major interests involved under the umbrella of civil society, with big businesses, economic and political interests and money involved. There is a need to find adequate legal solutions to change the system from within; otherwise, if we are serious in addressing the needs of people and their fundamental rights, it will be difficult to get states to recognize those needs and supersede their own narrow self-interest.

Finally, there is more and more interest in global heritage, and there are now new approaches to engaging and mobilizing that concern. Individuals and organizations are bypassing UNESCO and starting their own international agencies. The International Alliance for the Protection of Heritage in Conflict Areas (ALIPH), based in Switzerland, is a good example of a major foundation with expansive funding to work on heritage preservation and community-building worldwide, with many projects situated in the Middle East. We are seeing the proliferation of NGOs, whether in the United States or India, as well as across the Middle East and Europe. Endangered heritage has been a rallying point, as has Indigenous culture, the environmental movement and concerns for a more just, equitable and sustainable future.

As UN Secretary-General Antonio Guterres proposed, the United Nations must also change and focus on delivery, not process, on people, not bureaucracy. For World Heritage that may entail not inscribing sites on a list at all, but rather allowing communities to determine their own paths. Nations still marginalize and persecute their minorities, so simply appealing to sovereign entities offers no solution. The measure of international oversight that World Heritage affords still yields considerable value. The past practices of working closely with nongovernmental agencies, universities, and other institutions might be further reinvigorated, although those organizations lack UNESCO's reach and capacity to embrace the world and have their own agendas. In such a time of reduced circumstances, expanding networks and collaborations may prove the most expedient way forward.

6.6 Global Heritage Governance

C. Capdepon de Bigu-Poirrier · B. Choi · R. Negredo Fernández
I. Rupp · T. Willhalm
Brandenburg University of Technology Cottbus-Senftenberg, World Heritage Studies, Cottbus, Germany
e-mail: capdecar@b-tu.de; choiboyo@b-tu.de; negrerog@b-tu.de; ruppisab@b-tu.de; Tanja.Willhalm@b-tu.de

Since the adoption of the World Heritage Convention in 1972, UNESCO has been the leading agency for the creation of a global heritage framework. The Convention states that "(…) deterioration or disappearance of any item of the cultural or natural heritage constitutes a harmful impoverishment of the heritage of all the nations of the world", and that "protection of this heritage at the national level often remains incomplete because of the scale of the resources which it requires".

This government-centred framework implies that States Parties to the Convention are entirely responsible for its implementation. Hence, heritage protection relies on States Parties and their willingness to cooperate with one another and with UNESCO as a multilateral catalyst. Although States Parties are the primary World Heritage actors, there has been some, albeit limited, integration of the expert community and civil society into the operational structure of the Convention through the Advisory Bodies ICOMOS, ICCROM and IUCN, explicitly mentioned in the treaty. This inclusion provides some access for civil society, but it is not open-ended – the Advisory Bodies are numbered and named in the Convention.

In the decades following the Convention's drafting, the diversity of actors involved in international politics increased considerably. Civil society organizations positioned themselves to contest state power in many areas and claim broader participation in decision-making processes. However, the increasing involvement of non-state actors in global governance arrangements was a challenge. Civil society actors undertook to remind States Parties about legal compliance with conventions they had signed up to, thus pushing for more people-oriented, inclusive and transparent processes.

In different political areas, civil society platforms and non-governmental organizations have made a place for themselves at the table, becoming relevant actors that would need to be taken into account when drafting, executing, amending or abolishing policies. For example, Greenpeace and the World Wildlife Fund are making their voices heard on contemporary climate issues; the World Social Forum's deliberations are introducing possible structural reforms towards a more people-oriented world economy. Thousands of activists are contributing to the online documentation of human rights violations through social media. Although not using conventional frameworks, these examples highlight actors that have become influential players within a transforming system of global governance.

As in other areas, non-governmental organizations and civil society initiatives have been created in the World Heritage arena to fill the gaps of non-inclusive governance and provide critical support to the World Heritage Committee, World Heritage Centre and the wider system.

Only recently, however, has civil society been indirectly implicated as an actor in the World Heritage system. In 2002, the World Heritage Committee adopted the Budapest Declaration on World Heritage aiming to "ensure the active involvement of our local communities at all levels in the identification, protection and management of our World Heritage properties." The involvement of local communities was later formalized when a fifth "C" for "Communities" was added to the four other previously identified strategic objectives of the Convention – *credibility*, *conservation*, *capacity-building* and *communication*. These developments seem to affirm the role of communities in heritage governance and lend themselves to the perception of a potential opening for civil society involvement in the overall governance of the World Heritage framework. This principle was, however, not endorsed in the States Parties' approach to decision making.

We wanted to understand how in this context, independent heritage initiatives can influence the system, and more broadly their role in global heritage governance. We looked at the two most relevant civil society-driven initiatives that address the implementation of the World Heritage Convention.

World Heritage Watch is a German-based NGO that defines itself as the "global voice of civil society for world heritage." The NGO's main activities include two annual fora and a published report. They are intended to provide the World Heritage Committee and its advisory bodies with supplementary, civil society sourced and up-to-date information about World Heritage properties and their management. That information – if taken into account – is meant to enable the Committee to make better informed and less politically biased decisions. Therefore, the World Heritage Watch fora and report focus particularly on heritage sites that need urgent attention or are ignored by the Committee and Advisory Bodies.

World Heritage Watch facilitates its fora and produces its report in cooperation with members of its informal global network of civil society groups. It is important to note that the goal of World Heritage Watch is not to inspire civil society engagement with heritage but rather to work with groups that are not officially recognized and included in the decision-making processes but are engaged in heritage protection and management at the property level. The lobbying activities of World Heritage Watch are centred around gaining official access and recognition within the World Heritage system for civil society actors and their contributions to heritage protection. Of course, progress in this regard has been gradual and slow as the system is set up to exclude these actors.

World Heritage Watch also promotes solidarity between the various civil society groups in its global network. During the World Heritage Watch fora, these groups and organizations share information and expertise but also offer each other support. Thereby they can demystify the processes of the World Heritage system and increase public awareness of World Heritage. These clarifying activities are invaluable for the many civil society groups unfamiliar with the Convention's modalities and the somewhat hermetic processes of its implementation.

OurWorldHeritage is an emergent NGO presenting itself as a civil society platform. The main goal of this organization is to foster a public dynamic that is strong enough to have an impact on the World Heritage system, directed towards a more inclusive and less politically dominated implementation of the 1972 Convention. The initiators of OurWorldHeritage are mainly concerned about the politicization of the World Heritage Committee and State Parties dedicating their resources to the nomination process rather than protecting their heritage (nominated or not). This neglect of conservation endangers what the Convention is supposed to preserve. More systematic and inclusive consideration of diversity and human rights is needed to modernize the system. These would need to include individuals and groups of all profiles with particular emphasis on the interests of Indigenous communities.

The founders of OurWorldHeritage have demonstrated an impressive capacity to gather a substantial number of young and emerging professionals as volunteers for the initiative. This was possible thanks to their personal, professional and academic networks and strong presence on social media. They also emphasized creating communication channels for people of different native languages to reach the broadest possible section of the public. Additionally, the organization has successfully involved established heritage professionals, such as site managers and IT, tourism and industry actors.

Rather than directly confronting the World Heritage system with its deficiencies in a non-constructive way, OurWorldHeritage has been generating thought-provoking debates and collecting substantive recommendations. This project, running through 2021, should produce a significant set of recommendations to submit to political authorities responsible for the World Heritage system.

The core principle of the 2030 Agenda for Sustainable Development is "leaving no one behind", which means eliminating all forms of discrimination and injustice. Given the many excluded and marginalized communities in the World Heritage system, it is essential to consider how civil society actors can enhance the representation of communities in global heritage governance. The two examples of World Heritage Watch and OurWorldHeritage demonstrate that civil society actors can help make the political space more transparent. The two initiatives create spaces for heritage researchers, professionals, the public and anyone interested in heritage to counteract the discrimination, exclusion, domination and inequality that continuously emerges in the area of World Heritage. One of the main concerns of the initiatives is how politically biased and non-transparent decisions and processes impact the inclusiveness, equity and broader effectiveness of the World Heritage system and the global heritage governance it represents.

These two initiatives show why civil society organizations should be involved in shaping cultural heritage policy and illustrate how these organizations can expand their sphere of influence and promote democratic opportunities. Civil society actors in the World Heritage system can create the necessary leverage and power to mitigate the gap between the decision makers and those relegated from the decision-making process. Their work can build platforms where the voices of marginalized communities can be heard. They cast a light on the hidden dynamics of what goes on behind the scenes. As mediators, they put local communities' opinions and concerns on the international stage and create entry points for the advice of heritage experts that do not have access to committee deliberations. These initiatives act as a point of convergence for all those voices that are ignored in the decision-making process.

The analysed initiatives are a qualitative representation of two different platforms that contribute to the same greater goal. World Heritage Watch and OurWorldHeritage have broadened the inclusion of civil society in a field in which – paradoxically – non-state actors have barely had a voice. Hopefully, the coming decades will see an increased dynamic of these and similar initiatives that can build a more solid grid of connections between initiatives, governments, experts, UNESCO staff and civil society at large.

Civil society actors are challenging and working to transform the dichotomy of inclusion and exclusion in all stages of heritage management and decision-making. As emerging experts in heritage, we believe it is vital to act towards a unified goal, a shared aspiration: diversity and equity through networking and the building of coalitions. To compel UNESCO and its affiliated state actors to more inclusive policies, it is necessary to create, mobilize, strengthen and empower civil society along with a more credible and transparent bottom-up approach in responding to a diversity of social actors. The principles and the objectives of the 1972 Convention can only be achieved through active participation and the shared responsibility of civil society.

Civil society platforms related to World Heritage still have a way to go and broader audiences to attract. We raise the following questions in order to stimulate a better understanding of the role we can all play in the system:

- How do civil initiatives contribute to a fairer and more representative World Heritage system? How can they have an active role in correcting flaws and filling gaps?
- How can initiatives cooperate to build a stronger network that allows for less fragmented civil society representation in World Heritage governance?
- How can initiatives influence state decisions at a local, regional and international level?

The role of civil society within the heritage sector – and specifically within the World Heritage system – is an under-researched topic. It has not yet received the academic attention it deserves. Further research is necessary to provide insights into the current state of heritage governance and the roles of diverse actors and to suggest avenues that can lead to a more equitable future for global heritage governance.

Chapter 7
Urban Transformation and Related Conflicts at UNESCO World Heritage Sites

Christer Gustafsson and Matthias Ripp

Abstract Cities are constantly changing. Today, policy-makers all over the world are discussing how to create the conditions for developing green, healthy and safe cities where people meet and innovations are created. There is a great need to change to a climate-smart society with people at the centre. Urban cultural heritage is also constantly changing; however, the World Heritage Convention and its tools take their starting points in the preservation of monuments and ensembles. This introduction to section four presents the emerging scientific concept of urban transformation, relates it to conflicts at UNESCO World Heritage Sites (WHS) and discusses possible ways forward in the overall frame of this book.

Keywords Urban transformation · City · World heritage · Development processes

7.1 Conflicts Related to Urban Transformation

There is not one conflict but a bundle of diverse conflicts related to urban transformation and the implementation of the World Heritage Convention (WHC). These conflicts can be related to the (unequal) distribution of wealth, health, access to common goods and different generations of users (McPhearson et al., 2016). Global challenges, besides increased urbanisation, include climate change, war, terror, different and conflicting values and differences between global and local agendas and objectives or other parts of the context in which these conflicts may arise. Some of these conflicts are affecting sites on the World Heritage List (WHL), and others are more considered a result of inscription, however, the distinction between cause and

C. Gustafsson
Department of Art History, Uppsala University, Uppsala, Sweden
e-mail: christer.gustafsson@konstvet.uu.se

M. Ripp (✉)
Organisation of World Heritage Cities, Regensburg, Germany
e-mail: ripp.matthias@regensburg.de

M.-T. Albert et al. (eds.), *50 Years World Heritage Convention: Shared Responsibility – Conflict & Reconciliation*, Heritage Studies,
https://doi.org/10.1007/978-3-031-05660-4_7

Table 7.1 Examples of conflicts associated with urban transformation at World Heritage Sites

World Heritage site	Conflict	Type of conflict	Included dimension	Scope
Kairo (Smith, 2003)	Uncontrolled expansion of urban settlement in the vicinity of the pyramids	Visual integrity as well as the urban context of World Heritage Sites (WHS)	Social visual economic	Whole Kairo
Bamberg (Kremer & Lehmeier, 2009)	Over-tourism	Conflicts due to different needs of temporary and permanent users	Social economic	WH area and central parts of the city
Berlin WHS (Bernt & Holm, 2009)	Gentrification	Change in the urban and social fabric due to uncontrolled market developments	Social cultural	Specific areas
Krakow (Szromek et al., 2020)	Over-tourism gentrification	Change in the urban and social fabric due to uncontrolled market developments	Social economic cultural	WH area and beyond
Cusco (Bromley & Mackie, 2009)	Over-tourism gentrification displacement of informal trade	Conflicts due to different needs of temporary and permanent users	Social economic	The old town, gateway to Machu Picchu and Machu Picchu

Note: Table prepared based on authors' considerations and the references mentioned in the table

effect is far from easy. While developing a typology or presenting an ultimate collection of these conflicts is not within the focus of this introduction, some examples from different contexts around the world are provided in Table 7.1.

Inscription on the WHL is not always the (only) cause for these conflicts. Conflicts can be the result of different needs of temporary (e.g., tourists) and permanent (inhabitants) users of the city. World Heritage listing can lead to significant changes in the urban fabric and a range of conflicts between temporary and permanent users of the city (Siemer & Matthews-Hunter, 2017).

The example of gentrification can be observed in a number of WHS around the globe (Krakow, Rhodos, Amsterdam, Cusco, Shanghai, Visby, and several others) and is often caused by multiple factors, some of which can be related to general societal changes like an improved overall economic situation (Zuk et al., 2018). As a result of the unique history of Berlin as a divided city, the process of gentrification has developed at times and in patterns that are markedly different from other global cities (Siemer & Matthews-Hunter, 2017). In different districts, examples of "new-build gentrification" (Davidson & Lees, 2005) with the boom of luxury housing estates and "rental gentrification" (van Criekingen, 2010) with the displacement pressure resulting from the gap between long-term rental agreements and new contract rents are evident. In addition, elements of "tourism gentrification" (Gotham, 2005) with the transformation of rental housing into holiday flats can be observed,

Fig. 7.1 Protest sticker from an NGO against gentrification in Berlin. (Note: Image by Matthias Ripp, 2015)

with the stock of social housing reduced from 370,000 units in 1993 to less than 150,000 in 2012 (Holm, 2014) (Fig. 7.1).

Other conflicts that are related to urban transformation are, e.g., over-tourism, which is noticeable in many WHS such as Krakow, Cusco and Bamberg, and displacement effects on traditional crafts and business such as informal trade in Cusco. For many WHS, conflicts are a result of challenges of global significance like climate change. Urban WHS such as Regensburg are struggling with an increased number of days with extreme temperatures and the effects of disasters and severe weather like flooding. Other WHS are threatened by close proximity forest fires. These are not potential but existing life-threatening risks and events, and their prevention would often mean altering the historic urban fabric and deviating from dogmatic and strict rules of preservation for historic ensembles. Behind these specific cases, where, for example, walls to protect towns and cities from flooding or new trees in a WHS are opposed by local preservationists with reference to the WH listing, there are often conflicts between different values that are unfortunately rarely verbalised. Artistic values of monuments – sometimes even described as intrinsic values (Fusco Girard & Vecco, 2019) – conflict with use values (Smith, 2006). Khalaf (2021) suggests, in view of the global challenges that we face today, the concept of authenticity needs to be adapted, and more elements of a process should be incorporated into our World Heritage management approaches.

Similar issues arise with conflicts between the heritage sector and urban transformation and urban development as described by Gustafsson (2011), who suggests such conflicts can be overcome if stakeholders and decision-makers first enter a "trading zone" of discussion, from which all parties ultimately benefit. Many of the described conflicts have the potential to contradict the original objective of the World Heritage Convention, even if sometimes in an indirect way. Gentrification, as one example, can also lead to development pressure with the spread of more "heavy" urban functions like hotels, which can result in severe architectural interventions and potentially harm the heritage fabric and impact local societies (Okech, 2010).

Further examples in the following chapters of this book are connected to urban transformation that has occurred as a direct or indirect result of the designation of a city as a WHS or happened in parallel. These examples include conflicts in an African setting that have been identified as a result of colonialism and its long-term implications. Other examples are touristic mega-events, such as huge international sports events or the European Capitals of Culture, and particularly the festivalisation of heritage cities that led to the temporary rapid growth of tourism and tourism-related activities, which hitherto resulted in conflicts, for example, through congestion in the streets, rising prices for housing, etc. Planned revitalisation of urban areas is another field of practice where urban heritage is related to conflicts that can be rooted in touristification or gentrification. The effects of various processes of urban transformation have been analysed and described, for example, in McCormick's "Advancing Sustainable Urban Transformation", which serves as an analytical lens to describe and understand the continuous, complex and contested processes and dynamics in cities (McCormick et al., 2013) (Fig. 7.2).

Fig. 7.2 Tourists at the Bund in Shanghai. (Note: Image by Matthias Ripp, 2018)

7.2 The Role of Cultural Heritage and the World Heritage Convention

During the last decades, the understanding of cultural heritage has changed. In the UNESCO Convention concerning the Protection of the World Cultural and Natural Heritage 1972 (article 1), three different types of cultural heritage are mentioned: monuments, groups of buildings and sites (UNESCO, 1972). This rather "sectoral" understanding of UNESCO is focusing on different elements of cultural heritage, especially tangible heritage, and does not address, for example, the more multi-layered, versatile and complex urban heritage (Ripp & Rodwell, 2016) that many people encounter on a daily basis. Several other international conventions followed that have broadened the understanding of cultural heritage.

1975, the Resolution of Bruges: Principles Governing the Rehabilitation of Historic Towns, adopted by ICOMOS, included references to beauty, the human scale and social function (ICOMOS, 1975). In 2015, ICOMOS published a review to understand the impact of this influential Resolution: "A Future for Our Past. The 40th anniversary of European Architectural Heritage Year (1975–2015)" (Falser and Lipp, 2015). This publication represents the first comprehensive appraisal of the European campaign of 1975 as the most important and successful campaign of its time, with its recognition of the importance of urbanistic ensembles, plurality within the categories of historic monuments, citizen engagement, and, finally, and legal and administrative measures for monument protection.

1976, the UNESCO Recommendation Concerning the Safeguarding and Contemporary Role of Historic Areas (Nairobi Charter) addressed community identity and at the same time promoted the integration of historic areas into "the life of contemporary society (as) a basic factor in town planning and land development" (UNESCO, 1976).

1987, ICOMOS declared the integration of urban conservation into policies for socio-economic development and the participation of residents in the Charter for the Conservation of Historic Towns and Urban Areas (Washington Charter) (ICOMOS, 1987).

Then, in 2005, the "Faro Convention", officially titled the Council of Europe's Framework Convention on the Value of Cultural Heritage for Society, stated in Article 1, c (Aims of the Conventions), that "the conservation of cultural heritage and its sustainable use have human development and quality of life as their goal" (Council of Europe, 2005). Consequently, the understanding of cultural heritage has become wider and different definitions have been used. "Traditionally, planners viewed historic areas as a collection of monuments and buildings to be preserved as relics of the past, whose value was considered to be totally separate from their day-to-day use and city context" (Siravo, 2014, p. 161). This mainly materialistic approach to heritage is based on the physical appearance of monuments and a traditional interpretation of heritage and preservation as a mainly material science (Ripp, 2021).

UNESCO's Recommendation on the Historic Urban Landscape in 2011 finally supported a holistic view of heritage management, acknowledging the different urban layers and connections and at the same time beginning to integrate the goals of urban heritage conservation and those of social and economic development (UNESCO, 2011). The International Guidelines on Urban and Territorial Planning (UN-Habitat, 2015), the Pact of Amsterdam (European Commission, 2016a) and the Urban Agenda for the EU (European Commission, 2016b, 2019) and, finally, the European Commission's Cohesion Policy 2014–2020 (2014) all follow the concept of the Historic Urban Landscape.

The World Heritage Convention, on the other hand, was based on a more linear, monument or site-based understanding of preservation that was described by some authors as a preservationist paradigm following mainly material aspects (Holtorf, 2012). While contemporary theories of conservation are changing this notion (Viñas, 2002; Orbaşli, 2017), the scientific discussion connected to the implementation of the World Heritage Convention – for example, on the definition of the outstanding universal value (OUV), statements of significance or the emerging idea of attribute mapping – is still following the narrative of definition and separation rather than that of connecting and integrating different aspects into one holistic, systemic and procedural understanding of heritage. Slowly more systemic approaches are gaining popularity in the field of heritage (Barile & Saviano, 2015), and, in parallel, concepts like urban transformation can be beneficial for understanding and mitigating existing conflicts connected to the implementation of the World Heritage Convention. By narrowing down the implementation to maintaining the few attributes that are considered the core qualities of World Heritage sites, the complexity of the system of urban heritage and the processes of urban transformation can hardly be addressed in a comprehensive way.

While the scope of cultural heritage has broadened, it shifted from a single-monument and object-based approach to a more holistic understanding. This shift towards a more complex idea of heritage has stimulated systemic approaches that can better address higher levels of complexity and multi-level stakeholder approaches (Sacco et al., 2014). While a more linear and sectoral approach was prevalent in the beginning of monument preservation, policies and tools also followed such an approach. Laws to preserve monuments and methods to conserve the physical monuments in the best possible way have been rooted in this thinking. Later, when the concept of heritage became more complex (Ripp, 2018), for example, with the introduction of the ensemble and later the Historic Urban Landscape, there was a need for methods, tools and policies that reflected this complexity and were more related to systemic approaches.

7.3 Urban Transformation as an Emerging Scientific Concept to Address the Systemic Nature of Heritage

Urban transformation as an emerging scientific concept has the potential to analyse existing conflicts and expand the possible solutions beyond already existing interventions to broader systemic interventions that take greater account of the complexity of the users.

According to Anne Maassen and Madeleine Galvin (2019, p. 9), transformation—urban or otherwise—is commonly described "…using adjectives such as deep, far-reaching, radical, long-term, persistent and sometimes also as systemic and structural, irreversible, non-linear, non-incremental, complex (multi-scale, multi-actor, multi-level), and inherently contextual and political…". They suggest that urban transformation is an outcome of change processes in which (large) parts of cities change in fundamental ways. It involves "…overcoming inherited patterns of exclusion, neglect or risk across the various social, technical, and natural systems that make up the city…" (Maassen & Galvin, 2019, p. 9). The authors indirectly follow the notion that what we call a city is a system that consists of several subsystems. More precisely "Cities are complex systems that are made up of different entities that form the components of the system. These entities can be subjects, objects, processes, other fluid and changing developments, organisations and subsystems" (Ripp, 2021, p. 25). Processes of transformation are severely affecting these entities or their respective subsystems and lead to altered urban functions and new local needs and opportunities (Hölscher & Frantzeskaki, 2021).

Urban transformation research is an emerging interdisciplinary field with open boundaries that combines complex system studies and urban studies. It is a field that explores the patterns and dynamics of change that link cities and diverse socio-technical and social-ecological systems across levels and scales and develops new forms of intervention to foster the sustainability of these systems (Wolfram et al., 2016, p. 20). Rather than focusing on separate phenomena, defining and excluding them in their entirety, urban transformation research emphasizes the connections, interrelations and processes between them. Seven key factors that co-shape urban transformations have been recognised: agency, politics, capacity, policy, experiments, foresight and geography. The term "urban transformation" thus refers to the process and the outcome of changing the systemic configuration of urban areas (Maassen & Galvin, 2019, p. 9) and is mostly studied from the perspective of the sustainability performance or achievements of these areas.

The adoption of the 2030 Agenda for Sustainable Development (United Nations, 2015) alongside the New Urban Agenda (2016) set global standards of achievement in sustainable urban development, rethinking the way we build, manage, and live-in cities through collaboration between relevant stakeholders (UN-Habitat, 2016) to achieve safe, inclusive, resilient and sustainable cities (Maassen & Galvin, 2019, p. 2). Urban transformation and urban development are different. Urban development may be defined as "a process of synergetic integration and co-evolution among the great subsystems making up a city (economic, social, physical and environmental), which guarantees the local population a non-decreasing level of wellbeing in the long term" (Camagni, 2017, p. 272). While urban development is associated with a "project-approach", urban transformation may be more related to processes that occur suddenly– desired or not.

As the methods related to urban transformation are very diverse, it is beyond the scope of this paper to introduce or demonstrate the application of any of the methods or even to list all of them. The following are some relevant methods in relation to UNESCO World Heritage sites:

Governance and Planning, recognised as key methods for innovation and implementation of sustainable urban transformation (Wamsler et al., 2013). The planning process and the concept of governance highlight the critical roles of collaboration and engagement of stakeholders, particularly residents in urban areas (Radywyl & Biggs, 2013).

Collaboration and learning, it is only through collaborative action that urban sustainability projects can be effective, particularly when there are ambitious goals (McCormick et al., 2013). The international community has also responded with solutions that are related to the global sustainability agenda and the Sustainable Development Goals, for example, target 11.4 "safeguarding the world's cultural and natural heritage" to achieve Goal 11 in making "cities safe, inclusive, resilient and sustainable" (United Nations, 2015). In addition, the adoption of the Davos Declaration (Conference of Ministers of Culture, 2018) "Towards a high-quality Baukultur for Europe 2018" responded to societal needs and values of the built environment by enabling the assessment of Baukultur qualities of places and was fully supported and continues to be endorsed by international partner Europa Nostra.

7.4 Conclusion, Outlook

In the context of the described shift in the understanding of cultural and especially urban heritage, urban transformation is an approach that takes into account the complexity and systemic nature of urban heritage and can develop a deeper understanding of the conflicts and the ways in which they connect, interrelate and influence different parts of the urban heritage system. While many urban development projects still follow a more linear-thinking-based project management concept, a more systemic view and the introduction of methods based on systems thinking, and especially circular economy as well as circular business and governance models (Gravagnuolo et al., 2017; Stanojev & Gustafsson, 2021), would be more suitable to address urban complexity. A paradigm of rationalities (Nida-Rümelin, 2011) is often even prevalent in the Management Plans, which are the most prominent tools for World Heritage management. Data and measurable parameters, often quantified, form the basis for rather long-term defined objectives that are then to be met by the different parts of the local authorities. Tools and projects that followed this paradigm of rationalities have resulted in important findings and contributed greatly to professionalise and improve the management of WHS. These types of management tools are rooted in linear-thinking-based management by objective theories and do not offer enough flexibility to adapt to rapid challenges and therefore are not even fully implemented at many WHS after their design (Ripp & Rodwell, 2016). There are also Resource Manuals, such as the Management Guidelines for World Heritage Cultural Sites (UNESCO World Heritage Centre et al., 2013; Feilden & Jokilehto, 1993) and a Resource Manual on Managing Natural World Heritage (UNESCO World Heritage Centre, 2012) with an accompanying manual in 2013 for cultural properties. To facilitate resilience in WHS and achieve long-term objectives, we need a more systemic understanding of the existing interconnections and more flexibility (Ripp & Rodwell, 2016; Ripp, 2018) as well as more stakeholders involved in the decision-making processes (Gustafsson, 2011, 2019). To fully understand the system and processes that cultural heritage represents (Ripp, 2018), it is also necessary to consider the entities that are outside or contradict this rationality paradigm and are changing, fluid and unpredictable to some extent but most certainly occur in connection with urban transformation. These include, for example, values (Della Torre, 2013), emotions (Yusoff, 2019), motivation (Ripp & Hauer, 2017), sense of ownership (Boyd et al., 1996) and agency of heritage (Collins, 2007). They can be described as essential, even constitutional, to what we might call the human factor in heritage.

Focusing on connections makes processes clearer rather than narrowing the discussion to small sub-phenomena. Focusing on narrow (often disciplinary-led) phenomena rarely leads to the holistic and systemic understanding that is needed to fully embrace what is happening in terms of urban transformation. Drawing circles around specific urban phenomena, such as the preservation of a listed building, often misses the connection to other phenomena or systems (such as use, processes of climate change, social processes in the city, etc.) that are relevant and important if we look at the bigger picture. This is also necessary if cultural heritage is to be used to implement the UN SDGs, which are formulated as individual objectives, but which are all based on a systemic understanding of sustainable development (Ripp, 2021).

The key players need to be trained and enabled to implement a more systemic understanding. For examples of how exactly this can be done, the work of Otto Scharmer from MIT or the recent findings of consultants and coaches from the field of organisational development are valuable resources (Atwater & Pittman, 2006). Urban transformation has been revealed as a systemic transformation that includes transformation within people. Following the ideas of Scharmer and his Theory U Concept, true change in any system always involves change on a personal level (Scharmer, 2009). Therefore, a stronger focus on this aspect can be helpful to understand and influence urban transformation. The spirit of interdisciplinary and cross-sectoral work is equally relevant in academia to understand these urban systems and for urban practitioners on the ground to understand and address the urban transformation processes.

Future research should embrace transformation or change, even at its extremes such as emerging and total loss, as a natural process of any system. When linking urban transformation research to the SDGs and the new paradigm of resilience (SHELTER, 2020), it does not help to look at singular SDGs like Nr.11. All of the SDGs can be affected or part of urban transformation processes and conflicts; therefore, a systemic and holistic view is beneficial, as is explored, for example, in the SOPHIA (Social Platform for Holistic Heritage Impact Assessment) Project (https://sophiaplatform.eu/en).

The attitude – and this is more than a mere understanding – that every system includes the "birth" or emerging of new "things" or "entities" and their "death" or total loss can broaden perspectives and enhance understanding and empathy between other disciplines, experts and researchers that are embracing different views or paradigms.

Acknowledgements This text was prepared with outstanding assistance from Jonquille Clifford, Bath, U.K., jonquille.clifford@gmail.com.

References

Atwater, J. B., & Pittman, P. H. (2006). Facilitating systemic thinking in business classes. *Decision Sciences Journal of Innovative Education, 4*(2), 273–292.

Barile, S., & Saviano, M. (2015). From the management of cultural heritage to the governance of the cultural heritage system. In G. Golinelli (Ed.), *Cultural heritage and value creation* (pp. 71–103). Springer.

Bernt, M., & Holm, A. (2009). Is it, or is not? The conceptualisation of gentrification and displacement and its political implications in the case of Berlin-Prenzlauer Berg. *City, 13*(2–3), 312–324.

Boyd, B., Cotter, M., O'Connor, W., & Sattler, D. (1996). Cognitive ownership of heritage places: Social construction and cultural heritage management. *TEMPUS-ST LUCIA QUEENSLAND, 6*, 123–140.

Bromley, R. D., & Mackie, P. K. (2009). Displacement and the new spaces for informal trade in the Latin American City Centre. *Urban Studies, 46*(7), 1485–1506.

Camagni, R. (2017). Sustainable urban development: Definition and reasons for a research programme. In R. Capello (Ed.), *Seminal studies in regional and urban economics* (pp. 261–282). Springer. https://doi.org/10.1007/978-3-319-57807-1_13

Collins, J. (2007). The sounds of tradition: Arbitrariness and agency in a Brazilian cultural heritage center. *Ethnos, 72*(3), 383–407.

Conference of Ministers of Culture. (2018, January 20–22). *Davos declaration: Towards a high-quality Baukultur for Europe*. Conference of Ministers of Culture 20–22 January 2018. https://davosdeclaration2018.ch/media/Brochure_Declaration-de-Davos-2018_WEB_2.pdf

Council of Europe. (2005). *Council of Europe framework convention on the value of cultural heritage for society*. https://www.coe.int/en/web/conventions/full-list/-/conventions/rms/0900001680083746?module=treaty-detail&treatynum=199

Davidson, M., & Lees, L. (2005). New-build 'gentrification' and London's riverside renaissance. *Environment and Planning, A, 37*(7), 1165–1190.

Della Torre, M. (2013). Values and heritage conservation. *Heritage & Society, 6*(2), 155–166.

European Commission. (2014). *EU cohesion policy 2014–2020*. European Commission. https://ec.europa.eu/regional_policy/sources/docgener/informat/basic/basic_2014_en.pdf

European Commission. (2016a). *The pact of Amsterdam*. Netherlands Presidency of the Council of the European Union (EU), European Commission.
European Commission. (2016b). *Urban Agenda for the EU*. European Commission.
European Commission. (2019). *Urban agenda for the EU, multi-level governance in action. Regional and urban policy*. European Union. https://ec.europa.eu/regional_policy/sources/docgener/brochure/urban_agenda_eu_en.pdf
Falser, M., & Lipp, W. (2015). Eine Zukunft für unsere Vergangenheit. Zum 40. Jubiläum des Europäischen Denkmalschutzjahres (1975–2015). *Monumenta*, 3.
Feilden, B. M., & Jokilehto, J. (1993). *Management guidelines for World Cultural Heritage Sites* (First ed.). ICCROM.
Fusco Girard, L., & Vecco, M. (2019). Genius loci: The evaluation of places between instrumental and intrinsic values. *Bollettino Del Centro Calza Bini, 19*(2), 473–495.
Gotham, K. F. (2005). Tourism gentrification: The case of New Orleans' Vieux Carre (French Quarter). *Urban Studies, 42*(7), 1099–1121.
Gravagnuolo, A., Fusco Girard, L., Ost, C., & Saleh, R. (2017). Evaluation criteria for circular adaptive reuse of Cultural Heritage. *Bollettino Del Centro Calza Bini, 17*(2), 2017.
Gustafsson, C. (2011). *The Halland model. A trading zone for building conservation in concert with labour market policy and the construction industry, Aiming at Regional Sustainable Development*. University of Gothenburg. ISBN 978–91–7346-668-4.
Gustafsson, C. (2019). Conservation 3.0 – Cultural Heritage as a driver for regional growth. *SCIRES-IT: SCIentific RESearch and Information Technology, 9*(1), 21–32.
Holm, A. (2014). Berlin's gentrification mainstream. In M. Bernt, B. Grell, & A. Holm (Eds.), *The Berlin reader* (pp. 171–188). Transcript-Verlag. https://doi.org/10.14361/transcript.9783839424780.171
Hölscher, K., & Frantzeskaki, N. (2021). Perspectives on urban transformation research: Transformations in, of, and by cities. *Urban Transformations, 3*, Article 2. https://urbantransformations.biomedcentral.com/articles/10.1186/s42854-021-00019-z
Holtorf, C. (2012). The heritage of heritage. *Heritage & Society, 5*(2), 153–174.
ICOMOS. (1975). *The resolutions of Bruges: Principles governing the rehabilitation of historic towns*. ICOMOS. https://www.icomos.org/publications/93towns7k.pdf
ICOMOS. (1987). *Charter for the conservation of historic towns and urban areas*. ICOMOS. https://www.icomos.org/charters/charters.pdf
Khalaf, R. W. (2021). World Heritage on the move: Abandoning the assessment of authenticity to meet the challenges of the twenty-first century. *Heritage, 4*(1), 371–386.
Kremer, D. G. D., & Lehmeier, D. G. H. (2009). *Tourismus im UNESCO-Welterbe Bamberg*. Universität Bamberg. https://www.uni-bamberg.de/fileadmin/uni/fakultaeten/wiai_lehrstuehle/kulturinformatik/Publikationen/Kremer_Bericht_2010_01_28.pdf
Maassen, A., & Galvin, M. (2019). What does urban transformation look like? Findings from a global prize competition. *Sustainability, 11*(17), 4653. https://doi.org/10.3390/su11174653
McCormick, K., Anderberg, S., Coenen, L., & Neij, L. (2013). Advancing sustainable urban transformation. *Journal of Cleaner Production, 50*, 1–11. https://doi.org/10.1016/j.jclepro.2013.01.003
McPhearson, T., Iwaniec, D. M., & Bai, X. (2016). Positive visions for guiding urban transformations toward sustainable futures. *Current Opinion in Environmental Sustainability, 22*, 33–40.
Nida-Rümelin, J. (Ed.). (2011). *Praktische Rationalität: Grundlagenprobleme und ethische Anwendungen des rational choice-Paradigmas* (Vol. 2). Walter de Gruyter.
Okech, R. N. (2010). Socio-cultural impacts of tourism on World Heritage sites: Communities' perspective of Lamu (Kenya) and Zanzibar Islands. *Asia Pacific Journal of Tourism Research, 15*(3), 339–351.
Orbaşli, A. (2017). Conservation theory in the twenty-first century: Slow evolution or a paradigm shift? *Journal of Architectural Conservation, 23*(3), 157–170.
Radywyl, N., & Biggs, C. (2013). Reclaiming the commons for urban transformation. *Journal of Cleaner Production, 50*, 159–170.

Ripp, M. (2018). *Heritage as a system and process that belongs to local communities. Reframing the role of local communities and stakeholders*. Council of Europe/Faro Convention Workshop. https://rm.coe.int/heritage-as-a-system-and-process-that-belongs-to-local-communities-mr-/16807bc255
Ripp, M. (2021). *A metamodel for heritage-based urban development* [Doctoral dissertation]. Brandenburgische Technische Universität Cottbus-Sendtenberg.
Ripp, M., & Hauer, S. (2017). Communication model for built heritage assets going from knowing to identification. *Built Heritage, 1*(4), 1–10.
Ripp, M., & Rodwell, D. (2015). The geography of urban heritage. *The Historic Environment: Policy & Practice, 6*(3), 240–276.
Ripp, M., & Rodwell, D. (2016). The governance of urban heritage. *The Historic Environment: Policy & Practice, 7*(1), 81–108.
Sacco, P. L., Ferilli, G., & Tavano Blessi, G. (2014). Understanding culture-led local development: A critique of alternative theoretical explanations. *Urban Studies, 51*(13), 2806–2821. https://doi.org/10.1177/0042098013512876
Scharmer, C. O. (2009). *Theory U: Learning from the future as it emerges*. Berrett-Koehler Publishers.
SHELTER. (2020). *Sustainable Historic Environments holistic reconstruction through Technological Enhancement & community-based Resilience*. D.2.1. Resilience Structure. https://shelter-project.com/documents/deliverables/
Siemer, J., & Matthews-Hunter, K. (2017). The spatial pattern of gentrification in Berlin. *Prairie Perspectives: Geographical Essays, 19*, 49–57.
Siravo, F. (2014). Planning and managing historic urban landscapes. In F. Bandarin & R. Van Oers (Eds.), *Reconnecting the city: The historic urban landscape approach and the future of urban heritage* (pp. 161–175). Springer.
Smith, M. (2003). *Issues in cultural tourism studies*. Routledge.
Smith, L. (2006). *Uses of heritage*. Routledge.
Stanojev, J., & Gustafsson, C. (2021). Smart specialisation strategies for elevating integration of cultural heritage into circular economy. *Sustainability, 13*, 3685. https://doi.org/10.3390/su13073685
Szromek, A. R., Kruczek, Z., & Walas, B. (2020). The attitude of tourist destination residents towards the effects of overtourism—Krakow case study. *Sustainability, 12*(1), 228.
UNESCO. (1972). *Convention concerning the protection of the world cultural and natural heritage*. UNESCO.
UNESCO. (1976). *Recommendation concerning the safeguarding and contemporary role of historic areas*. UNESCO. https://atom.archives.unesco.org/u73c1
UNESCO. (2011). *Recommendation on the historic urban landscape*. UNESCO. https://whc.unesco.org/en/hul/
UNESCO World Heritage Centre. (2012). *Resource manual: Managing natural World Heritage*. UNESCO/ICCROM/ICOMOS. ISBN 978–92–3-001075-1. https://whc.unesco.org/document/117412
UNESCO World Heritage Centre, ICCROM, ICOMOS, & IUCN. (2013). *Resource manual: Managing cultural World Heritage*. UNESCO. https://whc.unesco.org/document/125839
UN-Habitat. (2015). *International guidelines on urban and territorial planning*. (HS/059/15E) UN-Habitat. https://www.uclg.org/sites/default/files/ig-utp_english.pdf
UN-Habitat III. (2016). *THE NEW URBAN AGENDA. The United Nations conference on housing and sustainable Urban development*. UN-Habitat.
United Nations. (2015). *Transforming our world: The 2030 Agenda for Sustainable Development. Seventieth session, Agenda items 15 and 116*. United Nations. http://www.un.org/ga/search/view_doc.asp?symbol=A/RES/70/1&Lang=E
van Criekingen, M. (2010). Gentrifying the re-urbanisation debate, not vice versa: The uneven socio-spatial implications of changing transitions to adulthood in Brussels. *Population, Space and Place, 16*(5), 381–394.
Viñas, S. M. (2002). Contemporary theory of conservation. *Studies in Conservation, 47*(1), 25–34.

Wamsler, C., Brink, E., & Rivera, C. (2013). Planning for climate change in urban areas: From theory to practice. *Journal of Cleaner Production, 50*, 68–81.
Wolfram, M., Frantzeskaki, N., & Maschmeyer, S. (2016). Cities, systems and sustainability: Status and perspectives of research on urban transformations. *Current Opinion in Environmental Sustainability, 22*, 18–25. https://doi.org/10.1016/j.cosust.2017.01.014
Yusoff, N. (2019). Cultural heritage, emotion, acculturation, ethnic minority, valence. *Journal of Ethnic and Cultural Studies, 6*(3), 53–63.
Zuk, M., Bierbaum, A. H., Chapple, K., Gorska, K., & Loukaitou-Sideris, A. (2018). Gentrification, displacement, and the role of public investment. *Journal of Planning Literature 2018, 33*(1), 31–44.

Chapter 8
Temporary Uses as a Toolkit for Heritage-Led Sustainable Urban Development

Mariko Ikeda

Abstract This paper reconsiders the possibilities for heritage conservation through everyday practices found in temporary uses as relevant and cost-effective tools in a constantly transforming urban environment, contributing to a more sustainable urban development. For this aim, three of the author's previous case studies of temporary uses in the city of Berlin are reconsidered from the perspective of heritage conservation through everyday practices and citizen participation. Berlin, with its rapidly changing urban environment since 1989, has been an experimental hub for countless temporary uses in a short period of time and therefore provides useful insights into the viability of temporary uses for urban heritage conservation from a variety of perspectives. This paper shows that temporary uses, especially ones that develop into permanent businesses, help to protect buildings from decay, revitalize neglected urban areas, contribute to the realization of the SDGs, and provide affordable spaces for cultural and social activities.

Keywords Urban heritage · Temporary use · Everyday practices · Ruin · Gentrification · Berlin

8.1 Introduction

Our urban heritage is endangered by several factors. One of the dangers, the vacancy of historic buildings that are an important part of our urban heritage, can be the result of deindustrialization, redevelopment projects, failed businesses, real estate speculation, negligence, and, more recently, economic pressures resulting from the ongoing COVID-19 pandemic. When these buildings are left vacant, they are at a

M. Ikeda (✉)
Faculty of Art and Design, University of Tsukuba, Tsukuba, Japan

Degree Program of Heritage Studies, Graduate School of Comprehensive Human Sciences, University of Tsukuba, Tsukuba, Japan
e-mail: marikoi@heritage.tsukuba.ac.jp

M.-T. Albert et al. (eds.), *50 Years World Heritage Convention: Shared Responsibility – Conflict & Reconciliation*, Heritage Studies,
https://doi.org/10.1007/978-3-031-05660-4_8

greatly increased risk of damage and decay. World Heritage Sites in urban areas are no exception from such developments. A look at the city of Edinburgh, with its Old and New Towns registered as World Heritage, shows that some historic buildings have been vacant for a long time (e.g., the Royal High School for 50 years) or have become vacant due to the recent economic situation in the COVID-19 pandemic (e.g., Jenners department store in 2021).

The best way to preserve these buildings is to use them. From a historical perspective, use and reuse were first and foremost responsible for preserving buildings in historic centres (Organization of World Heritage Cities et al., 2014, p. 53). English Heritage, which cares for over 400 historic sites in England, claims that the best way to protect buildings today is to keep them occupied, even if the use is on a temporary or partial basis (English Heritage, 2011). Such temporary uses, meaning the use of a building or a space for a limited time, have attracted the attention of urban planners and administrations over the last decades. This practice, also known as ‘meanwhile use’, is generally regarded as having advantages for both the tenants and the property owners. For the tenants, temporary use can be an attractive, low-cost way to experiment with a new business idea, start artistic and cultural activities, or set up non-profit projects. Centrally located properties can be rented for relatively low prices, and initial investment for these sites is often low due to the time-limited character of the venture (Oswalt et al., 2013). For property owners, including municipalities, temporary use of a site can contribute to the reduction of maintenance costs and to the preservation of a property’s value through cultural activities. Activities on the property can also prevent damage caused by vandalism or disuse and can contribute to a positive image of the site, which can attract new investors (Senatsverwaltung für Stadtentwicklung, 2007).

Berlin, the “capital of temporary uses” (Bodenschatz, 2011), is a showcase for the wide range and the dynamic nature of informal ways of preserving urban heritage. Many of Berlin’s temporary uses provide valuable insight into the conflicts facing heritage in urban spaces and their solutions. This paper reconsiders three of the author’s previously investigated case studies of temporary uses in Berlin from the perspective of heritage reuse and sustainable urban development. The paper specifically asks how these temporary uses contribute or have contributed to the preservation of neglected heritage in urban areas and how these temporary uses contribute to the achievement of sustainable development goals (SDGs) 8 (decent work and economic growth), 11 (sustainable cities and communities), and 12 (responsible consumption and production) (UN, 2015).

The range of case studies is important in the context of urban World Heritage Sites but especially for historic cities because ownership is much more widely spread in such diffuse heritage properties compared to other World Heritage Sites. According to UNESCO, “in a heritage city [...] the bulk of the historic buildings will be privately owned and many will be used for non-heritage purposes” (UNESCO et al., 2013, p. 14). This description is comparable to situation in Berlin, which is not a historic city in the World Heritage sense but has three World Heritage Sites and a large number of historic buildings.

8.2 Temporary Use, Heritage, and Sustainable Development

It can be argued that the idea of temporary use is, from the outset, a heritage-led sustainable way of development since it uses and reuses the available fabric. A study on 20 temporary uses in heritage buildings around the globe found that most of them were utilizing a former vacant structure that was in poor condition and that the changes made to the physical fabric were mainly positive and reversible (Tuohy Main, 2014). In other words, the temporary uses contributed to the maintenance of the buildings without making irreversible changes or damaging the fabric. Furthermore, Baum (2012, p. 30) notes that the reutilization of existing structures always encompasses an examination of their history, architecture, atmosphere, and context. This process can create a bond between the users and the property, fostering an identity related to the structure's history and its urban environment. Rellensmann (2010) sees, besides the prevention of decay and demolition through temporary uses, a sensitization of the users as well as visitors and investors to issues of heritage and preservation.

In a publication by the REFILL network (Refill, 2018, p. 12–13), a network of 10 European cities for the exploration of temporary uses, the political representatives of the participating cities conclude that temporary use has become a necessity to fight the social and environmental crisis. They note that the users "bring administrations in touch with active grassroots initiatives exploring new urban solutions." However, they also note that "it remains a challenge to adapt city governance into a form of decision-making that is less top-down and more based on co-creation with the people." This shows that although temporary use has become a more widely adopted tool for urban development over the last 20 years (at least in some European countries), there is still room for improvement in how temporary uses are implemented.

This is also mirrored by the disregard of temporary use as a possible long-term solution by official heritage bodies. English Heritage (2011, p. 9) notes that meanwhile uses "will allow prospective tenants to see the building in use and make it easier to attract long-term occupiers." According to Heritage Council Victoria (2013, p. 4), "temporary uses can be a good way to prevent deterioration until a long-term use is found." They mainly see temporary uses as a viable strategy for preventing further structural damage but not as a viable option for long-term heritage conservation itself, which would require "a more complex understanding of the site and its context" that can only be achieved by heritage professionals.

Literature specifically focused on the sustainability of temporary uses is very sparse. A study conducted in the Austrian city Graz suggests the reutilization of the many unused vacant buildings in the city through temporary uses to counter the ongoing loss of soil surface due to new constructions that can contribute to a higher risk of flooding, loss of soil functions, loss of biodiversity, and alteration of microclimates (Reitsamer, 2018).

Again, heritage bodies do not regard temporary uses as sustainable: temporary uses have the advantage of "increased prospect of a sustainable use – 'meanwhile' uses can make the space more attractive and vibrant, and increase awareness of the property" (English Heritage, 2011, p. 9). However, Bennett (2017, p. 28) analyses the current situation of empty stores in the UK and concludes that "meanwhile" use is becoming more and more the norm for commercial leases and therefore asks, "What is actually wrong with a sequence of short-term adaptive uses?" He identifies "urban law and policy's ruinphobia" as the main problem, where only long-term commitment is seen as a successful way of utilizing stores in the UK's town centres. On the other hand, Sandler (2016, p. 24) has found that decay can also be seen as a way to represent the history of buildings and sites more truthfully than restoration. In this "counterpreservation", occupants of counterpreserved buildings have infused decay with positive associations of social inclusiveness, freedom, and creativity. Finally, Oswalt et al. (2013, p. 376) emphasize the bottom-up approach of temporary uses by noting that they represent an urban development without financial resources, based on the utilization of the city. This point relates to SDG 12, responsible consumption and production (UN, 2015).

8.3 Berlin as an Experimental City for a Heritage-Led Sustainable Urban Development

8.3.1 Temporary Use and Its Emergence in Berlin

Historic buildings have played a vital role in the urban transformation of the inner city of Berlin (Holm & Kuhn, 2011). In West Berlin in the 1970s to 1980s, citizen initiatives launched preservation activities with the goal to rescue historic buildings that were slated for demolition due to large-scale redevelopment plans based on the concept of a car-friendly city. Furthermore, the housing shortage in the 1970s led to "rehab squatting" (*Instandbesetzung*), in which vacant and dilapidated houses were occupied and immediately renovated by the squatters (Holm & Kuhn, 2011). Some of the preserved buildings from this time, such as the "Künstlerhaus Bethanien", are still used as cultural spaces today. One of the buildings of the vacant hospital complex Bethanien in the Kreuzberg district was squatted in December 1971 to establish a youth centre with integrated housing. In the following years, several art groups achieved a reutilization of the hospital's main building as a centre for the arts, while it became a property of the city of Berlin and was put under monument protection (*Denkmalschutz*).

Later, in the years after the fall of the Berlin Wall in November 1989, the inner city of Berlin and particularly the districts of former East Berlin were characterized by a high number of vacant lots and buildings. This situation gave a variety of users the opportunities to occupy the buildings and establish temporary uses in the often historic and derelict building stock. During the 1990s, new music clubs, art houses, galleries, bars, urban gardens, and alternative living spaces invigorated the vacant buildings and lots (Senatsverwaltung für Stadtentwicklung, 2007). Many of these projects were initiated by participants of Berlin's vibrant subcultures, which had emerged in both parts of Berlin during the division and came together after the reunification. These projects emphasize the importance of urban heritage for creative uses and can be seen as a development where, as described by urban sociologist Jane Jacobs (1961, p. 188), "new ideas are born in old buildings." On the other hand, these developments in Berlin also signify the importance of creative and temporary uses for the urban heritage, which will be further explored in the following case studies.

These case studies were conducted for some of the author's previous research between 2010 and 2014 and exemplify how temporary uses, especially artistic, cultural, and creative activities in urban areas, can contribute to the preservation of heritage in a sustainable way. They also clearly show where possible conflicts in this practice lie.

8.3.2 *Case Study 1: Kunsthaus Tacheles: Ruin, Art House, Tourist Attraction*

This case study investigated the art facility "Kunsthaus Tacheles" with a focus on its history as a cultural squat, its history as a building, its role as a free space, and its development into a tourist attraction in a rapidly transforming district in the city centre of Berlin (Ikeda, 2014). The Kunsthaus Tacheles was a cultural facility in the Mitte district of Berlin and housed ateliers for many national and international artists. Furthermore, it had exhibition rooms, a theatre, a cinema, a bar, a music club, and an open-air area that was used for art installations and exhibitions.

The building was opened in 1909 as one of Berlin's largest department stores at the time. It was damaged during the Second World War, but it was still usable. After the war, it was used as office space, travel agency, storage, professional school, and cinema, among other uses. At the beginning of the 1980s, the building was partially demolished to make way for a new street. However, a small part of the building remained and was slated for demolition in 1990. To stop the demolition, the artist initiative Tacheles, consisting of artists from East and West Berlin, squatted the building.

The users achieved the registration of the Tacheles building as a monument (*Baudenkmal*) after the building was surveyed and found to be in a good structural condition. In this process, the artists rediscovered the eventful history of the building, which was located in a quarter that had a vibrant Jewish culture before the Second World War. The New Synagogue, once Germany's largest and most magnificent synagogue, is only a few 100 m away. The property changed hands throughout the 1990s and 2000s, and it became an object for speculation and unrealized construction projects. After a long battle with the property owners, the last artists left Tacheles in 2012, making way for a large-scale redevelopment project. In the future, the Tacheles building will be incorporated into the new development as a photography museum, which is supposed to keep the building's history, such as its war damage and the artist interventions, visible to the public (Hilburg, 2020).

In this case study, artists initially looking for an affordable place for their activities became the main actors behind the preservation of urban heritage and the rediscovery of its historical background as well as its opening up to the public. However, in the process of internal institutionalization and increased property speculation on the outside, the former temporary use faced problems of touristification and the loss of its original identity and authenticity. Furthermore, these problems became part of a greater process of gentrification in the surroundings of the reused property.

8.3.3 Case Study 2: Revitalization of Industrial Buildings for Music and Socio-Cultural Projects

This case study investigated three temporary uses in Berlin's Friedrichshain district in the years 2013 and 2014. The following three properties were investigated: A large former railway repair workshop, which was turned into a socio-cultural space with more than 30 different businesses, including concert halls, food stalls, and a theatre (RAW); a music club (Maria); and a sports and music location for young people (YAAM) (Ikeda, 2018).

The Friedrichshain district, particularly, a wide area along the Spree River, was heavily affected by deindustrialization after the fall of the Wall, which led to the closure of factories and railway properties. Another reason for vacant lots and buildings in this district was the construction boom of office buildings in other inner-city districts, such as Mitte, in the early 1990s. Because the inflated growth expectations for Berlin never materialized, a high number of development plans for new office spaces in Friedrichshain remained on hold for several years. This was a fortunate situation for informal users who were looking for an affordable space for temporary use in the inner city.

In the case of RAW, a single location is used by a multitude of individuals and organizations and offers a music school, a theatre, a circus, and night-time venues. Furthermore, through these activities, a closed and unused property in the inner city was made accessible to the public by local actors. In a similar way, the two smaller locations, YAAM and Maria, combine several uses in one location. During the day, YAAM facilitates sports activities, holds occasional market events, and provides food and drinks, and at night, it turns into a music club that generates the necessary revenues to finance its youth and cultural work. Maria, in addition to its main function as a music club and concert location, was used for theatre and other stage performances, art exhibitions, and as a recording studio. In addition, it was important for all three projects that they were located in spacious areas with an industrial

Fig. 8.1 Berghain, a world-famous techno club in a former power plant built in 1953, Berlin. (Note: Image by Mariko Ikeda, 2014, August 26)

character and few residential buildings because they held concerts and other activities that involved noise, which could have led to problems in a residential neighbourhood. Similar examples for temporary uses in vacant industrial buildings in Berlin that have become world-famous music clubs are Tresor and Berghain (Fig. 8.1).

8.3.4 Case Study 3: Economic Evaluation and Gentrification in a Residential Area Through Creative Industries

This case study analyzed the process of gentrification in the Reuter Quarter in the Berlin district of Neukölln in the years 2008–2014 (Ikeda, 2019). The Reuter Quarter is a densely populated residential quarter in the northern part of Neukölln with a substantial historic building stock dating from the early twentieth century. In the 2000s, the quarter was characterized by high unemployment, a high percentage of residents from foreign countries, and a high number of vacant stores and apartments.

Due to this situation, a so-called "Neighbourhood Management", a subordinate project of the national social urban program *Soziale Stadt* (social city), was launched in 2002 by the city government. The project aimed at improving the social environment through German language courses for foreign residents and street festivals to foster the formation of neighbouring communities. In addition to this, a district revitalization program commenced in coordination with a private organization in 2005. The Temporary Use Agency (TUA) was founded in 2005 by an architect and an urban planner. Their first project was to find tenants for some of the 130 vacant stores in Neukölln's Reuter Quarter. First, TUA recruited people who were looking for vacant spaces and supported the contractual negotiations with the owners, realizing 2-year rental contracts with no rent. These contracts had advantages for both sides: The tenants paid no rent but renovated the empty shops by themselves, and the property owners got a new user for the vacant shop who made the place more attractive and liveable, attracting customers, new tenants, and new residents to the area.

In the beginning, the applicants were primarily people related to art, handicrafts, or social projects. The purpose of the program was to encourage district improvement through matching supply and demand for the vacant stores. According to TUA, 56 contracts were formed, and more than 200 jobs were created in the Reuter Quarter between 2005 and 2007. TUA was partly financed by subsidies from Neighbourhood Management, but no subsidies were used for rents or investments for the spaces themselves. The goal was from the beginning to moderate rental agreements between real estate owners and tenants so that they are independent of subsidies. Most of the initial temporary users became permanent users with normal rental agreements after the 2-year contract had ended. The TUA was later renamed Coopolis and coordinated the temporary use of 150 vacant shops in four different neighbourhoods of the northern part of Neukölln. New fashion stores, sewing workshops, galleries, cafes, youth facilities, music clubs, and other venues changed the district's character, making it an attractive area for the creative industries and local initiatives.

8.4 Discussion of the Case Studies

The most important finding of these case studies is that temporary uses can become permanent fixtures and successful businesses that provide employment and a wide range of cultural and social services (SDGs 8 and 11). This was the case in all the analyzed temporary uses. Regardless of their success, case 1 (Tacheles) and case 2 (RAW) also show that as they grew into more mature organizations, the economic situation in the surrounding areas improved and attracted new investors, leading to conflicts with traditional ways of urban development. The artists of Tacheles had to make way for a large-scale redevelopment project. The users of RAW were more fortunate as they were incorporated into future development plans and acknowledged as important cultural and social projects. Furthermore, Tacheles and RAW exemplify the findings of Baum (2012): The users have become involved in the history and the architecture of the used buildings and even more in the context of the buildings and their environment. In both cases, formerly closed spaces were made available to the public, and their histories were rediscovered and publicized, putting the temporary uses in relation to their surrounding areas and the local communities.

In the first two case studies, the users were actively protecting the utilized buildings from decay by carrying out maintenance and repair work and by greening and beautification activities in the surrounding areas. Therefore, the temporary uses contributed to the realization of SDGs 11 ("sustainable cities and communities") and 12 ("responsible consumption and production") (UN, 2015). Furthermore, by providing affordable spaces for cultural or social activities, work opportunities were created (SDG 8) and the attractiveness of communities was increased.

Moreover, the users themselves generated most of the funding for the repair of the buildings and the improvement of the infrastructure through income-generating events, such as concerts and public readings, selling products and services, or catering. In short, these places did not rely on funding from the local government and were self-sustained operations (SDG 12). Therefore, in addition to their obvious short-term advantages identified in previous studies, temporary uses cannot be ignored as a viable long-term alternative to conventional strategies of urban development and heritage conservation as they are functioning businesses.

Case study 2 also showed that temporary uses, despite their success, face various problems that can threaten their existence. There are ways for the city administration to support temporary uses in such times, for example, by acting as an intermediary in the negotiation process for a renewed rental agreement or by providing a publicly owned property in which the temporary use can continue to operate in the case of the termination of a rental agreement. Therefore, temporary rental agreements, which were often desired in the beginning, became a limiting and problematic factor when the businesses grew and sought to become permanent establishments. The case study also made clear that the administration of Berlin did not have a consistent strategy to integrate these and other temporary uses into their long-term urban planning processes, although temporary uses had been acknowledged by the administration as an important factor for the city as early as 2001 (Senatsverwaltung für Stadtentwicklung, 2001) and pronounced as a tool for sustainable urban development in 2007 (Senatsverwaltung für Stadtentwicklung, 2007). The main reason for this lack of consistent strategy was the conflict with the city's original plan for this area to become a newly built large-scale media and office quarter. In practice, the city administration was merely exploiting temporary uses to create a positive "creative" image in times of economic hardship. This finding echoes the difficult relationship between temporary uses and administrative bodies pointed out by Refill (2018).

Case study 3 showed that a temporary use agency is an effective tool for the revitalization of a problematic residential district with a high number of vacant spaces. It allowed local entrepreneurs to establish new businesses with minimal investment, strengthening the local community and economy (SDGs 8 and 11).

Moreover, letting artists, creative entrepreneurs, social projects, and other business operators use vacant spaces for a below-market rent for a limited time protected properties from decay and vandalism until the situation in the district was attractive enough for tenants who could afford usual market prices. This is a temporary use practice favoured by heritage bodies such as English Heritage. However, it was not always necessary to attract new tenants, as many of the temporary uses were successful themselves and therefore could afford increased rent prices after the initial contract ended. The study also showed that this kind of planned revitalization can be a contributing factor to gentrification, making a once undesired quarter attractive for more affluent residents and displacing the original population due to rising rents. In this case, the temporary uses did not ultimately lead to a sustainable community (SDG 11) but a rapidly transforming quarter.

8.5 Conclusion

This paper has shown how temporary uses have been discussed regarding heritage preservation and sustainable urban development. Several previously studied temporary uses in Berlin have then been reconsidered from this perspective. The discussion above clarified the positive aspects of temporary uses for the preservation of vacant historic buildings and highlighted the aspects contributing to the achievement of sustainable development goals 8, 11, and 12. Through temporary uses, vacant old buildings are rediscovered, reutilized, and protected by the users and the spatial and financial scale of conservational activities is often on a small scale. However, those spaces can become "lived spaces" (in the sense of Armand Frémont's *espace vécu*), which are spaces of everyday life and social interrelationships that the users themselves create and are open to the public.

Furthermore, the identified problems could provide guidance for finding better ways to implement temporary uses into sustainable urban development strategies in a post-growth era. Especially in the context of shrinking cities and tighter public budgets, temporary use can become a viable strategy for a less formal and less costly way of utilizing and protecting our urban heritage. Involving grassroots initiatives that carry out voluntary preservation activities offers an additional sustainable way of heritage preservation. The involvement of local communities in heritage protection also makes the users consider the history of their heritage and how it shapes their identities (Fig. 8.2).

Finally, in the ongoing COVID-19 pandemic, the number of insolvent shops, bars, and restaurants due to lack of customers and mounting economic pressure is increasing, and we can expect to find our cities with numerous empty buildings in the near future. These spaces could benefit from temporary uses through opportunities for employment and community building, preserving the urban heritage, and increasing the resilience of communities.

Fig. 8.2 New West, a temporary use project and civic museum themed around club culture in Berlin in an "ordinary building" built after 1970s, Berlin. (Note: Image by Mariko Ikeda, 2014, November 10)

Acknowledgements This research was funded by JSPS KAKENHI Grant Number 19KK0020/ 12J01040/ 17J02079 and by JST SAKIGAKE Grant Number JPMJPR21R1.

References

Baum, M. (2012). Zukunftsfähigkeit braucht Wandlungsfähigkeit und Stabilität. In M. Ziehl, S. Oßwald, O. Hasemann, & D. Schnier (Eds.), *Second hand spaces. Über das Recyceln von Orten im städtischen Wandel* (pp. 28–35). Jovis.

Bennett, L. (2017). Forcing the empties back to work? Ruinphobia and the bluntness of law and policy. In J. Henneberry (Ed.), *Transience and permanence in urban development* (pp. 17–29). Wiley Blackwell. https://doi.org/10.1002/9781119055662.ch2

Berghain, a world-famous techno club in a former power plant built in 1953, Berlin. [Photograph] Mariko Ikeda. 2014, August 26.

Bodenschatz, H. (2011, October 6). Große Stadt–kleines Denken. *Der Tagesspiegel*. https://www.tagesspiegel.de/berlin/architekturdebatte-6-grosse-stadt-kleines-enken/4689198-all.html

English Heritage. (2011). *Vacant historic buildings. An owner's guide to temporary uses, maintenance and mothballing*. English Heritage.

Heritage Council Victoria. (2013). *Adaptive reuse of industrial heritage: Opportunities & challenges*. Heritage Council Victoria. https://heritagecouncil.vic.gov.au/wp-content/uploads/2014/08/HV_IPAWsinglepgs.pdf

Hilburg, J. (2020, September 15). The next Fotografiska will open in Berlin, courtesy of Herzog & de Meuron. *The Architect's Newspaper.* https://www.archpaper.com/2020/09/the-next-fotografiska-will-open-in-berlin-courtesy-of-herzog-de-meuron/

Holm, A., & Kuhn, A. (2011). Squatting and urban renewal: The interaction of squatter movements and strategies of urban restructuring in Berlin. *International Journal of Urban and Regional Research, 35*(3), 644–658. https://doi.org/10.1111/j.1468-2427.2010.001009.x

Ikeda, M. (2014). Cultural squatting, conflict involved, and role as free space in urban transformation: Case study of the Kunsthaus Tacheles in the former east district of Berlin. *Geographical Review of Japan Series A, 87*(3), 224–247. https://doi.org/10.4157/grj.87.224

Ikeda, M. (2018). Temporary use of vacant urban spaces in Berlin: Three case studies in the former Eastern inner-city district Friedrichshain. *Geographical Review of Japan Series B, 91*(1), 1–16. https://doi.org/10.4157/geogrevjapanb.91.1

Ikeda, M. (2019). Role of culture and consumption in the process of gentrification: Case study of the Reuter Quarter in the former West Berlin Neukölln. *Geographical Review of Japan Series B, 92*(1), 10–32. https://doi.org/10.4157/geogrevjapanb.92.10

Jacobs, J. (1961). *The death and life of great American cities*. Random House.

New West, a temporary use project and civic museum themed around club culture in Berlin in a 'ordinary building' built after 1970s, Berlin. [Photograph] Mariko Ikeda. 2014, November 10.

Organization of World Heritage Cities, City of Lyon, World Heritage Centre, France-UNESCO Convention, Council of Europe, The Getty Conservation Institute, & ICOMOS. (2014). *Developing historic cities – Keys for understanding and taking action*. UNESCO World Heritage Centre, France-UNESCO Cooperation Agreement.

Oswalt, P., Overmeyer, K., & Misselwitz, P. (Eds.). (2013). *Urban catalyst. Mit Zwischennutzungen Stadt entwickeln*. DOM Publishers.

Refill. (2018). *A journey through temporary use*. Refill.

Reitsamer, N. (2018). *Zwischennutzung – ein Mittel zur nachhaltigen Stadtentwicklung? Eine Analyse am Beispiel der Stadt Graz* [Master's thesis]. Karl-Franzens-Universität Graz. Unipub. https://unipub.uni-graz.at/obvugrhs/download/pdf/3494642?

Rellensmann, L. (2010). *Das Phänomen der Zwischennutzung unter besonderer Berücksichtigung der Denkmalpflege am Beispiel der jüngeren Berliner Stadtgeschichte* [Unpublished master's thesis]. Brandenburgische Technische Universität Cottbus-Senftenberg. https://www-docs.b-tu.de/fg-denkmalpflege/public/downloads/Rellensmann_Masterarbeit_Zwischennutzung.pdf

Sandler, D. (2016). *Counterpreservation: Architectural decay in Berlin since 1989*. Cornell University Press. https://doi.org/10.7591/cornell/9781501703164.001.0001

Senatsverwaltung für Stadtentwicklung. (2001). *Spreeraum Friedrichshain-Kreuzberg. Leitbilder und Konzepte*. Senatsverwaltung für Stadtentwicklung.

Senatsverwaltung für Stadtentwicklung. (Ed.). (2007). *Urban pioneers. Temporary use and urban development in Berlin*. Jovis.

Tuohy Main, L. (2014). *Temporary use: A potential strategy for historic buildings at risk* [MSc dissertation]. The University of Edinburgh. Era. https://era.ed.ac.uk/bitstream/handle/1842/9633/MSc%20Arch%20Cons%202014%20Lilian%20Tuohy%20Main.pdf

UN. (2015). *A/RES/70/1 – Transforming our world: the 2030 Agenda for Sustainable Development.* https://www.un.org/en/development/desa/population/migration/generalassembly/docs/globalcompact/A_RES_70_1_E.pdf

UNESCO, ICCROM, ICOMOS, & IUCN. (2013). *Managing cultural world heritage*. UNESCO World Heritage Centre. https://whc.unesco.org/document/125839

Chapter 9
Going Beyond Tourism Attraction in the Festivalisation of World Heritage Cities

Zachary M. Jones

Abstract Tourism has long been noted as a double-edged sword for World Heritage cities that can lead to a wide range of socio-economic benefits while also introducing many stresses that both physically damage sites and affect local communities through gentrification and other socio-economic changes. Festivals, events and cultural mega-events are often framed with a focus on growing tourism, but they can also provide unique opportunities to align heritage with Sustainable Development Goals. This chapter explores these dynamics by looking at three trends that the festivalisation of heritage cities can lead to: establishing and promoting heritage-based city images; spreading out events to reduce stresses; expanding traditional definitions of heritage through involving local communities. Several examples from across Europe that have hosted the Expo, European Capital of Culture (ECoC) and the UK City of Culture (UKCoC) demonstrate varying alignments with the Historic Urban Landscape (HUL) approach, Sustainable Development Goals (SDGs) and the UNESCO 5Cs to promote the sustainable development and inclusion of historic environments in broader city strategies. The chapter concludes by calling for a more integrated governance approach that can reframe approaches to go beyond just tourism attraction while anticipating and avoiding the potential range of risks of festivalisation.

Keywords Cultural heritage · Cultural mega-events · Festivalisation · Mass tourism · Sustainable development

---

Z. M. Jones (✉)
Politecnico di Milano, Milan, Italy
e-mail: zachary.jones@polimi.it

M.-T. Albert et al. (eds.), *50 Years World Heritage Convention: Shared Responsibility – Conflict & Reconciliation*, Heritage Studies,
https://doi.org/10.1007/978-3-031-05660-4_9

9.1 Rethinking Tourism in World Heritage Cities

Cultural heritage tourism has long been identified as a double-edged sword with potential positive and negative impacts on historic sites (Ashworth, 2000; MacCannell, 1976). Yet is unlikely that the founders of the World Heritage Convention could have predicted the exponential growth of global tourism and its impact on World Heritage sites over the last 50 years. 2018 witnessed 1.4 billion international tourist arrivals (UNWTO, 2019), and studies have estimated that up to 50% of the tourism in Europe is driven by cultural heritage (O'Brien et al., 2015). On the one hand, this growth has led to an increase in conservation funding as well as supporting local economies and development (Nuryanti, 1996; Strauss & Lord, 2001), while on the other it has too often led to a loss of authenticity and introduced stresses that physically damage sites (Timothy & Nyaupane, 2009). One increasingly common approach that cities have used to promote themselves and grow tourism has been through festivalisation strategies (Quinn, 2009; Santa-Cruz & López-Guzmán, 2017). Such approaches typically aim at rebranding cities, focusing on culture and cultural heritage as attractors to provide a new identity, and are often used by many post-industrialised cities (Bianchini & Parkinson, 1993). Festivalisation strategies can range from smaller week-long cultural events, music or film festivals and biennales to year-long cultural mega-events such as the European Capital of Culture (ECoC) and other Capital/City of Culture (CoC) programmes worldwide (Green, 2017; Jones, 2020). García and Cox (2013) found that most host cities experienced a tourism boost during the year hosting the ECoC and long-term tourism numbers surpassed those of the event within 5 years. One prominent example is the Matera-Basilicata 2019 ECoC, which saw overnight visits nearly triple in the 4 years between the awarding and hosting of the event (Ponzini et al., 2020a).

This chapter goes beyond a tourism-centric view of the festivalisation of heritage-rich cities by looking at the practical ways that event organisers, city planners and heritage experts can reframe these events to align with other long-term strategies and aims, including those of the Historic Urban Landscape (HUL) approach, Sustainable Development Goals (SDGs) and the UNESCO 5Cs (i.e., Credibility; Conservation; Capacity-building; Communication; Communities). Section 9.2 discusses the recent trend of World Heritage cities, particularly in Europe, turning towards festivalisation strategies, the potential threats this introduces and the opportunities it can present to connect WH properties with wider city visions and plans. Sections 9.3, 9.4 and 9.5 then use several examples that highlight how cultural mega-events can be used to establish heritage-based city images, spread out events to reduce stresses on key historic areas and expand traditional definitions of heritage. Section 9.6 concludes by examining some negative cases due to a missing cohesion between heritage and festival strategies and calls for more integrated tourism management that can help in seeking out more sustainable tourism models that can contribute to long-term goals and strategies. Many of the examples presented are derived from previous research that was carried out within a PhD (Jones, 2020) and as part of the HOMEE Research Project (Heritage Opportunities/threats within Mega-events in Europe), funded by the JPICH 2017 Heritage in Changing Environments Joint Call (Ponzini et al., 2020b). Previously studied as in-depth case studies, the examples have been specifically selected to demonstrate ways that festivalisation strategies can and should go beyond purely pro-growth tourism objectives to contribute to long-term sustainable development.

9.2 World Heritage Cities and Festivalisation

Within Europe, many cities with a World Heritage-recognised urban quarter or district have hosted the ECoC. In fact, more than one-third of all host cities from 1985 until 2020 have successfully bid for and hosted this cultural mega-event (see Fig. 9.1). Without including the other cities in Europe or globally that have bid for this or other types of mega-events, such events have clearly become attractive for

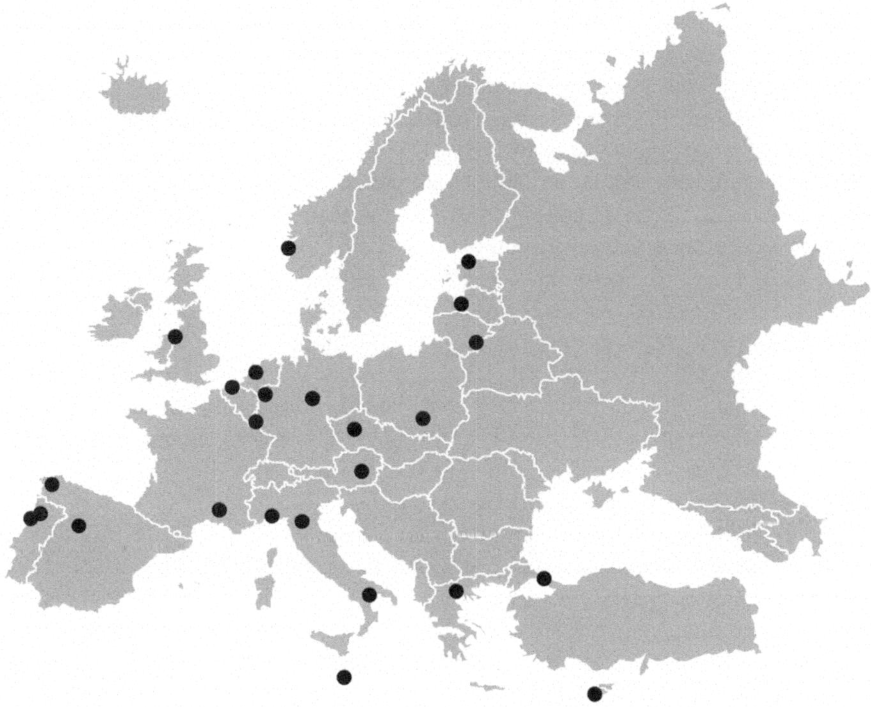

Fig. 9.1 Location of ECoC host cities from 1985 to 2020 that also contain an urban area recognised as a World Heritage Site. (Note: Author elaboration based on Cultural Mega-Events by Jones, 2020, reproduced by permission of Taylor & Francis Group)

World Heritage cities. The upcoming Olympic host cities of Paris 2024 and Milan-Cortina 2026 reveal the ongoing continuation of this trend as areas within the Paris, Banks of the Seine World Heritage Site will host a number of competitions along with the Palace and Park of Versailles. The Ancient Roman Arena in the City of Verona World Heritage Site will also serve as the site of the closing ceremony of the 2026 Olympics. With such events poised, or at least intended, to attract mass tourism, there is a need for event organisers and city decision makers to anticipate such issues while also seeking to align with broader sustainable and heritage management goals.

There are a range of threats to sustainability that these mega-events can introduce. First, mass tourism can overwhelm sensitive heritage spaces that are not designed to handle such high numbers of visitors (Zubiaga et al., 2019). Such high visitor numbers typically lead to increases in pollution from traffic or littering. Beyond physical threats, intangible changes can also occur, ranging from gentrification to general increases in costs adjusted for tourists rather than locals, affecting sites' authenticity as well as the meanings of heritage (Gravari-Barbas, 2018). Beyond local impacts, global tourism has an increasing carbon footprint, and tourism transport alone represents 5.3% of total global emissions (UNWTO, 2019), introducing many long-term climate change-induced threats to World Heritage sites (Fatorić & Seekamp, 2017). Considering these issues, particularly in light of the recent weaknesses to an overreliance on tourism revealed by the pandemic, city decision makers and event organisers must seek out other relevant uses of large events and festivals in World Heritage cities, particularly those that can link to the Sustainable Development Goals and other UNESCO-defined aims.

Though there may not be an obvious alignment between cultural mega-events and the Historic Urban Landscape (HUL) approach promoted by UNESCO, a deeper investigation into the ways these events come to be planned within heritage-rich cities highlights the potential to meet many of the promoted aims. HUL aims to reintegrate historic areas of cities into wider activities and plans, to work against them becoming mere stage sets for tourists (Bandarin & Van Oers, 2012, 2015; Veldpaus et al., 2013). The approach aims to break away from historic district/quarter thinking that isolates urban heritage and instead embraces the historic city in its entire complexity, integrating tangible and intangible aspects. Cultural mega-events can represent a key moment for cities to develop new planning and governance strategies (Jones, 2019); for World Heritage cities, this can mean seeking out new ways of promoting or activating urban heritage spaces. CoC events have also been shown to play a critical role in cultivating public participation (Tommarchi et al., 2018) and could be harnessed as part of wider community engagement tools for urban heritage management or to address dissonant heritage narratives, as is discussed in the following sections. For these reasons, and as previously argued in Jones and Ponzini (2018), there are existing alignments between the use of mega-events and the management of historic urban environments. Considering that World Heritage cities have embraced hosting cultural mega-events in the recent past and near future while the pandemic has left cities questioning typical tourism models, this chapter presents several examples of how events can go beyond pro-growth tourism strategies to align with broader physical, social and economic sustainability goals.

9.3 Establishing Heritage-Based City Images Through Cultural Events

The ECoC program began in 1985 and has become especially popular over the past several decades for its perceived ability to help cities establish new identities based on cultural production or its existing cultural heritage offer. Glasgow 1992 was notably the first host city to demonstrate how a de-industrialised city could use the event to establish a new city image and begin to shift trends of decline (García, 2005; Mooney, 2004). Other cities have since used the event to bring about tangible or intangible transformations and establish themselves as cultural destinations. Two such examples are the Genoa 2004 European Capital of Culture and the Hull 2017 UK City of Culture (UKCoC), one of the national programmes inspired by the ECoC. Both cities invested significantly in their historic built environment to promote themselves at a national and international level on the basis of their cultural heritage. Genoa has since seen sustained tourism growth in the 15 years following the event, at a higher rate than the national average, with a growing share of visitors attending heritage sites and museums (Jones, 2020). Notably, the city's unique system of urban palaces and streets was recognised as a World Heritage Site in 2006, one of the main goals of city decision makers (Jones, 2022). Meanwhile, Hull used the event to promote a heritage education through the arts approach that made the city's history and heritage accessible to a wide range of local citizens and visitors (Tommarchi & Bianchini, 2022). The event also saw a huge boost to the city's tourism, with over six million visits recorded during the year (CPPI, 2018), and the city plans to continue investing in its cultural heritage in the years to come through the Yorkshire Maritime City Project. Both cities utilised the events to carry out ambitious heritage-focused urban regeneration schemes resulting in a significant impact on the historic areas of these cities while connecting to broader city development and growth by embedding the city's heritage within long-term plans and strategies.

Regarding the UNESCO 5Cs, the ECoC and UKCoC supported conservation, capacity-building and communication of heritage values to local and international audiences. The events brought significant funding that allowed both Genoa and Hull to complete wide-reaching urban regeneration and conservation programs. The particular governance of the CoCs also required a range of actors to collaborate together to complete works (Tommarchi et al., 2018), increasing local capacity-building even if, as in the case of Genoa, such networks were not retained over the long term (Jones, 2019). In terms of SDGs, there were noticeable improvements to Innovation and Infrastructure (SDG 9) as well as Good Jobs and Economic Growth (SDG8), particularly regarding the tourism sector in Genoa over time. The city has also largely avoided concerns of gentrification over the years due to its particular urban structure in the historic city centre that has retained a diverse socio-economic mix (Briata, 2010). While it is too early to observe long-term effects in the case of Hull, the UKCoC has established a new approach to promoting and investing in the city's cultural heritage as a driver of growth and development, locating heritage as a key element in future plans.

9.4 Spreading Out Event Locations to Reduce Localised Stresses

One of the most immediate and direct threats that mass tourism poses to cultural heritage is the physical damage that can be caused through environmental impact (Coccossis, 2016; Richards, 2018). Such issues are aggravated when visitors focus on a few key 'must-see' areas, often World Heritage sites, which creates an imbalance of isolated tourism effects. Festivals, CoCs and mega-events can be used to address such acute problems by diffusing events across urban and peri-urban areas to promote alternative and under-recognised areas of cities, demonstrating the wider historic contexts of World Heritage sites. In this way, event programmes can be specifically designed to help reduce stresses on already highly visited areas of cities and to draw visitors and locals alike to new areas. Such processes can be observed during the 2015 Milan Expo and the transversal Expoincittà programme that was carried out in tandem across the metropolitan region.

Earlier festivals in Milan had already established a tradition of spreading out events to alternative or previously overlooked quarters of the city, initially through the popular Design Week as well as other events like Piano City Milano or Fashion Weeks that have led to the formerly industrial neighbourhoods of Lambrate and Tortona becoming recognised as the city's cultural districts (Armondi & Bruzzese, 2017). As the Expo site was located outside of the city limits in a largely rural area, the Expoincittà programme introduced thousands of small- and micro-scale cultural events throughout the city of Milan. The programme provided gallery spaces for exhibits or small public spaces and squares for performances. Rather than using the prominent or highly visible spaces of the city, the programme utilised many 'leftover' spaces, activating them through cultural activities and attracting visitors to parts of the city they might not normally visit in order to participate in such events (Di Vita, 2022). Expoincittà has since been rebranded as YesMilano! and continues to schedule and host events and festivals across the city to help diversify the range of offerings and attractors.

Another example is the aforementioned Matera-Basilicata 2019 ECoC, noteworthy for its significant increase in tourism before hosting the event. Event organisers developed five 'alternative' pathways stemming out from The Sassi and the Park of the Rupestrian Churches of Matera World Heritage Site to encourage visitors and locals to explore areas beyond the historic city centre and the fragile Sassi area. Unfortunately, such efforts were not tightly linked with the spatial organisation of events throughout the year of celebration, with only occasional events taking place along these new pathways. Without stronger incentives, many tourists were unaware of the city's broader cultural and natural heritage and typically followed the pre-existing touristic routes within the city centre (Ponzini et al., 2020b). Though these new alternative pathways may not have been activated to their full potential during the event itself, they may yet become part of the ECoC's legacy as it introduced a rethinking of such spaces and potential ways to move through the city. In the future, these may yet be utilised and integrated with future strategies to manage the touristic impact on the historic centre. Such an approach can help to create more Sustainable Cities and Communities (SDG 11) and work towards Reduced Inequalities (SDG 10) by spreading out the potential benefits of tourism to reach the areas of cities too often overlooked by tourists and left behind by more central areas in terms of the spread of economic benefits.

9.5 Expanding Traditional Definitions of Heritage Through Cultural Mega-Events

One of the emerging ways that cities have come to utilise cultural mega-events has been to highlight alternative heritage spaces or narratives that may have been previously overlooked or considered too challenging to confront. Events like the Essen for the Ruhr 2010 ECoC emphasised the region's rich industrial heritage, concentrating a significant portion of the event around the Zollverein Coal Mine Industrial Complex in the Essen World Heritage Site. Such trends take an important step towards more locally framed understandings and valuing of heritage that go beyond the cleaned-up versions presented for tourists (Aygen, 2013; Barthel, 1996). Such endeavours frequently invite local communities into these processes, one of the precepts of the 5Cs, and can represent an important step in reconnecting inhabitants with their own heritage (UNESCO, 2007). Two such examples can be seen in the Wrocław 2016 and Pafos 2017 ECoCs, where the events were used to address the difficult histories of both cities. The Polish city of Wrocław is noteworthy for undergoing a complete population exchange following the Second World War. Previously known as the German city of Breslau, the city used the 2016 ECoC as an opportunity to recognise and highlight its multi-cultural, multi-lingual past rather than trying to ignore it, dealing with themes of emigration and immigration within the event (Sanetra-Szeliga, 2022). A key part of the programme was awarding microgrants to local citizens and organisations to stage events in or requalify alternative spaces, giving them the opportunity to define new spaces of value within the city. The microgrant scheme was deemed so successful that it has been continued as part of the ECoC legacy (Sanetra-Szeliga, 2022).

The 2017 Pafos ECoC also provided the opportunity to address the city's, and more broadly the entire country's, difficult history of Turkish-Cypriot heritage spaces and memories. Rather than focus exclusively on the city's existing World

Heritage Site, the candidature phase for the ECoC was a bottom-up process that provided the opportunity for local volunteers along with previously displaced Turkish-Cypriot residents to be involved to a much greater degree in the forming of the proposal and addressing dissonant heritage narratives (Dova et al., 2019). The eventual programme included a number of restoration projects aimed at the historic centre area of Ktima and included several Turkish-Cypriot sites that had been abandoned. Works included renovations and urban improvements to the Attikon and Palia Ilektriki cultural centres along with pedestrianising larger portions of the historic centre, making it more accessible. As Pafos was already a well-established tourism destination, notable for its seaside resorts, the ECoC went beyond a purely tourism-driven approach to begin focusing more on the city's cultural heritage for the benefit of local inhabitants (Dova et al., 2022). Public participation has taken on a leading role within CoC events in recent years (Tommarchi et al., 2018), and though these processes can encounter many challenges, both Wrocław and Pafos were successful in broadening conceptions of heritage that are meaningful for locals and not overly simplified or sanitised for the sake of tourists. In this way, such approaches begin to work towards creating Peace and Justice (SDG 16) by addressing past wounds and difficulties, bringing together diverse groups of society that have been in conflict.

9.6 Risks and the Need for Long-Term Governance and Management

This chapter has demonstrated the potential for cultural mega-events to go beyond a mere pro-growth model of touristification of heritage spaces and cities, showing some of the ways they can be used to align with the aims of the HUL approach, SDGs and the UNESCO 5Cs. While no means an exhaustive list, the examples presented here overview some of the ways that CoCs, cultural mega-events and other festivals can be oriented to maximise the benefits for World Heritage sites and host cities. Yet ensuring that such events do not threaten World Heritage cities requires adequate preparation, planning and governance of these processes from the bidding throughout the legacy phases. One of the surprising issues detected during the case study research conducted as part of the HOMEE Project was that none of the host cities had developed specific tourism management plans in preparation for hosting a cultural mega-event. This finding revealed a clear imbalance of these events being utilised to attract a wide range of visitors without event organisers or city decision makers actively preparing for the inevitable tourism-related stresses that such events would introduce. This distance between heritage management and tourism is by no means new (Du Cros, 2001) but is an issue that should be addressed in the planning and organisation of future events within heritage-rich cities.

While the chapter has largely focused on the potential positive outcomes, disconnects between event planning and the development of long-term strategies and policies can indeed lead to negative outcomes. In the case of the Matera-Basilicata 2019 ECoC, interviews with representatives from the foundation that organised the events, the city municipality and the ministry of culture revealed that each of these entities considered the issue of tourism and tourists themselves to be outside their official purview (Ponzini et al., 2020b). Only the regional tourism body actively focused on tourism issues, and their main priority was to grow tourism as much as possible, particularly as the region of Basilicata is the second-least visited region in the entire country of Italy (ISTAT, 2019). A disconnected approach that overlooks critical issues like tourism can lead to cities being woefully unprepared for mass tourism and the wide range of physical and socio-economic impacts, and the city of Matera found itself ill-prepared to manage tourism flows during the peak tourism season. The case of the Liverpool 2008 ECoC is another cautionary example where the much-celebrated event interacted little with the city's World Heritage Site and instead aligned with future urban growth and development in city strategies (Jones, 2017; West, 2022). The long-term effects of this approach led to the unfortunate delisting of the Maritime Mercantile City World Heritage Site in 2021, only the third ever site to lose its status. While the delisting was by no means a direct result of the 2008 ECoC, the example of Liverpool stands as a severe lesson of the risks of festivalisation strategies that do not prioritise heritage or link to long-term Sustainable Development Goals.

Just as the examples cited in this paper have demonstrated the potential for these events to positively rethink the heritage–tourism dynamic, cities must work towards more multi-disciplinary governance and management of events, tourism and heritage in order to anticipate a wide range of issues and work towards more sustainable approaches. As many of the events' elements are planned early on during the bidding phase, it is crucial for heritage experts to be involved within the organising committees to avoid eventual tourism stresses while also including greater public participation to consider under-recognised local heritage spaces that relate to the context of globally recognised World Heritage sites. A more integrated governance approach can be the key to ensuring that event plans are embedded within longer-term city strategies rather than framing them as one-off tourism-centric events. Naturally, the COVID-19 pandemic introduced an unexpected global shift in tourism trends. This disaster has demonstrated that the complete lack of tourism can be just as damaging to historic cities as mass tourism (UNESCO, 2020) as local economies dependent upon visitors have struggled to survive, and tourism-centric approaches may threaten the future viability of heritage-led urban regeneration or promotion schemes. While the future of global tourism remains unclear, and it may take several years until tourism returns to pre-pandemic levels, this situation can provide an opportunity for heritage cities to reflect and rethink the role of tourism and how it can interact with and support heritage going forward. While the future of cultural events, festivals and mega-events are just as uncertain, as they adapt to post-COVID scenarios, they can continue to serve as tools to explore and experience heritage through new lenses that focus on and involve local communities in ways that align with and can help achieve long-term sustainable goals.

References

Armondi, S., & Bruzzese, A. (2017). Contemporary production and urban change: The case of Milan. *Journal of Urban Technology, 24*(3), 27–45. https://doi.org/10.1080/10630732.2017.1311567

Ashworth, G. J. (2000). Heritage, tourism and places: A review. *Tourism Recreation Research, 25*(1), 19–29. https://doi.org/10.1080/02508281.2000.11014897

Aygen, Z. (2013). *International heritage and historic building conservation: Saving the world's past*. Routledge. https://doi.org/10.4324/9780203083567

Bandarin, F., & Van Oers, R. (2012). *The historic urban landscape: Managing heritage in an urban century*. Wiley Blackwell.

Bandarin, F., & Van Oers, R. (Eds.). (2015). *Reconnecting the city: The historic urban landscape approach and the future of urban heritage*. Wiley.

Barthel, D. (1996). Getting in touch with history: The role of historic preservation in shaping collective memories. *Qualitative Sociology, 19*(3), 345–364.

Bianchini, F., & Parkinson, M. (1993). *Cultural policy and urban regeneration: The West European experience*. Manchester University Press.

Briata, P. (2010). Living or leaving? Regeneration policies in multi-ethnic contexts in North Italy. In P. Ache & M. Ilmonen (Eds.), *Space is luxury* (pp. 590–604). Aalto University.

Coccossis, H. (2016). Sustainable development and tourism: Opportunities and threats to cultural heritage from tourism. In L. F. Girard & P. Nijkamp (Eds.), *Cultural tourism and sustainable local development* (pp. 65–74). Routledge. https://doi.org/10.4324/9781315258720

CPPI. (2018). *Cultural transformations: The impacts of Hull UK City of Culture 2017. Preliminary outcomes evaluation*. University of Hull. Culture, Place and Policy Institute.

Di Vita, S. (2022). Cultural events and heritage policy for the Milan Expo 2015: Experimental intersections between mega-event and city. *European Planning Studies, 30*(3), 499–513.

Dova, E., Sivitanidou, A., & Balasis, E. (2019). Grafting culture onto urban space: The exploration of community-based practices towards sustainability in the case of Pafos, Cyprus. In M. A. Segantini (Ed.), *Tangible – Intangible heritage(S) – Design, social and cultural critiques on the past, present and the future* (pp. 290–301). AMPS Research Organization.

Dova, E., Sivitanidou, A., Anastasi, N. R., & Tzortzi, J. G. N. (2022). A mega-event in a small city: Community participation, heritage and scale in the case of Pafos 2017 European Capital of Culture. *European Planning Studies, 30*(3), 457–477. https://doi.org/10.1080/09654313.2021.1959721

Du Cros, H. (2001). A new model to assist in planning for sustainable cultural heritage tourism. *International Journal of Tourism Research, 3*(2), 165–170. https://doi.org/10.1002/jtr.297

Fatorić, S., & Seekamp, E. (2017). Are cultural heritage and resources threatened by climate change? A systematic literature review. *Climatic Change, 142*(1–2), 227–254. https://doi.org/10.1007/s10584-017-1929-9

García, B. (2005). Deconstructing the city of culture: The long-term cultural legacies of Glasgow 1990. *Urban Studies, 42*(5), 841–868. https://doi.org/10.1080/00420980500107532

García, B., & Cox, T. (2013). *European Capitals of Culture: Success strategies and long-term effects*. European Parliament Committee on Culture and Education.

Gravari-Barbas, M. (2018). Tourism as a heritage producing machine. *Tourism Management Perspectives, 25*, 173–176. https://doi.org/10.1016/j.tmp.2018.01.004

Green, S. (2017). *Capitals of Culture: An introductory survey of a worldwide activity*. Prasino.

ISTAT. (2019). *19 Turismo*. Italian National Institute of Statistics. https://www.istat.it/it/files//2019/12/C19.pdf

Jones, Z. M. (2017). Synergies and frictions between mega-events and local urban heritage. *Built Heritage, 1*(4), 22–36. https://doi.org/10.1186/BF03545655

Jones, Z. M. (2019). Policy and Practice Integrating cultural events and city agendas: Examples from Italian/UK practice. *Town Planning Review, 90*(6), 587–599. https://doi.org/10.3828/tpr.2019.38

Jones, Z. M. (2020). *Cultural mega-events: Opportunities and risks for heritage cities*. Routledge.
Jones, Z. M. (2022). Short-term gains and long-term challenges to learning from mega-event planning in the city of Genoa. *European Planning Studies, 30*(3), 437–456. https://doi.org/10.1080/09654313.2021.1959720
Jones, Z. M., & Ponzini, D. (2018). Mega-events and the preservation of urban heritage: Literature gaps, potential overlaps and a call for further research. *Journal of Planning Literature, 33*(4), 433–450. https://doi.org/10.1177/0885412218779603
MacCannell, D. (1976). *The tourist: A new theory of the leisure class*. University of California Press.
Mooney, G. (2004). Cultural policy as urban transformation? Critical reflections on Glasgow, European city of culture 1990. *Local Economy, 19*(4), 327–340. https://doi.org/10.1080/0269094042000286837
Nuryanti, W. (1996). Heritage and postmodern tourism. *Annals of Tourism Research, 23*(2), 249–260. https://doi.org/10.1016/0160-7383(95)00062-3
O'Brien, G., O'Keefe, P., Jayawickrama, J., & Jigyasu, R. (2015). Developing a model for building resilience to climate risks for cultural heritage. *Journal of Cultural Heritage Management and Sustainable Development, 5*(2), 99–114. https://doi.org/10.1108/JCHMSD-06-2013-0021
Ponzini, D., Bianchini, F., Georgi, J., & Sanetra-Szeliga, J. (Eds.). (2020a). *Mega-events and heritage: The experience of five European cities*. International Cultural Centre. http://www.tau-lab.polimi.it/homee-mega-events-and-heritage-the-experience-of-five-european-cities/
Ponzini, D., Jones, Z. M., D'Armento, S., Scandiffio, A., Bianchini, F., & Tommarchi, E. (2020b). *Urban heritage and mega-events: The case of Matera-Basilicata European Capital of Culture 2019*. Transnational Architecture and Urbanism. http://www.tau-lab.polimi.it/matera-report-file.pdf/
Quinn, B. (2009). Festivals, events and tourism. In T. Jamal & M. Robinson (Eds.), *The SAGE handbook of tourism studies* (pp. 483–503). Sage.
Richards, G. (2018). Cultural tourism: A review of recent research and trends. *Journal of Hospitality and Tourism Management, 36*, 12–21. https://doi.org/10.1016/j.jhtm.2018.03.005
Sanetra-Szeliga, J. (2022). Culture and heritage as a means to foster quality of life? The case of Wrocław European Capital of Culture 2016. *European Planning Studies, 30*(3), 514–533.
Santa-Cruz, F. G., & López-Guzmán, T. (2017). Culture, tourism and world heritage sites. *Tourism Management Perspectives, 24*, 111–116. https://doi.org/10.1016/j.tmp.2017.08.004
Strauss, C. H., & Lord, B. E. (2001). Economic impacts of a heritage tourism system. *Journal of Retailing and Consumer Services, 8*(4), 199–204. https://doi.org/10.1016/S0969-6989(00)00023-0
Timothy, D. J., & Nyaupane, G. P. (2009). *Cultural heritage and tourism in the developing world: A regional perspective*. Routledge.
Tommarchi, E., & Bianchini, F. (2022). A heritage-inspired cultural mega-event in a stigmatized city: Hull UK City of Culture 2017. *European Planning Studies, 30*(3), 478–498.
Tommarchi, E., Hansen, L. E., & Bianchini, F. (2018). Problematising the question of participation in Capitals of Culture. *Participations, 15*(2), 154–169. https://www.participations.org/Volume%2015/Issue%202/10.pdf
UNESCO. (2007). *Proposal for a 'Fifth C' to be added to the Strategic Objectives*. UNESCO. https://whc.unesco.org/en/documents/8896
UNESCO. (2020). *Monitoring World Heritage site closures*. UNESCO. https://en.unesco.org/covid19/cultureresponse/monitoring-world-heritage-site-closures
UNWTO. (2019). *International tourism highlights 2019 edition*. UNWTO. https://www.e-unwto.org/doi/book/10.18111/9789284421152
Veldpaus, L., Pereira Roders, A. R., & Colenbrander, B. J. F. (2013). Urban heritage: Putting the past into the future. *The Historic Environment: Policy & Practice, 4*(1), 3–18. https://doi.org/10.1179/1756750513Z.00000000022
West, T. (2022). Liverpool's European Capital of Culture Legacy Narrative: A selective heritage? *European Planning Studies, 30*, 534–553.
Zubiaga, M., Izkara, J. L., Gandini, A., Alonso, I., & Saralegui, U. (2019). Towards smarter management of overtourism in historic centres through visitor-flow monitoring. *Sustainability, 11*(24), 7254. https://doi.org/10.3390/su11247254

Chapter 10
Sustainable Urban Heritage vs Heritage Orthodoxy

Dennis Rodwell

Abstract Against the backdrop of recent crises, such as the Covid-19 pandemic and various armed conflicts, this chapter highlights the unique vulnerabilities of urban heritage and how these have been exacerbated by an over-simplified construction and commodification of heritage in the official discourse. The chapter reflects on issues and contentions about the definition of urban heritage and the prevailing focus on listing sites and monumental values, highlighting the frequent contrast between community narratives and outstanding universal value. Aside from challenges to its existence, urban heritage must also find ways to mediate between environmental, social, and economic pressures that World Heritage status imposes and the over-riding principles and objectives of the SDGs and other ambitious policies.

This chapter proposes that addressing all these challenges is linked to a shift toward an inclusive human- and environment-focused definition of urban heritage, which leaves room for community statements of values, not just OUV. This chapter presents the case for recognizing urban heritage's compendium of values to consolidate its sustainability and spread and minimise the risks. In short, it is a call to move away from heritage orthodoxy and toward sustainable urban heritage.

Keywords Authenticity · Community · Continuity · Inheritance · Resource · Sustainability

10.1 Introduction

The Covid-19 pandemic has served to highlight the inherent vulnerability of historic cities to heritage orthodoxy. Whereas the 1972 World Heritage Convention did not anticipate heritage branding, commodification, or mass tourism, the simplistic, abstracted, and carefully distilled definitions of cultural heritage in the Convention

D. Rodwell (✉)
Independent Researcher and Consultant, St. Boswells, Scotland, UK
e-mail: dennis@dennisrodwell.co.uk

M.-T. Albert et al. (eds.), *50 Years World Heritage Convention: Shared Responsibility – Conflict & Reconciliation*, Heritage Studies,
https://doi.org/10.1007/978-3-031-05660-4_10

coupled with the ongoing understatement of the compendium of values in successive editions of the UNESCO Operational Guidelines (UNESCO, 1972a, 2021), effectively distances urban heritage from the communities that are its primary, secure, long-term custodians and stakeholders. There is a lack of attention to the commitment under Article 5 of the Convention (UNESCO, 1972a), expanded in the contemporaneous 1972 UNESCO Recommendation concerning the Protection, at National Level, of the Cultural and Natural Heritage (UNESCO, 1972b). Ongoing challenges include synchronisation with today's global agendas, including the 2030 United Nations Sustainable Development Goals (SDGs) (United Nations, 2015). Such discordances contribute to the widespread *reductio ad absurdum* that urban heritage is object-focused and justifies its survival primarily as a raw material for high-end urban regeneration and tourism. Vulnerabilities include displacement of communities and gentrification, aggressive *contemporary* interventions, financial downturns, and pandemics. The sustainability of urban heritage demands a far more substantive foundation. This chapter interrogates *urban heritage* as the manifestation of continuously inhabited places of everyday human as well as often closely defined cultural significance, and presents the case for the recognition of the compendium of social, cultural, economic, and environmental values beyond those recognised within mainstream heritage orthodoxy, to the objective of spreading and minimising risks and vulnerabilities and reinforcing the sustainability of urban heritage.

10.2 Context in Time

The concept for the 1972 World Heritage Convention evolved through the 1960s from a coalescence of interests. These included the commitment set out in the UNESCO constitution to the conservation and protection of the world's cultural *inheritance* (UNESCO, 1945); the foundation in 1948 of the International Union for Conservation of Nature and Natural Resources (IUCN); and the establishment in 1965 of the International Council on Monuments and Sites (ICOMOS). In the aftermath of the destruction of monuments and cities during the Second World War, the context in time recognised the cumulative effects of neglect and decay, alongside peacetime threats posed by rapid social and economic changes coincidental with the ascendancy of new ideas in architecture and urban planning. Together with the advent of the environmental movement, key words at the time included *protection* and *conservation*; the concept of *sustainability* was subsumed for the natural world; and the popularisation of *sustainable development* awaited the Brundtland Report (Brundtland Commission, 1987). The need for awareness and education was a key driver for the Convention.

In the intervening half century, the 1972 starting points of neglect and decay have remained, augmented by accelerating socio-economic changes. Additionally, there has been a resurgence of destructive armed conflicts, including in the Balkans

(1990s), the wider Middle East and North Africa (MENA) region (2000s and ongoing), and Ukraine. These challenges have been compounded by emergent and intensified phenomena of *urban transformations*, including: mass tourism; the commodification of heritage allied to promotion and prioritisation of its *economic* value; the gentrification of historic neighbourhoods and associated socio-economic displacements; the primacy attached to *contemporary* interventions in the built environment (ICOMOS, 1964; UNESCO, 2005 and 2011); the fashion for tall buildings allied to a disregard for their impact on land values within and adjoining historic areas; augmented by actual and projected impacts of climate change. Whereas the ramifications of these multiple challenges are variable by location and time, the UNESCO brand has focused many of these in World Heritage Sites. This all imposes severe challenges on the objectives of responsibility, reconciliation, and sustainability.

10.3 The World Heritage Convention and Heritage Orthodoxy

10.3.1 Urban Heritage, Authenticity, and Integrity

Interpreted from the definitions under Article 1 of the 1972 Convention, urban heritage is categorised under *groups of buildings* and focused on the *tangible* (UNESCO, 1972a). This presents challenges in the context of the key word *authenticity*, as is made clear in successive editions of the UNESCO Operational Guidelines from January 1987 through July 2019: "historic towns which are still inhabited and which, by their very nature, have developed and will continue to develop under the influence of socio-economic and cultural change, a situation that renders the assessment of their *authenticity* [this author's emphasis] more difficult and any conservation policy more problematical" (UNESCO, 2019, Annex 3, 14 (ii)). This, notwithstanding the 1994 Nara Document on Authenticity (ICOMOS, 1994), which expanded the understanding of *authenticity* beyond materials and substance to include form and design, use and function, traditions, techniques and management systems, location and setting, and spirit and feeling (summarised at UNESCO, 2021, paras 79–86). The key word *integrity* also presents challenges. Defined in the Operational Guidelines as "a measure of the wholeness and intactness of the … cultural heritage and its attributes", with the stipulation that "relationships and dynamic functions present in cultural landscapes, historic towns or other living properties essential to their distinctive character should also be maintained" (UNESCO, 2021, paras 87–89), the context is constrained by reference to *outstanding universal value* and *significance*, concepts which are challenged by the dynamics of inhabited towns. The premise of heritage orthodoxy is that *authenticity*, *integrity*, and *distinctive character* (the essence of *identity*) are determined by physical attributes. Further, whereas "transmission to future generations" features in the

1972 Convention (UNESCO, 1972a, Article 4), there is dissonance between the 2011 UNESCO *Recommendation on the Historic Urban Landscape* (UNESCO, 2011) and the 2011 ICOMOS *Valletta Principles* (ICOMOS, 2011). The former insists on *layering*, an ambiguous concept, not least in re-iterating the insistence in the 2005 Vienna Memorandum on interventions being *contemporary* (UNESCO, 2005, Article 21); the latter insists on *continuity*, more closely allied to transmission.

10.3.2 Urban Heritage and Orthodoxy

Focus on *monumental values* expressed in materials and substance has roots in European heritage orthodoxy. In Anglo-centric philosophy, it is signalled in the 1877 Manifesto of the Society for the Protection of Ancient Buildings (1877). At the global level, it underscores interpretations of the Venice Charter (ICOMOS, 1964), ICOMOS's founding doctrinal text, envisaged as conditioning *universal* parameters for protection and conservation, and largely unchallenged in mainstream orthodoxy until the Nara Document (ICOMOS, 1994). Exceptions include successive editions of the Burra Charter, 1979 onwards (ICOMOS-Australia, 2013). Whereas the 2003 UNESCO Intangible Cultural Heritage Convention (UNESCO, 2003) counter-balanced the 1972 Convention, the text of the 2011 UNESCO Recommendation is limited in its understandings of *urban heritage* and *values*, *conservation* and *authenticity*, anticipates circumscribed definitions and categories of *tangible* and *intangible* heritage as selected objects and manifestations, and derives from normative approaches (UNESCO, 2011).

Notwithstanding some doctrinal loosening, for example, the 2005 Faro Convention (Council of Europe, 2005) and the 2014 ICOMOS Florence Declaration (ICOMOS, 2014) together with early implementations of the Historic Urban Landscape approach beyond Europe – in Cuenca, Ecuador; Ballarat, Australia; and trials of "HUL Quick Scan" in Indonesia – core philosophy, training, and practice prioritise specialists as the instigators of the heritage discourse, positioning citizens and communities as adherents to narratives constructed by others.

The mainstream European-derived approach is manifest in the definition formulated in the context of India: "[Urban Heritage] Refers to the built legacy of the city's history and includes protected and unprotected monuments, individual and groups of buildings of archaeological, architectural, historic and cultural significance, public spaces including landscapes, parks and gardens, street layout defining identifiable neighbourhoods or precincts, which together identify the visual, spatial and cultural character of the city" (National Institute of Urban Affairs, 2015, p.68). Parallel expression infuses the manifesto of the President of the ICOMOS International Committee on Historic Cities, Towns and Villages (CIVVIH) (Echter, 2020).

10.3.3 World Heritage, the World's Heritage, and the Operational Guidelines

November 2022 sees the 50th anniversary of the adoption of the World Heritage Convention (UNESCO, 1972a). Whereas the Preamble to the Convention addresses "assuring the conservation and protection of the world's heritage" *as well as* those "parts of the cultural and natural heritage [that] are of outstanding interest", the intervening decades-long focus on the World Heritage List has overshadowed States Parties' over-arching commitment to the collectivity of the cultural and natural heritage in their territories, as itemised under Article 5 and signalled in Article 12 of the Convention, and expanded upon in the parallel 1972 UNESCO Recommendation (UNESCO, 1972b), a largely overlooked document in the UNESCO archive that is intended to underpin the 1972 Convention.

Focus on the List has assisted the disproportionate promotion of a highly selected group of heritage properties to the prejudice of the advancement of comprehensive global heritage conservation and sustainable management and contributed to the increasing politicisation of outcomes at successive sessions of the World Heritage Committee.

The UNESCO *Operational Guidelines for the Implementation of the World Heritage Convention*, which have gone through over a score of revisions since the first edition in June 1977, are regularly updated; most recently, in July 2021. Consistently, they focus on the processes for inscription, the monitoring and periodic reporting of properties in the List, and celebration of the brand (UNESCO, 2021). In their selective approach to the provisions and commitments in the Convention, from inclusive to exclusive, they are effectively mistitled.

10.4 Urban Heritage

10.4.1 The Heritage Construct and Outstanding Universal Value

Today's *heritage construct* emerged around the turn of the third and fourth quarters of the twentieth century. Previously, *heritage* was *inheritance* and understood holistically; it was not split into *cultural* and *natural*, *tangible* and *intangible*, nor subject to maximal value judgements and what Marc Askew has described as UNESCO's "fetishism for making lists" (Askew, 2010, p. 32).

Importantly, processes of inclusion into lists of heritage are simultaneously processes of exclusion, of people as well as places. In today's interdisciplinary field of heritage studies, heritage is understood "as a social and political construct", in which "heritage results from a selection process, often government-initiated and supported by official regulation" (Labadi & Logan, 2016: foreword). Selection processes are top-down, not bottom-up, and the protection of heritage is generally assumed to be atypical and exceptional, largely determined by specialists, and expensive. As Laurajane Smith argues, the dominant *authorised heritage discourse* "constitutes the idea of heritage in such a way as to exclude certain actors and interests from actively engaging with heritage", framing audiences as passive recipients of the authorised meaning of heritage and creating significant barriers to "the social and cultural roles that it may play" (Smith, 2006).

Mainstream concepts of heritage confer value based on the perspective of an educated elite. This can exclude both long-established as well as incoming communities within historic cities. Narratives constructed to evidence *outstanding universal value* constitute carefully edited intensifications of the authorised heritage discourse. "We connect people to their heritage" headlines the mission statement of Edinburgh World Heritage Trust (Edinburgh World Heritage, n.d.). Urban populations are not homogenous. Such statements imply that the manifold constituent communities are not connected to *their* heritage: multiple heritages are not recognised; non-adherents to the discourse are excluded; only one narrative is legitimised.

Discordance between the UNESCO narrative of *outstanding universal value* and a host community is well-illustrated in the case of Paramaribo, the capital of Suriname, a country that gained its independence in 1975 (Fig. 10.1). The Historic Inner City of Paramaribo was inscribed in the World Heritage List in 2002, highlighting its Dutch colonial and Christian heritage (UNESCO, n.d.). The complexity of the ethnic, religious, social, and cultural diversity of the historical as well as present-day communities in Paramaribo is not reflected in the UNESCO synthesis; this ignores, for example, the main synagogue and the assemblage of Modernist buildings from the 1950s and 1960s (Strik & Lambert, 2018).

Fig. 10.1 Paramaribo, Suriname. (*Note.* Image by Dennis Rodwell 2018)

The inner city is currently under-occupied and in poor condition, lacking the animation that characterised its historical multi-functionality. There is a shortfall of support for the heritage of the country's colonial past in political and governmental circles, filtering through to omissions in heritage education, professional and craft skills training, and interpretation material, all underscored by lack of community support. The monumentalisation of this World Heritage Site, factored into *contested heritage*, offers little intellectual access and challenges its sustainability; the spectrum of *values* is in serious need of expansion. Labadi has raised important questions concerning alignments of *outstanding universal value* with values perceived by today's communities, as well as dissonances between the identification of normative values pre-inscription and the long-term conservation and management of sites post-inscription (Labadi, 2013).

10.4.2 What Is Urban Heritage?

The unique identity of any historic city is a conjunction of people, place, and time. Just as natural heritage sites cannot survive as ecosystems without wildlife, historic cities are contingent on human functionality. An integrated approach to urban heritage is not simply a question of the restoration of buildings, ensembles, and public spaces. It subsumes an understanding of the dynamics of everyday life and timelines of socio-economic continuity in the communities that host and animate a quantum and diversity that extends far beyond prescribed definitions of tangible and intangible cultural heritage. This distinguishes it fundamentally from built heritage.

Urban heritage, comprehended generically as embracing a legion of globally diverse living historic cities and urban districts, in which citizens have and continue to conduct their daily lives in complex and dynamic relationships with a heterogeneity of physical environments, is a highly complex field that fits uneasily into heritage orthodoxy. The human factor – the synergy between the miscellany of human activities and the myriad of physical places – is missing, as is the strategic vision to position urban heritage mainstream in the geography of urban planning (Ripp & Rodwell, 2015 and 2016).

The safeguarding and transmission of urban heritage are dependent on continuity of functionality. The key words *authenticity* and *integrity* are struggling to be considered, let alone appraised, in relation to continuity in the socio-cultural life of historic cities and their neighbourhoods (Brown-Saracino, 2010). The 2011 UNESCO *Recommendation* omits to expand *authenticity* and *integrity* to embrace communities and neighbourhoods (UNESCO, 2011). Anthropologists and sociologists are not incorporated into the heritage discourse.

A major global threat to urban heritage manifests from its simplistic construction and commodification as *heritage*, coupled with the interpretation of heritage *values* in monetary terms, whether for tourism, gentrification, or other. This has prejudiced people-based interests, whether of habitation, commerce, or the broader community, as well as visitor understandings of cities' culture and their heritage. This challenges the sustainability of historic cities, including their resilience to resist major downturns, whether financial, such as the 2007–2008 global crisis, or pandemics, Covid-19, 2020 onwards.

10.5 Changes in Values and Attitudes

The pre-millennium trajectory of heritage selectivity allied to orthodox processes of designation is interrogated by today's agendas of sustainability and climate change, which place a broader onus on the "3Rs" of sustainability: *reduce*, *recycle*, and *reuse*. The assumption that closely defined *heritage* determines identity is also called into question by statements such as "The unlisted buildings enshrine the human stories, the memories of the community. They are the real heritage. It is they that determine the sense of identity, of place, and of belonging. These are the places where the historic environment is at the heart of sustainable communities" (Goodey, 2007).

In India, the shoots of intellectual independence from the orthodox monumentalist approach to urban heritage are emerging. As a 2013 position papers states, "(an) important dimension of urban heritage in India is its living character, where the past is very much part of the present lives of the people; as an evolving cultural resource in which continuity and change are deeply embedded" (India Institute for Human Settlements, 2013, p. 2).

A longstanding champion of an inclusive approach to India's urban heritage is Professor A. G. Krishna Menon, who argues that "the nascent field of urban conservation in India offers the potential to review the dominant paradigms of urban planning and develop more context-specific and appropriate strategies for tackling the problems of Indian urbanisation" (Menon, 2017, p. 34).

For this, Menon recommends revisiting the pioneering approach demonstrated by Patrick Geddes in the reports he produced for cities in India in the period 1915–1919 (Tyrwhitt, 1947). Regarding cities as organic systems, each a unique human artefact in its equally unique local and regional environment rather than simply an example of an abstract typology, Geddes insisted on the need for comprehensive historical, geographic, biological, climatic, sociological, economic, cultural, and institutional insight and knowledge, and on nurturing the shoots of innovation and creativity rather than restraining the evolution of a city based on its roots at some historical moment in time (Geddes, 1915).

Expansion of the concept of *values* in heritage management remains work in progress (including Avrami et al., 2019). This reinforces the view that, in rhythm with twenty-first-century agendas, an inclusive compendium of the values that citizens attribute to urban heritage needs to be articulated and embraced, one that is cross-disciplinary, unconstrained by bureaucratic and academic silos, and encapsulated under headings that include

- *community* – all social values and relationships;
- *resource* – in multiple senses, including environmental capital/embodied energy;
- *usefulness* – including ongoing adaptation and creative reuse; and
- *cultural* – broadly defined, especially as recognised and appreciated by inhabiting communities.

This supports the thesis that there are no limits to the appreciation of *heritage* once there is respect and recognition for all sectors and age groups in any given society (Rodwell, 2015).

10.6 Urban Transformations

Change and transformation are the normative state of cities. They may not be intended, desired, or beneficial. World Heritage stands at the apex of today's commodification of *heritage* that adds value to buildings, visitor experiences, countless merchandise, and much else.

The growth of the global tourist industry has precipitated major changes in patterns of use, notably in cities in the World Heritage List. The procedures for World Heritage inscription and monitoring neither anticipate nor include provisions to oversee changes of functionality, including the supplanting of the core original function of cities as places of residence.

Venice is a prime example. A city whose early-1970s population of over 130,000 represented a balance between the number of households and the number of dwellings in the city, today counts less than 60,000 residents, mostly in peripheral areas of the city; the 22 million visitors in a year have assumed precedence (Fig. 10.2).

Whereas cruise ships are the most visible sign of this transformation, the population drop attracts scant attention from the media or from the World Heritage Committee. With a near tourist mono-culture in the city centre, the economic impact of the Covid-19 pandemic could not be more severe.

Heritage listings, once perceived as a burden, deflating property values and rent returns, now attract a *heritage premium*. In today's volatile financial markets, heritage properties provide investment security, fuelling gentrification and loss of community diversity (Rodwell, 2018).

A comparison of the cities of Zamość, Poland, and Sibiu, Romania, the first treated as a *monument*, inscribed as a *group of buildings* and with the policy objective of decanting the established population to the city's outskirts, the second not inscribed, and with a revitalisation programme prioritising the existing community, is informative (Rodwell, 2010).

The heritagisation of cities imposes severe penalties on their authenticity as well as their viability. *Authenticity* invoked through normative, selected cultural attributes only embraces the defining characteristics of an established city superficially. *Transformations* that sever the connections and continuity between place and

Fig. 10.2 Venice and its Lagoon was inscribed in the UNESCO World Heritage List in 1987. (*Note.* Image by Dennis Rodwell 2015)

people can transform functioning historic cities into variants of Disneyland, sites whose resilience and viability are severely challenged in times of pandemic and associated closure to travel and tourism.

10.7 Culture, Heritage, and the Sustainable Development Goals

In the lead-in to 2015 and the definition of the SDGs, an initial aspiration was to introduce *culture* as the fourth dimension of sustainable development, complementing *environment*, *society*, and *economy* (Brundtland Commission, 1987). Comprehended holistically, *culture* is an inclusive, cross- and inter-sectoral concept that embraces all fields of human activity and endeavour, conventionally sectioned into discrete disciplines and fields, nominally the arts, humanities, and sciences, and encompassing all interests, pursuits, and occupations that employ the word *culture* (Williams, 1981 and 1988). In the event, consensus was not reached between competing claims, the opportunity to position *culture* as the common feature that binds human engagement across all 17 SDGs and 169 targets was missed, and *cultural heritage* features explicitly and implicitly only to a limited extent. As such, the potential of *culture* to impact coherently across the sustainability agenda is seriously constrained.

The dilemma facing the UNESCO World Heritage system in its attempt to align itself with the United Nation's Sustainable Development Agenda is set out in Larsen and Logan (2018), which clearly identifies the challenge of mediating between the environmental, social, and economic pressures that World Heritage status imposes on highly selected properties and the over-riding principles and objectives of the SDGs (Rodwell, 2021). The same handicap impacts the policy guidance issued by ICOMOS (2021).

10.8 The Challenge

The premise of heritage orthodoxy is selective survival according to exacting conservation standards formulated from philosophies founded on European models that were intended to have *universal* application. Two main pillars for this are the World Heritage Convention (UNESCO, 1972a) and the Venice Charter (ICOMOS, 1964), both dating from the third quarter of the twentieth century. Across the intervening decades, the bibliography and compass of charters and homologous texts have expanded dramatically (Ripp & Rodwell, 2015: appendix). At the same time, as we approach the second quarter of the twenty-first century, the global agendas of environmental protection, sustainability, and climate change have assumed centre stage. Selective survival is no longer the responsible option; the canvas has enlarged dramatically, and emphasis on human-centred approaches has accelerated. Heritage orthodoxy has yet to position itself centrally in this new global reality.

Understood as continuously inhabited places, urban heritage – the major challenge that conservation theorists and practitioners face in this twenty-first century – does not fit well into the cultural heritage definitions in the 1972 Convention, into successive editions of the Operational Guidelines or with narrow interpretations of the key words *authenticity* and *integrity*, or with heritage orthodoxy generally. Urban heritage in the World Heritage List faces intensified challenges, the principal of which are encapsulated by its *commodification* and transformations that fundamentally affect its functionality and securitisation (Rodwell, 2019).

Recognising the challenge is the essential precursor to addressing it. For this, a step-change is needed to move beyond simplified linear cause and effect models of interventions and comprehend each and every historic city as systems with multiple sub-systems, all in continuous motion. This requires close cooperation and partnerships with disciplines and interests that have not traditionally been associated with the heritage field and have often been seen as adversaries (Ripp & Rodwell, 2015 and 2016).

10.9 The Way Forward

November 2022 will celebrate the 50th anniversary of the World Heritage Convention, not of the World Heritage List. The Convention was a phenomenon of its time and has served us well. To date, however, there has been only a selective activation of the Convention's provisions, and there is much in reserve. Specifically, State Parties' obligations to the *world's heritage* under the Convention's Preamble and Article 5 now need to be prioritised, in conjunction with a focus on the 1972 Recommendation (UNESCO, 1972b).

In tandem, the UNESCO Operational Guidelines – whose successive editions are long overdue a root and branch overhaul – should be re-formulated away from their unique focus on World Heritage to incorporate the many themes, agendas, and associated UNESCO Conventions, Recommendations, and other post-1972 issues and documents which are not currently integrated. Some, such as the 2011 Recommendation on the Historic Urban Landscape (UNESCO, 2011), have yet, a decade later, to be mentioned at all.

There is a complementary need to embrace an inclusive human- as well as environment-focused definition of urban heritage that recognises the spectrum of values inherent in it – summarised above as *community*, *resource*, *usefulness*, and *cultural* broadly defined.

Moving on, inclusive statements of values, not just of outstanding universal value, but ones that position people in their communities at the apex of the hierarchy of stakeholders, need to be developed further, alongside understandings of *authenticity* and *integrity* that embrace people in their communities.

10.10 Conclusion

Urban heritage constitutes the oldest and most historic parts of our cities, and heritage orthodoxy has an important ongoing role to play. At the same time, a prerequisite for the sustainability of historic cities is to secure balanced futures for them founded on continuity of their *raison d'être* as multi-varied and multi-functioning human habitats. Historic cities were not settled and constructed as *heritage*, as a raw material for high-end urban regeneration and tourism. Such is a hi-jacking of their *authenticity* and *integrity*, in contradiction of the drivers for and underlying ethos of the 1972 World Heritage Convention. The collapse of international travel and tourism for much of 2020 and continuing into 2022 has highlighted the vulnerability of relying on a single sector of the economy to justify and support another. The heritage sector suffers an abundance of risks at the best of times. The Covid-19 pandemic has served as a timely warning for the heritage sector that it must broaden its approach, spread and limit its risks and vulnerabilities, revisit the roles that urban heritage has and can continue to perform, and reinforce its sustainability. The "eggs in one basket" approach leaves it too exposed and fraught with danger.

References

Askew, M. (2010). The magic list of global status: UNESCO, World Heritage and the agenda of states. In S. Labadi & C. Long (Eds.), *Heritage and globalisation* (pp. 19–44). Routledge.
Avrami, E., Macdonald, S., Mason, R., & Myers, D. (Eds.). (2019). *Values in heritage management: Emerging approaches and research directions*. Getty Conservation Institute.
Brown-Saracino, J. (2010). *A neighborhood that never changes: Gentrification, social preservation, and the search for authenticity*. University of Chicago Press.
Brundtland Commission. (1987). *Our common future: The World Commission on Environment and Development*. Oxford University Press.
Council of Europe. (2005). *Council of Europe Framework Convention on the value of cultural heritage for society (Faro convention)*. Council of Europe.
Echter, C. -P. (2020). *Reflections of an expert: Key priorities with regards to urban heritage*. https://www.youtube.com/watch?v=ZENc3XYKmtg
Edinburgh World Heritage. (n.d.). *Edinburgh World Heritage: Vision, mission, values*. https://ewh.org.uk/about-us/our-history/vision-mission-values/
Geddes, P. (1915). *Cities in evolution*. Williams and Norgate.
Goodey, F. (2007). *Institute of Historic Building Conservation (IHBC) Summer School*. [Conference Presentation]. IHBC 2017 Summer School, Liverpool, England.
ICOMOS. (1964). *The Venice charter. International charter for the conservation of monuments and sites*. ICOMOS.
ICOMOS. (1994). *The Nara document on authenticity*. ICOMOS.
ICOMOS. (2011). *The Valletta principles for the safeguarding and management of historic cities, towns and urban areas*. ICOMOS.
ICOMOS. (2014). *The Florence declaration on heritage and landscape as human values*. ICOMOS.
ICOMOS. (2021). *Heritage and the sustainable development goals: Policy guidance for heritage and development actors*. ICOMOS.
ICOMOS-Australia. (2013). *The Burra charter. The Australia ICOMOS Charter for places of cultural significance*. ICOMOS-Australia.
India Institute for Human Settlements. (2013). *Socio-economics of urban heritage*. https://doi.org/10.24943/seuh.iihspp.2013
Labadi, S. (2013). *UNESCO, cultural heritage, and outstanding universal value: Value-based analyses of the world heritage and intangible cultural heritage conventions*. AltaMira Press.
Labadi, S., & Logan, W. S. (Eds.). (2016). *Urban heritage, development and sustainability: International frameworks, national and local governance*. Routledge.
Larsen, B. L., & Logan, W. (Eds.). (2018). *World Heritage and sustainable development: New directions in world heritage management*. Routledge.
Menon, A. G. K. (2017). The rationale for reviewing current concepts of urban planning and developing new ones in India. *Built Heritage, 1*(3), 34–43. https://doi.org/10.1186/BF03545673
National Institute of Urban Affairs. (2015). *Urban heritage in Indian cities: Compendium of good practices*. National Institute of Urban Affairs. https://www.niua.org/pearl/sites/default/files/books/GP-IN4_HERITAGE.pdf
Ripp, M., & Rodwell, D. (2015). The geography of urban heritage. *The Historic Environment: Policy & Practice, 6*(3), 240–276. https://doi.org/10.1080/17567505.2015.1100362
Ripp, M., & Rodwell, D. (2016). The governance of urban heritage. *The Historic Environment: Policy & Practice, 7*(1), 81–108. https://doi.org/10.1080/17567505.2016.1142699
Rodwell, D. (2010). Comparative approaches to urban conservation in central and Eastern Europe: Zamość, Poland, and Sibiu, Romania. *The Historic Environment: Policy & Practice, 1*(2), 116–142. https://doi.org/10.1179/175675010X12662480109199
Rodwell, D. (2015). The limits of heritage: What limits? In K. Jagodzińska & J. Purchla (Eds.), *The limits of heritage* (pp. 25–40). International Cultural Centre.
Rodwell, D. (2018). "Gentry"? Heritage conservation for communities. *Change Over Time, 8*(1), 74–100. https://doi.org/10.1353/cot.2018.0004

Rodwell, D. (2019). Urban heritage, communities, and environmental sustainability. In H. Oevermann & E. Gantner (Eds.), *Securing urban heritage: Agents, access, and securitization* (pp. 62–79). Routledge.
Rodwell, D. (2021). Book review of Larsen & Logan (2018). *Context, 168*, 50–51.
Smith, L. (2006). *Uses of heritage*. Routledge, Taylor & Francis Group.
Society for the Protection of Ancient Buildings. (1877). *Manifesto of the Society for the Protection of Ancient Buildings*. SPAB.
Strik, F., & Lambert, A. (2018). *Paramaribo: De Modernistische Architectuur van ir.* Peter J Nagel. LM Publishers.
Tyrwhitt, J. (Ed.). (1947). *Patrick Geddes in India*. Lund Humphreys.
UNESCO. (1945). *Constitution*. UNESCO.
UNESCO. (1972a). *Convention concerning the protection of the world cultural and natural heritage*. UNESCO.
UNESCO. (1972b). *Recommendation concerning the protection, at national level, of the cultural and natural heritage*. UNESCO.
UNESCO. (2003). *Convention for the safeguarding of the intangible cultural heritage*. UNESCO.
UNESCO. (2005). *Vienna memorandum on "World Heritage and contemporary architecture – Managing the historic urban landscape"*. UNESCO World Heritage Centre.
UNESCO. (2011). *Recommendation on the historic urban landscape*. UNESCO World Heritage Centre.
UNESCO. (2019). *Operational guidelines for the implementation of the World Heritage Convention*. UNESCO World Heritage Centre.
UNESCO. (2021). *Operational guidelines for the implementation of the World Heritage Convention*. UNESCO World Heritage Centre.
UNESCO. (n.d.). *Historic inner city of Paramaribo*. UNESCO. http://whc.unesco.org/en/list/940
United Nations. (2015). *2030 Sustainable development goals*. United Nations. https://sdgs.un.org/goals
Williams, R. (1981). *Culture*. Fontana.
Williams, R. (1988). *Keywords*. Fontana.

Chapter 11
The Politics of Shared Heritage: Contested Histories and Participatory Memory Work in the Post-Colonial Urban Landscape

Jan Küver

Abstract Shared heritage is a concept that serves to address cultural ties between countries or people that emanate from colonial history, including conflicts and contestations as well as connections and commonalities. This contribution evaluates the potential of shared heritage to work as a tool for a transformative heritage management practice through exploring the post-colonial heritage landscape of Iringa, Tanzania. The historical dynamics of colonialism have left various tangible and intangible traces throughout Iringa Town and Region. Combining ethnographic and historical methods, this paper examines historical narratives of different social groups, representations of these trajectories in the regional museum, and community responses to buildings and sites of colonial origin in the cityscape. In line with UNESCO's Historic Urban Landscape (HUL) approach, observed applied conservation activities are discussed in the light of local development processes. I argue that shared heritage can serve as a viable concept to grapple with the colonial legacy vested in the HUL while at the same time using the discursive energy provided by these conflicts to support the cultural, social, and economic development of communities.

Keywords Shared heritage · Colonial history · Historic urban landscape · Iringa · Tanzania

---

J. Küver (✉)
University of Iringa, Iringa, Tanzania

fahari yetu Tanzania, Iringa, Tanzania
e-mail: jkuever@fahariyetu.net

M.-T. Albert et al. (eds.), *50 Years World Heritage Convention: Shared Responsibility – Conflict & Reconciliation*, Heritage Studies,
https://doi.org/10.1007/978-3-031-05660-4_11

11.1 Introduction

This paper builds on a line of inquiry from my PhD thesis (Küver, 2021b) in developing the theoretical implications of the concept of *shared heritage* and its positioning within the heritage discourse. Among different readings of the concept, shared heritage serves to address cultural ties between countries or people that result from colonial history, seeking to negotiate conflicts and contestations as well as connections and commonalities emanating from this historical legacy. The evaluation is done through the lens of the post-colonial heritage landscape of Iringa, Tanzania – i.e., efforts of conserving elements in this landscape – as a particular case of shared heritage.

Iringa is a medium-sized town of about 150,000 inhabitants, and the surrounding administrative region has approx. one million inhabitants (United Republic of Tanzania, 2013) in the Southern Highlands of Tanzania. Iringa harbours a diverse heritage landscape with various natural and cultural attractions, yet it is particularly known for the history of the rise and fall of the Hehe chiefdom during the second half of the nineteenth century, which culminated in a grim war of anti-colonial resistance against the German colonial conquest.

This paper investigates a bundle of applied heritage conservation and management activities that were part of the cultural heritage conservation and management initiative *fahari yetu Tanzania*, a programme that I established and coordinate in Iringa myself. fahari yetu – a Swahili term translating to "our pride" – combines academic research, historical restoration, museum exhibitions, professional capacity building, community outreach, and tourism commodification into a holistic heritage management practice (http://fahariyetu.net). In the following sections, I will show how colonial history emerges as shared heritage from the case, discuss community responses to the applied conservation of "shared" remnants of this historical legacy, and lay out concluding reflections of working towards a shared Historic Urban Landscape.

11.2 Theories and Methods

Shared heritage allows for various theoretical readings and applications in institutional policy and practice. First of all, the idea of sharing is inherent to the concept of heritage as a universal cultural archive or inventory to be made accessible for different people and cultures of the world as proclaimed in the UNESCO Conventions (1972; 2003). As such, it can be applied in transnational contexts to create new narratives of a common history, such as in the shared heritage programmes established by the national governments of France (Bibliothèque Nationale de France, n.d.) and the Netherlands (Netherlands Cultural Heritage Agency, n.d.), which document and depict the historical connections of the two countries with other countries and places around the globe and how these connections have contributed to shaping contemporary French and Dutch culture. Beyond national histories, the acknowledgement of transcultural dynamics transfers the concept to heritage communities emerging from the diversity of contemporary migration societies. This view recognises that heritage – whether places, landscapes, practices, or collections – is frequently connected with and valued by multiple and diverse communities who share a collective responsibility for its care and safeguarding (ICOMOS, 2020). From this perspective, shared heritage is not about identifying an original or rightful owner but seeks to elicit stories from objects, buildings, or sites, stories that are attached to the knowledge, skills, and values of different users and are passed on and transmitted between people and generations.

Furthermore, shared heritage is intentionally provocative and explores the ideas of cooperation against its counterpoints, contestation, and resistance (ICOMOS, 2020). In some cases, cultural features and values have been forced upon populations and resisted rather than collectively adopted. Thus, the value of places representing these features can be contested, leading to debates about their conservation. In this vein, a fourth reading of shared heritage addresses the historical injustice of colonialism and current post-colonial discourses that critically examine this heritage from the perspective of the colonized. This requires intensive cooperation and research in dialogue between representatives of both former colonizing and colonized cultures. Research and practice dealing with African colonialism and its ramifications in Africa and Europe use the term shared heritage in mainly two respects (Vanhee, 2016, p.6): The first is in reference to cultural property of colonial origin in African countries. These are buildings, monuments, and sites that were originally designed by metropolitan architects but are now appropriated by post-colonial users. The second, conversely, is in reference to cultural property that was created by Africans and collected by Europeans. This critical museum discourse also includes human remains that are now in European museum collections. Both, buildings of European origin in Africa and ethnographic objects of African origin in Europe, are called shared heritage because people feel that they say something about Africa as well as about Europe (Vanhee, 2016, p.6).

The quest for a shared heritage theory and practice can be grounded in the *heritage for sustainable human development* paradigm as articulated in the *Cottbus Declaration* of 2012 and a number of subsequent publications (Albert, 2015; Albert et al., 2013). This paradigm advocates a critical pragmatism, which integrates two preceding theoretical approaches, the *Institutionalized Heritage Discourse* and *Critical Heritage Studies*. While the institutionalized UNESCO framework aims at the pragmatic identification, classification, and conservation of global heritage resources (UNESCO, 1972, 2003), Critical Heritage Studies is an ideological critique of this institutionalization that exposes its underlying conceptual biases and asymmetrical power relations (Smith, 2006). The above-named readings of shared heritage mirror this integration of structural-pragmatist with critical-constructivist approaches towards a transformative practice driven by community-based actors.

However, the theoretical underpinnings of the presented readings of shared heritage and implications for their integration seem to be in an early stage of development and thus require further substantiation. The investigated case of Iringa bears references to all of them. While the articulation and interpretation of its historical perspectives is a meaningful addition to the global heritage archive, which obviously touches on the common history between Tanzania and Germany as well as other countries, the third and the fourth reading of shared heritage provide the most constructive conceptual frame for the case analysis. Both are meaningfully supported and integrated through the *Historic Urban Landscape* (HUL) concept. Premised on "the dynamic nature of living cities", the HUL is regarded as a configuration of material and immaterial elements that refer to the past of different groups and communities and the history of contact between them, seeking the "integration of historic urban area conservation, management, and planning strategies into local development processes" (UNESCO, 2011). The HUL focus ties in well with the idea of shared heritage as a transcultural thinking space emerging from a diverse community of contemporary users with diverging interests in conservation and representation. Furthermore, in a post-colonial setting, the HUL in many cases contains or is even characterised by architectural and memorial remnants from the colonial past whose present interpretation and representation are accompanied by controversies and conflicts. This paper is mainly an evaluation of such colonial remnants in the Iringa HUL, which also brought forward references to the debate on displaced cultural property and human remains.

In terms of methodology, the paper was inspired by Setha Low's (2016) *ethnography of space and place*. Low's (2016, p. 36) approach lends itself to utilizing ethnography in heritage studies and linking it with other fields concerned with space, place, and territory – such as urban studies and architecture. Low (2016, p. 68) supposes that space is socially constructed through structures of race, class, and gender, and transformations and contestations of space occur through people's interactions, memories, and feelings. Embodied by the people inhabiting them, spaces have intersecting "trajectories" of their social construction (Low, 2016, p. 149–150). The concept serves to access the trajectories of the Historic Urban Landscape of Iringa with its contestations and examine historical sites and material objects in relation to people, stories, and conflicts through various data sources, including life story interviews, observations, and visitor testimonies. Moreover, assessing shared heritage touches on the relationship between heritage and history. According to Lowenthal (1998, p. x), investigating history as heritage work is not an inquiry into the past aiming to know what actually happened, but rather borrowing from historical inquiry to enliven historical study and interpretation. The paper implements this methodological notion by complementing ethnographic approaches with conventional historical inquiry into written sources, original diaries from past protagonists, and archive documents.

11.3 Colonial History as Shared Heritage

Iringa is particularly known for the history of the rise and fall of the Hehe chiefdom in the course of the booming slave and ivory trade during the second half of the nineteenth century. As a stronghold of anti-colonial resistance during the Hehe wars from 1891 to 1898, Iringa also played a prominent role in the history of German East Africa. In this section, I unfold this history through the entangled perspectives of two different social groups who contributed to building up Iringa Town and Region.

The indigenous perspective is represented in the story of Mpangile Wangimbo, which is set at the crossroads between Iringa's old days as a powerful chiefdom and a new era of colonial administration (Küver, 2021a). Mpangile is born around 1870 (Nigmann, 1908, p. 20) during the tribal wars marking the rise of the Hehe chiefdom under his father, Chief Munyigumba. During the reign of his famous older brother, Chief Mkwawa, he becomes a Hehe warrior and fights in the war of resistance against the Germans during the early 1890s. In an effort to "divide and rule", the Germans try to harness Mpangile's popularity to undermine Mkwawa's influence and install him as "native chief" in the newly established Iringa Town on Christmas Day 1896. Only 2 months later, they publicly execute him at the gallows, based on allegations that he was aiding his fugitive brother Mkwawa in the anti-colonial resistance. In the context of the current shared heritage discourse, Mpangile's story specifically touches on current negotiations about the provenance and possible restitution of cultural property and human remains between Tanzania and Germany. Archival records show that Mpangile's head was taken to and kept in the *Museum for Völkerkunde* in Berlin (Brockmeyer et al., 2020, p.129–130). It is, therefore, part of the prominent "Hehe case" that also includes the skull of his brother Mkwawa, which was restituted under British administration in 1954, and the skull and bones of their father Chief Munyigumba, which German records claim were removed from the gravesite in Iringa during the war (Brockmeyer et al., 2020, p.129–130).

The second perspective is that of what is remembered of the *Schutztruppe* in Iringa, which I trace through the story of the Hans Poppe family. The so-called 'protection troops' were deployed in Iringa in the course of the war against Mkwawa and consisted of German officers and African soldiers, the *Askari*. The story begins with German officer Max Poppe coming to Iringa as a Schutztruppe officer during the First World War. In Iringa, he engages in a relationship with the daughter of one of his Askari comrades, and his son Hans is born and grows up with his mother's family in Iringa. Hans serves in the British and independent Tanzanian military police before he is killed in a border clash with Idi Amin's Ugandan forces in 1971. Two of his sons follow in his footsteps and become pilots in the Tanzanian military. Instead of serving their government, they use their position in a failed coup d'etat against president Nyerere in the early 80s and are sent to prison (Mwakikagile, 2010, p. 693–690). After being pardoned a decade later, they become wealthy businesspeople after the country's shift towards a capitalist economy during the 1990s.

11.4 Shared Heritage Sites and the Community

The historical dynamics of colonialism have left various tangible and intangible traces throughout Iringa Town and Region. The examination of three prominent aspects of this shared heritage from the fahari yetu case serves as a lens to magnify the entanglement of the historical trajectories introduced above.

11.4.1 Iringa Boma – The Building

The term "shared heritage" was first commonly used in the context of architectural heritage (Vanhee, 2016, p. 6). Accordingly, a notable number of German colonial buildings and monuments are listed in Tanzania's national cultural register (Kamamba, 2017, p. 320). One among these buildings is the old German hospital in Iringa, whose historical restoration and re-opening as *Iringa Boma – Regional Museum and Cultural Centre* has been the central measure of the fahari yetu programme. The building was designed by the German administration in 1914 and built by Askari soldiers just before the outbreak of WWI, designed to serve as a hospital for the growing European population in Iringa (Tanzania National Archives, G7/191, n.d.). After the war, the British colonial government made it the regional administrative headquarters, a use that was kept by the Tanzanian government after the country's independence. fahari yetu took over the building from the District Commissioner in 2014 and performed restoration works from 2015 to 2016 before re-opening it in June 2016 (Fig. 11.1).

> Some people have the notion that – because this is a German building – what is presented inside must be the German version of the history. That is why they are reluctant to embrace the Boma. (Deborah, exhibition coordinator, personal interview in December 2018)

The critical question was how the local community would receive the new Boma museum and cultural centre. Laurajane Smith (2006, p. 81) asserted that all heritage is uncomfortable to someone and is thus necessarily contested. Applied to the Boma, we have observed that restoring the building has brought a submerged

Fig. 11.1 Iringa Boma during British administration (Troll, 1934, left), after restoration in 2016 (right). (*Note.* Photograph (right) by fahari yetu Tanzania. Printed with permission)

post-colonial consciousness back to the surface. In his comparative investigation of coastal Swahili townscapes, Heathcott (2013, p. 22) employs the term *instabilities of heritage* to describe how dynamics of social inequality can undermine the economic development potential of historic preservation by creating resentment in the community against perceived nostalgia of colonial history. Applied to the case of Iringa, the exposition of the Boma through restoration brought the buildings' colonial association back into public awareness, and it was resented by parts of the community as an effort of reconstructing colonial nostalgia for tourist consumption. The former Boma exhibition coordinator remembers a visitor who contrived a conspiracy theory according to which the Boma restoration was to be seen as an attempt of the Germans to re-colonize the Hehe land. Whether or not this is a rare individual view, local people have repeatedly expressed the notion that the Boma is a German place until today and that we had restored something German with value for European foreigners.

As a consequence, we had to find and engage in the right efforts to make the people own the place. Such measures included advertising the conference room, which has become a popular venue for wedding committee meetings among the long-established population of Iringa; promoting the Boma Café, which has become a meeting point for the local chapter of the ruling political party in Tanzania; developing upstairs workshop and office facilities, which have drawn in various cultural artisan groups and local businesses; and convening cultural events and art exhibitions, which have garnered a following among musicians, artists, and expatriates. All these activities and services brought community actors with their interests to embrace and appropriate the space, and Iringa Boma now provides a safe and inclusive public space that offers educational, recreational, and business opportunities for various local constituencies.

11.4.2 The Exhibition

The exhibition consists of an introductory section and five thematic rooms, each with a specific theme: "Iringa history", "Iringa worship and healing", "Iringa culture and ethnography", "interactive display", and "explore Iringa Region". In the community perception of the exhibition, the above-mentioned instabilities of heritage became evident. Most notably, community members contested the way colonial history and anti-colonial resistance were narrated in Room 1:

> My feeling was that the exhibition avoids to show the true nature of the colonial relationship. On some panels, it sounds like it was a partnership between the Hehe and the Germans. But colonialism was never a partnership but always a forced and unlawful appropriation of land and people. Even if it was a long time ago, the people cannot honestly leave that experience of violent oppression and humiliation behind and be OK with it. (Clara, exhibition visitor, personal interview in August 2020)

Clara's concern about not sensitively reflecting the injustice and brutality of the colonial relationship in the exhibition reminds us that shared heritage remains a highly contested idea. Indeed, many voices speaking from the side of the former colonized – in the academic discourse and beyond – are rejecting the term "shared" in relation to colonial history and heritage. Mirroring the partnership statement in particular, van Beurden (2018) admonishes that the use of the seemingly neutral "shared heritage" suggests an equilibrium that pre-emptively erases the context of inequality in which cultural exchange took place in the colonial system. Yvonne Owuor (2020) vividly raises the question of how to deal with the brutality experienced by Africans in the course of colonial history as a shared experience and forcefully dismisses the shared history concept as yet another neo-colonial instrument of cultural appropriation through levelling historical power asymmetries.

Instabilities of heritage also surfaced in the local community perception of the Boma museum in general. First of all, we realized that many people in the community did not share the European notion of a museum as an exhibition of things from the past. Second, most of those who were actually aware of the idea seemed to understand the museum exhibition as a European concept, as a place meant for foreigners and tourists. Luntumbue (2015, p. 17; as cited in Vanhee, 2016, p. 6) rejects the idea that the colonial past would constitute a shared history, reminding us that history is always written from a specific viewpoint. His reminder allows us to interpret the community understanding as a perception of the museum writing history and culture from a colonial point of view, which excludes the local perspective.

Our approach to counter the local perception was to actively involve the indigenous community in exhibition design. In the course of the Coronavirus outbreak in Tanzania in spring 2020, the Boma launched a workshop series on traditional healing and the use of medicinal plants, which were conducted by well-known local healers. The performative workshops successfully broke with the Eurocentric exhibition concept and strengthened the museum's acceptance within the community. The same applies to a new exhibit showcasing folktales from the rural communities of Iringa Region, which was under installation by the time of writing this paper.

11.4.3 Reaching Out into the Cityscape

Another ongoing fahari yetu key activity is the integration of the Boma with the surrounding Historic Urban Landscape through the development of an international standard history trail. The trail development includes the restoration and enhancement of specific target sites. The target sites include the remaining building structures of the old German military station, which was built as the first building of Iringa Town in 1896. Today, the dilapidated main building serves as a storage facility for the Iringa central police, and the surrounding barracks accommodate police officers with their families. fahari yetu has proposed to restore and refurnish parts of the property as a historical hotel and guesthouse.

In front of the military station stands the Maji Maji Memorial. The Maji Maji war was an armed rebellion of a united front of different ethnic groups against German colonial rule in the south-eastern part of German East Africa from 1905 to 1907. The memorial was erected by the German station commander after the war to honour the Askari soldiers from Iringa who fell in the fight against the Maji Maji rebels. Despite being right there in the city centre for more than a hundred years, most people in the community apparently do not know its meaning. In order to raise awareness, fahari yetu renovated the monument in cooperation with the central police in summer 2021. Whether this measure will prove successful or not, the question of whose history the monument tells remains. Is it German heritage because it was built by the German colonial administration? Is it Askari heritage because it commemorates fallen Askari soldiers with their names and ethnic origin? Or is it anti-colonial resistance heritage because it reminds us of the lethal fight against the oppressors? The example shows the lines we have to think along when we imagine shared heritage, that it is shared from ambivalent and diverse perspectives.

Located just across from the memorial is the old market, which was built by the Germans in the early 1900s as an effort of relocating the commercial centre of Iringa Region to the new German town settlement. The original building structure is still intact yet obscured on all sides with iron sheet-covered shop frame constructions. For the case of the market in the Old Stone Town of Zanzibar – a similar example in which a market originally engineered by colonial forces was re-designed in a makeshift fashion after independence – Heathcott (2013, p. 24–25) observes how conservation officials decry the ramshackle additions and emphasize architectural form over social utility and human creativity. Similarly, my own as well as my colleagues' ideas for rehabilitating the market in Iringa envisage its dismantling to restore the visibility of the colonial structure. At the same time, we are well aware that the makeshift additions serve the livelihoods of many shop-owners and petty business operators, and any intervention with this business microcosm would stir up serious contention and conflict.

Heathcott (2013, p. 35) proposes the concept of *investment parity* to help reconcile such conflict of interest. Investment parity advocates a linked development process where investments in restoring "historic" neighbourhoods are matched by similar investments in "non-historic", especially low-income neighbourhoods. The idea was reflected in our discussions with local government officials who emphasized that heritage conservation projects should ensure immediate socio-economic counter value. In the case of the military station, replacing the run-down garrison barracks with the construction of a modern residence building could create such value and give the conservation positive PR in the local community. For the case of the market, a similar replacement with a new building to accommodate the petty traders living off the makeshift additions was suggested (Fig. 11.2).

Yet, from the local perspective, the question of why colonial sites should be preserved remains. How can the people be brought on board in conserving this historical landscape? fahari yetu's response is to ensure investment parity through connecting the urban history trail with Chief Mkwawa's late nineteenth century Hehe capital in Kalenga, by then an industrious town fortified by an impressive stone wall. Today there is only a village left, home to the Mkwawa memorial museum in which the famous skull of the Chief is displayed since its restitution in 1954. The reconstruction of a part of the old stone fort wall would support the museum in evoking Kalenga's pre-colonial glory. Such reconstruction would connect well with the restoration of the German military station in town, both representing political and military power and its transformation over time.

Fig. 11.2 Old market in 1908 (upper row left) and 2019 (upper row right), inside the market in 2016 (lower row left and right). (*Note.* Photograph upper row left, by Manfred Baumann. Printed with permission. Photographs lower row left and right, by fahari yetu Tanzania. Printed with permission)

11.5 Concluding Reflection – Towards a Shared Historic Urban Landscape

First of all, the case has shown that the controversy around the shared heritage concept is yet to be resolved. The disempowerment and dispossession of the colonized cannot be undone by branding what happened "shared history" (Vanhee, 2016, p. 7). But the applied conservation and representation of buildings and sites can signify that disempowerment and dispossession as integral parts of their history. An appropriate representation can only be realized by putting in place modes of inclusive community participation, modes that require full participation in both the creative process and decision making (Vanhee, 2016, p.7.). Under this prerequisite, shared heritage can become a useful tool of sensitive confrontation with the past and forge a shared understanding from which to investigate the post-colonial urban landscape with its diverse perspectives.

The production of shared heritage requires a careful examination of the conditions in which a building or site was created and how its use and meaning transformed over time, with consideration given to the agency of all those involved and implicated. The highlighted sites and representations all show that colonial history carries ambiguous connotations across different social constituencies and that its memory cannot easily be harnessed in a collective gaze. Neither should a collective gaze be the aim, but rather the multivocal articulation of different perspectives, whether it be historical narratives and experiences, representations in museum exhibitions, interpretations and usage options surrounding historical sites, or associations of local culture with physical remnants of colonialism. From this point of view, shared heritage serves as a viable concept and practice to grapple with the colonial legacy vested in the Historic Urban Landscape, to embrace the ambiguity and multivocality of this legacy, and to transcend the common notion of exclusive ownership of heritage along the colonial divide. It can thus be regarded as an element of a transformative practice, which addresses the injustice of the colonial past but at the same time uses the discursive energy provided by these conflicts to support the cultural, social, and economic development of communities.

Of course, the results of this paper call for further research on shared heritage as an evolving practice. The case has shown that there is a multitude of voices beyond the colonizer and the colonized emanating from shared heritage, a diverse range of perspectives further research should strive to explore in full. It has also become evident that the Historic Urban Landscape of Iringa is being haunted by expatriated cultural collections and human remains, calling for joint provenance research, restitution, and exhibition projects. Lastly, further research needs to tackle the issue of positionality in developing the concept of shared heritage, addressing such questions as who is comfortable speaking of shared heritage and why and how we can utilize these positions in creating sustainable development output from heritage.

References

Albert, M.-T. (Ed.). (2015). *Heritage studies: Volume 4. Perceptions of sustainability in heritage studies* (Vol. 4). De Gruyter.

Albert, M.-T., Bernecker, R., & Rudolff, B. (Eds.). (2013). *Heritage studies: Volume 1. Understanding heritage: Perspectives in heritage studies* (Vol. 1). De Gruyter.

Bibliothèque Nationale de France. (n.d.). *Patrimoines Partagés*. BNF. https://heritage.bnf.fr/indexen.html

Brockmeyer, B., Edward, F., & Stoecker, H. (2020). The Mkwawa complex: A Tanzanian-European history about provenance, restitution, and politics. *Journal of Modern European History, 18*(2), 117–139. https://doi.org/10.1177/1611894420909033

Heathcott, J. (2013). Historic urban landscapes of the Swahili coast: New frameworks for conservation. In United Nations Educational, Scientific and Cultural Organization (Ed.), *Swahili historic urban landscapes: Report of the historic urban landscape workshops and field activities on the Swahili coast in East Africa 2011–2012* (pp. 20–39). UNESCO.

ICOMOS. (2020). *Shared cultures, shared heritage, shared responsibility*. ICOMOS. https://www.icomos.org/en/77-articles-en-francais/72051-18-april-2019-theme-shared-cultures.

Kamamba, D. M. (2017). Shared cultural heritage between Germany and Tanzania. In L. Reyels (Ed.), *Humboldt Lab Tanzania. Objekte aus kolonialen Kriegen im Ethnologischen Museum, Berlin. Deutsch-tansanische Perspektiven* (pp. 306–329). Dietrich Reimer Verlag.

Küver, J. (2021a). European footsteps in the land of the chief. In Y. Aixelá-Cabré (Ed.), *Africa in Europe and Europe in Africa: Memories, cultural heritage and sociopolitical experiences* (pp. 85–110). Peter Lang Publishing.

Küver, J. (2021b). *Shared history, urban landscapes, and local arts and crafts: Perspectives on cultural heritage as a means of sustainable community development in Iringa, Tanzania* [unpublished PhD thesis]. Brandenburg Technical University Cottbus-Senftenberg.
Low, S. M. (2016). *Spatializing culture: The ethnography of space and place*. Routledge.
Lowenthal, D. (1998). *The heritage crusade and the spoils of history*. Cambridge University Press.
Luntumbue, T. M. (2015). Rénovation au Musée de Tervuren: questions, défis et perspectives. *L'art Même, 65*, 15–17.
Mwakikagile, G. (Ed.). (2010). *Nyerere and Africa: End of an era* (5th ed.). New Africa Press. (Original work published 2002).
Netherlands Cultural Heritage Agency of the Ministry of Education, Culture and Science. (n.d.). *Shared cultural heritage*. NCHA. https://english.cultureelerfgoed.nl/topics/shared-cultural-heritage
Nigmann, E. (1908). *Die Wahehe.: Ihre Geschichte, Kult-, Rechts-, Kriegs- und Jagdgebräuche*. Ernst Siegfried Mittler.
Owuor, Y.A. (2020). Derelict shards: The roaming of colonial phantoms. Online keynote address given at the International Conference: Colonialism as Shared History – Past, Present, Future, 7–9 October 2020, Berlin. *CODESRIA Bulletin Online*, No. 10. https://codesria.org/IMG/pdf/10-_y_owuor_codbul_online_1_.pdf
Smith, L. (2006). *Uses of heritage. Heritage studies*. Routledge.
Tanzania National Archives. G7/191: Regierungsbauten in Iringa, 1905–1914. TNA. Dar es Salaam.
Troll, C. (1934). *Hauptquatier der britischen Administration* [Photograph]. Deutsche Fotothek. http://www.deutschefotothek.de/documents/obj/71648976.
UNESCO. (1972). *Convention concerning the protection of the world cultural and natural heritage*. UNESCO.
UNESCO. (2003). *Convention for the safeguarding of the intangible cultural heritage (ICHC)*. UNESCO.
UNESCO. (2011). *Recommendation on the historic urban landscape (HUL)*. UNESCO.
United Republic of Tanzania. (2013). 2012 population and housing census: Population distribution by administrative areas. URT. Dar es Salaam.
van Beurden, S. (2018). The pitfalls of 'shared heritage'. *How to move on with Humboldt's legacy? Rethinking ethnographic collections*. https://blog.uni-koeln.de/gssc-humboldt/the-pitfalls-of-shared-heritage/
Vanhee, H. (2016). On shared heritage and its (false) promises. *African Arts, 49*(3), 1–7.

Chapter 12
UNESCO World Heritage and Cultural Property Protection in the Event of Armed Conflict

Friedrich Schipper

Abstract Bamiyan, Palmyra, Timbuktu are examples of iconic toponyms of ancient civilization that have been used as headlines in international media coverage of the Islamist extremists' war – including the Taliban, Daesh, and Ansar Dine – against World Cultural Heritage as listed by UNESCO. Further, more sites are being inscribed on UNESCO's list of *World Heritage in Danger* or referred to in the reports of ICOMOS' *Heritage at Risk programme* because of the imminent threat posed by current armed conflict, for example, in Afghanistan, Armenia and Azerbaijan, Iraq, Libya, Mali, Syria, Yemen, and other countries struck by war or the violent consequences of political instability. Nowadays, war, terrorism, vandalism, and iconoclasm pose an equal threat to cultural heritage. This poses a series of questions to heritage as well as conflict studies.

Keywords Cultural property · Armed conflict · Blue Shield

12.1 Introduction

Cultural heritage has always fallen victim to war (Dolezalek et al., 2021). Movable cultural property has been taken as war trophies or war booty and has been claimed as war reparation (Ferstman & Goetz, 2020). Immovable cultural property has been damaged or destroyed either as collateral damage during hostilities, such as loss during pillage or as an intentional target.

In modern history, both World Wars caused massive destruction and, even more, the intended annihilation of cultural heritage, the "Looting of Loewen" (Born & Störtkuhl, 2017) or the "Baedeker Blitz" (Rothnie, 1992) are some of the most poignant and pervert examples of warfare against cultural property and infrastructure

F. Schipper (✉)
Theresian Military Academy, Austrian Armed Forces & Austrian National Committee of the Blue Shield, Vienna, Austria
e-mail: friedrich.schipper@blueshield.at

M.-T. Albert et al. (eds.), *50 Years World Heritage Convention: Shared Responsibility – Conflict & Reconciliation*, Heritage Studies,
https://doi.org/10.1007/978-3-031-05660-4_12

in this period. At the same time, these two World Wars also witnessed the first military countermeasures like the Imperial German Army "Kunstschutz" (Kott, 1997) or the US Army's "Monuments Men" (Edsel, 2009) as militarized expert task forces.

Modern-day civilization's response to this vast destruction of culture in war resulted in joint work on creating normative measures of International Humanitarian Law (IHL) striving for the prevention or at least mitigation of the loss of heritage due to military action. The 1935 Roerich Pact – the *Treaty on the Protection of Artistic and Scientific Institutions and Historic Monuments* – and the 1954 Hague Convention – the *Convention for the Protection of Cultural Property in the Event of Armed Conflict* – were the main legal instruments born in this era (Schipper & Frank, 2013).

About two decades later, the 1972 World Heritage Convention (UNESCO 1972)– the *Convention Concerning the Protection of the World Cultural and Natural Heritage* – was the result of a new way of thinking about heritage in an era of renewed international cultural cooperation as a reaction to modern globalization. At that time, the academic community dealing professionally with heritage paid little attention to the formerly dominant aspects of heritage protection in armed conflict. The World Heritage Convention itself hardly touches upon this topic. In the many proxy wars of the Cold War period against the backdrop of the nuclear arms race of the two superpowers, the issue of the destruction of cultural property faded from the military, juridical, and political agenda.

The shift from the ideological wars of the Cold War period to the identity wars of the post-Cold War period – against the backdrop of the nuclear disarmament deals – resulted in a renaissance of aggression against culture and its material expressions. Again, the international community reacted by jointly working on a Second Protocol 1999 to the Hague Convention 1954 (Toman, 2009). From a European viewpoint, the Yugoslav Wars 1990–1999 may be considered as decisive drivers for this diplomatic initiative and legal development (Abtahi, 2015).

The Second Protocol entered into force in 2004, about a year after the looting of the Iraq Museum in Baghdad, which is considered by the heritage community as the "Solferino"[1] momentum for the protection of cultural heritage in the event of armed conflict. The event and political reaction triggered worldwide public outrage and even enhanced anti-US resentments, which continues to have an after-effect to this day. But it also raised awareness of the fact that – regardless of the development of International Humanitarian Law – the armies of all contributing nations to the so-called "Coalition of the Willing" lacked any expertise on how to implement cultural property protection in planning and operations effectively, and what role cultural property protection could play in civil–military cooperation in general and in a post-conflict period in terms of peacebuilding and reconciliation in particular (Gooren, 2007). Regardless, a contemporary survey would have revealed the same lack of competence in almost all armed forces worldwide.

[1] This is a reference to the battle of Solferino, which led to the establishment of the Red Cross.

The past decade even witnessed an increased targeting of cultural property in a variety of armed conflicts, as well as terror attacks and iconoclastic vandalism. Regarding the regions of the Middle East and Northern and Western Africa, this phenomenon in part corresponds to the sparking-off of the so-called "Arab Spring" and the subsequent and prolonged armed conflicts and political instabilities with the above-mentioned main perpetrators (Schipper, 2011). This dramatic development is reflected by enhanced public and media awareness.

At the same time, the governmental and inter-governmental sectors actively respond to these challenges and improve their capabilities and capacities. The International Criminal Court (ICC) launched a first trial exclusively dealing with violations against the 1954 Hague Convention regarding the destruction of heritage in Timbuktu, *The Prosecutor v. Ahmad Al Faqi Al Mahdi. ICC-01/12-01/15* (Casaly, 2016).[2] By passing a resolution for establishing the *United Nations Multidimensional Integrated Stabilization Mission in Mali* (MINUSMA), the UN Security Council integrated CPP explicitly in the mandate of a UN PK (Peace Keeping) mission for the first time, UN Security Council Resolution 2531, 29 June 2020 (Schorlemer, 2020).[3] Furthermore, armed forces worldwide are currently taking steps to implement measures of military CPP as defined by the 1954 Hague Convention and its two protocols (Kila, 2012; O'Keefe et al., 2016; Schipper, 2015; Schipper & Eichberger, 2010).

Nevertheless, iconoclastic strikes against cultural heritage have increasingly been carried out through the past 20 years by Islamist actors, which are all listed as terrorist groups by the international community. These strikes are aimed at and may finally result in the cultural cleansing of whole regions as a central part of their overall strategy to reach hegemony over certain geographical and historical realms (Smith et al., 2016). In addition, the looting of and illicit trafficking in cultural property generates a source of income for these groups, as the relevant body of the UN, UNODC, clearly indicates (Musu, 2021).[4]

As asserted by the World Heritage Committee in its 41st session in Krakow on 2–12 July 2017 and published in its Decision 41 COM 7, the threat to cultural heritage caused by armed conflict is also seriously increasing through the use of World Heritage properties for military purposes as well as through illicit trafficking of cultural property and wildlife for the purpose of financing belligerent non-state groups (UNESCO, 2017, sec. 8 & 12).

Taking up this WHC Decision, the 200 participants from more than 30 countries, gathered at the Royal Castle of Warsaw, on the occasion of the International Conference on Reconstruction "The Challenges of World Heritage Recovery" (UNESCO, 2018, 2) addressed amongst other issues

---

[2] See https://www.icc-cpi.int/mali/al-mahdi for more information about the ICC trial of *Ahmad Al Faqi Al Mahdi*

[3] See https://minusma.unmissions.org/en/mandate-0 for more information about UN Security Council Resolution 2531 (United Nations, 2020).

[4] See https://www.unodc.org/unodc/en/organized-crime/intro/emerging-crimes/trafficking-in-cultural-property.html for more information about trafficking cultural property.

> the growing impact of armed conflicts and disasters on important cultural and natural heritage places, including World Heritage properties, which in recent years have resulted in their widespread destruction on a scale similar to that of World War II, notably within historic urban areas and archaeological sites" (§3) and they continued to condemn "in the strongest terms, the numerous intentional attacks on cultural properties and in general the perpetration of all policies of 'cultural cleansing' aimed at erasing diversity, inciting sectarian violence and preventing the affected population from realizing their human rights, including cultural rights" (§4) as phrased out in the subsequent "Warsaw Recommendation on Recovery and Reconstruction of Cultural Heritage.[5]

The following topics are, in any case, major issues of legal, political, diplomatic, military, and academic consideration, research, and development.

12.2 Targeting Cultural Property: Reasons and Justifications, Effects, and Consequences

Damage to or destruction of cultural property may occur during any military action caused by any of the conflict parties as collateral damage. This must be distinguished from the intentional targeting of cultural property. The causes, reasons, and motives for the intentional targeting of cultural property remain the most important basic question in cultural property protection (CPP), in particular in relation to the effects on its direct stakeholders and the consequences for communities and societies.

Descriptions of particular cases, including more or less reliable reconstructions of the course of events and documentation of the extent of destruction, are available for many or even most of the scenarios of the past 30 years. Nevertheless, we lack empiric data on the *why* (Bevan, 2006). This *why* may be easy to address with a few superficial and hypothetical words that can fit most scenarios. But it is difficult to identify the tactical reasons or strategic rationales more precisely behind such hostile actions that may be directed against iconic cultural or religious buildings as well as ordinary cemeteries. Nonetheless, those scholars, who have examined this phenomenon, suggest that it is common in events of ethnic cleansing, posing the equivalent of a cultural cleansing (Schorlemer, 2016).

This striking explanation obviously fits what we know about the targeting and destruction of cultural property during the Yugoslav Wars or the on-off armed conflict between Armenia and Azerbaijan. However, the destruction of Islamic cultural property in Mali, for example, was seemingly not related to ethnic cleansing, but the evident iconoclasm was related to other causes. The destruction of Islamic heritage by Islamists is about the interpretation of righteous Islam and the definition of Islam. And the destruction of the remains of ancient civilizations, e.g. in Palmyra, hardly fits the ethnic cleansing explanation as the temples of former civilizations are

[5] For more information about the Warsaw Recommendation on Recovery and Reconstruction of Cultural Heritage, see https://whc.unesco.org/document/168799

not part of contemporary practices of any religious or ethnic group (Alshehaby, 2020). The Doha Conference of Ulama on Islam and Cultural Heritage, Doc. CL/CH/THS/2001/CD/H/1, explored the realm of cultural heritage protection in Islam's Sunnah (UNESCO et al., 2001).[6]

Recently, Johan Brosché et al. (2017) have proposed a "typology of motives" that I would refer to as four categories of cultural property-related attacks allowing for systematic empiric research:

- attacks related to conflict goals, in which cultural property is targeted because it is connected to the issue the warring parties are fighting over,
- military-strategic attacks, in which the main motivation is to win tactical advantages in the conflict,
- signalling attacks, in which cultural property is targeted as a low-risk target that signals the commitment of the aggressor, and
- economic incentives where cultural property provides funding for warring parties.

While the current intentional targeting of cultural property certainly has its general context in the above-mentioned identity wars, the hermeneutical bracket constituted by cultural cleansing and ethnic cleansing is insufficient as an overall explanation to help face future challenges of cultural property protection. The proposal by Johan Brosché et al. or similar recent ideas may serve as a renewed basis for military, legal, sociological, and other research.

12.3 Safeguarding of and Respect for Cultural Property

The safeguarding of and respect for cultural property are the two main principles of cultural property protection defined by the 1954 Hague Convention according to Articles 3 and 4 and further developed in Articles 5 and 6 of the 1999 Second Protocol.

Safeguarding of cultural property refers to all measures to prepare in peacetime for the protection of cultural property situated within a state territory against the foreseeable effects of an armed conflict. This includes all civil as well as military measures in peacetime and, in particular, all measures that both segments of the governmental sector – civil and military – have to undertake as a joint effort. That means that safeguarding cultural property is the strategic dimension of civil–military coordination. Regarding the military side, this requires the integration of cultural property protection on a doctrinal level (national security or defence doctrine) to enable any military policy-making or strategic planning. Practically, these issues

[6] For more information about the Proceedings of the Doha Conference, see http://unesdoc.unesco.org/images/0014/001408/140834m.pdf

should fall into political science, where cultural property protection and safeguarding is contemporarily not on the agenda of any prominent research programme.

Respect for cultural property refers to all measures to protect it by refraining from any use of the property and its immediate surroundings or the equipment used for its protection that would likely expose it to destruction or damage in the event of armed conflict. This also includes refraining from any act of hostility directed against such property. As these measures apply to the conduct of armed conflict, which means military operations, they fall within the responsibility of the military. In consequence, cultural property protection has to be established within the military on a normative basis, which means on the level of a corresponding directive and, as such, it is an issue of military science (Rush, 2012).

From doctrine to directive, from political to military science, the safeguarding of and respect for cultural property are the most basic principles for successfully implementing cultural property protection, and a comprehensive research approach is required to foster cultural property protection in the military as well in civil–military cooperation.

12.4 The Instruments of Special and Enhanced Protection

Cultural property protected under the regime of 1954 Hague Convention and its two protocols shall not just be listed and communicated to the depository of the Convention, which is UNESCO, but also marked with a protective emblem, the so-called Blue Shield (UNESCO, 1954, Art. 6, 16 & 17). Nevertheless, it is a widely ignored or even unknown fact that cultural property may also be granted Special Protection according to Chap. II of the 1954 Hague Convention[7] or Enhanced Protection according to Chap. III of the Second Protocol (UNESCO, 1954; UNESCO, 1999a, 1999b).[8] The regime of Enhanced Protection was introduced in the Second Protocol as the regime of Special Protection was by then evaluated as not having met the expectations of its authors and considered an unsuccessful instrument. However, whether the regime of Enhanced Protection may be evaluated as having been more successful so far is doubtful, at least. Therefore, it remains an open and important question, why these two legal protection instruments fail to meet the expectations of the juridical and diplomatic community that developed and negotiated them.

[7] For information about the International Register of Cultural Property under Special Protection, see http://www.unesco.org/new/fileadmin/MULTIMEDIA/HQ/CLT/pdf/Register2015EN.pdf (UNESCO, 2015).

[8] For information about the List of Cultural Property under Enhanced Protection, see http://www.unesco.org/culture/1954convention/pdf/Enhanced-Protection-List-2017_EN.pdf (UNESCO, 1999a)

12.5 The Tools of Registers/Inventories

Accurate, complete, accessible, and secure registers or inventories of cultural property are the basic formats of information about such property, which are necessary for all kinds of professional administration and management and protection. Such lists usually form a key element in national heritage legislation and consequently also in International Humanitarian Law regarding the management and protection of heritage. Beyond the basic norm to keep such lists, there are also standards, like ICOM's or the International Council of Museums' Object ID, for how to identify and record cultural goods or how to keep a priority list for the purpose of emergency evacuation of a collection (Thornes et al., 1999; Yasaitis, 2005).[9] Cultural property protection in the sense of the 1954 Hague Convention and its Protocol is also based on the principle of listing protected property, see Art. 12 regarding the International Register as laid out in the Regulations for the Execution of the Convention for the Protection of Cultural Property in the Event of Armed Conflict.

In the event of armed conflict, such lists become the basis of no-strike lists by the military. These no-strike lists, which are compiled and maintained by target officers of the air force or artillery, contain all kinds of elements that bombing or shelling must avoid for humanitarian or other reasons, e.g. hospitals.

The concern that such lists may also be subject to abuse and turned into strike or target lists has been raised and discussed since their adoption. Listing or not listing cultural property may therefore be considered an ethical, legal, and practical dilemma. Nevertheless, the experience of past conflicts shows that such lists are a necessary condition for the respect of cultural property, as it cannot or can hardly be identified by the military during operation (Stone, 2013).

12.6 Cultural Property Protection and Peacekeeping Operations

Peacekeeping comprises activities intended to create conditions that favour lasting peace. Evidence shows that peacekeeping reduces civilian and battlefield deaths, as well as reduces the risk of renewed warfare. Within the United Nations (UN), there is a general understanding that peacekeepers monitor and observe peace processes in post-conflict areas and may assist ex-combatants in implementing peace agreement commitments that they have undertaken. Such assistance may come in many forms, including confidence-building measures, power-sharing arrangements, electoral support, strengthening the rule of law, and economic and social development. Accordingly, UN peacekeepers (often referred to as Blue Berets or Blue Helmets

[9] See https://icom.museum/en/resources/standards-guidelines/objectid/ for more information about ICOMs Object ID list

because of their light blue berets or helmets) can include soldiers, police officers, and civilian personnel (Powles & Negar Partow, 2015).

As all International Humanitarian Law automatically and without exception applies to all peacekeeping missions, the Conventions for the protection of heritage, in particular the 1954 Hague Convention and its two Protocols, also apply. Therefore, cultural property protection is and has always been an implicit element of every Peacekeeping Operation. Nevertheless, history and reality show that cultural property protection has de facto been a neglected aspect of most Peacekeeping Operations, with cultural property protection usually not being explicitly mentioned as a task in the mission mandates issued by the UN Security Council. As a consequence, most Peacekeeping Operations have no specialized personnel commissioned to conduct measures of cultural property protection in the area of responsibility and no cultural property protection measures set in their rules of engagement. Furthermore, no cultural property protection content is provided in the context of on-mission training for its personnel. In the long term, this also results in cultural property protection being ignored even in the context of the lessons-learned process (Micewski & Sladek, 2002). UNIFIL and MINUSMA may be referred to as exceptions to this otherwise defective reality of Peacekeeping Operations. Meanwhile, the United Nations Department of Peace Operations, as the entity providing political and executive direction to all Peacekeeping Operations, offers at least a compact online course on cultural property protection that has been created in cooperation with UNESCO.[10] Nevertheless, the development of integral cultural property protection in teaching and training (pre-deployment and on-mission), as well as operations, remains a desideratum in the context of Peacekeeping Operations (Petrovic, 2018).

12.7 Blue Shield and the Role of Expert Organizations

In 1996, in the context of the efforts leading towards what would become the Second Protocol to the Hague Convention, the International Committee of the Blue Shield (ICBS) was established as a joint committee of the UNESCO-affiliated cultural heritage expert organizations – International Council of Museums (ICOM), International Council on Monuments and Sites (ICOMOS), International Council of Archives (ICA), and International Federation of Library Associations (IFLA) – with the task to protect the world's cultural heritage from threats such as armed conflict as well as natural and man-made technical disasters. Originally intended as the "cultural equivalent of the Red Cross",[11] its name derives from the distinctive emblem used to signify cultural sites protected by the 1954 Hague Convention. In the very

[10] See https://cdn.peaceopstraining.org/course_promos/unesco_pcp/unesco_pcp_english.pdf for more information about online courses on cultural property protection (UNESCO, 2020).

[11] See https://theblueshield.org/about-us/history/ for more information about Blue Shield International

same year, the first national committees of the Blue Shield were established to perform the operative work in the sense of ICBS.

ICBS was introduced to the 1999 Second Protocol as an expert organization and consultative body to UNESCO, namely to its Committee for the Protection of Cultural Property in the Event of Armed Conflict. Its IHL-related mandate derives from this reference in Article 27.3 of the Second Protocol.

In 2008, the national committees joined to make up the Association of National Committees of the Blue Shield (ANCBS) in order to coordinate operative work worldwide. In 2016, both ICBS and ANCBS merged to become what is now Blue Shield International.

Today, Blue Shield has developed into a network of expert organizations that offers competence in CPP to national governments and armed forces or to UN peacekeeping operations (PKO), including but not limited to awareness-raising initiatives and lobbying activities for implementing cultural property protection and training, fact finding in conflict areas, and emergency response. It shares this field with a few other organizations that have developed specialized expertise in the field of cultural property protection in armed conflicts, such as International Centre for the Study of the Preservation and Restoration of Cultural Property (ICCROM), ICORP (International Scientific Committee on Risk Preparedness), DRMC (Disaster Risk Management Committee), and others. As a consequence, proposals for guidelines, standard-setting instruments and training curricula etc. have been frequently developed and published over the past decade, and many more initiatives in response to current crises and disasters have been performed (Stone, 2019).

While such activities supporting cultural property protection on the level of expert organizations have steadily increased over the past 25 years, there is still an urgent need for a common policy leading to internationally accepted standards in trainings and missions.

12.8 Outlook: World Heritage in Danger

Armed conflict and natural disasters may pose major threats to World Heritage sites. Several of these sites are on UNESCO's list of World Heritage in Danger and consequently referred to in the reports of ICOMOS' Heritage@Risk programme.[12] According to paragraphs 179–180 of the Operational Guidelines of the World Heritage Convention, dangers can be a. "ascertained", referring to specific and proven imminent threats, or b. "potential", when a property is faced with threats which could have negative effects on its World Heritage values. In the case of armed conflict, dangers are always classified as "potential" (UNESCO, 2019). Inscribing a site on the List of *World Heritage in Danger* allows the World Heritage Committee,

[12] See https://www.icomos.org/en/what-we-do/risk-preparedness/heritage-at-risk-reports for more information about ICOMOS's Heritage at Risk Reports.

among other things, to allocate immediate assistance from the World Heritage Fund to the endangered property.

To decide whether a World Heritage site is endangered or not, an assessment mission has to be carried out. The assessment of World Heritage is definitively a core task of ICOMOS. Its Heritage@Risk programme was endorsed by its General Assembly in 1999. The aim of this programme's reports is to identify endangered monuments and sites, present typical case studies and trends, and share suggestions for solving individual or global threats to our cultural heritage. Nevertheless, the assessment of World Heritage sites in areas of armed conflict goes beyond the usual challenges posed to an expert team and requires military-compatible skills. Blue Shield is well-prepared for such fact-finding missions in conflict areas, though it does not necessarily have expertise on World Heritage. While World Heritage is increasingly becoming a target of higher priority, particularly in the Islamist extremists' war against heritage, the creation of a joint task force, the development of a legal basis for such a task force, and guidelines for joint missions are the best practice consequences. It is furthermore a necessity in terms of peacebuilding because cultural heritage – while being a target in armed conflict – may at the same time also constitute an essential resource for reconciliation (Stone, 2012; Tandon et al., 2021; Walters et al., 2017).

References

Abtahi, H. (2015). *The protection of cultural property in times of armed conflict: The practice of the international criminal tribunal for the former Yugoslavia*. Routledge.

Alshehaby, F. (2020). Cultural heritage protection in Islamic tradition. *International Journal of Cultural Property, 27*(3), 291–322.

Bevan, R. (2006). *The destruction of memory*. Reaktion Books.

Born, R. & Störtkuhl, B. (Eds.). (2017). *Apologeten der Vernichtung oder »Kunstschützer«? Kunsthistoriker der Mittelmächte im Ersten Weltkrieg*. Böhlau Verlag.

Brosché, J., Legnér, M., Kreutz, J., & Ijla, A. (2017). Heritage under attack: Motives for targeting cultural property during armed conflict. *International Journal of Heritage Studies, 23*(3), 248–260.

Casaly, P. (2016). Al Mahdi before the ICC: Cultural property and world heritage in international criminal law. *Journal of International Criminal Justice, 14*(5), 1199–1220.

Dolezalek, I., Savoy, B., & Skwirblies, R. (Eds.). (2021). *Beute. Eine Anthologie zu Kunstraub und Kulturerbe*. Matthes & Seitz.

Edsel, R. M. (2009). *The greatest treasure hunt in history. The story of the Monuments Men*. Center Street.

Ferstman, C., & Goetz, M. (Eds.). (2020). *Reparations for victims of genocide, war crimes and crimes against humanity. Systems in place and systems in the making*. Brill.

Gooren, R. (2007). Civil-military cooperation in protecting cultural heritage. *Forum Kulturgüterschutz, 10*, 50–54.

Kila, J. (2012). *Heritage under siege. Military implementation of cultural property protection following the 1954 Hague Convention*. Brill.

Kott, C. (1997). Die deutsche Kunst- und Museumspolitik im besetzten Nordfrankreich im Ersten Weltkrieg. Zwischen Kunstraub, Kunstschutz, Propaganda und Wissenschaft. *Kritische Berichte. Zeitschrift für Kunst- und Kulturwissenschaften, 25*(2), 5–24.

Micewski, E., & Sladek, G. (Eds.). (2002). *Protection of cultural property in the event of armed conflict. A challenge in peace support operations*. National Defence Academy Vienna.
Musu, C. (2021). Cultural heritage protection, illicit antiquities, and the international fight against terrorism financing. In B. Romiti & A. Folliero (Eds.), *The safety and security of cultural heritage in zones of war or instability* (NATO Science for Peace and Security Series – E: Human and Societal Dynamics) (Vol. 153, pp. 15–31). IOS Press.
O'Keefe, R., Péron, C., Musayev, T., & Ferrari, G. (2016). *Protection of cultural property: Military manual*. UNESCO.
Petrovic, J. (2018). The cultural dimension of peace operations: Peacekeeping and cultural property. In A. H. Campbell (Ed.), *Global leadership initiatives for conflict resolution and peacebuilding* (pp. 84–115). IGI Global.
Powles, A., & Negar Partow, N. (Eds.). (2015). *United Nations peacekeeping challenge: The importance of the integrated approach*. Routledge.
Rothnie, N. (1992). *The Baedeker Blitz. Hitler's attack on Britain's historic cities*. Ian Allan Publishing.
Rush, L. W. (2012). *Archaeology, cultural property, and the military*. Boydell Press.
Schipper, F. (2011). The Arab spring and blue shield. Protecting archaeological sites in conflict zones. *Res Militares, 11*(2), 4–5.
Schipper, F. (2015). Kulturgüterschutz im Kontext des humanitären bzw. Kriegsvölkerrechts. Eine historische und systematische Überschau bis heute. In J. Lang (Ed.), *Montecassino 1943 – Rettung vor der Zerstörung. Nationaler und internationaler Kulturgüterschutz* (pp. 249–263). Eigenverlag.
Schipper, F., & Eichberger, H. (2010). The protection of cultural property in the event of armed conflict. The cultural property protection officer as a liaison between the military and the civil sector. *Present pasts. Journal of the Heritage Studies Research Group at the Institute of Archaeology, UCL, 2*(1), 169–176.
Schipper, F., & Frank, E. (2013). A concise legal history of the protection of cultural property in the event of armed conflict and a comparative analysis of the 1935 Roerich pact and the 1954 Hague convention in the context of the law of war. *Archaeologies. Journal of the World Archaeological Congress, 9*(1), 13–28.
Schorlemer, S. v. (2016). *Kulturgutzerstörung. Die Auslöschung von Kulturerbe in Krisenländern als Herausforderung für die Vereinten Nationen* (The United Nations and Global Change, Vol. 11). Nomos Verlag.
Schorlemer, S. V. (2020). Military intervention, the UN security council, and the role of UNESCO. The case of Mali. In A.-M. Carstens & E. Varner (Eds.), *Intersections in international cultural heritage law* (pp. 82–104). Oxford University Press.
Smith, C., Burke, H., Leiuen, C. d., & Jackson, G. (2016). The Islamic State's symbolic war: Da'esh's socially mediated terrorism as a threat to cultural heritage. *Journal of Social Archaeology, 16*(2), 164–188.
Stone, P. G. (2012). Human rights and cultural property protection in times of conflict. *Journal of Heritage Studies, 18*(3), 271–284.
Stone, P. (2013). *War and heritage. Using inventories to protect cultural property*. (Conservation Perspectives.) The Getty Conservation Institute. https://www.getty.edu/conservation/publications_resources/newsletters/28_2/war_heritage.html
Stone, P. G. (2019). Protecting cultural property in the event of armed conflict: The work of the blue shield. In M. Dumper (Ed.), *Contested holy cities: The urban dimensions of religious conflicts* (pp. 147–168). Routledge.
Tandon, A., Harrowell, E., & Selter, E. (2021). *PATH – Peacebuilding assessment tool for heritage recovery and rehabilitation*. ICCROM.
Thornes, R., Dorrell, P., & Lie, H. (1999). *Introduction to object ID: Guidelines for making records that describe art, antiques and antiquities*. Getty Information Institute.

Toman, J. (2009). *Cultural property in war: Improvement in protection: Commentary on the 1999 second protocol to the Hague convention of 1954 for the protection of cultural property in the event of armed conflict*. UNESCO.
United Nations. (2020). *Mandate: The security council unanimously approved resolution 2531 on 29 June 2020, renewing the mandate of MINUSMA for another year.* UN MINUSMA. https://minusma.unmissions.org/en/mandate-0.
UNESCO. (1954). *Protocol to the convention for the protection of cultural property in the event of armed conflict 1954*. UNESCO. https://unesdoc.unesco.org/ark:/48223/pf0000082464_mul
UNESCO. (1972). *Convention concerning the protection of the world cultural and natural heritage*. UNESCO.
UNESCO. (1999a). *List of cultural property under enhanced protection.* UNESCO. http://www.unesco.org/culture/1954convention/pdf/Enhanced-Protection-List-2017_EN.pdf
UNESCO. (1999b). *Second protocol to the Hague convention of 1954 for the protection of cultural property in the event of armed conflict 1999.* UNESCO. http://portal.unesco.org/en/ev.php-URL_ID=15207&URL_DO=DO_TOPIC&URL_SECTION=201.html
UNESCO. (2015). *International register of cultural property under special protection.* UNESCO. http://www.unesco.org/new/fileadmin/MULTIMEDIA/HQ/CLT/pdf/Register2015EN.pdf
UNESCO. (2017) *Decision: 41 COM 7 state of conservation of the properties inscribed on the World Heritage List.* UNESCO. https://whc.unesco.org/en/decisions/6940/
UNESCO. (2018). *Warsaw recommendation on recovery and reconstruction of cultural heritage*. UNESCO.
UNESCO. (2019). *Operational guidelines for the implementation of the World Heritage Convention*. UNESCO.
UNESCO. (2020). *Protection of cultural property: Online course for the military, police, and law enforcement.* UNESCO. https://cdn.peaceopstraining.org/course_promos/unesco_pcp/unesco_pcp_english.pdf
UNESCO, ISESCO, ALESCO & OIC. (2001). *Proceedings of the Doha conference of Ulamâ on Islam and cultural heritage*. UNESCO. https://unesdoc.unesco.org/ark:/48223/pf0000140834
Walters, D., Laven, D., & Davis, P. (2017). *Heritage and peacebuilding*. The Boydell Press.
Yasaitis, K. E. (2005). Object ID: A model of global collaboration. *Journal of Museum Management and Curatorship, 20*(1), 21–39.

Chapter 13
Palmyra: Bridging Past and Future

Zeina Elcheikh

Abstract Targeting and destroying Syria's cultural heritage have become a common concern, especially in the case of Palmyra. The ruined city enjoyed a significant position in the country's history and bore a large share of the violence in Syria's protracted tragedy. Since 2014, militants of ISIS (Islamic State in Iraq and Syria) have committed many acts of terrorism, claiming thousands of lives and shattering others, looting antiquities and destroying historic sites, including World Heritage sites. In 2015, ISIS extremists destroyed Palmyra's major monuments, and, since then, this destruction has been the focus of debates on the academic and professional levels and in the media. This chapter has two parts. The first part briefly reviews Palmyra's long history, in which monuments have been subject to selective narratives and official instrumentalisation. The second part looks at the current debates on Palmyra's heritage loss in the light of the actual conflict, in which the local community has been unheeded. This chapter suggests that future efforts need to (re)consider the role of local communities in heritage debates and the right(s) to their heritage to bridge the discontinuity between the past and the future caused by the terrorism and conflict.

Keywords Palmyra · Heritage targeting · ISIS · Media · Local communities · Human rights

13.1 Introduction

Since its destruction in 2015, Palmyra has been, and continues to be, present in the debates on Syrian heritage. The emergence of the Islamic State in Iraq and Syria (ISIS) as a new actor in the protracted Syrian conflict added a new dimension to heritage targeting and damage. This trend is not new. Conflicts and wars have been

Z. Elcheikh (✉)
University of Stuttgart, Stuttgart, Germany
e-mail: elcheikh_zeina@yahoo.com

M.-T. Albert et al. (eds.), *50 Years World Heritage Convention: Shared Responsibility – Conflict & Reconciliation*, Heritage Studies,
https://doi.org/10.1007/978-3-031-05660-4_13

part of human history, and the advance of the arms industry has caused more death and more significant damage. In recent years, armed conflicts have shown how heritage sites have been reconfigured as "material impediments and transformed into proxies for ideological adversaries" (Meskell, 2018, p. 186–187). Indeed, Palmyra was no longer just a historical site but a symbol of the Syrian state authority and the international community, upon which the militant group wanted to impose its resentment. The shock at the broadcasted atrocity in the oasis was immediately framed within a discourse of terrorism, barbarism and iconoclasm. Moreover, raised voices called for lessons to be learned from other conflicts with iconoclastic strife (Korsvoll, 2021; Papaioannou, 2015).

The (Western) media coverage has focused on the ISIS destruction of pre-and non-Islamic sites. However, these atrocities were not the only ones committed by the extremist group. The ISIS militants have also targeted sites of other communities, including those of other Islamic denominations, to eliminate opposing beliefs by removing physical evidence from the landscape (Jones, 2018). The radical group produced and propagated visual materials of their atrocities as a performance intended to re-enact historical iconoclasm and obliterate the historical memory of local communities (Harmanşah, 2015) and even the actual presence of these communities. Nada Al Hassan, the Chief of the Arab States at the UNESCO World Heritage Centre, stated that "the inhabitants of Palmyra have been in large part, if not entirely, destroyed, and their voice is silenced" (Margit, 2018).

The destruction in Palmyra has underlined the value placed upon World Heritage sites, especially in the case of armed conflicts. It also prompts the question of how to deal with these sites when they are partially or entirely destroyed. Through the medium of images, opposing publics have also been produced (Joselit, 2020). Several initiatives – framed within a traditional (Western) account – to (digitally) reconstruct the damage in Palmyra emphasise materiality as heritage evidence. At the same time, later changes, even destruction and absence, have been part of Palmyra's long narrative. Besides, Syrian authorities have planned to re-open the historic site in a broader push to revive tourism in the war-torn country (Margit, 2018).

The calls and initiatives tend to be detached from priorities on the ground and the local communities' concerns about peace and socio-economic response and recovery. UNESCO's work to protect culture and preserve heritage amidst armed conflicts contributes to the resilience of communities and the reduction of disaster risks, resonating across many Sustainable Development Goals (SDGs) set up for the Agenda 2030 (UNESCO, 2018), but little consideration has been given to local communities and to Syrian experts as main stakeholders in the preservation of this cultural heritage resource (Al Quntar, 2018). Consequently, the danger of a binary rose in a dilemma that appears to advocate either caring about ancient monuments and objects or the affected people (Al-Azm, 2018, p. 101).

Through media and the various initiatives and calls, it has been considered that Syria's heritage will have a healing and reconciliation power in the post-war era to come. However, and especially in the case of conflicts, heritage also needs to be viewed from a human rights perspective since heritage is important in itself and in relation to its human dimension (Bennoune, 2016). Thus, conceptualising heritage through the prism of civil rights rescues its preservation from commodification, bureaucratic calcification and destructive extremist bigotry and reasserts the community's social, economic and environmental wellbeing as integral to the heritage discourse (Rabbat, 2016, p. 272).

Palmyra became a terrain on which several agencies and interests vied for control. Long before the actual Syrian war and the terror attacks of ISIS militants, the oasis has been a field for marching armies and travellers and archaeologists exploring its monuments and ruins. Within the broad viewpoint of "competitive archaeology" (Corbett, 2014), the interest in Palmyra kept showing how different powers have strived for antiquities and used them to create specific narratives and national identities.

Before addressing the actual debates on Palmyra, a short review contextualises the interest in the site, from its (re)discovery in the seventeenth century to the contemporary conflict.

13.2 The Ruins in the Syrian Desert: Heritage and Narratives

Palmyra, or Tadmor, enjoyed the availability of water and arable soil and a strategic location in the Syrian desert. It became a modest centre in the early stages of Roman rule, with a population mainly of Aramaean and Arab origins (Burns, 2017, p. 236). Palmyra reached its heyday when its commercial activities extended along the Silk Road. While incorporated into the Roman province of Syria, Palmyra continued to enjoy sovereignty, especially under Queen Zenobia, who became a provocation to Rome's domination in the region. Aurelian captured and destroyed Palmyra, which his successors reshaped on a smaller scale.

When Christianity gained strength in Palmyra, buildings for the new form of worship started to appear. Early in the fifth century, the Temples of Bel and Baalshamin were turned into churches (Browning, 1979; Burns, 2009). By the end of the Umayyad reign in the eighth century, the oasis was governed from Homs. When the Abbasids moved the capital to Baghdad, Tadmor started to decline (Browning, 1979, p. 51). After being devastated by earthquakes and by the Timurids, it was rebuilt but not to its former grandeur. The Druze ruler Fakhr Al-Din Al-Maani II used Palmyra to control the desert areas, and he reconstructed the castle that had been previously erected under the Mamluks in the seventeenth century.

During the Islamic period, the Temple of Bel's cella was turned into a mosque and used as such until the site's massive clearance in the late 1920s. In the ancient monuments, the locals sought shelter from the frequent nomad Bedouins' invasions. Many of the materials used for the dwellings were taken from other decaying structures and cemented with mud. Defacing sculptures has also been reported (Frank & Brownstone, 1986, p. 140). Although Palmyra never disappeared, its "rediscovery" by European travellers would bring the world's attention to the spectacle of ruins in the desert – as though it only came into being in the seventeenth century.

In 1678, English traders in Aleppo ventured out across the desert to reach the ruins but returned soon after to escape robbers. In 1691, William Halifax headed a new group, including the Dutch artist Gerard Hofstede van Essen, who made a detailed drawing of the ruins. Four years later, Halifax published his account, "Relation of a Voyage to Tadmor". In the mid-eighteenth century, two Englishmen, Robert Wood and James Dawkins, visited the site and documented the most significant monuments. They recruited the Italian architect Giovanni Battista Borra, who made minutely detailed drawings published in Wood's and Dawkins' "The Ruins of Palmyra Otherwise Tedmor in the Desart" [sic], in 1753. The book had a swift and widespread impact on architectural taste in Europe and became a sensation from London to St. Petersburg. A Paris edition reached the Russian court. The French enclave immediately named St. Petersburg "Palmyra of the North" or "Northern Palmyra" (Rostovtzeff, 1932, p. 122) and equated Catherine the Great with legendary Queen Zenobia – who became a motif in operas, literature and arts.

Palmyra was regarded as the most prosperous, luxurious, elegant and "romantic" of Syria's towns (Rostovtzeff, 1932, pp. 120–121). Its picturesque remains were (re) discovered during a romantic period of European history, marked by travel, exploration and the growing emotional effect of classicism. Intriguingly, illegal antiquities trafficking intensified in parallel to the excitement about the ruined town and the new finds – an interest that never ceased.

The twentieth century witnessed excavations at Palmyra's urban centre, temples and primary monuments. Numerous Palmyrene art objects went to several museums abroad. After the First World War, Palmyra attracted experts from various disciplines to work on the site. In 1929, the French started the extensive excavation works for which the local community had to be moved to a new, French-built village next to the site. Rostovtzeff (1932, p. 127) recalled that "the main temple of Palmyra was freed from the hundreds of poor huts built on and around it". During the Second World War, Syria was still under the French administration. At that time, with the agreement of the Vichy French, the Germans started using several airfields in the Syrian territories against British operations, including Palmyra.

By the end of the French Mandate in 1946, the Syrian Directorate General of Antiquities and Museums (DGAM) undertook the enormous task of coordinating excavations supported by international institutions. Communities regarded the interest in the relics of the past as a tool of cultural imperialism imposed by the West, and it was later wielded as an instrument in service of the Syrian government's plan to impose power and national identity. The growing interest in the past had been "institutionalized by a reification of the past into governmental branding" (Al-Manzali, 2016). The ruins of Palmyra, its theatre and Queen featured on Syrian banknotes and postage stamps and remained one of Syria's main tourist destinations for decades. The thriving leisure industry and the facilities created around it in the oasis secured additional incomes for the locals.

In 1980, UNESCO declared Palmyra a World Heritage Site – a measure of the success of a given state in assuring international assistance in protecting heritage sites within its territory and a successful brand for tourism marketing. Despite

Palmyra's long narrative, worthy of inclusion on its own merit, it was inscribed as "Rome's expansion in and engagement with the East" (UNESCO World Heritage Centre, n.d.) without considering its continuity under the Arabs and Muslims.

13.3 Heritage Destruction and Preservation: Instrumentalisation and Propaganda

Since the beginning of the Syrian conflict in March 2011, UNESCO has tried to draw more attention to the potential damage that is threatening the country's cultural heritage. This rich legacy includes six sites that have been listed as World Heritage, including Palmyra.

In April 2013, the ISIS militant group expanded into Syria. It declared its new caliphate in June 2014, adopting the name of Islamic State, which has been criticised by Muslim scholars and communities alike. The radicals appointed themselves as defenders of true Islam. The iconoclastic destruction of religious images and monuments went hand-in-hand with the group's efforts to establish an Islamic state. The ISIS extremists' claims of their (religiously motivated) targeting of monuments were based on a fundamentalist interpretation of Islam. In March 2015, UNESCO called for global action to face the systematic destruction of archaeological sites by the ISIS militants and stated that the destruction should be considered a war crime against humanity's heritage and cultural cleansing. In Syria and Iraq, the group demolished several pre-Islamic sites and artefacts, believing that such structures were idolatrous. On the other hand, profits from extensive looting networks helped to fund the group's armed operations.

In May 2015, the ISIS militants reached Palmyra and captured parts of the historic citadel. A few weeks later, the world watched the radicals destroying the Temple of Baalshamin after detonating a massive quantity of explosives inside the building. The Roman theatre, which had hosted the International Festival of Palmyra since 1992, became the stage for a dreadful act when ISIS child executioners slaughtered Syrian soldiers. The atrocities also extended to civilians in Palmyra. In August 2015, ISIS fighters beheaded the renowned archaeologist Khaled al-Assad, who worked for decades in Palmyra. Posthumous tributes were paid to him in Syria and abroad. But what of the thousands of nameless residents of Palmyra?

The oasis received internally displaced Syrians from other conflict zones. When the hostilities reached the town, people were more concerned about seeking refuge and fulfilling their basic needs. Most of them left Palmyra when ISIS took over in May 2015, and those who remained fled to other cities when the fighting became too severe. Palmyra remained under the control of the terrorist group until the Syrian government reclaimed the city in April 2016. The ISIS militants occupied Palmyra again in December 2016 and attacked other monuments. The Syrian forces liberated Palmyra for a second time in March 2017 (Abdulkarim, 2020).

The attacks on the monuments of Palmyra were perhaps the most publicised of the extremists' actions, with videos of the temples, statues and museum exhibits being demolished virally circulating on social media. The ISIS atrocities unleashed an international outcry, and UNESCO and other organisations have strongly condemned their acts of terror against World Heritage sites. Several initiatives were set into motion with much enthusiasm to document the damage and recollect the past. Raised voices called for the world to learn lessons from other conflicts with iconoclastic strife – namely, the Taliban's destruction of Bamiyan's colossal statues of Buddha in 2001. However, it is worth noting that Palmyra was a listed World Heritage site for decades before the ISIS atrocities, unlike the Buddhas of Bamiyan that were listed two years after their destruction.

Despite Syria being considered one of the most dangerous countries in the world due to the conflict and Palmyra being littered with mines, the perilous situation on the ground did not prevent groups of foreign experts from travelling to witness, assess and document the damage. In all its types, the media reported continuously and with varying degrees of emotion on the new state of ruination in Palmyra. Furthermore, several Western countries organised special temporary and virtual exhibitions dedicated to Palmyra and its memory and printed corresponding catalogues. Numerous conferences and colloquia, organised and held outside Syria, widely addressed the future of Palmyra as they selectively showcased its past. Furthermore, despite the sanctions on the country imposed by the same powers advocating the safeguarding of Syrian heritage and the international community's failure to end the calamity of millions of Syrians in and outside the country, renowned institutions, such as Le Louvre in Paris, offered asylum to artefacts under threat (Jones, 2015). Immediately after the destruction, not only voices from the political, academic and creative scenes were raised, but heated discussions about the future of Palmyra and its reconstruction also commenced.

Italian experts painstakingly restored sculptures to "erase the act of violence" (Di Donato & Said-Moorhouse, 2017), and digital technologies allowed several teams to resurrect vanished monuments and (virtually) replicate parts of the lost heritage. Experts from Oxford and Harvard Universities and the Institute for Digital Archaeology (IDA) rebuilt a two-thirds scale replica of Palmyra's Arch of Triumph. It was made in Italy from Egyptian marble. The reproduction was displayed in a high-profile political event on Trafalgar in London as an "act of solidarity" (Boyle, 2016), and unveiled by the then Mayor of London, Boris Johnson, as part of World Heritage Week, before heading to New York and Dubai and eventually, according to the plan, to be relocated in Palmyra to be displayed next to the ruins of the original arch (Bacchi, 2016).

After Palmyra's recapture in 2016, the Mariinsky Orchestra, founded in 1783 under Catherine the Great, came from the (other) Palmyra of the North. The orchestra played music as a celebration at the same theatre in which ISIS child executioners staged their offence a few months earlier. The Russian president, Vladimir Putin, praised the liberation of Palmyra in a broadcast from a video screen on the main stage. According to Putin, the concert was dedicated to the sufferers of "international terrorism", which he termed a "terrible evil" (Harding, 2016).

Since the beginning of the conflict in Syria, Russia has supported the Syrian administration of President Bashar al-Assad in the country, first politically and later on a military level. Some Syrian–Russian partnerships started to promote archaeological and heritage restoration projects in the war-torn country and put a basic touristic infrastructure in place, including a project to rehabilitate Syria's ancient city of Palmyra. However, these endeavours have been considered to be mainly for propaganda purposes, using Syrian archaeologists as their representatives to add archaeological legitimacy, and as Russia's attempts to expand its influence and to gain a foothold in the country's vital sectors in Syria, including antiquities and tourism sectors (Bacchi 2016; Margit, 2018; Hardan, 2021).

Voices from organisations, such as the Association for the Protection of Syrian Archaeology (APSA), claimed that a rapid and exhaustive reconstruction of Palmyra would also be immoral since it would be serving as indirect rehabilitation for a government broadly accused of massacring its people, whose forces had also been accused of overseeing looting and damaging heritage sites with indiscriminate bombings (Bacchi, 2016).

On the other hand, the replica of the triumphal arch displayed in London has sparked a fierce debate on its political and archaeological implications since the ruins must not be turned into a fake replica of the monuments' former glory (Jones, 2016; Bacchi, 2016). The restoration of Palmyra as programmed by the Syrian government and supported by Russian and other European and Western leaders has been regarded as both ethically and scientifically wrong and amounting to nothing more than "transforming Palmyra into a pro-Bashar theme park" (Bacchi, 2016).

With Palmyra's devastating present and indeterminate future, its glorious past has become a nostalgic refuge. However, endeavours to undo the damage cannot erase the vicious event that caused it. With thousands of images of the destruction and millions of search engine results for the keywords "Palmyra" and "ISIS", forgetting what happened is out of the question.

13.4 Discussion

The violence inflicted through ISIS terrorist attacks has punctuated the oasis in both spatial and temporal terms. Destruction has become a chapter in Palmyra's long narrative, and the destroyed monuments have, themselves, become historical documents. There is an increasing focus on later alteration and transformations of objects and monuments, even on their destruction and absence. The Buddhas of Bamiyan, constantly cited for the similarity of their fate with that of Palmyra, illustrate this case well. The Bamiyan Valley was inscribed on the World Heritage List after the destruction of the Buddhas by the Taliban in 2001. The physical and material absence of the Bamiyan statutes did not reduce their significance, and the debates around them are still ongoing, even after two decades. The focus on materiality often oversimplifies intangible values. Palmyra embodied many identities and influences. So, if Palmyra is to be reconstructed, which one of its several pasts should be restored, and who would decide?

Today, Western governments and institutions would spend money and share expertise to rebuild Palmyra and rescue its artefacts, but not lift the sanctions or accommodate refugees. To say that traumatised people are disregarded in favour of monuments may sound like an overstatement, yet it is not entirely detached from colonial attitudes and local governments' attempts to create and enforce a national identity.

In this regard:

> instead of promoting a better understanding of the way in which Syrian cultural heritage resources and their fate are mobilised for political or sectarian gains, a concern with an idealistic reconstruction of Palmyra de-sensitizes audiences and other experts alike to the context that gives Palmyra its significance: heritage reconstruction as a reconciling and unifying role in post-conflict Syria, with the Syrian people (not monuments) at the core of the reconstruction process (Al Quntar, 2018).

Despite the calls and attempts to raise awareness about a shared and common (world) heritage, it is not always evident whether all people and communities have received that message and have conceded and prioritised it in the same way. In the World Heritage Convention from 1972, UNESCO defined World Heritage sites as "heritage of mankind as a whole" – a Eurocentric perception based on an all-encompassing and universal view. Local communities are not monolithic in what concerns their heritage. They appreciate their past and its relics differently.

Arabs, in general, wrote about archaeology and ancient history to deploy their narrative in the fight against colonialism (Corbett, 2014, p. 124). A new literary genre, the historical novel, introduced a fresh way of thinking about the past, so Arabic readers could imagine that they were the heirs of a drama that stretched back to antiquity and celebrated old heroes. Beirut journalist, Salim Al-Boustani, wrote an early novel, Zenobia, in 1871. The protagonist was the queen of Palmyra, "whose name deeply rooted Syrian identity and historic Syrian womanhood" (Reilly, 2019, p. 77). Although Palmyra's population probably spoke other languages in addition to Arabic, they are usually called "Arabs" in official Syrian historical narratives. Mustafa Tlass, the former long-serving Syrian minister of defence and amateur historian, wrote Zenobia queen of Palmyra to emphasise the Arab identity of the oasis. The former Syrian president, Hafez Al-Assad, saw himself as a saviour of Arabness and the latest fortress in opposition to the West and its imperialism, squared off against an official interpretation and instrumentalisation of ancient history. Zenobia's rebellion against Rome made her a patriotic symbol in Syria (Sahner, 2014, pp. 134–135), and Palmyra was charged with the secular Arab nationalism propagated by the Syrian administration.

For decades, the use of archaeological heritage to promote national harmony did not enjoy significant success. Consequently, heritage became a target for those who wanted to destroy those states (Jones 2018, p.53). The mindless destruction of monuments and artefacts and targeting local communities by ISIS was meant to turn these areas into the groups' endeavours to establish their Islamic Caliphate. With the increasing violence and the rising toll of victims and displaced persons, voices criticised international organisations and institutions for caring more about antiquities when people were dying, being forcibly displaced and losing their homes (Westcott, 2016). Nada Al Hassan mentioned that a large part if not all of Palmyra's inhabitants have been devastated, and their voice is silenced, and through the debates of reconstructing Palmyra and reviving tourism, there it is possible that the victims and their trauma would be trivialised (Margit, 2018).

In the light of the protracted Syrian conflict, the growing focus on the destruction of heritage in the country, including Palmyra, has been a blessing and an affliction alike. On the one hand, awareness of the importance of cultural heritage has been raised as a common wealth that brings components of the war-torn country together. On the other hand, the sanctions posed on the country and the fact that several organizations, including UNESCO, work with governmental bodies rather than individuals has created another area of conflict.

Therefore, a more comprehensive approach needs to be taken into account, one which includes not only heritage sites from a material point of view and perspective of experts but also local communities. Through democratic citizenship and through education (SDG4), the sense of continuity and belonging is supported. Moreover, a peaceful environment (SDG16) and a sustainable community (SDG11) could be established through the (active) involvement of the locals in the future plans to bring life back to the oasis and its ruins once their basic needs (SDG1) are fulfilled. Such an approach needs to reconsider heritage preservation as a civil right (Rabbat, 2016) instead of an ideal imposed from above. Accordingly, the significance of cultural heritage will continue to be both a message from the past and a pathway to the future (Bennoune, 2016).

13.5 Concluding Thoughts

Throughout millennia, the fortune of Palmyra rose a little and fell a little. In 2015, Palmyra was demolished in a demonstration of power, and the debates on its future are no less significant than those demonstrations of power. Over the centuries, the local communities have perceived the ruins as continuity rather than something frozen in time. They dwelled in monuments and reused ancient fragments in new buildings. The residents of Palmyra have loved their oasis, literally their home, and they are part of its long account, as much as the ruins and their ancient builders.

Today, with the loss and damage caused by war and terrorism in Palmyra, the pressing question "what should be done next?" It is not easy to separate people's cultural heritage from the people themselves, and a balance needs to be struck between universal and local values of cultural heritage. The oasis could serve as an example for how this balance might be struck.

The world is watching Palmyra. In the context of the 50th anniversary of the World Heritage Convention, it is necessary to encourage revisions on how to adopt approaches which aim to integrate cultural heritage and its safeguarding in the life of local communities. For those who experienced the damage of war and reign of terror, basic living requirements need to be fulfilled, along with education to achieve a peaceful environment and a sustainable community based on democratic citizenship and rights.

By (re)considering heritage and its preservation as a civil right, the way is paved for a more comprehensive approach that actively includes and involves local communities, helping them build more resilient societies and strengthen their awareness of their own heritage in the post-conflict era to come.

References

Abdulkarim, M. (2020). Palmyra: From war and destruction to rehabilitation. In M. Nagaoka (Ed.), *The future of the Bamiyan Buddha statues* (pp. 113–121). Springer. https://doi.org/10.1007/978-3-030-51316-0_7

Al-Azm, A. (2018). Preserving heritage in Syria. In P. Newsom & R. Young (Eds.), *Post-conflict archaeology and cultural heritage: Rebuilding knowledge, memory and community from war-damaged material culture* (pp. 91–105). Routledge.

Al-Manzali, M. (2016, October 2). *Palmyra and the political history of archaeology in Syria: from colonialists to nationalists.* Mangal Media. https://www.mangalmedia.net/english//palmyra

Al Quntar, S. (2018). Palmyra and expert failure. *International Institute for Asian Studies*. The Newsletter No. 80 Summer 2018. https://www.iias.asia/the-newsletter/article/palmyra-expert-failure

Bacchi, U. (2016, April 20). *Palmyra Arch in London: 'Unethical' reconstruction of 'Disneyland' archaeology criticized.* International Business Times. https://www.ibtimes.co.uk/palmyra-arch-london-unethical-reconstruction-disneyland-archaeology-criticised-1555659

Bennoune, K. (2016, October 25). *Cultural heritage is a human rights issue*. UNESCO. https://en.unesco.org/news/karima-bennoune-cultural-heritage-human-rights-issue

Boyle, D. (2016, April 20). *Arch of Triumph rises again in London's Trafalgar Square after being destroyed by Isil.* The Telegraph. https://www.telegraph.co.uk/news/2016/04/19/palmyras-arch-of-triumph-rises-again-in-londons-trafalgar-square/

Browning, I. (1979). *Palmyra*. Noyes Press.

Burns, R. (2009). *Monuments of Syria*. I.B.Tauris.

Burns, R. (2017). *Origins of the colonnaded streets in the cities of the Roman East*. Oxford University Press.

Corbett, E. (2014). *Competitive archaeology in Jordan: Narrating identity from the Ottomans to the Hashemites*. University of Texas Press.

Di Donato, V., & Said-Moorhouse, L. (2017, February 17). *Palmyra treasures restored after ISIS hammer attack*. CNN. https://edition.cnn.com/2017/02/17/europe/palmyra-treasures-restored-isis/index.html

Frank, I. M., & Brownstone, D. M. (1986). *The Silk Road: A history*. Facts on File Publishers.

Hardan, M. (2021, June 22). *Russia gains foothold in Syria's Palmyra through archaeological restoration*. Al-Monitor. https://www.al-monitor.com/originals/2021/06/russia-gains-foothold-syrias-palmyra-through-archaeological-restoration#ixzz72nfoz0tK

Harding,L.(2016,May5).*PalmyrahostsRussianconcertafterrecapturebySyrianforces*.TheGuardian. https://www.theguardian.com/world/2016/may/05/palmyra-amphitheatre-hosts-russian-concert-after-recapture-by-syrian-forces

Harmanşah, Ö. (2015). ISIS, heritage, and spectacles of destruction in the global media. *Near Eastern Archaeology, 78*(3), 170–177.

Jones, C. W. (2018). Understanding ISIS's destruction of antiquities as a rejection of nationalism. *Journal of Eastern Mediterranean Archaeology and Heritage Studies, 6*(1-2), 31–58.

Jones, J. (2015, November 20). *Asylum for artefacts: Paris's plan to protect cultural treasures from terrorists*. The Guardian. https://www.theguardian.com/artanddesign/jonathanjonesblog/2015/nov/20/asylum-artefacts-paris-terrorists-louvre-isis

Jones, J. (2016, April 11). *Palmyra must not be fixed. History would never forgive us*. The Guardian. https://www.theguardian.com/artanddesign/jonathanjonesblog/2016/apr/11/palmyra-isis-syria-restored-3d-printers-vandalism

Joselit, D. (2020, July 22). *How Western nations use the Arch of Palmyra and other symbols of Islamic heritage as political tools*. Artnet News. https://news.artnet.com/art-world/david-joselit-heritage-and-debt-1895594

Korsvoll, N. H. (2021, January 21). Bamiyan to Palmyra: Modern Islamist iconoclasm in war. *The Art of Value*. https://theartofvalue.blog/2021/01/29/iconoclasm-in-war/

Margit, M. (2018, September 11). *Assad regime courts tourists to Palmyra*. The Media Line. https://themedialine.org/top-stories/assad-regime-courts-tourists-to-palmyra/

Meskell, L. (2018). *A future in ruins: UNESCO, World Heritage, and the dream of peace*. Oxford University Press.

Papaioannou M. (2015, November 2015). The Buddhas of Bamiyan, Timbuktu, Palmyra: What next? *Cambridge International Law Journal*. http://cilj.co.uk/2015/11/20/the-buddhas-of-bamiyan-timbuktu-palmyra-what-next/

Rabbat, N. (2016). Heritage as a right: Heritage and the Arab Spring. *International Journal of Islamic Architecture, 5*(2), 267–278. https://doi.org/10.1386/ijia.5.2.267_1

Reilly, J. A. (2019). *Fragile nation, shattered land: The modern history of Syria*. I.B. Tauris.

Rostovtzeff, M. (1932). *Caravan cities: Petra, Jerash, Palmyra, Dura* (D. T. Talbot Rice, Trans.). Clarendon Press.

Sahner, C. (2014). *Among the ruins: Syria past and present*. Hurts.

UNESCO. (2018). *Culture for the Agenda 2030*. UNESCO. http://www.unesco.org/culture/flipbook/culture-2030/en/Brochure-UNESCO-Culture-SDGs-EN2.pdf

UNESCO World Heritage Centre. (n.d.). *Site of Palmyra*. https://whc.unesco.org/en/list/23

Westcott, T. (2016, April 4). *Forget about antiquities; what about Palmyra's civilians?* The New Humanitarian. https://www.thenewhumanitarian.org/feature/2016/04/04/forget-about-antiquities-what-about-palmyra-s-civilians

Chapter 14
Countering the Narratives of Destruction: Textual Evidence and the Tradition of Heritage Preservation in Islam

Azeez Olaniyan and Akeem O. Bello

Abstract The relationship between Islam and cultural heritage preservation continually comes under scrutiny. This is because of vicious attacks on heritage and artefacts by groups laying claim to Islamic tenets and texts to justify their action. Thus, under this pretext, heritage sites and cultural actors and icons, are eliminated. Why is this so? What is the position of Islam on heritage preservation? How can the narratives of destruction be countered? This chapter interrogates these questions. It argues that the narratives of destruction derive from poor interpretations of the texts and traditions of Islam in respect of cultural heritage. Several monuments in the Islamic world also predate the establishment of Islam. The study brings out textual facts and traditions to counter the narratives of violent elements such as Islamic State in Iraq and Syria (ISIS), Al-Qaida, Boko Haram, Islamic State in West Africa Province (ISWAP) and Al-Qaida in the Islamic Maghreb (AQIM).

Keywords Preservation · Destruction · Culture · Heritage · Islam

14.1 Introduction

Heritage exists in both natural and cultural forms, and the understanding of this dichotomy is a major step in coming to terms with its importance both in historical and contemporary discourses. The demarcation has been well clarified in articles 1 and 2 of the World Heritage Convention of 1972 adopted in Paris. For instance, article 1 considers cultural heritage to include monuments, groups of buildings and sites inherited from past generations, maintained in the present for the benefit of

A. Olaniyan (✉)
Department of Political Science, Federal University Oye Ekiti, Oye-Ekiti, Nigeria
e-mail: azeezolaniyan@gmail.com

A. O. Bello
Department of Political Science, Bamidele Olumilua University of Education, Science and Technology, Ikere-Ekiti, Nigeria
e-mail: bello.akeem@bouesti.edu.ng

M.-T. Albert et al. (eds.), *50 Years World Heritage Convention: Shared Responsibility – Conflict & Reconciliation*, Heritage Studies,
https://doi.org/10.1007/978-3-031-05660-4_14

future generations, while article 2 classifies natural heritage as including geological and physiographic formation, natural features, and natural sites (UNESCO, 1972). Thus, it is obvious that while natural heritage occurs as a gift of nature to humanity with little input from man, cultural heritage is a manifestation of human ingenuity, creativity, and practices. While the two are particularly important to humanity, cultural heritage has, however, attracted more attention because it has come under more serious and sustained attacks in recent times. Therefore, discourses on heritage preservation have tended to focus more on cultural forms.

Cultural heritage comes in tangible and intangible forms. While the tangibles are the physical manifestations such as buildings, sites, monuments, paintings, sculptures and manuscripts, the intangibles are found in customs, mores, myths, values, and practices that define a people or society and have evolved over several generations. Cultural heritage is crucial to humanity because of its "existential, ideological, commercial and educational" values (Nilson & Thorel, 2018). It is basically a manifestation of the past and encompasses unique treasures of inestimable value bequeathed to humanity by ancient civilisations. As a legacy, it connects the present with the past and projects into the future. Heritage has proven to be part of the wealth of a nation and thus crucial to human existence, identity, and memory.

However, the destruction of monuments has featured across time and space because, in the words of Rachel Bokkem (2017) "destruction of physical or intangible artefacts that embody the ideas, beliefs, and characteristics of past societies is a well-tested means of control and power". Thus, monuments, artefacts and landscapes have come under destructive intentional attacks. The need for the preservation of these valuable treasures has emerged as one of the major concerns of the global community. This particularly informs the establishment of the World Heritage Convention, which remains the most successful international legal instrument that propagates the protection of world cultural heritage (UNESCO, 2012).

Despite the concerted efforts of global bodies and agencies, World Heritage sites continually face vicious attacks. In fact, in recent times, a trend has emerged whereby Islam has been employed by terror groups to perpetrate violence against cherished World Heritage. Specifically, what could be regarded as the most brutal threat to World Heritage is the emergence of terror groups whose main agenda is the elimination of monuments and World Heritage, citing Islamic texts and injunctions as their authority for such actions (Isakhan & Zarandona, 2017). Through their violent actions, priceless World Heritage sites have been ruined. This destruction has been seen across nations like Nigeria and Mali in Africa and Syria and Iraq in the Middle East etc. As a result of this, some pertinent questions arise: what Islamic authority informs this violent destruction? What are the instances of such destruction globally? What can be done to reverse the narratives of destruction by the "Islamic" militants? This chapter interrogates these questions.

The chapter is divided into six sections. This first section introduces the topic; the second engages in historical and theoretical analysis of the relationship between religion and heritage; the third focuses on instances of heritage destruction by Islamist terrorists; the fourth examines their justification for doing so; the fifth examines the counterevidence to their narratives; the sixth provides a conclusion.

The study is qualitative and relies heavily on published texts such as book chapters, journals, newspapers, magazines, the Qur'an, Hadith and internet sources.

14.2 Religion and Cultural Heritage

Since time immemorial, religion has played a prominent role in society. It has left indelible imprints on landscapes and fostered social relations, cultures and practices (Zeybek & Arslan, 2017). Indeed, a great number of what has come to us today as icons and monuments of inestimable value are inspired by religious belief, and the builders or creators saw the task of building or creating such monuments as their religious duty. Thus, important monuments such as the great pyramids of Giza, the Sphinx of Giza, Taj Mahal, Lumbini, Borobodur, Angkor Wat, Stonehenge, the Babri Mosque in Ayodhya, the Buddhist images at Bamiyan, the mosques and Churches in Bosnia, the Djinguereber mosque and the treasured manuscripts in Timbuktu and several others are all connected to religious beliefs and knowledges of the people of the period in which they were made (Singh, 2008; Smith, 2015). In essence, it is difficult to separate religion from cultural images, which have come down to the present generation as heritage (Isakhan & Zarandona, 2017).

On the other hand, however, religion has featured prominently in the destruction of cultural heritage across time and space. According to Zeybek and Arslan (2017, p.3), history is replete with instances of the destruction of cultural heritage due to fanatic iconoclasm or the collateral effects of armed conflicts. In ancient Egypt, King Akhenaten, in his authority to introduce a new way of worship to his kingdom, ordered the destruction of all the old Egyptian gods (called Amun) and sent royal officials to chisel out every reference to the gods, including images, paintings, temples, tombs and cartouches (*Encyclopaedia*, 2018). The Roman Emperor Theodosius wanted to eradicate all vestiges of non-Christian society in Alexandria town and ordered the demolition of the Temple of Serapis (Zeybek & Arslan, 2017). In the same vein, the defacing of the great Sphinx of Giza was alleged to have been carried out by a Muslim conqueror who regarded Egyptian monuments as products of infidels that must be eviscerated (Giradi, 1995). In the Bible, the destruction of the Canaanites, including their landscape, arts and artefacts by the Israelites was justified by claims of instruction from God who gave them the permission to "drive out all the inhabitants of the land before you, and destroy all their idols, and destroy all their molten images, and quite pluck down all their high places" (*King James Bible*, 1769/2017, Number 33:35). In 1299 CE, the Somnath Temple in Gujarat was invaded and destroyed by the Delhi Sultanate armies, based on Islam's aversion to structures and images considered to be idols (Eaton, 2001).

In contemporary times, the Chinese Cultural revolution of 1966–1976, in a bid to foster atheism, witnessed the destruction of many iconic religious and secular images, artefacts and structures that bore semblance to China's traditional past, which the Mao Zedong regime described as impediments to development. In 1992, Hindu zealots, led by Vishva Hindu Parishad and Bajrad Dal, destroyed the

430-year-old Babri Masji Ayodhya mosque in a show of intolerance towards the Muslim group (Tully, 2002). In South Korea, several Buddhist shrines and monuments have been set on fire by a group of people identified as Pentecostal Christian fundamentalists (Wells, 2000). In Brazil, several traditional religious devotees and their monuments, edifices and images have come under vicious attacks and been destroyed by a group of Pentecostal zealots called "Soldiers of Jesus" (McCoy, 2019).

These examples reveal a strong relationship between religion and cultural heritage. The relationship operates on two opposing levels: symbiotic and adversarial. On the first level is the reality that religion inspires monuments, which in return help to uplift the image and status of religious beliefs. That is why it is difficult to separate religion from most of the ancient cultural heritage, because, as argued by Singh (2008, p. 2), heritage sites are "places where the spirit of nature and culture meet and are additionally symbolized and maintained by people's attachment to rituals performed there." However, on the other hand, religion has been used as a cover to attack cultural materials. But those structures that are dubbed as idols or heathenish are products of other religious beliefs, symbolizing a clash of religious beliefs. The clash, however, reveals a deliberate attempt by one religion to obliterate the vestiges of another.

To a large extent, discourses on the destruction of cultural heritage can be subsumed under the rubric of intentional and unintentional perspectives of conflict and the rational choice theoretical perspective of terrorism. Arthur Westing (2008) argued that environmental destruction is one of the consequences of war, but that the consequences can be intentional or unintentional. Since the environment plays a crucial role in war, combatants do deliberately consider the environment in their planning and tactics (Olaniyan & Okeke-Uzodike, 2020). In the same vein, religious zealots employ tactics that result in the deliberate destruction of cultural heritage, which is also part of the living environment, as a way of erasing the vestiges of what they consider inferior religion. Rational choice theory identifies terrorism as a deliberate act in which the perpetrators are conscious of their actions (Caplan, 2006; Anderton & Carter, 2005; Olaniyan, 2017). The theory argues that terrorism is a deliberate action that involves a high level of organisation, planning and calculation, meaning that the perpetrators of the destruction in the guise of terrorism knew what they were doing. They make deliberate decisions to destroy and plunder because they consider these artefacts as contrary to their interpretation of beliefs. It is within the context of intentional destruction as a rational choice that the current attacks by terror groups on cultural heritage in Islamic societies are situated.

14.3 Tales of Destruction

The Muslim world has come under the severe grip of terrorism, and casualties have seldom been human beings alone. There are tales of horror concerning materials as well, one of which is the wanton destruction of monuments, artefacts, culture and landscapes. The activities of the terrorists show a determination to obliterate history and the past from the Islamic world. The following are a few examples that will suffice to demonstrate this point. In March 2001, the Taliban militants blew up the 1500-year-old Bamiyan Buddhas statues in Afghanistan, to the outrage of the world (*Asia News*, 2009). In 2012, al-Qaida and Ansar Dine militants vandalized hundreds of ancient mausoleums, which were the resting places of Timbuktu's legendary 333 saints and destroyed part of the iconic Djinguereber mosque in Timbuktu (Smith, 2015; Bello, 2020). In addition, 4203 pieces of the famous Timbuktu manuscripts from the Ahmed Baba Research Centre and many other libraries were destroyed by the same terror group (Smith, 2015; Williams, 2018). In March 2015, ISIS destroyed the Sufi shrines situated near Tripoli (Kingsley, 2015). In late December 2016, ISIS terrorists destroyed part of the Roman amphitheatre in the ancient city of Palmyra, Syria (Bokkem, 2017), and beheaded Khaled Al-Asaad, the custodian of the Palmyra statues, for his refusal to show the terrorists the location of hidden statues (Hubbard, 2015). In 2009, a Taliban militant group targeted and destroyed much of Pakistan's Buddhist artefacts left from the Buddhist Gandhara civilisation (*Asia News,* 2009). On 7 February 2012, militants stormed the national museum of the Maldives and destroyed Buddhist artefacts (Wright, 2012). In Syria, much of the country's cultural heritage was destroyed by ISIS militants, such as the minaret of the Great Mosque in Aleppo, Al-Madina Souq, Krak des Chevailers, Khsruwiyah Mosque, Lion of Al-lat, the Temple of Bel and Baalshamin, the Palmyra Temple, the Arch of Triumph, the Monastery of St Elian and the Armenian genocide Memorial Mosque as well as several ancient sculptures in the city of Raqqa (Tastekin, 2017). In Iraq, ISIS destroyed much of the cultural heritage in the areas it controlled, which included 28 religious buildings, Shiites tombs, mosques, shrines and churches, the ancient city of Nimrod, the walls of Nineveh and the iconic Mosul library and museum (Khalid, 2015).

14.4 Justifying Destruction

Curiously, there are virtually no Quranic texts that directly sanction the destruction of cultural artefacts. There are, however, examples of traditions of the Prophet and his immediate followers regarding cultural artefacts, concerning which two major references can be made. The first says: "Shun the abomination of idols and shun the world that is false" (Holy Quran, 2021, 22:30). This verse is more of an admonition to believers to avoid idol worshipping. It did not outrightly sanction the destruction of such objects. However, it has become one of the means of justification for modern-day militants to destroy statutes. For example, after the destruction of Palmyra artefacts by ISIS militants, a justification that was posted in *Dabiq,* its propaganda magazine, reads thus: Baal is a false divinity for which people sacrificed their children as indicated in the book of Jeremiah (Old Testament). But by the Grace of Allah, soldiers of the Caliphate destroyed it (Isakhan & Zarandona, 2017).

This line of thought is also reflected in the dastard killing of Khalid al-Asaad, the custodian of Palmyra statutes, during the siege of ISIS on Syria. As reported by Bel Trew (2016),

> The blood-soaked headless corpse of Khaled al-Asaad, 81, a former chief of the UNESCO world heritage site, was strung up on traffic lights by ISIS. His severed head was placed between his feet, and next to it was a sign accusing the antiquities expert of being an "apostate" and "director of idolatry" (Trew, 2016).

The second reference in the Quran contains the story of how Prophet Ibrahim destroyed idols in his father's temple. The story goes that Prophet Ibrahim launched an overnight raid on his father's shrine, axed every statue, except the biggest one upon which he hung the axe. By daybreak, he was confronted, and the following exchanges ensued:

> … He said to his father and his people: 'What do you worship? They said, 'We worship idols, and to them we are ever devoted.' He said: 'Do they hear you when you call on them? Or do they benefit you or do they harm you?' They said: 'Nay but we found our fathers doing so (Holy Quran, 2021, 26:69–74)

Again, the Abraham story can be interpreted as more of a message for people to stop worshipping idols. However, the audacity of Abraham to confront his father and the entire leaders of the religion of the time, albeit successfully, has become a source of inspiration to modern-day militants in the Muslim world.

In the tradition of the Prophet, there is evidence of the outright removal of idols from society. The most popular of these Prophetic actions was the removal of idols and images from the Kaaba after the conquest. The story goes that upon the conquest of Mecca, the Prophet ordered the removal and destruction of all idols that were kept in the Kaaba (Fachrudin, 2015). Although this was done to prove monotheism and prevent idol worshipping in what was considered a sacred place, it has become a major reference point for militants. For example, Isakhan & Zarandona (2017) reported how an ISIS militant justified the destruction of monuments in Mosul, Iraq, by saying "The Prophet Muhammad shattered the idols with his own honourable hands when he conquered Mecca. The Prophet Muhammad commanded us to shatter and destroy statues. This is what his companions did later on, when they conquered lands".

To the terrorists, the destruction of statues and killing of custodians of culture is justified on the pretence of preventing idolatry and worshipping of other deities. It is also, in their estimation, a way of purifying Islam and supporting the establishment of a so-called Islamic state. To the terrorists, the management and sustenance of traditional monuments, artefacts, traditional medicine, and other forms of traditions are considered acts of paganism and idol worship. They equally preach against tourism to historical sites on the grounds that they encourage social vices like fornication, adultery, and corruption (Bello, 2020). To the extremists, ancient manuscripts, shrines, and tombs also represent mixing pre-Islamic beliefs with Islam, which is a heresy that must be stopped.

In addition to the above, the Hadith (which contains the sayings of the Prophet) also recorded how the Prophet preached against idols and the keeping of images. In the Hadith, the Prophet was reported to have said: "I have been sent to join ties of relationship (with kindness and affection), to break the idols and to proclaim the oneness of Allah (in a manner that) nothing is to be associated with him" (Sahih Bukhari, Book 04, Number 1812). By and large, these verses and traditions have served as the basis for militants' onslaught against cultural artefacts.

14.5 Evidence of Heritage Preservation

Evidence from the preceding section suggests a strong aversion and rejection of any statues, monuments and images because they are regarded as symbolizing idols. However, in the Islamic world, there are monuments, beliefs and practices of cultural values that have been preserved and even become central parts of the Islamic religion. These tend to nullify or modify the provisions of the verses and the tradition quoted above. Some of these include the following:

(i) **The Preservation of the Kaaba**

The Kaaba is the cuboid structure in the heart of Mecca city that Muslim faithful turn to while observing their five daily prayers and circumambulate during pilgrimage. It is regarded as the holiest place to the Muslims, and every believer of the faith is enjoined to visit the place at least once in their lifetime. This is clearly stated in the Quran thus:

> God has made the Kaaba, the Sacred House, an asylum of security, Hajj, and 'Umrah (pilgrimage) for mankind... (Holy Quran, 2021, 5:97).

The Kaaba predated the formal institutionalization of Islam because, according to Khudair (2021), it was built by the Prophet Abraham, but later turned into a house of idols. But upon the formation of Islam, the building was preserved and integrated into the religion. The Kaaba is an iconic heritage.

(ii) **The Black Stone in the Kaaba**

The Black Stone (al-hajar-al-aswaj) is part of the Kaaba building. The people kiss this stone during the annual pilgrimage. After the establishment of Islam as a religion, this stone was preserved and became a place of visit during pilgrimage to the Holy Site. Some pilgrims always struggle to touch it, just like the Prophet and the Khalifs did (Fachrudin, 2015). From all indications, the Black Stone is a preserved heritage with a long historical value from antiquity.

(iii) **The Preservation of Safa and Marwa**

Safa and Marwa are two small, enclosed mountains adjacent to the Kaaba, between which pilgrims travel back and forth seven times during the circumambulating of the Kaaba as part of the pilgrimage rituals. Allah is quoted in the Quran thus:

> Behold! Safa and Marwa are among the Symbols of Allah. So, if those who visit the House in the Season or at other times, should compass them round, it is no sin in them. And if

> anyone obeyeth his own impulse to good, – be sure that Allah is He Who recogniseth and knoweth (Holy Quran, 2021, 2:158).

The Kaaba and Safa and Marwa are pre-Islamic monuments that were preserved for centuries until incorporated into Islam much later (Al-Harbi, 2020).

(iv) **The Death and Preservation of the Pharaoh's Body**

The death and preservation of the Pharaoh's body is another example of heritage preservation in Islam as stated in the Quran thus: 10 verses 90–92 as follows:

> We brought the tribe of Israel across the sea, and Pharaoh and his troops pursued them out of tyranny and enmity. Then, when he was on the point of drowning, he [Pharaoh] said: "I believe that there is no god but Him in Whom the tribe of Israel believes. I am one of the Muslims"; "What, now! When previously you rebelled and were one of the corrupters? *Today we will preserve your body so you can be a Sign for people who come after you.* Surely many people are heedless of Our Signs (Holy Quran, 2021, 10:90–92).

According to Maurice Bucaille (2003), the body of the Pharaoh was recovered from the sea in 1908 and kept in an Egyptian museum. According to Hossam Mahdy (2019), the Pharaoh's preservation and recovery were not only to serve as signs and proof of past civilisations but also to drive home the importance of archaeology as a means of authenticating history, which explains cultural heritage preservation.

(v) **The Preservation of the Quran (The Islamic Holy Book)**

The Quran is regarded as the greatest heritage to the Muslim faithful. According to Afzal Iqbal (1967 p142), the words of the Quran were collected as they were revealed to the Prophet, committed to memory by the early Muslims, recorded in writing by scribes and later ordered to be compiled by the first Caliph, Abu Bakri. The originality of the holy book has been preserved for over a thousand years. The Quran has served as a link between the period before Mohammad and now. That is what heritage preservation is all about.

(vi) **Preservation of Al-Hijri pre-Islamic Site**

In 2008, the al-Hijri archaeological site became listed as a World Heritage site in Saudi Arabia. This follows long years of Saudi Arabia's preservation of the place. The significance of this is that the site is a pre-Islamic site, which was mentioned in the Quran as quoted below:

> Surely the people of al-Hijr[45] also rejected the Messengers, calling them liars. We also gave them Our Signs, yet they turned away from them. They used to hew out houses from the mountains[46] and lived in security. Then the Blast caught them in the morning and whatever they had been earning proved of no avail (Holy Quran, 2021, 15:80–84).

According to the above reference, al-Hijri site is considered to belong to people who were regarded as enemies of God because they rejected messengers sent to them during their time on earth (Mahdi, 2019). The site is not considered a place of worship in Islam, but it was preserved by Saudi Arabia, the seat of the religion of Islam, which illustrates the compatibility of Islam with heritage preservation.

14.6 Towards Countering the Narratives of Destruction

The preceding two sections demonstrated the apparent contradiction in the theory and practice of cultural heritage preservation in Islam. Section three presented clear Quranic and Prophetic directives against the making and veneration of statues and monuments. Section four highlighted evidence of the preservation of some monuments and objects as part of Islam. How do we reconcile the contradiction? One way to do this, which is lost to the terrorists, is to adopt Aziz Fachrudin's argument that a distinction exists between *timthal* (mere statutes for decoration) and *sanam* (statutes for worship). Echoing the views of reformists like Muhammad Jadul-Haq (former Grand Sheikh of Al-Azrar Mosque in Egypt) and famous Muslim thinker Muhammed Umara, Fachrudin argues that statues for decoration are allowed, provided they are not worshipped (Fachrudin, 2015). This seems to be the reason why monuments like the Pyramid and the Sphinx were allowed to stand even when Egypt was conquered by the Muslims. The same reason explains the presence of numerous ancient monuments in the Middle East and other Muslim-dominated areas, even long after the entrenchment of Islam in those areas.

These arguments can assist in educating innocent and unsuspecting potential recruits of terrorism in line with SDG 4, which centres on education. The violent attacks on heritage by extremists and terrorists operating under the banner of Islam relies on a poor level of education on Islamic tenets. The indoctrination of people into their fold also relies on ignorance and self-centred beliefs. To counter the violent campaign, there is the need to apply the principles of SDG 4 as a bloodless and effective strategy of discouraging the recruitment of community members into their folds, largely in Muslim countries. Therefore, the instruments needed to counter the narratives of violence by terrorists are the proper education of the community dwellers concerning Islamic principles on cultural heritage preservation and the promotion of the sociocultural inter-relationship between Muslims and non-Muslims. Mary Kaldor, in a conference, is reported to have said, "very often, it is the local people destroying cultural heritage, either because of their beliefs, ignorance or for economic reasons" (Elrha, 2016). This major challenge to the preservation of cultural heritage can only be addressed through the purposeful education of the people.

In the same vein, Gharib argues that the built environment is part of Islamic tradition and that its preservation is part of the necessities of Islamic law (Gharib, 2017). Furthermore, in the opinion of Yusuf al-Qaradawi (cited in Gharib, 2001), a deliberate violation of the environment also constitutes a crime against the principles of Islam. Thus, the terrorists' arguments for violence against the built environment are erected on fraudulent foundations, and the terrorists, according to Mahmud Shahbanaz (2016, p.1), are nothing but "isolated extremists dreaming up their own versions of Islam". It should be noted that SDG 11 is focused on making cities resilient and sustainable. The heritage sites are important components of the environment. Terrorism has emerged as one of the greatest threats to heritage preservation in many human environments in the contemporary world and particularly in Muslim societies. In line with SDG 11, ensuring the sustainability of cultural heritage is of urgent importance. There appears to be a concerted effort to obliterate the long-cherished heritage from Islamic cities and settlements. This is a threat to history, identity and the economy of these areas. Abdul Nasir Khan, curator of Taxila Museum at Islamabad, Pakistan, avers that "Militants are the enemies of culture" (U.S. Committee of the Blue Shield, 2009). Nothing could be truer than this statement. The synchronization of SDG 4 and 11 will help to make the environment resilient and sustainable. All efforts that assist in the preservation of the treasure of humanity must be intensified.

14.7 Conclusion

There is a strong relationship between religion and cultural heritage. The relationship is both symbiotic and adversarial. While religion has assisted in the creation of monuments and artefacts of cultural value, it has also been employed to launch vicious attacks on monuments. Therefore, the current onslaught on cultural heritage by some people laying claim to the Islamic religion represents an instance of religion playing an adversarial role in the cultural heritage discourse. Even though some texts endorse the destruction of statues and monuments, Islamic tradition also shows evidence of the preservation of some of these icons, some of which have been incorporated into the Islamic religion.

This chapter shows that terrorists operating in the Muslim world can be effectively checkmated by amplifying the evidence of heritage preservation in Islam. This will dissuade people from being recruited into the extremist fold because many believe they are joining extreme for religious purposes.

References

Al-Harbi, M (2020, May 20). In pictures: The story of a Muslim ritual with roots dating back 5,000 years. *Al-Arabiya News*. https://english.alarabiya.net/features/2018/01/06/IN-PICTURES-The-story-of-a-Muslim-ritual-that-goes-back-5-000-years

Anderton, C. H., & Carter, J. R. (2005). On rational choice theory and the study of terrorism. *Defence and Peace Economics, 16*(4), 275–282.

Asia News. (2009, November 27). Taliban trying to destroy Buddhist art from the Gandhara period. *Asia News*. http://www.asianews.it/news-en/Taliban-trying-to-destroy-Buddhist-art-from-the-Gandhara-period-16984.html

Bello, A. (2020). Countering Boko Haram insurgency: Interrogating culture destruction attempts and culture conservation efforts in North-eastern Nigeria. *Journal of College of Education, 2*(1), 12–20.

Bokkem, R. V. (2017). History in ruins: cultural heritage destruction around the world, *Perspectives on history, A newsmagazine of America Society of History*, April. Retrieved on 31 January 2021 from: https://www.historians.org/publications-and-directories/perspectives-on-history/april-2017/history-in-ruins-cultural-heritage-destruction-around-the-world

Bucaille, M. (2003). *The Bible, the Qu'ran and science: The Holy Scriptures examined in the light of modern knowledge* (7th Rev. Expended ed.). Tahrike Tarsile Qur'an.
Caplan, B. (2006). Terrorism: The relevance of the rational choice model. *The Political Economy of Terrorism, 128*(1/2), 91–107.
Eaton, R. M. (2001). Temple desecration and Indo-Muslim states. *Frontline,* 5 January. http://www.columbia.edu/itc/mealac/pritchett/00islamlinks/txt_eaton_temples2.pdf
Elrha. (2016). *The importance of communities in protecting heritage in conflict zones*. Elrha. https://wwwelrha.org/project-blog/importance-communities-protecting-heritage-conflict-zones/
Encyclopaedia. (2018). Akhenaten. https://www.encyclopedia.com/people/literature-and-arts/english-literature-1499-biographies/akhenaten#3446400012
Fachrudin, A. A. (2015, April 2). Does Islam permit the destruction of all statutes? *The Jakarta Post.* https://www.thejakartapost.com/news/2015/04/02/does-islam-permit-destruction-all-statues.html
Gharib, R. Y. (2017). Preservation of built heritage: An Islamic perspective (1). *Journal of Cultural Heritage Management and Sustainable Development, 7*(3), 1–17.
Giradi R. (1995, November 9). Sphinx nose. *Baltimore Sun.* https://www.baltimoresun.com/news/bs-xpm-1995-11-09-1995313086-story.html
Holy Quran. (2021). The Qur'an (Abdullah Yusuf Ali, Trans.). India: Kitab Bhavan Publishers.
Hubbard, B. (2015, August 20). Syrian expert who shielded Palmyra antiquities meets a grisly death at ISIS' hands. *The New York Times.* https://www.nytimes.com/2015/08/20/world/middleeast/isis-palmyra-syria-antiquities-scholar-beheaded.html
Iqbal, A. (1967). *The culture of Islam: An analysis of its earliest pattern.* Institute of Islamic Culture.
Isakhan, B., & Zarandona, J. G. (2017, June 4). Erasing history: Why Islamic States is blowing up ancient artefacts. *Conversation.* https://theconversation.com/erasing-history-why-islamic-state-is-blowing-up-ancient-artefacts-78667
Khalid, E. T. (2015, February 13). Iraq churches, mosques under ISIL attack. *Al-Shorfa.* https://en.wikipedia.org/wiki/Destruction_of_cultural_heritage_by_the_Islamic_State
Khudair, D. A. (2021, July 17). The history of the Kaaba. *Arab News.* https://www.arabnews.com/node/1895886/saudi-arabia
Kingsley, P. (2015, March 12). ISIS vandalism has Libya fearing for its cultural treasures. *The Guardian.* https://www.theguardian.com/world/2015/mar/07/isis-destroy-libya-cultural-treasures
Mahdy, H. (2019). Is conservation of cultural heritage halal? Perspectives on heritage values rooted in Arabian-Islamic values. In E. Avrami, S. Macdonald, R. Mason, & D. Meyers (Eds.), *Values in heritage management: Emerging approaches and research directions* (pp. 127–140). Getty Conservation Institute.
McCoy, T. (2019, December 8). 'Soldiers of Jesus': Armed neo-Pentecostals torment Brazil's religious minorities. *Washington Post.* https://www.washingtonpost.com/world/the_americas/soldiers-of-jesus-armed-neo-pentecostals-torment-brazils-religious-minorities/2019/12/08/fd74de6e-fff0-11e9-8501-2a7123a38c58_story.html
Nilsen, T., & Thorell, K. (2018). Introduction. In C*ultural heritage preservation: The past, the present and the future.* Halmstad University Press. http://www.diva-portal.org/smash/get/diva2:1224014/FULLTEXT01.pdf
Olanıyan, A. (2017). *Boko Haram insurgency and the spectre of female suicide bombings in Nigeria.* Paper presented at the international conference by the IAFOR International Conference on Social Sciences at the Crowne Plaza, Dubai Festival City, United Arab Emirates 26–28 February, http://papers.iafor.org/papers/iicssdubai2017/IICSSDubai2017_31174.pdf
Olaniyan, A., & Okeke-Uzodike, U. (2020). When two elephants fight: Insurgency, counter-insurgency and environmental destruction in northeastern Nigeria. *Journal of Contemporary African Studies, 38*(3), 1–18.
Shahnawaz, M. (2016, January 4). 8 rules of engagement taught by Prophet Muhammed. *Ilmfeed.* https://ilmfeed.com/8-rules-of-engagement-taught-by-the-prophet-muhammad/

Singh, R. P. B. (2008). The contestation of heritage: The enduring importance of religion. In B. Graham & P. Howard (Eds.), *Ashgate Research Companion to Heritage & Identity* (pp. 125–141). Ashgate Pub.
Smith, A. V. (2015, May 27). Timbuktu's Djinguereber mosque: A history of cities in 50 buildings, day 5. *The Guardian.* https://www.theguardian.com/cities/2015/mar/27/timbuktu-djinguereber-mosque-history-cities-buildings
Tastekin, F. (2017, January 12). Return to Aleppo: A squandered hope. *Al-Monitor.* https://www.al-monitor.com/originals/2017/01/turkey-syria-return-to-aleppo-squandering-legacy.html
Trew, B. (2016, April 6). Family describe horror at city custodian's beheading. *The New York Times.* https://www.thetimes.co.uk/article/family-describe-horror-at-killing-of-citys-custodian-5flgkj7z7
Tully, M. (2002, December 5). Tearing down the Babri Masjid. *BBC News.* http://news.bbc.co.uk/2/hi/south_asia/2528025.stm
UNESCO. (1972). *Convention concerning the protection of the world cultural and natural heritage.* UNESCO.
UNESCO. (2012). *World Heritage, humanity's gift to future.* UNESCO. https://whc.unesco.org/en/activities/487/
US Committee of the Blue Shield. (2009). *Taliban suffocate Pakistan Buddhist heritage.* https://uscbs.org/news/taliban-suffocate-pakistan-buddhist-heritage/
Wells, H. L. (2000). Korean temple burnings and vandalism: The response of the society for Buddhist-Christian Studies. *Buddhist-Christian Studies, 20*, 239–240. https://doi.org/10.1353/bcs.2000.0035
Westing, A. (2008). The impact of war on the environment. In B. S. Levy & V. W. Sidel (Eds.), *War and Public Health* (pp. 69–86). Oxford University Press.
Williams, S. (2018). *Damage to Timbuktu's cultural heritage worse than first estimated reports UNESCO mission.* https://en.unesco.org/news/damage-timbuktu%E2%80%99s-cultural-heritage-worse-first-estimated-reports-unesco-mission
Wright, T. (2012, February 11). Islamism set stage for Maldives coup. *Wall Street Journal.* https://www.wsj.com/articles/SB10001424052970203824904577215050907606404
Zeybek, O., & Arslan, M. (2017). Examining cultural heritages harmed by religious fanaticism: Sample of the Palmyra Ancient City. *Bartın Orman Fakültesi Dergisi, 19*(2), 1–10. https://doi.org/10.24011/barofd.322903

Chapter 15
The Role of Heritage in Post-War Reconciliation: Going Beyond World Heritage Sites

Lorika Hisari, Kristen Barrett-Casey, and Kalliopi Fouseki

Abstract It is widely acknowledged that reconciliation and sustainable development are processes that necessitate involvement from local, national and international actors. However, with the attention of international actors overwhelmingly focused on World Heritage sites, this chapter seeks to examine the potential consequences of the disparity in treatment between those sites on the World Heritage List and those that are not but are still significant for their local communities. Kosovo and Iraq are the two cases we use to explore the role, use and treatment of heritage in post-war recovery and reconciliation and how this is affected by World Heritage status. Through an examination of heritage as a political process, we can approach a more in-depth understanding of how heritage shapes and reshapes the politics of post-war memory, inter-community relations, and the extent to which the international community uses World Heritage in these communities to mandate their own politics of remembrance. We argue that heritage can have a "pacifying" role and contribute to peacebuilding, but this will need active, transformative actions from UNESCO which go beyond the Convention and, if possible, beyond politically influenced decision-making. This chapter seeks to fill a gap in the literature of how the local, national and international interact in the post-war environment, as well as the true impact of potential inequalities created by World Heritage.

Keywords World Heritage Status · Post-war heritage · Reconciliation · Kosovo · Iraq · World Heritage Convention

L. Hisari (✉) · K. Fouseki
UCL Institute for Sustainable Heritage, London, UK
e-mail: lorika.hisari.19@ucl.ac.uk; kalliopi.fouseki@ucl.ac.uk

K. Barrett-Casey
Compass Archaeology, London, UK
e-mail: k.b.casey8@gmail.com

M.-T. Albert et al. (eds.), *50 Years World Heritage Convention: Shared Responsibility – Conflict & Reconciliation*, Heritage Studies,
https://doi.org/10.1007/978-3-031-05660-4_15

15.1 Introduction

One of the opening observations of the 1972 UNESCO Convention is that heritage is "increasingly threatened with destruction" (UNESCO, 1972), and nowhere is this clearer than in war. Increasingly, heritage has been seen as an essential "tool" of post-war recovery and reconciliation, particularly in aiding the achievement of the United Nation's 17 Sustainable Development Goals (SDGs). Some have gone further and suggested that World Heritage status itself is an enabler of post-war redevelopment (Kalman, 2017; Saldin & Forbes, 2018; Alsalloum & Brown, 2019; Alsalloum, 2019).

The World Heritage Convention has contributed significantly to the protection of heritage sites. The recognition of values at the universal level has sublimated the understanding of heritage from a common point of view in order to build a sense of shared responsibility and respect. However, war provides a particularly challenging context for heritage as it is often the target of destruction, which indicates the political nature of heritage. Whilst heritage, and UNESCO, have been seen by researchers and practitioners alike as central to the reestablishment of societies that have undergone conflict and destruction, few have moved beyond such theories to examine what this could look like on the ground. What role can UNESCO play in post-conflict reconciliation? How can the achievements of the Convention be built on to enable sustainable redevelopment in such vulnerable and damaged societies?

To begin a conversation around these questions, this chapter undertakes a comparison of two very different post-conflict societies which have interactions with UNESCO. The first is in Kosovo, where there are four World Heritage sites. The second is Mosul, Iraq, where there is the "Revive the Spirit of Mosul" project but no World Heritage sites. Through such comparison, we assess which has the greatest potential to enable more effective post-conflict reconciliation, as well as observing potential negative consequences, and ultimately what role UNESCO can have in this process, particularly in contributing to the 2030 Sustainable Development Goals. One of our core arguments is that guidance on how to capture and deal with the socio-political, local dynamics will be particularly useful for future stakeholders working on post-war reconstruction through heritage.

This analysis is conducted in light of ongoing academic discourses on the UNESCO conventions, with particular focus on the World Heritage Convention and its future implications discussing sustainable conservation, responsibility and reconciliation in the efforts for building peace in post-war/post-conflict societies as "a permanent ideal and aspiration, as well as a right and a duty, and a foundation for sustainable development" (UNESCO, 2013).

15.2 UNESCO, World Heritage Status and the Post-War Context

There are three issues for post-conflict societies trying to demonstrate OUV. The first is the obvious difficulty in selecting which criteria are relevant. The second is the cost (in time and money) of the nomination process. The third is the potential lack of political or economic influence that such societies have over the World Heritage selection process. In cases of the destruction and reconstruction of heritage, a fate often witnessed in conflict, sites also struggle to demonstrate other requirements such as integrity, authenticity, adequate buffer zone, protection and management places. These additional requirements have, aside from the context of post-conflict societies, been the subject of decades-long debate by scholars and practitioners, with alternative, more inclusive requirements being suggested, such as "continuity" (Khalaf, 2018).

This state of affairs surrounding World Heritage status causes several issues and impediments to the achievement of reconciliation. The biggest danger is that it will ultimately lead to a two-tier system of heritage internationally, in which World Heritage status with all its economic benefits will only be achieved by those who are politically influential and capable of meeting (what are considered ill-defined and unscientific) criteria and affording the costs of nomination. The heritage of those countries too poor or too unstable to meet these demands will inevitably be left behind, with the risk being that their heritage will be forgotten and neglected. One of the biggest points of contention, then, in the context of post-conflict reconciliation, is what happens if this occurs within the same historic urban environment. Would the World Heritage status of one site negatively affect another site in the same environment? Could this lead to animosity between communities, where one's heritage is "valued" and the other's is not?

The issues which lead to conflict are necessarily political. A former UNESCO project leader summarized the potential problem of UNESCO's policy of neutrality "UNESCO is very much a political organization and less concerned with long-lasting results on the ground. [They] neglect the root causes of the conflict, which, for me, was the most important thing" (anonymous personal communication). Whilst acknowledging the great work done in developing policies more appropriate for post-conflict societies, such as the Historic Urban Landscape Framework, we must now find ways UNESCO can bridge the gap between theory and practice in these societies so they can play a meaningful role in sustainable development and reconciliation, which addresses the roots of conflict.

15.3 The Role of Heritage in a Post-War Context

There is a call for cultural heritage to be recognized "as a crucial element of the recovery process immediately following the end of an armed conflict, and not be considered a luxury to await attention later" (Stanley-Price, 2007, p. 1). The current literature demonstrates an emerging discourse of what has been identified to be one of the most critical issues of our time by asking "how can heritage contribute to peacebuilding?" as well as the belief that "heritage can provide solutions" (Walters et al., 2017, p. 1).

The use of heritage for reconciliation is a subject of debate in some contexts, especially where reconciliation is still seen "as a challenging and threatening process" because the "genuine reconciliation will mean some compromise, or at least the rehumanisation of old enemies" (Hamber & Kelly, 2005, cited in Vos, 2015, p. 726). The discourse also revolves around the distractive question of whether heritage *heals* or *hurts*. Although, as noted by Giblin (2014, p. 515), critical academic views of post-conflict heritage typically cite "a lack or failure of healing", the author suggests a reconsideration of post-conflict cultural healing with the developmental question of "how" actors symbolically engage and not "whether" they should engage in post-conflict healing heritage.

Post-war recovery is mainly driven by the international community, making the reconstruction and the project itself a complex matter that becomes part of political processes, especially when "there is a rush to reconstruct and excavate abroad" (Plets, 2017, p. 20). These processes confirm or create new identities involving organizations that might have different aims and agendas and have little chance for success if they are imposed on a society by potentially defining new (ideal) landscapes (Legnér, 2018; Higueras, 2013). Moreover, without adequate engagement, there is a risk that the local population and their heritage sites will be exposed to "further waves of violence and iconoclasm" (Isakhan & Meskell, 2019, p. 1193). This is in complete opposition to the intention of post-war reconstruction, which should, among others things, aim for "… creating a peaceful environment that will prevent a relapse into violence" (Barakat, 2007, p, 29)

Moreover, as a historic urban environment is a complex and dynamic system in constant change and transformation (Fouseki & Nicolau, 2018), any effort on the ground for reconstruction and revitalization through heritage requires synergies between stakeholders as well as periodical assessments of value change. UNESCO's 2011 *Recommendation on the Historic Urban Landscape* recognizes the complexity of the historic urban system and calls for a socio-spatial approach towards urban heritage conservation. Despite a greater focus on the "material" and "visual" aspects of the historic urban landscape (Fouseki, 2019), the Recommendation is a critical milestone for shifting heritage management from a "freezing in time" approach to integrating the complexity of its values into management frameworks.

15.4 World Heritage Status and Post-War Kosovo – The Case of the Church of the Holy Virgin of Levisha/Ljeviša in the Historic Center of Prizren

Four heritage sites of the Serbian Orthodox Church (SOC) in Kosovo were inscribed on the UNESCO World Heritage List as "Medieval Monuments in Kosovo", with Serbia as a state party, in 2004 and 2006, respectively five and seven years after the war ended. The inscribed sites were Deçan/Dečani Monastery, the Patriarchate of Peja/Peć Monastery, the Church of the Holy Virgin of Levisha/Ljeviša and the Graçanica/Gračanica Monastery (UNESCO, n.d.). In 2006, the four heritage sites were inscribed on the List of World Heritage in Danger and remain on the list until the present day (https://whc.unesco.org/en/list/724/).

A careful look at the State of Conservation reports confirms that the threatening factors have remained literally the same over the 14-year course of reporting (2007–2021) (UNESCO, n.d.). Civil unrest is one of the indicated threats in the annual reports from 2007 up to 2021. According to the Draft Decision: 44 COM 7A.33 (2021), the World Heritage Committee decided to retain the Medieval Monuments in Kosovo on the List of World Heritage in Danger until its 45th session in 2022 (https://whc.unesco.org/en/soc/4033).

During this period (2004–2021), the socio-political dynamics (in Kosovo) on the ground changed, which also affected cultural heritage conservation. After the declaration of independence in 2008, 44 heritage sites, mainly orthodox churches and monasteries of the SOC in Kosovo, as well as the historic urban core of Prizren, were granted a legal conservation and protection status through the establishment of (special) protective zones (Hisari & Fouseki, 2020). The zones are also included in strategic documents such as Kosovo's Spatial Plan 2010–2020+ and National Strategy for Cultural Heritage 2017–2027.

The Church of the Holy Virgin of Levisha/Ljeviša is located in the Historic Center of Prizren, an urban core where the multi-layered and diverse heritage buildings and sites best illustrate its long history and cultural and religious diversity since ancient times (Management Plan for the Historic Center of Prizren, 2020, pp. 1–8). This long history of diversity represents the spirit of the place today. Evidence for this includes the three main religious heritage buildings in the Historic Center (Sinan Pasha Mosque, Saint George Cathedral of the SOC in Kosovo and Our Lady of Perpetual Succour Catholic Cathedral) that are open and functional for believers and visitors, and the city's ethnic composition (Agjencia e Statistikave të Kosovës, 2011). The Historic Center of Prizren is a delineated area (Fig. 15.1) protected by a specific law adopted in 2012 (Law No. 04/L-066 on Historic Center of Prizren) (Hisari & Fouseki, 2020). The natural and cultural elements of the historic area are enshrined in the urban fabric, with over 100 cultural monuments (Law No. 04/L-066 on Historic Center of Prizren, Appendix II) creating a distinct urban heritage.

It includes the Church of the Holy Virgin of Levisha/Ljeviša. The Management Plan for the Historic Center of Prizren (2021–2030) has been drafted with the support of the Ministry of Culture, Youth and Sports and Municipality of Prizren, incorporating provisions of the Law on Historic Center of Prizren (2012).

The UNESCO listed heritage building of the Church of the Holy Virgin of Levisha/Ljeviša is located on the north-western side of the Historic Center and has a delineated buffer zone for its "effective protection" (UNESCO, n.d.) , according to the *Operational Guidelines* (UNESCO, 2019, p. 30) (Fig. 15.1). This is a living residential area with traditional houses and other heritage buildings in the vicinity of the Levisha/Ljeviša Church. A road beside the monument serves as a connection to other parts of the city, thus making it an integrated part of the Historic Center and

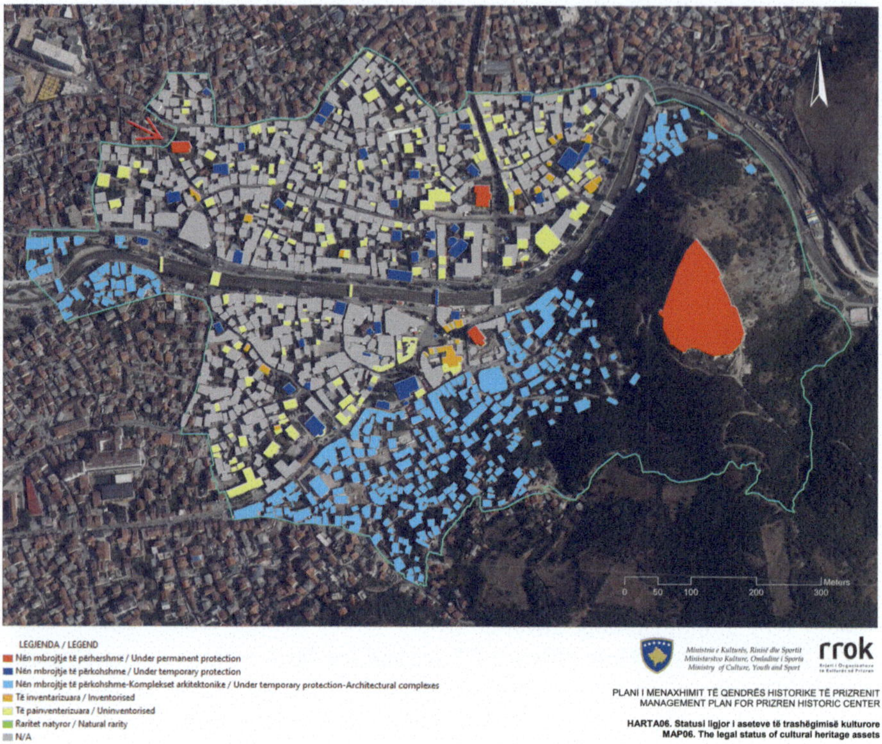

Fig. 15.1 Protective zone of the Historic Center of Prizren with marked cultural heritage buildings under legal protection, including the WHS Church of the Holy Virgin of Levisha/Ljeviša (northwestern side, in red). (Note. Map, by E. Toska, 2017. Source: Management Plan for the Historic City of Prizren (2021–2030), Ministry of Culture, Youth and Sports, Republic of Kosovo, 2020, p. 22 (WHS Church of the Holy Virgin of Levisha/Ljeviša, indicated with an arrow by L. Hisari)

the wider urban area. Current observations indicate that people and vehicles pass the monument on a daily basis. There is a Kosovo Police booth in front of the monument for its safeguarding (Fig. 15.2). More importantly, the Church of Levisha/Ljeviša provides religious services on specific days (TV Klan Opinion, 2020).

In reality, the buffer zone of the World Heritage site of the Church of Levisha/Ljeviša now presents a fraction of a wider protective area of the whole historic environment with potential impacts on the effective conservation and management of the site. As a consequence, there might also be socio-political implications in the future. For instance, in the high-level dialogue between Kosovo and Serbia, which the European Union (EU) started to facilitate in 2011, as an ongoing political and technical process for the normalization of relations (Strategic Communications EEAS, 2020) cultural heritage is expected to be on the negotiating table. Further alternatives that include "territorial solutions" in Kosovo, such as enclaves, extraterritoriality or dual sovereignty – condominium, are considered as possible options (Surlić & Novaković, 2020). Are these suggestions well-informed by the reality on

Fig. 15.2 Church of the Holy Virgin of Levisha/Ljeviša and the surrounding area. (*Note*. Image by L. Hisari, 2021)

the ground? How will they impact, for example, a multi-ethnic historic urban environment, such as the Historic Center of Prizren? Will they contribute to inter-community post-war reconciliation and peace or induce further segregation and new rounds of conflict? These are the questions that must be considered and discussed.

15.5 UNESCO Reconstruction Projects and the Socio-Political Dynamics of Reconciliation: The Case of Revive the Spirit of Mosul

UNESCO's "Revive the Spirit of Mosul" project in Iraq, whilst contemporary, ongoing and incomplete, represents several issues which have accompanied UNESCO's similar projects previously. Although the project has the potential to not only reconstruct what has been lost but also "build back better" through the management of change and continuity (Khalaf, 2020), we want to highlight the necessity of a people-centred, if not people-led, approach, the need for continuous dialogue with the local population, and the consequences of neglecting these elements, which can already be seen in the project.

Mosul certainly has international importance; it was here at the al-Nouri mosque in 2014 that the Islamic State declared their Caliphate, and through the destruction of the same mosque three years later that they were declared to be defeated. However, UNESCO's competition to design the reconstruction of al-Nouri mosque has been met with outrage and upset, including accusations that Iraqis have been almost systematically excluded from the process. The winning Egyptian firm has chosen a modernist, cubist style, along with new changes in purpose that include a mixed-sex school and a public garden. This has been one of the main points of contention, with some saying this design belongs more in the UAE than in the architecturally unique Mosul. Professor Ihsan Fethi has been one of the most outspoken critics. In his open letter, he describes this competition as "un-necessary and flawed… without any real consultation with all members of the"Technical Committee "which was treated more like a formality rather than a real partner in taking important decisions" (Cambridge Heritage Research Centre Bulletin, May 2021, p. 12).

Revive the Spirit of Mosul is concentrated on very few landmark sites, namely, Al-Nouri Mosque and its Al-Hadba Minaret, the Nabi Younis Shrine, Al Saa'a Church and Al Tahera Church. Whilst UNESCO and the EU want to reconstruct houses and schools in the Old City of Mosul and Basra, interviewed heritage professionals who have worked on the project highlighted that focusing overwhelmingly on landmark sites risks obscuring other sites that are important to different communities in Mosul.

The potential risk is that this neglect and prioritization could inadvertently prolong the conflict through heritage management, and, as such, we should reflect not only on the nature of the conflict and its causes but also on how much this is caused by the dynamics of international intervention. Other participants spoke of how projects in Syria and Yemen have been affected, if not ceased, because of the attention the landmark sites of Mosul have attracted. With $100 million, "Revive the Spirit of Mosul" has the biggest budget that UNESCO has ever managed, with a large amount of funding received from the UAE. In interviews, participants emphasized how much decision-making power the funders had over which sites were chosen and how the politics of post-conflict societies generally make heritage management difficult. As one anonymous participant described:

> There is a lot of work in Iraq, most of which was neglecting Kurdistan, as well as the south of Iraq. This is why five out of the 20 million Euros of the European Union project on Mosul are addressing the restoration of historic houses in the Old City of Basra. We tried our best to cover the different areas of the country. The south of Iraq is in turmoil, to the extent that the government is resigning because of corruption or the shortage of basic services. So, we always try to look at that part of the country as well. But it is obvious that the interest and focus of UNESCO is on Mosul.

As such, one of the neglected problems in post-conflict heritage research has been *who* has the most influence over projects and, therefore, on the future management, memory and identity issues that will shape these societies for generations, with the potential upset and animosity this could cause.

However, from a theoretical perspective, such projects present positive opportunities for redevelopment and reconciliation. Greater community engagement and even leadership can be harnessed by focusing on the issues of *who*, which would allow these projects to capture the new values and meanings generated through the destruction and reconstruction of heritage and provide meaning-making and purpose, and, in turn, advancing the process of reconciliation. One method, suggested by one of our interview participants, was to pay for and teach reconstruction skills to local people through projects, which they could then reinvest in their own properties. A strategy like this should be explored further to integrate heritage into broader urban development.

15.6 Discussion

The World Heritage Convention has contributed significantly to the protection of heritage sites. The recognition of values at the universal level has sublimated the understanding of heritage from a common point of view in order to build a sense of shared responsibility and respect. However, war provides a particularly challenging context for heritage as it is often the target of destruction, which indicates the political nature of heritage. For instance, whilst memory and heritage are important for individual and group identity and for their sense of belonging (Apaydin, 2020), the approach of international organizations, such as UNESCO, "honors political agreements where actors are unquestionable and uncontested, while at the same time it dismisses the memories arising from conflicts, which remain unrecognized unless those memories are included in national narratives" (Jaramillo, 2015, p. 200). Moreover, heritage in a post-war period can be easily used in a misleading or destructive way, especially when the involved international actors fail to adequately engage with the local population. Furthermore, this lack of adequate engagement risks the local population and their heritage sites being exposed to further waves of violence, for example, when heritage is utilized to support state power and nationalism or when political involvement acts as "strategic manipulation" corresponding to various agendas and aims (Isakhan & Meskell, 2019; Hadžić & Eaton, 2017; Plets, 2017, p. 22).

The values are firstly associated with local communities, the people who use, live in or by the respective heritage site and are most commonly the parties that are directly affected by the consequences of war or conflict, and who need reconciliation for sustainable recovery and peacebuilding. Chapter VI of the *Operational Guidelines* (UNESCO, 2019) particularly mentions the component of communities. However, the question remains, how much, if at all, is this important element of the Convention implemented in practice? The challenges of such research in complex post-war circumstances must be recognized; however, the lack of research-based decision-making and its impacts on the ground should also be emphasized. The two cases – the "freezing" of threatening factors that keep four World Heritage sites on the Danger List despite changes on the ground in Kosovo and the lack of meaningful engagement with stakeholders, particularly the local population, in a post-war reconstruction in Iraq – demonstrate the underemphasized importance of an in-depth understanding of the local context in a post-war period. Thus, we question how accurately UNESCO and the status of "World Heritage in Danger" can capture these vital elements on the ground? Can and should UNESCO have a role in reconciliation beyond "high politics"? What could this look like?

As Nikander & Zirl (2016) point out, "it will be crucial that the international community recognizes the equal and universal value of the heritage of the different communities in Kosovo, for example through the inclusion of sites representing the cultural heritage of all communities on the UNESCO World Heritage list. Such international recognition could well contribute to the communities taking pride in Kosovo's cultural heritage and eventually losing sight of to whom it belongs" (Nikander & Zirl, 2016).

During war, there are dramatic changes at a rapid pace, which make the role of heritage in post-war conservation and reconstruction even more critical. Given the pace of change and transformation, and the nature of the Convention, which may imply a "freezing" in time for heritage sites, there may be some challenges for those planning to work with heritage for post-war reconstruction. We argue that heritage can have a "pacifying" role (contributing to peacebuilding), but this would need active, transformative actions from UNESCO, which move beyond the Convention, and, moreover, beyond politically influenced decision-making.

15.7 Conclusions

The 1972 Convention must be celebrated for preserving the sanctity of the post-war universalist ideas upon which UNESCO was founded and which have been threatened by the targeting and destruction of heritage ever since. However, to continue to play an effective role in reconciliation, which can help achieve the 2030 Agenda, UNESCO must bridge the gap between its policy and practice of community engagement.

As an intergovernmental organization, UNESCO is by its very nature limited in its ability to engage with local communities. Further, the fast-changing nature of post-conflict societies presents unique challenges to working on the ground, not least of which is the danger and instability of these societies limiting the effectiveness of operations.

In both cases, we can see the realities of these struggles and the theoretical limitations of capturing the rapid new dynamics through UNESCO's conventions, policies and frameworks. For example, the ongoing high-level political dialogue between Kosovo and Serbia for the normalization of relations, which includes heritage as a topic, may need to be informed by the reality on the ground as it is directly related to sustainable conservation and, moreover, to sustainable peacebuilding. This is not necessarily the case with the ongoing efforts, and this affects the lives of people surrounding these heritage sites in multi-ethnic historic urban environments. With Mosul, there is heavy international involvement, not only from the UN and the EU but also countries like the UAE, Egypt and Saudi Arabia, whose level of influence has led some to accuse them of systematically excluding Iraqis from the "Revive the Spirit of Mosul" project.

An immediate question that arises from this chapter is *how* heritage plays a role beyond policy in reconciliation. It is not inevitable that a World Heritage status or a UNESCO project can enable such a process, nor that it will be sustainable. The idea of sustainable reconciliation must be built from the bottom up rather than imposed from the top down, and the change of dynamics must be recognized. As such, one recommendation is the establishment of open dialogues led by the people who have been directly affected by conflict, where communities can work together. Instead of "keeping" the WHS on the Danger List without considering the changes on the ground or planning and implementing the projects, a slight reassignment of role could enable UNESCO to mediate these conversations, providing expert advice and resources to community-led projects and processes.

References

Agjencia e Statistikave të Kosovës. (2011). *Regjistrimi i Popullsisë, Ekonomive Familjare dhe Banesave në Kosovë 2011. Popullsia sipas gjinisë, etnicitetit dhe vendbanimit.* ASK. [online]. https://ask.rks-gov.net/media/1613/popullsia-sipas-gjinis%C3%AB-etnicitetit-dhe-vendbanimit.pdf

Alsalloum, A. (2019). Rebuilding and reconciliation in Old Aleppo: The Historic Urban Landscape perspectives. In R. A. Pereira & F. Bandarin (Eds.), *Reshaping urban conservation* (pp. 57–77). Springer.

Alsalloum, A., & Brown, A. (2019). Towards a heritage-led sustainable post-conflict reconciliation: A policy-led perspective. *Sustainability, 11*(6), 1686.

Apaydin, V. (2020). *Critical perspectives on cultural memory and heritage: Construction, transformation and destruction.* UCL Press.

Barakat, S. (2007). Postwar reconstruction and the recovery of cultural heritage: Critical lessons from the last fifteen years. In N. Stanley-Price (Ed.), *Cultural heritage in postwar recovery: Papers from the ICCROM Forum held on October 4–6, 2005. ICCROM Conservation Studies* (pp. 26–39).ICCROM.

Cambridge Heritage Research Centre. (2021, 4 May). *Bulletin: Cambridge Heritage Research Centre.* https://www.heritage.arch.cam.ac.uk/documents/2021-5-4-bulletin

Fouseki, K. (2019). Re-conceptualizing 'historic urban landscapes'. In V. Bharne & S. Trudi (Eds.), *Routledge companion to global heritage conservation* (pp. 26–38). Routledge.

Fouseki, K., & Nicolau, M. (2018). Urban heritage dynamics in 'heritage-led regeneration': Towards a sustainable lifestyles approach. *The Historic Environment: Policy & Practice, 9*(3–4), 229–248.

Giblin, J. D. (2014). Post-conflict heritage: Symbolic healing and cultural renewal. *International Journal of Heritage Studies, 20*(5), 500–518. http://www-tandfonline-com.

Hadžić, L., & Eaton, J. (2017). Rebuilding the broken: Regional restoration camps as a meeting platform in the Western Balkans. In D. Walters, D. Laven, & P. Davis (Eds.), *Heritage and peacebuilding* (pp. 205–220). Boydell & Brewer.

Higueras, A. (2013). Aid and reconstruction of heritage in the context of post-conflict societies. *Archaeologies.*, *9*(1), 91–105.

Hisari, L., & Fouseki, K. (2020). Post-war cultural heritage preservation in Kosovo: Rethinking the implementation of Ahtisaari Plan Annex V. *Heritage.*, *3*(1), 98–115.

Isakhan, B., & Meskell, L. (2019). UNESCO's project to "Revive the Spirit of Mosul": Iraqi and Syrian opinion on heritage reconstruction after the Islamic State. *International Journal of Heritage Studies, 25*(11), 1189–1204.

Jaramillo, C. (2015). Memory and transitional justice: Toward a new platform for cultural heritage in post-war Cyprus. *Santander Art and Culture Law Review, 1*(2), 199–220.

Kalman, H. (2017). Destruction, mitigation, and reconciliation of cultural heritage. *International Journal of Heritage Studies, 23*(6), 538–555.

Khalaf, R. W. (2018). A proposal to apply the historic urban landscape approach to reconstruction in the World Heritage context. *The Historic Environment: Policy & Practice, 9*(1), 39–52.

Khalaf, R. W. (2020). Cultural heritage reconstruction after armed conflict: Continuity, change, and sustainability. *The Historic Environment: Policy & Practice, 11*(1), 4–20.

Law No. 04/L-066 on Historic Center of Prizren. (2012). https://gzk.rks-gov.net/ActDetail.aspx?ActID=2836

Legnér, M. (2018). Post-conflict reconstruction and the heritage process. *Journal of Architectural Conservation, 24*(2), 78–90.

Ministry of Culture, Youth and Sports. (2020). *Management Plan for the Historic Center of Prizren (2021–2030).* https://www.mkrs-ks.org/repository/docs/PMQHP-eng.pdf, 1.1, 1–8.

Nikander, P., & Zirl, V. (2016, October 5). Building blocks for peace: Cultural heritage protection in Kosovo. *Organization for Security and Co-operation in Europe.* https://www.osce.org/magazine/271966

Plets, G. (2017). Violins and trowels for Palmyra: Post-conflict heritage politics. *Anthropology Today., 33*(4), 18–22.

Saldin, M., & Forbes, C. (2018). The role of cultural heritage in building peace and reconciliation: Conflict, disaster and the future of heritage. *Historic Environment, 30*(3), 76–83.

Stanley-Price, N. (2007). The thread of continuity: Cultural heritage in postwar recovery. In N. Stanley-Price. (Ed.), *Cultural heritage in postwar recovery: Papers from the ICCROM Forum held on October 4–6, 2005. ICCROM Conservation Studies.* ICCROM

Strategic Communications EEAS. (2020, July 15). Dialogue between Belgrade and Pristina. *European Union External Action Service.* https://eeas.europa.eu/headquarters/headquarters-homepage_en/349/Dialogue%20between%20Belgrade%20and%20Pristina

Surlić, S. & Novaković, I. (2020). *Serbian cultural and religious heritage in Kosovo from Ahtisaari's special zones to the final status.* Working Group of the NCEU for Chapter 35. Institute for Territorial Economic Development. http://www.lokalnirazvoj.org/upload/Publication/Documents/2020_06/2020_Serbian_Cultural_and_Religious_Heritage.pdf

TV Klan Opinion. (2020, September 30). *Opinion- Opinionzbulon Prizrenin ndryshe.* [Video]. Youtube. https://www.youtube.com/watch?v=tiTa5ZTyG4U

UNESCO. (1972). *Convention concerning the protection of the world cultural and natural heritage.* UNESCO.

UNESCO. (2011). *Recommendation on the historic urban landscape, including a glossary of definitions.* UNESCO. https://whc.unesco.org/uploads/activities/documents/activity-638-98.pdf

UNESCO. (2013). High-level Discussion 2 – Background note. *Peace and reconciliation: How culture makes the difference.* UNESCO. Peace and Reconciliation | United Nations Educational, Scientific and Cultural Organization (unesco.org)

UNESCO. (2019). *Operational guidelines for the implementation of the World Heritage Convention.* UNESCO.

UNESCO. (n.d.). *Medieval Monuments in Kosovo*. https://whc.unesco.org/en/list/724/documents/

Vos, C. (2015). Debating the reconciliatory use of heritage. European post-monumentalism versus regional national-monumentalism. *International Journal of Heritage Studies. Vol., 21*(7), 716–733. [online] Available from: Taylor & Francis Online. http://www-tandfonline-com [Accessed 18 February 2020]. p: 726.

Walters, D., Laven, D., & Davis, P. (Eds.). (2017). *Heritage and peacebuilding*. Boydell & Brewer Press.

Chapter 16
Fighting Terrorist Attacks Against World Heritage – An Integrated Approach

Sabine von Schorlemer

Abstract This contribution will examine intentional acts of terrorist groups and organised networks directed against cultural heritage as a challenge for international law, e.g., by creating grey zones. In particular, it will be asked to what extent criminal law enforcement can be identified as a *missing link* in the system of the legal protection of cultural property, including World Heritage. It is seen as necessary to strengthen criminal sanctions for possession and sale of illegally trafficked cultural objects. In that respect, it will be argued, the Nicosia Convention on Offences Relating to Cultural Property, adopted by the Council of Europe in 2017, may give new impetus to criminal law as a tool in the fight against offences against cultural property, helping to reduce lawless areas. Cultural heritage protection, it is emphasised, requires a robust, legally integrated approach, including criminal prosecution for plundering, smuggling, and destruction.

Keywords Organised networks · Terrorism · Illicit trade · Criminal law · Nicosia Convention · Complementarity

16.1 Introduction

As the EU's 2020 Security Union Strategy emphasised, "[t]rafficking in cultural goods has [...] become one of the most lucrative criminal activities, a source of funding for terrorists as well as organised crime" (European Commission, 2020).

S. von Schorlemer (✉)
Saxon State Minister for Higher Education, Research and the Fine Arts (retd.); Chair of International Law, European Union Law and International Relations; UNESCO Chair in International Relations, Technische Universität Dresden, Dresden, Germany
e-mail: sabine.schorlemer@tu-dresden.de

M.-T. Albert et al. (eds.), *50 Years World Heritage Convention: Shared Responsibility – Conflict & Reconciliation*, Heritage Studies,
https://doi.org/10.1007/978-3-031-05660-4_16

Given the fact that the so-called Islamic State of Iraq and Syria (ISIS/Daesh) set up a system of illicit trafficking in art and antiquities as a source of steady income, we are facing today interrelated terrorist crimes – plundering and smuggling on the one hand and destruction on the other. "[V]arious episodes prove that artefacts have been stolen and smuggled abroad and the revenues used to buy weapons, and that cultural heritage has been destroyed or desecrated in order to weaken the resistance of the enemies through the mortification and humiliation of their culture" (Committee on Offences Relating to Cultural Property, 2016, p. 3). This leads to a "vicious circle" with new attacks and new loss of cultural property (Committee on Offences Relating to Cultural Property, 2016, p. 3; Brown, 2017).

Therefore, it will not suffice to protect World Heritage (WH) sites against physical attacks. As the 1972 Convention Concerning the Protection of the World Cultural and Natural Heritage (1972) does not deal with criminal law sanctions and, generally, prosecutions and convictions for cultural objects-related crimes are weak (Munnelly, 2021, p. 3), it will be asked to what extent stronger criminal law enforcement can be identified as a *missing link* in the system of legal protection of cultural heritage.

Cultural heritage protection, it will be argued, needs to be approached in a broader legal framework designated to combat serious international crimes – both destruction and trafficking – and to provide appropriate law enforcement for both of them.

16.2 Intentional Attacks by Terrorist Groups: Avoiding Lawless Areas

Iconic WH is at risk, especially when terrorist groups gain influence on the ground, be it in peacetime or in wartime. Intentional acts of terrorist groups and organised networks[1] directed against cultural heritage present a challenge for international law, particularly due to the existence of grey zones.

16.2.1 The Relevance of International Humanitarian Law (IHL)

Generally, armed conflict and terrorist activities threaten the integrity of cultural objects and resources. *In times of armed conflict,* cultural property and its surroundings have to be respected by non-state armed groups, provided that the State where the WH is located has ratified the IHL treaties. In these cases, non-state actors have

[1] See the United Nations Security Council Consolidated List at: https://www.un.org/securitycouncil/content/un-sc-consolidated-list

to refrain from using cultural property in ways which might destroy or damage the property (Convention for the Protection of Cultural Property in Armed Conflict, 1954, arts. 4 and 19; Protocol Additional to the Geneva Conventions of 12 August 1949, and Relating to the Protection of Victims of Non-International Armed Conflicts (Protocol II), 1977, arts. 1 and 16).

In cases of serious violations of the Second Protocol to the Hague Convention, 1954, as of 26 March 1999 (1999, hereinafter SP; for details; von Schorlemer, 2004), States, in whose territory the alleged offender is present, are obliged to either extradite or submit the person to its competent authorities (SP, arts. 15, 17, 18). Further, mutual legal assistance in connection with investigations or criminal or extradition proceedings is required (SP, art. 19). However, being ratified by only 84 State Parties (as of November 2021), the Second Protocol is not universally accepted.

16.2.2 The UNESCO Declaration Concerning the Intentional Destruction of Cultural Heritage

In the case of *peacetime destruction*, IHL does *not* apply, and, consequently, alleged offenders may not be prosecuted for war crimes. This may create a legal loophole: For example, when the Taliban destroyed the Buddha statues in Afghanistan in 2001, they were in control of large parts of the state's territory – no armed conflict was taking place.

Against these terrorist acts, the UNESCO General Conference adopted the Declaration concerning the Intentional Destruction of Cultural Heritage (UNESCO, 2003) at its 32nd session on 17 October 2003. Applying to tangible and intangible, movable and immovable objects, the Declaration emphasises the international community's commitment to fight against intentional destruction in any form (including terrorist acts).

States are called upon to "take all appropriate measures [...] to establish jurisdiction over, and provide effective criminal sanctions against, those persons who commit, or order to be committed, acts of intentional destruction of cultural heritage of great importance for humanity" (UNESCO, 2003, para. VII). In addition, States are encouraged "to take all appropriate measures, in accordance with international law, to cooperate with other States concerned with a view to establishing jurisdiction over, and providing effective criminal sanctions against, those persons who have committed or have ordered to be committed acts [...] and who are found present on its territory, *regardless of their nationality and the place where such act occurred*" (e.g. UNESCO, 2003, para. VIII (2), emphasis added).

Thus, the 2003 Declaration is the first UNESCO standard-setting document that aims to establish universal jurisdiction for cultural heritage crimes and to strengthen cooperation in the field of criminal sanctions. Still, the Declaration is a non-binding document, which has had limited effect (Lenzerini, 2003).

16.2.3 The Role of Criminal Law Prosecution

In the light of ISIS/Daesh terrorist acts, Security Council (SC) Resolution 2347 from 24 March 2017 reminded UN member states that they are obliged to bring perpetrators to justice whenever the latter direct "unlawful attacks against sites and buildings dedicated to religion, education, art, science or charitable purposes, or historic monuments", which may constitute a war crime (UN Security Council, 2017, para. 4). Furthermore, Resolution 2347 emphasised that illicit trafficking of cultural property may constitute a serious crime (Resolution 2347, para. 9).

Unequivocally, criminal law prosecutions are powerful legal instruments for fighting terrorist acts against WH, theft, and illegal trafficking.

Still, *international* criminal jurisdiction is but complementary: As the Rome Statute sets forth, the International Criminal Court in The Hague will consider a case admissible only when a State which has jurisdiction over it "is unwilling or unable genuinely to carry out the investigation or prosecution" (Rome Statute, 1998, art. 17, para. 1(a)). Legal requirements – such as ratification of the Rome Statute, clear evidence, and state cooperation with the Court – are often difficult to achieve.

Hence, *national* criminal law prosecution of deliberate offences against cultural property and international cooperation for that purpose remain of utmost importance. However, domestic jurisdictions often fail to prosecute and punish individual perpetrators (Wierczyńska & Jakubowski, 2020), not least because of the transnational character of the crimes.

16.2.4 Grey Zones and Terrorist Acts: The Transnational Character of Crimes

The Council of Europe (CoE) warned: "The black market trade in antiquities, art and artefacts by unscrupulous dealers who do not care about the illicit provenance of such cultural objects can end up funding corruption, terrorism, violence and other crimes." (Council of Europe, 2017b, para. 3; Mottese, 2018; Campbell, 2013). Furthermore, referring to INTERPOL, the UN Secretary-General explained that "as a result of the transnational nature of crimes related to culture, the involvement of several national jurisdictions leads to gaps and loopholes which are exploited by individuals and organized groups" (Guterres, 2017, p. 12).

Legal differences in national jurisdictions are well known by the art trade and used extensively. As experts outline: "The transnational nature of illicit activities is because experienced thieves and smugglers are well aware of the legal differences between countries and seek to exploit gaps or weaknesses in the national laws to increase profits from their wrongdoing and lower the chances of being caught." (Oñate, 2018, p. 240; see also Ulph, 2019; Munnelly, 2021, p. 6).

For example, a precious piece illegally excavated in Syria can be bought legally at an auction in Germany, thus acquiring a "spotless provenance" that helps it to be resold: Therefore "awareness that legal loopholes of this kind quickly turn into dangerous security threads [sic] is badly needed" (Groß, 2018, p. 52). Moreover, the amount of information given about when and where art and antiquities are transferred is "shockingly low. Dealers and buyers still prefer confidentiality and concealment" (Fincham, 2019, p. 334; Kulturstiftung der Länder, 2018).

Against this backdrop, already in the 1980s, the CoE had drafted the European Convention on Offences relating to Cultural Property (1985), the so-called Delphi Convention. This treaty was opened for signature on 23 June 1985. Unfortunately, however, it never entered into force. Apparently, the Delphi Convention failed due to objections regarding practical enforcement and because other organisations, "especially UNESCO, UNODC, UNIDROIT, and the EU", were seen as more efficient in combatting illegal art trade (Bieczyński 2017, p. 262).

But efforts continued.

16.3 The Nicosia Convention on Offences Relating to Cultural Property (2017): An Integrated Approach

The Nicosia Convention on Offences Relating to Cultural Property (Council of Europe, 2017a, hereinafter NC) was adopted on 3 May 2017 by the CoE Committee of Ministers, with ratification being possible since May 2017.[2] The elaboration of this new legal instrument was carried out in close cooperation with UNESCO, INTERPOL, UNODC, OSCE, and the EU.

As will be argued, the relevance of the Nicosia Convention for WH is huge.

Germany has been an active member in the drafting process of the Nicosia Convention. The First Meeting of the Committee on Offences Relating to Cultural Property, which was responsible for preparing the draft Convention (PC-IBC), taking place from 31 May to 1 June 2016, unanimously elected Hans-Holger Herrnfeld (Germany) as the Chair of the Committee.

16.3.1 Scope and Objectives

The Nicosia Convention focuses on transnational crimes, expressing concern that terrorist groups are deliberately destroying cultural heritage and using the illicit trade of cultural property as a source of financing their activities (NC, preamble, para. 7). In doing so, the treaty takes up a two-pronged approach (Bieczyński, 2017), addressing the increasing number of illegal acts connected to cultural property and "the concomitant destruction of the world's cultural heritage" (Blake, 2020, p. 176).

[2] For ratifications see: https://www.coe.int/de/web/conventions/full-list/-/conventions/treaty/221/signatures?p_auth=mm3sfMXa

The Convention's primary purposes are to "prevent and combat the destruction of, damage to, and trafficking of cultural property by providing for the criminalisation of certain acts." Furthermore, it aims to strengthen "crime prevention" and the criminal justice response to all criminal offences relating to cultural property (NC, art. 1).

The Convention applies to tangible movable or immovable heritage, designated or listed in accordance with the 1970 Convention on the Means of Prohibiting and Preventing the Illicit Import, Export and Transfer of Ownership of Cultural Property and the 1972 UNESCO Convention concerning the Protection of the World Cultural and Natural Heritage (NC, art. 2, para. 2 (a) and (b)). The adoption of definitions from UNESCO treaty law was criticised by some, as it may bring disadvantages for yet undiscovered objects (Gottlieb, 2020). However, when archaeological objects are unearthed, they may fall well within the scope of application of the Nicosia Convention, provided they are located in designated areas.

16.3.2 Value-Added: Criminalising Offences

Each Party of the Nicosia Convention shall ensure that not only "unlawful destruction" but also "damaging of movable or immovable cultural property" constitute a criminal offence (NC, art. 10, para. 1 (a)). As the Explanatory Report sets forth, this article has been drafted "mindful of the egregious demolitions at major cultural sites by terrorist groups such as those in Mali, Syria and Iraq" (Council of Europe, 2017b, para. 65).

Another obligation to criminalise under domestic law concerns intentional, "unlawful removal, in whole or in part, of any elements from movable or immovable cultural property" when designed for trafficking (NC, art. 10, para. 1 (b)). This may also include parts of WH, for example, "statues, frescoes and mosaics" (Council of Europe, 2017b, para. 68), when being cut up, dismembered, and seriously damaged by removing objects, e.g. from archaeological sites or cultural landscapes (Palmyra, Ancient Villages of Northern Syria, Samarra Archaeological City, Erbil Citadel, Babylon, etc.).

Other state obligations concern, inter alia, theft, unlawful excavation, and removal of movable cultural property; illegal exportation and illegal importation of cultural property (NC, arts. 3–6), but also the acquisition of stolen, illegally excavated, or illegally exported cultural property; the placing on the market of such property and the falsification of documents (NC, arts. 7–9).

The centrepiece of the treaty requires State Parties to establish jurisdiction over all of the offences mentioned above (NC, art. 12) which includes providing the judiciary in their countries with the necessary competencies to try cases of referred criminal offences and pronounce judgment on infringements, irrespective of whether these are committed on their territory, ships, aircraft, or by one of their nationals.

16.3.3 *Criminal Sanctions and the Art Trade*

Consensus is growing that it is necessary to strengthen criminal sanctions for possession and sale of illegally trafficked cultural objects. In the understanding of the CoE, the use of criminal sanctions is to be seen as "a means of last resort", in line with the principle of *ultima ratio* (Council of Europe, 2017b, para. 111). More specifically, sanctions have to be "effective, proportionate and dissuasive" and shall "take into account the seriousness of the offence" (NC, art. 14, para. 1 and 2).

In determining the sanction, the fact that the offences were committed either by persons "abusing the trust placed in them in their capacity as professionals" (e.g. "restorers, conservators, curators, auctioneers and dealers" (Council of Europe, 2017b, para. 95)) or by "a public official tasked with the conservation or the protection of movable or immovable cultural property" may be taken into consideration as aggravating circumstances (NC, art. 15 (a) and (b)). The same applies when the offence was committed within the framework of a criminal organisation or when the perpetrator is recidivistic (NC, art. 15 (c) and (d)).

Importantly, each Party shall ensure that legal entities, e.g. companies and associations or auction houses, are liable for criminal actions performed for their benefit by someone holding a "leading position" within the legal person (NC, art. 13, para. 1). For example, when a person, having the power to represent the legal person or to make decisions on behalf of it (NC, art. 13, para. 1 (a) and (b)), commits a criminal offence, then the liability of the legal person may be – dependent on the Party's decision – criminal, civil, or administrative (NC, art. 13, para. 3).

Sanctions against legal persons include, among others, temporarily or permanently disqualifying them from commercial activities or placing them under judicial supervision. In addition, the possibility of seizure and confiscation of the proceeds of the offences, or property whose value corresponds to such proceeds, are relevant measures as well (NC, art. 14, para. 3; Council of Europe, 2017b, para. 91). In this respect, the Convention vastly improves the implementation of existing cultural heritage law, making legal protection of cultural property more efficient at the same time.

16.3.4 *International Cooperation and Law Enforcement*

The Nicosia Convention fosters international cooperation to fight crimes that are destroying the World's Cultural Heritage and will be able to strengthen inter-organisational cooperation between the CoE and UNESCO. Effective cross-border cooperation is foreseen in investigating, prosecuting, and sentencing persons suspected of committing offences listed in the Convention.

As regards terrorist attacks against WH, the relevance of Art. 21 should be highlighted: Each Party shall cooperate "to the widest extent possible" to prevent and fight the "intentional destruction of, damage to, and trafficking of cultural property"

(NC, art. 21). In particular, parties should "facilitate co-operation" to protect and preserve cultural property "in times of instability or conflict" (NC, art. 21 (c)).

Action may include the establishment of safe havens as well, where "foreign movable cultural property endangered by such situations of instability or conflict can be safely stored, conserved and protected" (Council of Europe, 2017b, para. 126).

Additionally, the exchange of information regarding identification, seizure, and confiscation of cultural property, and contributions to international data collections (such as the Interpol Database on Stolen Works of Art) by "interconnecting national inventories or databases" are valuable instruments (NC, art. 21 (b)) for the work of police and law enforcement authorities.

16.3.5 Awareness-Raising

In general, deterrence to commit cultural heritage crimes can be enhanced by a better understanding of the profound direct and indirect damage certain acts cause to cultural property. We may safely assume that the more people consider cultural property as an integral part of national, European, and universal heritage, "the more difficult it will be for someone to commit an offence against them and the easier it will be to provide an effective means of protection" (Mottese, 2018, p. 133).

Awareness is essential: As Lazare Eloundou Assomo (Director of Culture and Emergencies, UNESCO) stressed at the Berlin Conference on Cultural Heritage and Multilateralism in November 2020, awareness-raising among the public as well as educational measures regarding younger people are crucial, for the "absence of ethics" is a fundamental cause of cultural heritage crimes (Eloundou Assomo, 2020). Along these lines, it is to be welcomed that the NC promotes awareness-raising campaigns by States addressed to the general public about protecting cultural property and the dangers posed by the crimes against it (NC, art. 20 (g)).

16.3.6 Relevance of Monitoring

Generally, monitoring is a significant element of international legal cooperation.

The Committee of the Parties foreseen by the NC is a body of State representatives with a mandate to monitor the implementation of the Convention, assisted by the Secretariat of the CoE. The Committee may "express an opinion on any question concerning the application" and make "specific recommendations to Parties concerning the implementation" (NC, art. 24, para. 3 (b) and (c)). Regrettably, a periodical state reporting mechanism – in order to provide required, sufficient, and conducive information to the Committee – is missing. As a result, the only way to gather information is left to voluntary cooperation and exchange of information (NC, art. 24, para. 3 (b); art. 24, para. 2).

Moreover, the Committee is not a standing body, either, but a body that will meet (only) at the request of at least a third of parties or of the Secretary General of the CoE. As a result, robust monitoring of the implementation of the Nicosia Convention has to be assured.

16.4 Critical Appraisal

The Nicosia Convention is a multifaceted instrument, helping States – as former Secretary General Thorbjørn Jagland emphasised – to combat criminal acts "effectively" (Council of Europe, 2016). Regarding its future acceptance, two important legal aspects are highlighted briefly:

16.4.1 Harmony of the NC with Existing Cultural Heritage Law

First of all, the Nicosia Convention fully respects not only binding SC resolutions (e.g. Res. 2199 (2015a); Res. 2253 (2015b); Res. 2322 (2016)) but also universally applicable UNESCO Culture Conventions already ratified by UNESCO Member States, e.g. the 1970 Convention or the UNIDROIT Convention (1995).

In addition, the Convention respects existing EU law, such as, for example, the EU Regulation 2019/880, adopted in early 2019. The Regulation is binding to all EU Member States and applies to cultural property originating from States outside the EU (European Union, 2019, art. 1, para. 2). Similar to the Nicosia Convention, the Regulation addresses organised crime, in particular "where such illicit trade could contribute to terrorist financing" (European Union, 2019, art. 1, para. 1).

The Regulation establishes a system of import licences required for certain categories of cultural objects (Annex, Part B, e.g. archaeological finds or items removed from monuments and sites when more than 250 years old), supported by documents and information providing evidence that the cultural objects have been exported in accordance with the laws and regulations of that country (European Union, 2019, art. 4, para. 4; Peters 2020). Coming close to NC regulations, the "new EU Import Regulation similarly relies on the importers' documentation that should support the lawful ownership history (provenance) before an object can be imported" (Campfens, 2020, p. 272).

Thus, it is essential to note that the Nicosia Convention "harmoniously coexists" with existing treaty law and instruments dealing with matters that the Convention also covers (Bieczyński, 2017, p. 270). The Convention complements international heritage law in a valuable way.

16.4.2 *Reflections on State Sovereignty and Ratification*

Second, the Nicosia Convention is a pragmatic and workable instrument to deal with the challenges of protecting our common heritage from plunderers, thieves, traffickers, dealers, and terrorists.

Therefore, ratification is a logical next step, according to German officials (Bundesregierung, 2019), not least because parts of the 2016 "Kulturgutschutzgesetz" coincide with obligations of the Nicosia Convention (Haas, 2021).

Once the treaty enters into force (provided that five ratifications, acceptances, or approvals exist) – other non-member states may be invited to accede to the Convention (NC, art. 28, para. 1). In that respect the Convention's rigorous openness for non-Member States is to be welcomed, for only by achieving wide acceptance and ratification of treaty law is it possible to create a close-knit international legal network to eradicate legal loopholes, which are readily exploited.

The Conference "Act for Heritage" that took place from 24 to 26 October 2019 in Nicosia promoted acceptance and ratification, additionally aiming to enhance stakeholder cooperation in the field. However, as it stands, the NC has been signed by just nine and ratified by four States. As the Head of Culture and Cultural Heritage Division of the Council of Europe emphasised, there is a "need to act […] [and] many more countries [need] to ratify this convention quickly" (Council of Europe, n.d.).

Experts are afraid that – similarly to the Delphi Convention – the new Convention might fail due to objections and concerns related to loss of state sovereignty, for "when criminal law is involved there is usually a reaction against the 'creeping assault' on domestic sovereignty" (Mottese, 2018, p. 140; Bieczyński, 2017).

It should be stressed, therefore, that the Nicosia Convention does not call into question the principle of state sovereignty. State Parties have the option to make far-reaching reservations at the time of signature or when depositing the instrument of ratification, acceptance, approval, or accession (NC, art. 30). The possible reservations may concern not only unlawful excavation and removal, but also illegal importation, destruction and damage, aiding or abetting and attempt, and jurisdiction (NC, art. 4, para. 2, art. 5, para. 2, art. 10, para. 2, art. 11, para. 3, art. 12, para. 3). In doing so, the Nicosia Convention gives treaty parties the liberty to pursue their own legal ways and solutions compatible with the framework provided by the Nicosia Convention.

Moreover, incursions on state sovereignty are unlikely to occur because most of the obligations in the Nicosia Convention have already been included in binding SC resolutions, paving the way for anti-terrorist state action. Accordingly, the SC (indirectly) encourages the ratification of the NC, as did the UN General Assembly, when it invited "Member States to consider becoming parties to the Convention" (UN General Assembly, 2018, para. 18).

16.5 Concluding Remarks

The Nicosia Convention may give new impetus to criminal law as a tool in the fight against offences against cultural property, helping also to reduce lawless areas regarding WH.

Cultural heritage protection requires a robust, legally integrated approach, including criminal prosecution. In that respect, the Nicosia Convention fulfils the promise given in the Bonn Declaration on World Heritage of the World Heritage Committee at its 39th session (Bonn, Germany) in June 2015, when State Parties committed themselves "to strengthen their national legislation and practice for the protection of cultural and natural heritage, also by introducing more effective measures to combat illicit trafficking and illegal trade of cultural properties" (World Heritage Committee, 2015, para. 27).

There is hope therefore that the 50th anniversary of the world-renowned World Heritage Convention will bring progress to the acceptance of the Nicosia Convention as a "missing link", contributing to the overall implementation of international cultural heritage law.

References

Bieczyński, M. M. (2017). The Nicosia Convention 2017: A new international instrument regarding criminal offences against cultural property. *Santander Art and Culture Law Review, 3*(2), 255–274. https://doi.org/10.4467/2450050XSNR.17.031.8432

Blake, J. (2020). Trafficking in cultural property where cultural heritage law and the international fight against transnational organized crime coincide. In A.-M. Carstens & E. Varner (Eds.), *Intersections in International Cultural Heritage Law* (pp. 157–180). Oxford University Press. https://doi.org/10.1093/oso/9780198846291.003.0007

Brown, J. (2017, July 4). The 2017 Blood Antiquities Convention – Protecting cultural property through criminal law. *GroJIL-Blog*. https://grojil.org/2017/07/04/the-2017-blood-antiquities-convention-protecting-cultural-property-through-criminal-law/

Bundesregierung. (2019). *Bericht der Bundesregierung zum Stand der Unterzeichnung und Ratifizierung europäischer Abkommen und Konventionen durch die Bundesrepublik Deutschland für den Zeitraum März 2017 bis Februar 2019* (Drucksache 19/10411). Deutscher Bundestag. https://dserver.bundestag.de/btd/19/104/1910411.pdf

Campbell, P. B. (2013). The illicit antiquities trade as a transnational criminal network: Characterizing and anticipating trafficking of cultural heritage. *International Journal of Cultural Property, 20*(2), 113–153. https://doi.org/10.1017/S0940739113000015

Campfens, E. (2020). Whose cultural objects? Introducing heritage title for cross-border cultural property claims. *Netherlands International Law Review, 67*(2), 257–295. https://doi.org/10.1007/s40802-020-00174-3

Committee on Offences Relating to Cultural Property. (2016). *Background paper on the Council of Europe Convention on Offences Relating to Cultural Property* (PC-IBC (2016) 02_en). Council of Europe. https://rm.coe.int/168070b4ff

Council of Europe. (1985, June 23). *European Convention on Offences relating to Cultural Property* (ETS No.119). Council of Europe.

Council of Europe. (2016, March 2). *"Blood antiquities": New Council of Europe initiative to combat illicit trafficking in cultural property* [Press release]. https://rm.coe.int/1680721000

Council of Europe. (2017a, May 19). *Council of Europe Convention on Offences relating to Cultural Property* (CETS No.221). Council of Europe.

Council of Europe. (2017b). *Explanatory report to the Council of Europe Convention on Offences relating to Cultural Property* (Council of Europe Treaty Series – No. 221). Nicosia: Council of Europe.

Council of Europe. (n.d.). *Offences relating to Cultural Property, Nicosia Convention – Long version* [Video]. https://www.coe.int/en/web/culture-and-heritage/cultural-property

Eloundou Assomo, L. (2020, November 16-18). *Closing Panel: Synergies for strengthened multilateralism in heritage protection – Challenges and perspectives* [Conference contribution, Video]. Cultural Heritage and Multilateralism: Regional and International Strategies for the Protection of Cultural Heritage, International Online-Conference, Berlin, Germany. https://cultural-heritage-and-multilateralism2020.com/event

European Commission. (2020). *Communication from the Commission on the EU Security Union Strategy* (COM/2020/605 final). European Commission. https://eur-lex.europa.eu/legal-content/EN/TXT/HTML/?uri=CELEX:52020DC0605&from=FR

European Union. (2019). *Regulation (EU) 2019/880 of the European Parliament and of the Council of 17 April 2019 on the introduction and the import of cultural goods* (O.J. (L 151) 1). European Union.

Fincham, D. (2019). The Blood Antiquities Convention as a paradigm for cultural property crime reduction. *Cardozo Arts & Entertainment Law Journal, 37*(2), 299–336.

United Nations General Assembly. (2018, December 13). *Return or restitution of cultural property to the countries of origin* (Resolution 73/130). U.N.Doc. A/RES/73/130.

Gottlieb, Y. (2020). Attacks against cultural heritage as a crime against humanity. *Case Western Reserve Journal of International Law, 52*, 287–330.

Groß, S. W. (2018). Can cultural property protection be an effective counter-terrorism instrument? *Journal of Art Crime, 19*, 47–60.

Guterres, A. (2017). *Report of the Secretary-General on the implementation of Security Council resolution 2347 (2017)* (S/2017/969). Security Council. https://digitallibrary.un.org/record/1323116/files/S_2017_969-EN.pdf

Haas, V. (2021). Kapitel 8 Datenschutz, gemeinsames Verfahren, Zoll. In K. von der Decken, F. Fechner, & M. Weller (Eds.), *Kulturgutschutzgesetz* (1st ed., pp. 509–528). Nomos; Facultas; Schulthess.

ICRC. (1977). *Protocol additional to the Geneva Conventions of 12 August 1949, and Relating to the Protection of Victims of Non-International Armed Conflicts (Protocol II)* (1125 U.N.T.S. 609). International Committee of the Red Cross.

Kulturstiftung der Länder. (2018). *Verbundname: Illegaler Handel mit Kulturgut in Deutschland. Verfahren zur Erhellung des Dunkelfeldes als Grundlage zur Kriminalitätsbekämpfung und -prävention am Beispiel antiker Kulturgüter, Akronym: ILLICID; Teilvorhaben (TV3): Antike Kulturgüter aus dem östlichen Mittelmeerraum: Identifizierung, Klassifizierung und Dokumentation von in Deutschland gehandelten Objekten: Schlussbericht zu Nr. 3.2, BNBest-BMBF 98* (Förderkennzeichen: 13N13645). Kulturstiftung der Länder, Stiftung bürgerlichen Rechts. https://doi.org/10.2314/KXP:1664575944

Lenzerini, F. (2003). The UNESCO Declaration concerning the Intentional Destruction of Cultural Heritage: One Sstep forward and two steps back. *The Italian yearbook of international law online, 13*(1), 131–145. https://doi.org/10.1163/221161303X00074

Mottese, E. (2018). Preventive measures in the Council of Europe Convention on Offences relating to Cultural Property: An Overview. *Santander Art and Culture Law Review, 4*(2), 121–142. https://doi.org/10.4467/2450050XSNR.18.021.10375

Munnelly, A. (2021). Compliant or complicit? Security implications of the art market. In *European Union, Institute for Security Studies*, Brief/23, 1-8. https://www.iss.europa.eu/sites/default/files/EUISSFiles/Brief_23_2021.pdf

Oñate, S. (2018). Treaty law and the protection of cultural property: Recent developments. *University of Vienna Law Review*, 2, 230–246 Seiten. https://doi.org/10.25365/VLR-2018-2-2-230
Peters, R. (2020). Nationalism versus internationalism: New perspectives beyond state sovereignty and territoriality in the protection of cultural heritage. In A.-M. Carstens & E. Varner (Eds.), *Intersections in international cultural heritage law* (pp. 364–388). Oxford University Press. https://doi.org/10.1093/oso/9780198846291.003.0016
Ulph, J. (2019). Legal perspectives on heritage crime: Reviewing due diligence measures for the art market. In V. Mitsilegas, S. Hufnagel, & A. Moiseienko (Eds.), *Research handbook on transnational crime* (pp. 356–382). Edward Elgar Publishing. https://doi.org/10.4337/9781784719449.00032
United Nations. (1998, July 17). *Rome Statute of the International Criminal Court* (2187 U.N.T.S. 3). United Nations.
UNESCO. (1954). *Convention for the protection of cultural property in the event of armed conflict*. UNESCO.
UNESCO. (1970). *Convention on the means of prohibiting and preventing the illicit import, export and transfer of ownership of cultural property*. UNESCO.
UNESCO. (1972). *Convention concerning the Protection of the World Cultural and Natural Heritage*. UNESCO.
UNESCO. (1999). *Second protocol to the Hague Convention of 1954 for the protection of cultural property in the event of armed conflict 1999*. UNESCO.
United Nations Security Council. (2015a, February 12). Resolution 2199, U.N.Doc.S/RES/2199.
United Nations Security Council. (2015b, December 17). Resolution 2253, U.N.Doc.S/RES/2253.
United Nations Security Council. (2016, December 12). Resolution 2322, U.N.Doc.S/RES/2322.
United Nations Security Council. (2017, March 24). Resolution 2347, U.N.Doc.S/RES/2347.
UNESCO. (2003). *Declaration concerning the Intentional Destruction of Cultural Heritage* (Doc. 32 C/Res. 33). UNESCO. https://unesdoc.unesco.org/ark:/48223/pf0000133171.page=68
von Schorlemer, S. (2004). Legal changes in the regime of the protection of cultural property in armed conflict. *Art Antiquity and Law, 9*(1), 43–77.
Wierczyńska, K., & Jakubowski, A. (2020). The Al Mahdi case from punishing perpetrators to repairing cultural heritage harm. In A.-M. Carstens & E. Varner (Eds.), *Intersections in International Cultural Heritage Law* (pp. 133–156). Oxford University Press. https://doi.org/10.1093/oso/9780198846291.003.0006
World Heritage Committee. (2015). *The Bonn Declaration on World Heritage*. UNESCO. https://whc.unesco.org/en/sessions/39COM

Chapter 17
Climate Change and World Heritage: An Introduction

Claire Cave

Abstract The rapid acceleration of science and technology has enabled people to make unprecedented changes to their environment and to alter the global climate. The changing climate, together with biodiversity loss, now pose significant threats to people and their heritage. This chapter provides an introduction to the impacts that climate change is having on World Heritage and how those impacts are being addressed. It considers the conflict that can be created between interventions to protect against climate change and the conservation of heritage values. Effective on-site management is an important tool in addressing climate change impacts and should be supported by states parties together with local engagement and national and international collaboration. World Heritage sites should not be viewed in isolation from their surrounding environment, and a strong World Heritage Climate Change policy is required to guide future management and implementation of the World Heritage Convention.

Keywords Climate change · Mitigation · Adaptation · Monitoring · Adaptive management

17.1 A Global Challenge

The adoption of the UNESCO *Convention Concerning the Protection of the World Cultural and Natural Heritage* (otherwise known as the World Heritage Convention) by the General Conference of UNESCO in 1972 occurred at a time of increasing political and public awareness of global environmental problems. That year, the UN hosted the first major intergovernmental conference on international environmental issues, which produced a declaration that proclaimed, "We see around us growing evidence of man-made harm in many regions of the earth: dangerous levels of

C. Cave (✉)
School of Archaeology, University College Dublin, Dublin, Ireland
e-mail: Claire.cave@ucd.ie

M.-T. Albert et al. (eds.), *50 Years World Heritage Convention: Shared Responsibility – Conflict & Reconciliation*, Heritage Studies,
https://doi.org/10.1007/978-3-031-05660-4_17

pollution in water, air, earth and living beings; major and undesirable disturbances to the ecological balance of the biosphere; destruction and depletion of irreplaceable resources; and gross deficiencies, harmful to the physical, mental and social health of man […]" (UN, 1972, Chap. 1.3). Climate change was not on the global agenda. The World Heritage Convention acknowledged the scale of the environmental challenges by noting the need to protect the world cultural and natural heritage not only from neglect but from the formidable phenomena of damage or destruction caused by changing social and economic conditions (UNESCO, 1972, Preamble). Fifty years later, the magnitude of global environmental problems has not diminished, and climate change has become one of the most significant and fastest-growing threats to people and their heritage.

The 2015 Paris Agreement, a legally binding international treaty on climate change under the 1992 United Nations Framework Convention on Climate Change (UNFCCC), aims to limit global warming to well below 2, preferably to 1.5 °C, compared to pre-industrial levels, i.e., relative to the period 1850–1900. To achieve this temperature goal, the treaty sets out objectives in line with the UNFCCC to stabilise greenhouse gas concentrations in the atmosphere "at a level that would prevent dangerous anthropogenic (human induced) interference with the climate system" (UN, 1992, Art. 2). The concentration of greenhouse gases (GHGs) is driven by human activities such as burning fossil fuels for electricity, heat, and transportation, producing methane and nitrous oxide through farming activities including management of livestock and use of fertilisers, and through deforestation and land-use change. Human-induced warming reached approximately 1 °C above pre-industrial levels in 2017 and is increasing at 0.2 °C per decade (Allen et al., 2018). The effects of heightened GHGs include an increase in land and ocean temperatures, resulting in more frequent heatwaves, an increase in the frequency and intensity of heavy precipitation events at the global scale and an increase in extreme weather events. These effects intensify regional droughts and water stress, flooding events and storms, and cause a reduction in sea-ice, glaciers, and ice sheets and an increase in sea levels. The associated risks include forest fires, expansion of desert terrain, declining ocean productivity, biome shifts, and the spread of invasive species, pests, and diseases (Hoegh-Guldberg et al., 2018). Already the impacts are affecting human health and wellbeing through reduced access to safe drinking water, reduced crop yields and food security, and socio-economic losses related to damaged infrastructure and industry. Climate change is not something that is happening in the future but is a phenomenon that is being felt by societies all around the world today.

The impacts of climate change also contribute to the biodiversity crisis, which similarly affects human wellbeing. A global assessment report of biodiversity and ecosystem services estimates that the natural extent of ecosystems has decreased by 47% due to human activities and continues to decline by at least 4% per decade (IBPES, 2019, p. 24). Land-use change is one of the major drivers of habitat loss, leading to an escalation in extinction rates such that approximately 25% of animal and plant species are now threatened with extinction (IPBES, 2019, p. 24). Other direct drivers of biodiversity loss include direct exploitation of species by humans, pollution, invasive alien species, and climate change. The loss of habitats and species impedes the capacity of ecosystems to provide services which benefit humanity. Such services include the supply of water, food, and soil maintenance; regulatory services including pollination, flood control, and carbon sequestration; and cultural services such as spiritual experience, cultural identity, and recreation. Biodiversity loss and climate change are both driven by human economic activities and mutually reinforce each other. Neither will be successfully resolved unless both are tackled together, and both will jeopardise progress in achieving the UN 2030 Sustainable Development Goals (SDGs) (Pörtner et al., 2021).

17.2 The Need for Change

Climate change impacts World Heritage sites not only directly by triggering shifts in habitats or through damage caused by incidents of extreme weather but also indirectly by the exacerbation of existing stresses such as unsustainable use, development pressures, and ineffective management. These threats are often interrelated, increasing the vulnerability of World Heritage sites. Furthermore, the destabilisation of social and environmental conditions caused by climate change and biodiversity loss will impact peoples' ways of life and their relationship with World Heritage. For example, people may be forced to migrate, and their former interaction with heritage sites and the associated cultural knowledge will dissipate. As well as peoples' dislocation from culturally important places, the distinctive dynamic of living World Heritage sites and their long-term maintenance and sustainability will be affected. Climate change is forcing change at an unprecedented scale across economic, environmental, political, and social spectrums. It poses a severe challenge to current conservation strategies and traditional heritage policies.

The general outlook for cultural and natural World Heritage is not positive. IUCN identified climate change as the most prevalent current threat and the largest potential threat to natural World Heritage sites (Osipova et al., 2020, p. viii). The same quantitative analysis has not been carried out for cultural sites, but the World Heritage Committee in 2019 urged all states parties to "step up action toward better understanding the climate vulnerability of World Heritage properties and put in place adaptation strategies that strengthen the resilience of properties" (UNESCO WHC, 2019, Para. 18.). However, cultural and natural heritage should not be considered only as a passive victim of natural and human-induced disasters but also as a tool that can be used proactively to develop and foster resilience and mitigate the threat of climate change and other stressors. Natural World Heritage sites protect large intact ecosystems, and this rich biodiversity offers carbon storage, soil stabilisation, water preservation, and flood prevention (Osipova et al., 2014). By its very nature, World Heritage encourages a sustainable approach to its stewardship and is often the product of an age-old interaction between humans and their environment. As iconic places, World Heritage sites have the potential to set standards in best practice conservation in tackling the impacts of climate change through planning, adaptation, and mitigation strategies. Rather than relying on top-down policies, engagement at the community level offers the opportunity for bottom-up commitment, to support and raise awareness of the deep and rapid shifts in human behaviour needed to address climate change. Apart from the need for a public commitment at global, national, and local levels to achieve such potential, it is important that issues of conflict, both likely and existing, regarding how World Heritage is managed, protected, and used in the face of climate change are recognised and addressed.

The World Heritage Convention has focussed heavily on the tangible aspects of heritage, but it is essential to recognise that heritage is more than individual structures and sites. Heritage exists within a human environment supported by an intangible dimension. Intangible heritage encompasses intangible "practices, representations, expressions, knowledge, skills" and associated artefacts and spaces. These expressions are transmitted from generation to generation, are constantly recreated in response to interactions with the changing environment, and give communities and groups a sense of identity and continuity (UNESCO, 2003, Art. 2). Linking intangible with cultural heritage, therefore, identifies cultural heritage as a cultural process, a product of traditions and shared beliefs and values that influence the attitudes, behaviour, and habits of people. This would indicate that heritage conservation should be understood as management of change to enable continuity in an ever-changing world.

The need to reduce GHG emissions to net zero, as highlighted in the Paris Agreement, demands change. It requires societal transformation involving fundamental reform of our way of living, land and water use, consumption patterns, and production processes. This, in turn, requires a cultural shift to adapt behaviours to accommodate nature friendly, sustainable, and climate-resilient development. As indicated by the 2030 Agenda for Sustainable Development and the 2015 Paris Agreement, the global community has the appetite to endorse ambitious and systematic targets for change, but whether it will enact the changes required remains to be seen. Looking to the future of the World Heritage Convention, we must examine how the implementation of the Convention can support such change while avoiding conflict.

17.3 Addressing Climate Change Impacts on World Heritage

Disasters are occurring more frequently. People and heritage and are increasingly exposed because of unplanned and rapid urbanisation, the decline of ecosystems, and poor land management. These, in turn, are compounded by factors such as weak governance, poor administration, and poverty. Most disaster risk is now climate related (UNDP, 2002). Long-term planning and prevention in the form of heritage risk preparedness is being promoted by both national and international organisations to help reduce the risks to heritage sites. The World Heritage Committee adopted a strategy for reducing risks from disasters at World Heritage sites in 2007 (UNESCO WHC, 2007), and the UN Office for Disaster Risk Reduction (UNISDR) produced the Sendai Framework for Disaster Risk Reduction in 2015. These documents emphasise a shift from disaster management to disaster prevention and mitigation. The importance of preventing new risk, reducing existing risk, and strengthening resilience as well as global, regional, national, and local collaboration and participation is highlighted. This is an important message as government funding in disaster management is typically skewed towards response and recovery rather than prevention and mitigation (de Vet et al., 2019). However, the increased scale and frequency of disasters is acting as a catalyst to promote a change in culture. In 2021, the Australian government announced substantial investment in resilience and mitigation following the 2019/2020 Australian megafires, which caused an estimated economic cost of 100 billion Australian dollars (Libatique, 2021). The Australian government has traditionally only dedicated 3% of disaster spending towards prevention (de Vet et al., 2019).

Local management interventions at heritage sites could be very effective at reducing climate sensitivity and improving resilience. For example, wildfires in the Tasmanian Wilderness, Australia, and the Ouadi Qadisha (the Holy Valley), Lebanon, although on the increase because of increased lightning strikes and droughts, respectively, are also influenced by the loss of traditional practices such as Aboriginal patch burning in Tasmania (Styger et al., 2018) and traditional cultivation systems and land management in Lebanon (Centre G.F.M., 2010). The loss of these practices has allowed a build-up of live and dead vegetation, which acts as ready fuel when a fire ignites. Recognition of these factors is an important element of risk preparedness and prevention. However, as the fire at Notre Dame in April 2019 has demonstrated, risk preparedness may be a balancing act between preservation of heritage values and safety. The cathedral's ancient oak attic, where the fire started, did not have a firewall or a sprinkler system in place because of concerns about how they would impact the integrity of the historic structure. Greater input from personnel from a range of disciplines could have helped in the risk assessment and questioned the assumption of the low risk of fire versus damage to integrity (Tannous, 2019).[1]

The extent of a disaster depends on the ability of the affected community to cope using available resources. Therefore, the identification of both hazards and vulnerabilities is important in risk preparedness. The diversity of World Heritage sites, e.g., monumental, urban, agricultural, archaeological, geological, aesthetic, and biodiverse, makes it very difficult to provide guidelines for vulnerability assessment. In answer, the Climate Vulnerability Index or CVI has been developed as a rapid assessment tool by John Day and Scott Heron of James Cook University. Critically it is based on the risk assessment approach and assesses both the "Outstanding Universal Value (OUV) vulnerability" and the "Community vulnerability" to climate change. The community vulnerability considers the economic, social, and cultural dependencies of the community (local residents and visitors,

[1] Nevertheless, because disaster management plans were in place the firefighters were prepared and knew how to protect the stain glass windows of Notre Dame from the fire and which works of art to rescue and in which order (Lesté-Lasserre, 2020).

both national and international) on the World Heritage and the capacity of the community to adapt (https://cvi-heritage.org/about). The emphasis on community vulnerability helps to ensure the inclusion of diverse groups of participants with different perspectives and to increase the level of awareness of the potential impacts of climate change. It is of mutual benefit if residents and other stakeholders appreciate the adaptive capacity required to cope with climate change. Ultimately, the aim of the tool is to make it possible to downscale climate scenarios to inform site management, regardless of the type of site, and to provide a practical and transparent approach to ensure wide participation and repeatability over time. Pilot projects are in the process of testing the tool's applicability across regions and states parties.

In terms of climate change policy, mitigation refers to the measures and activities that are put in place to reduce GHG emissions or enhance the sinks of such gases, e.g., forests and wetlands are carbon sinks in that they absorb more carbon than they release (UN, 1992; Sesana et al., 2018). Similarly, adaptation in terms of climate change refers to "adjustments in ecological, social, or economic systems in response to actual or expected climatic stimuli and their effects or impacts" (UNFCCC, 2021). There has been a greater focus on mitigation strategies in climate change research and policy than on adaptation (Sesana et al., 2018). In the field of cultural heritage, for example, reducing the GHG emissions of sites is typically associated with improving the energy efficiency of buildings, reducing the carbon footprint associated with the production and transport of building materials and encouraging and reducing the production of waste (Sesana et al., 2019). Comparable to the risk preparation strategy for Notre Dame, however, there is a delicate balance to be made between refurbishing or retrofitting historic buildings and preserving the cultural values. The conflict exists where the heritage value may be compromised when, for example, historical elements are removed and replaced in the name of energy efficiency or where features are affected, such as wall paintings covered by insulation. Consequently, heritage values can be seen as a barrier to mitigation strategies, and, to counteract this perception, it would be useful to have accessible examples of high profile, built heritage that have been effectively refurbished without compromising their integrity and authenticity (Sesana et al., 2019; Department of Culture Heritage & the Gaeltacht, 2019). As a first step, the ICOMOS Climate Change and Heritage Working Group (ICOMOS CCHWG, 2019) have put together a comprehensive outline of cultural heritage actions which support win-win scenarios where the safeguarding of heritage values is compatible with climate mitigation and adaptation strategies.

Adaptation solutions may also cause problems for the preservation of the authenticity and integrity of heritage sites. For example, the construction of shelters and roofs over monuments to protect them from adverse environmental conditions could impact the visual integrity of the sites and the authenticity of their form and design, materials, and location and setting. At the Megalithic Temples of Malta WHS, protective shelters were installed over three archaeological sites in response to a serious structural collapse caused by exposure to temperature fluctuations, rainwater, salts, and anthropogenic pollution (Cassar et al., 2018). Considering the impacts on integrity and authenticity, temporary, lightweight shelters were raised, which could be easily removed without impacting the surroundings and were designed in such a way as to maximise passive environmental control. An unexpected side effect was that visitors reported that the protection from the sun and rain made the site visit more comfortable and that the diffused light effect enhanced their experience (Becherini et al., 2016). However, the most important lessons learned were that environmental monitoring was required to observe whether the shelter improved the situation and to inform decisions about further adaptation strategies and that the need for a shelter depends on the unique circumstances of a particular site (Becherini et al., 2016; Cassar et al., 2018). To disseminate information about adaptation strategies at World Heritage sites, ICOMOS has partnered with Google and CyArk to produce an innovative online project "*Heritage on the Edge*" which "tells the story of climate change" at five diverse cultural World Heritage sites from Africa, Europe, South America, and South Asia. Using 3D models and infographics, the case studies highlight the climate change pressures and adaptation strategies happening at the sites and outline how straightforward approaches such as monitoring and maintenance can maximise conservation efforts (Google Arts & Culture, 2020).

Monitoring to understand change at heritage sites forms the basis of adaptive management. The adaptive management approach incorporates monitoring into a system of evaluation and revision, which allows for continuous updating of the management plan in line with changing circumstances and an expanding knowledge base (Cave & Negussie, 2017). Adaptive management is applicable to both cultural and natural sites. The Operational Guidelines for the implementation of the WH Convention promote adaptive management through a "cycle of planning, implementation, monitoring, evaluation and feedback" (UNESCO, 2019, Para 111c). This reflects the need to manage for change, not only for climate change but for other external pressures such as environmental degradation, urbanisation, and rising social and economic inequalities. While World Heritage sites are typically nominated with fixed boundaries, they are increasingly understood in the context of linkages to their surrounding environment, both in terms of their physical setting and the environmental, social, and economic sustainability of the wider area. This makes management planning a more challenging and complex process, from identifying the myriad factors that could impact a site to employing participatory approaches and communicating across different jurisdictions and administrative systems within and beyond heritage site boundaries. Limited resources and lack of finances create further obstacles. Perry (2019, p.4) advocates a triage approach, together with the forward-looking method inherent in adaptive management, to deal with the "wicked problem" of climate change. Triage is a method of prioritisation, where guidelines help to establish how scarce resources should be allocated to maximise the conservation of those attributes that might otherwise disappear. The "wicked problem" refers to the difficulty of managing the uncertainty of how climate change will impact a specific site together with the changing demands of politics, stakeholders, external threats, and competing public interests (Perry, 2019, p. 4). The important point is that there is no conclusion, all solutions are provisional, and managers must continually monitor the impact of their interventions so that they can further improve or adapt them to changing climate and environments (both biological and socio-economic) (Perry & Falzon, 2014).

With an urgent need to collect data to monitor change but limited funding, citizen science approaches offer heritage site managers an opportunity to better manage threatened heritage. Citizen science is a means to engage the public in the collection of scientific data to support long-term environmental monitoring. The process has some drawbacks, such as the need to supply training to ensure data is collected to the required standard, the need to manage the data collected, and the possibility of biases where certain times or locations are more popular with members of the public than others for example. However, technology in the form of smartphones and mobile data and free software such as Gmail, social media, and WordPress makes the process widely accessible and allows people to submit observations along with photos and videos easily. Furthermore, if developed properly, a citizen science partnership gives the public the opportunity to be proactively involved in protecting their heritage, raises awareness among the public of the impacts of climate change and biodiversity loss, and helps to build consensus on how to address the threats (Dawson et al., 2020; Donnelly, et al., 2014). Long-term, meaningful involvement of the public requires quality engagement in co-creation, monitoring, and evaluation as the project progresses (European Commission, 2017).

17.4 Outlook

In looking forward to the next 50 years of the World Heritage Convention, the priorities are related to how change is managed. Change may include loss of heritage sites and loss of Outstanding Universal Value (OUV), one of the principal concepts under the Convention. Climate change threatens the very existence of heritage sites, particularly terrestrial sites in vulnerable locations such as coastal areas exposed to increased erosion and sea level rise. Changing climate may also cause the displacement of the values that make up OUV, such as the agricultural and cultivated species in cultural landscapes and biological communities and threatened species in terrestrial and marine parks. Simultaneously, heritage sites are part of dynamic, socio-cultural processes and the attributes which communities value may change with the impacts of increasing stressors from climate change and biodiversity loss. Therefore, the Convention is faced with managing OUV in a world of fast-paced change while also recognising the principle of equity and the respective vulnerabilities and capabilities of states parties. Where diaster does occur, building resilience and capacity for disaster risk management are essential as highlighted in the Warsaw Recommendation on Recovery and Reconstruction of Cultural Heritage (UNESCO, 2018).The principles and strategies for implementing the WH Convention are contained in the Operational Guidelines, which are regularly revised to incorporate new knowledge and concepts in the context of heritage values and conservation. The evolving nature of these guidelines allows the Convention to adapt to change. However, a comprehensive World Heritage policy on climate change is required, together with a critical evaluation of how the tools and procedures of the Convention can continue to be effective and implemented. Collaboration is needed at global, national, and local levels. The WH Convention needs to pursue continued and meaningful interaction with other global multilateral environmental and cultural agreements at the level of the Secretariat and the states parties to exploit opportunities for synergistic activities and increased coordination between conventions in tackling the current crises. This collaboration will assist states parties in recognising the contribution that World Heritage sites can make in addressing national targets under these agreements, including the Paris Agreement. Furthermore, greater collaboration among governing authorities at a national level will support adaptation and mitigation strategies which are sympathetic to heritage values and help avoid maladaptive and conflicting policies. A poorly managed tourist industry, for example, creates many problems at heritage sites; tourism is also estimated to account for approximately 8% of global greenhouse gas emissions and, therefore, directly influences climate change (Lenzen et al., 2018). Emphasis should be placed on the conservation and effective management of existing World Heritage sites at a local level and the role World Heritage can play in generating changes in human behaviour in favour of nature-friendly, climate-resilient, sustainable development.

References

Allen, M. R., Dube, O. P., Solecki, W., Aragón-Durand, F., Cramer, W., Humphreys, S., & Zickfeld, K. (2018). Framing and context. In: *Global Warming of 1.5 C. An IPCC Special Report on the impacts of global warming of 1.5°C above pre-industrial levels and related global greenhouse gas emission pathways, in the context of strengthening the global response to the threat of climate change, sustainable development, and efforts to eradicate poverty.* Intergovernmental Panel on Climate Changed. https://www.ipcc.ch/site/assets/uploads/sites/2/2019/05/SR15_Chapter1_Low_Res.pdf

Becherini, F., Cassar, J., Galea, M., & Bernardi, A. (2016). Evaluation of the shelters over the prehistoric Megalithic Temples of Malta: Environmental considerations. *Environmental Earth Sciences, 75*(14), 1–13.

Cassar, J., Cefai, S., Grima, R., & Stroud, K. (2018). Sheltering archaeological sites in Malta: Lessons learnt. *Heritage Science, 6*(36). https://doi.org/10.1186/s40494-018-0201-6

Cave, C., & Negussie, E. (2017). *World Heritage conservation: The World Heritage Convention, linking culture and nature for sustainable development*. Routledge.

Center, G. F. M. (2010). Forest fire threat in Qadisha Valley. In *Lebanon: Precautionary action to prevent damage or destruction of the UNESCO World Heritage site*. Global Fire Monitoring Center (GFMC).

Dawson, T., Hambly, J., Kelley, A., Lees, W., & Miller, S. (2020). Coastal heritage, global climate change, public engagement, and citizen science. *Proceedings of the National Academy of Sciences, 117*(15), 8280–8286.

Department of Culture, Heritage and the Gaeltacht. (2019). *Built & Archaeological heritage climate change sectoral adaptation plan*. Government of Ireland.

de Vet, E., Eriksen, C., Booth, K., & French, S. (2019). An unmitigated disaster: Shifting from response and recovery to mitigation for an insurable future. *International Journal of Disaster Risk Science, 10*, 179–192. https://doi.org/10.1007/s13753-019-0214-0

Donnelly, A., Crowe, O., Regan, E., Begley, S., & Caffarra, A. (2014). The role of citizen science in monitoring biodiversity in Ireland. *International Journal of Biometeorology, 58*(6), 1237–1249.

European Commission. (2017). *LAB-FAB-APP: Investing in the European future we want*. The Lamy Report; European Commission, Brussels.

Google Arts & Culture. (2020). *Heritage on the edge – Discover how people around the world are using technology to protect their cultural sites against climate change*. Retrieved June 12, 2021 from https://artsandculture.google.com/project/heritage-on-the-edge

Hoegh-Guldberg, O., Jacob, D., Taylor, M., Bindi, M., Brown, S., Camilloni, I., Diedhiou, A., Djalante, R., Ebi, K. l., Engelbrecht, J., Guiot, F., Hijioka, Y., Mehrotra, S., Payne, A., Seneviratne, S. I., Thomas, A., Warren, R., & Zhou, G. (2018). Impacts of 1.5°C global warming on natural and human systems. In: *Global Warming of 1.5°C. An IPCC Special Report on the impacts of global warming of 1.5°C above pre-industrial levels and related global greenhouse gas emission pathways, in the context of strengthening the global response to the threat of climate change, sustainable development, and efforts to eradicate poverty*. Intergovernmental Pannel on Climate Change (IPCC). Retrieved from https://www.ipcc.ch/site/assets/uploads/sites/2/2019/06/SR15_Chapter3_Low_Res.pdf

ICOMOS Climate Change and Cultural Heritage Working Group. (2019). *The future of our pasts: Engaging cultural heritage in climate action*. International Council of Monuments and Sites (ICOMOS).

IPBES. (2019). *Summary for policymakers of the global assessment report on biodiversity and ecosystem services of the Intergovernmental Science-Policy Platform on Biodiversity and Ecosystem Services*. Intergovernmental Science-Policy Platform on Biodiversity and Ecosystem Services (IPBES). Retrieved from https://ipbes.net/sites/default/files/inline/files/ipbes_global_assessment_report_summary_for_policymakers.pdf.

Lenzen, M., Sun, Y. Y., Faturay, F., Ting, Y. P., Geschke, A., & Malik, A. (2018). The carbon footprint of global tourism. *Nature Climate Change, 8*(6), 522–528.

Lesté-Lasserre, C. (2020). Scientists are leading Notre Dame's restoration—And probing mysteries laid bare by its devastating fire. *Science, 367*(6483), 1182–1187.

Libatique, R. (2021). Australian government establishes new agency to address climate change risks. *Insurance Business Australia*. Retrieved June 10, 2021 from https://www.insurancebusinessmag.com/au/news/environmental/australian-government-establishes-new-agency-to-address-climate-change-risks-254362.aspx

Osipova, E., Wilson, L., Blaney, R., Shi, Y., Fancourt, M., Strubel, M., Salvaterra, T., Brown, C., & Verschuuren, B. (2014). *The benefits of natural World Heritage: Identifying and assessing ecosystem services and benefits provided by the world's most iconic natural places*. IUCN.

Osipova, E., Emslie-Smith, M., Osti, M., Murai, M., Åberg, U., Shadie, P. (2020). *IUCN World Heritage Outlook 3: A conservation assessment of all natural World Heritage sites*. IUCN.

Perry, J. (2019). Climate change adaptation in natural World Heritage sites: A triage approach. *Climate, 7*(9), 105.

Perry, J., & Falzon, C. (2014). *Climate change adaptation for natural World Heritage sites: A practical guide* (Vol. 37). UNESCO.

Pörtner, H. O., Scholes, R. J., Agard, J., Archer, E., Arneth, A., Bai, X., Barnes, D., Burrows, M., Chan, L., Cheung, W. L., Diamond, S., Donatti, C., Duarte, C., Eisenhauer, N., Foden, W., Gasalla, M. A., Handa, C., Hickler, T., Hoegh-Guldberg, O., … Ngo, H. T. (2021). *IPBES-IPCC co-sponsored workshop report on biodiversity and climate change*. IPBES & IPCC. https://doi.org/10.5281/zenodo.4782538

Sesana, E., Gagnon, A. S., Bertolin, C., & Hughes, J. J. (2018). Adapting cultural heritage to climate change risks: Perspectives of cultural heritage experts in Europe. *Geosciences, 8*(8). https://doi.org/10.3390/geosciences8080305

Sesana, E., Bertolin, C., Gagnon, A. S., & Hughes, J. J. (2019). Mitigating climate change in the cultural built heritage sector. *Climate, 7*(7), 90.

Styger, J., Marsden-Smedley, J., & Kirkpatrick, J. (2018). Changes in lightning fire incidence in the Tasmanian Wilderness World Heritage Area, 1980–2016. *Fire, 1*(3), 38. https://doi.org/10.3390/fire1030038

Tannous, W. K. (2019). The fire of Notre Dame: Economic lessons learned. *WIT Transactions on The Built Environment, 190*, 51–63.

UN. (1972). Declaration of the United Nations Conference on the Human Environment. *Report of the United Nations Conference on the Human Environment Stockholm, 5–16 June 1972*. UNESCO.

UN. (1992). *United Nations framework convention on climate change*. UN.

UNDP. (2002). *A climate risk management approach to disaster reduction and adaptation to climate change*. UNDP Expert Group Meeting on Integrating Disaster Reduction, Havana. Retrieved June 12, 2021 from https://www.oecd.org/derec/undp/47871337.pdf

UNFCCC. (2021). *What do adaptation to climate change and climate resilience mean?* United Nations Climate Change. Retrieved May 13, 2021 from https://unfccc.int/topics/adaptation-and-resilience/the-big-picture/what-do-adaptation-to-climate-change-and-climate-resilience-mean

UNESCO. (1972). *Convention concerning the protection of the world cultural and natural heritage*. UNESCO.

UNESCO. (2003). *Convention for the safeguarding of the intangible cultural heritage*. UNESCO.

UNESCO. (2018). *Warsaw recommendation on recovery and reconstruction of cultural heritage*. UNESCO. Retrieved from http://whc.unesco.org/en/events/1442/.

UNESCO. (2019). *Operational guidelines for the implementation of the World Heritage Convention*. UNESCO.

UNESCO World Heritage Committee (WHC). (2007). *Strategy for reducing risks from disasters at World Heritage properties* (Doc. WHC-07/31.COM/7.2). UNESCO. Retrieved from https://whc.unesco.org/archive/2007/whc07-31com-72e.pdf

UNESCO World Heritage Committee (WHC). (2019). *Pressing conservation issues, UNESCO World Heritage Committee Decisions* (Doc. WHC -19/43, COM/7.2). UNESCO.

Chapter 18
The Climate Crisis, Outstanding Universal Value and Change in World Heritage

William P. Megarry

Abstract Climate change is the greatest threat facing global natural and cultural heritage. All World Heritage (WH) properties will be impacted over the coming century, and our ability to adapt will often be limited. Yet climate change was a threat never envisioned by the drafters of the World Heritage Convention (WHC). This chapter considers how concepts central to the WHC may need to adapt to a rapidly changing world, to reflect three uncomfortable realities of the climate crisis and its impacts on heritage sites. Firstly, climate change is and will continue threatening and invalidating the Outstanding Universal Value (OUV) of many properties, and there may be little we can do to stop this. Secondly, climate change knows no borders and existing mechanisms may need to be rethought to reflect this. Thirdly, these challenges will, like climate change, disproportionately impact marginalised and indigenous communities in the Global South. It is suggested here that more precise and explicit guidance, which considers local climate modelling and an inclusive approach to values, within the existing proactive mechanisms of the WHC Operational Guidelines would result in a more consistent consideration of climate change impacts at WH properties, that reflects the spirit of the WHC.

Keywords World heritage convention · Climate change · Managing change · Climate justice · Climate vulnerability

18.1 Introduction

It has been 50 years since world leaders established the World Heritage Convention (WHC). This visionary document sought to protect places of Outstanding Universal Value, acknowledging for the first time the universal value of some places. At this

W. P. Megarry (✉)
Archaeology and Palaeoecology, School of Natural and Built Environment, Queen's University Belfast, Belfast, UK
e-mail: w.megarry@qub.ac.uk

M.-T. Albert et al. (eds.), *50 Years World Heritage Convention: Shared Responsibility – Conflict & Reconciliation*, Heritage Studies,
https://doi.org/10.1007/978-3-031-05660-4_18

key point of reflection, our global heritage faces a new threat which was not foreseen by the authors of the 1972 Convention. The impacts of anthropogenic climate change are global and transcend the more localised threats to heritage identified in the original document. Either directly or indirectly, all World Heritage properties are or will be impacted by climate change. As a heritage community, it is imperative that we use this significant anniversary to critically evaluate our ability to respond to this challenge. This reflection should start with our founding documents, including the WHC and its subsequent Operational Guidelines (most recent edition 2019). The utility of these documents to World Heritage management within the climate crisis is explored in this chapter, which opens with an overview of the current state-of-the-art in research and policy before focusing on the need to adopt a more values-based approach to understanding impacts, vulnerabilities and adaptation planning that is more sympathetic to inevitable change.

18.2 Climate Change and Heritage

Recent years have seen an increasing focus within the heritage community on the impacts of climate change on heritage. The IUCN World Heritage Outlook 3, published in 2020, noted that one-third of properties are in danger (Osipova et al. 2020). In the same year, the 20th ICOMOS General Assembly overwhelmingly voted to declare a climate and ecological emergency, building on escalating actions, including the publication of their 2019 Future of our Pasts (FooP) report (ICOMOS, 2019). Within the cultural heritage sector, the quantity of papers in high-ranking journals and special editions focused on the subject are increasing, reflecting a growing understanding of the urgency of the climate crisis (Fatorić & Seekamp, 2017; Sesana et al., 2021). There has also been increased public interest, with media attention focused on damage and loss at iconic sites such as Venice and Rapa Nui, and increasing interest and concern about the loss of local heritage (Megarry & Hadick, 2021). These examples emphasise the impacts of climate change on cultural heritage and the immense communicative power of these special places to stress urgency and promote action (Rockman & Maase, 2017).

Impacts on cultural heritage are both direct and indirect. Fatoric and Seekamp (2017) provide an excellent overview of the themes and modalities of these studies, which have varied from focusing on certain subsets of heritage or regions (Brooks et al., 2020; Hollesen et al., 2018; Perez-Alvaro, 2016; Reimann et al., 2018); to taking a broader impact-focused approach as outlined in both the ICOMOS FooP report (2019) and Sesana et al. (2021). Indirect impacts have been less well researched, perhaps due to their inherently complex, multifaceted and commonly regional or site-specific manifestations, where climate change often acts as a risk multiplier for existing stressors. The ICOMOS FooP dedicates substantial discussion to some of the "cross-cuttin" issues, which include equity and climate justice and the relationship between climate action and sustainable development (ICOMOS, 2019. The inclusion of heritage issues into climate policy has been less common, with some notable exceptions (Daly, 2019; Fluck, 2016).

18.3 The World Heritage Convention and Climate Change

Identifying and preparing for climate change impacts are already embedded, both explicitly and implicitly, within the WHC and its Operational Guidelines. The text of the WHC identifies a range of threats and specific impacts to heritage in Article 11.4, which introduces the concept of the *"List of World Heritage in Danger*". This is a reactive mechanism for supporting and providing assistance to properties and State Parties and reflects *"serious and specific dangers"*. It includes an indicative list of examples, which (unsurprisingly) omits climate change (WHC 11.4). Climate change was first mentioned in the 1997 revision of the Operational Guidelines for the implementation of the WHC, which included it as a potential environmental pressure in the "*Format and content of nominations*" section (Article 64). It remained in this section until 2005 when this guidance was placed in a separate annex (Annex 5). It only returned to the main text in the 2017 version where it is explicitly mentioned in Articles 111 and 118, which discuss the importance of environmental, heritage and strategic impact assessments to "ensure the long-term safeguarding of the Outstanding Universal Value, and the strengthening of heritage resilience to disasters and climate change" (UNESCO, 2019). It appears again in Article 239, which addresses international assistance, and, while not explicitly named in the guidelines, climate change is often included in periodic reporting and state of conservation reports and in secondary guidance documents. These mechanisms allow for both the proactive (within the nomination process, state of conservation (SOC) reports and in periodic reporting) and reactive (reactive missions, List of WH in Danger) inclusion of climate impacts at sites. An increasing number of WH properties are including climate impacts within these mechanisms, for example, in the SOC reporting. Often, threats like wildfire or extreme weather may be identified without specifically considering the underlying cause, climate change. However, the lack of explicit and specific guidance on climate vulnerability assessment means that the extent of its inclusion (or omission) can vary from site to site and can reflect local or national factors, including capacity, politics or even just the priorities of an individual assessor.

The primary policy document for WH sites and climate change remains the *2007 UNESCO Policy Document on the Impacts of Climate Change on World Heritage Properties* (UNESCO, 2008). In many ways, this document was ahead of its time, stressing global urgency and acknowledging wide-ranging direct and indirect impacts on properties. Its utility was (by design) limited to some key issues, including identifying synergies between WH policy with other key climate drivers, proposing key research needs and addressing legal questions and alternative mechanisms for heritage decision and policy makers. Current efforts by advisory committees to update this document are at an advanced stage.

The WHC and its Operational Guidelines are visionary and seminal documents for global heritage management and its mechanisms have evolved to include considerations of climate change. While these can be effective for many sites, the assessment of climate impacts at many properties remains largely reactive. This poses a challenge for conservation where proactive adaptation is required. One major problem here is that the inclusion of climate impacts alongside other threats creates a false equivalence, which undermines the scale of the climate crisis and its impact on heritage sites. In many cases, local action or adaptation will not be enough to address impacts that must be addressed at a global scale and in line with existing policies and drivers. There is a need for clear and explicit guidance for all WH properties and not just those deemed at immediate threat from climate change. Climate change is not an acute threat affecting selective WH properties. It is now the single largest threat to all cultural and natural heritage sites across the planet. To illustrate this point, the next section will address some specific challenges to the implementation of the WHC in a climate emergency.

18.4 Climate Change and the Challenge of Outstanding Universal Value

By design, the WHC and its Operational Guidelines are primarily focused on establishing and preserving the Outstanding Universal Value (OUV) of WH properties, and this OUV is substantiated through attributes which align with the ten criteria outlined in the Operational Guidelines. These attributes are entwined with complementary concepts of integrity and authenticity, which must be met and maintained to protect the WH status of properties and are outlined in a statement of OUV (SOUV). While wider sites or property values may change over time, the SOUV is set in stone and cannot be easily changed. The Operational Guidelines for the WHC stress the importance of including and acknowledging a range of values in the SOUV at the point of inscription and that these need to be protected. Understanding the significance, attributes and associated values of heritage properties is central to their preservation and conservation, yet concepts of significance and values are multifaceted and can have a range of meanings as outlined in The Burra Charter (Australia ICOMOS, 2013). This plurality of significance and values echoes the influential Nara Convention, which emphasises the cultural specificity of values (ICOMOS, 1994). This inclusive approach is central to more recent drafts of the WHC Operational Guidelines and has largely been embraced by more recent WH sites, which make explicit connections between heritage values and local communities in the SOUV, reflecting wider drivers for sustainable development.

This SOUV is not just the key to inscription but also an essential part of the ongoing monitoring of sites, as threats to the OUV of a property constitute a threat to its WH status. Given the desire to protect OUV, integrity and authenticity, and, in turn, the site inscription, the values contained within the SOUV are closely protected. A SOUV has both strengths and weaknesses when faced with climate change. On the one hand, they can be seen as immovable and inflexible. On the other hand, an inclusive SOUV can allow for broader interpretations that acknowledge a wider range of values and aligns them with the criteria of the WHC, which will prove more resilient to inevitable change. A more restrictive SOUV, which builds a case for OUV based solely on material or historical authenticity and integrity and does not consider wider social values, will be less flexible and less able to adapt to the impacts of climate change.

So why is climate change such a threat to the OUV of WH properties, and why do we need to consider these conflicts now? The remainder of this section will explore three key (and sometimes crosscutting) themes central to this dynamic.

18.4.1 Climate Change Is and Will Continue Impacting the Inscription Criteria and the SOUV of WH Properties

Climate change both directly and indirectly impacts both the physical fabric, authenticity and integrity of WH properties and other aesthetic, historic, scientific, social and spiritual values. This is inevitable, and many properties can be saved, but a minority cannot. In the climate emergency, we must prepare to manage both changes to the authenticity and integrity of properties and the loss of others (Perry, 2019). Climate change is a global issue which necessitates a global response, and while local adaptation efforts may reduce the severity of some impacts, the majority of properties and landscapes are going to change. The Great Barrier Reef and other World Heritage coral reefs are examples of this dynamic (Fig. 18.1) (Heron et al., 2017). In these cases, threats to the SOUV will mean that, through no fault of their own and with no capacity or ability to adapt, properties may be interpreted by the WH centre as being at risk and added to the List of WH Sites in Danger or even removed from the WH list. The OUV of other properties are intricately connected to their geography and climates, which will likely change over the coming century. Cultural landscapes will likely bear the brunt of these changes, especially those which rely on production, such as the Coffee Cultural Landscape of Colombia or the Champagne Hillsides, Houses and Cellars in France (Silva, 2017). In these situations, there is a need to manage changes to the OUV of properties and accept the unavoidable impacts on their authenticity and integrity.

Fig. 18.1 Coral Bleaching, Lizard Island, Great Barrier Reef, before (March 2016) & after (May 2016). (*Note*: Image by The Ocean Agency/Ocean Image Bank)

18.4.2 Climate Change Does Not and Will Not Respect Property Boundaries

Boundaries are an integral part of establishing and protecting OUV. Understanding the geographical context of a property is crucial to understanding its cultural significance. Article 88 of the Operational Guidelines stresses that, in order to accurately show a property's integrity, the extent of a WHS must be of sufficient size to 'ensure the complete representation of the features and processes that convey the property's significance' (UNESCO, 2019, 27). Buffer zones can also be used to add an extra level of protection. While nomination dossiers will often consider these boundaries within wider regional or national contexts, considering regional or national threats such as earthquakes, wildfires or storms relevant to site management, future climate impacts are rarely explicitly presented. In reality, the impacts of climate change are rarely restricted to the inscribed boundaries of WH properties. Direct impacts such as change of sea level, storminess or regional climate will alter large areas and result in changes at properties which cannot be adapted to. In extreme cases, particularly at coastal sites or properties on small islands, the structure of the landscape will change as sea levels rise, physically altering inscribed boundaries. An example of this is Levuka in Fiji, where the location of the signing of the 1874 Deed of Cession is eroding due to rising sea levels. In other cases, indirect impacts outside property boundaries like floods or wildfires will affect management and conservation practices as well as site sustainability. Recent fires in Australia brought this into sharp relief where threats to the Gondwana Rainforests of Australia WH property resulted in an official response from UNESCO noting the potential impact on site OUV. We must now consider a future where threats may come from outside of WH properties and where established boundaries need to be reassessed to reflect the impacts of climate change. Central to this new reality is working with climate scientists and obtaining downscaled climate models for properties and their wider landscapes.

18.4.3 This Threat to OUV Will, Like Climate Change, Disproportionately Affect the Global South

The issue of in-country heritage professionals and State Parties being unable to prevent changes to or loss of OUV due to climate change will be particularly acute in the Global South. This climate inequality has been acknowledged by UNESCO in their 2017 *Declaration of Ethical Principles in relation to Climate Change* (2017), which stressed the need for global solidarity and action. More recently, the 2019 Future of our Pasts report emphasised the inequality inherent in the climate crisis from a cultural heritage perspective (ICOMOS, 2019), noting that while climate change is predominantly caused by the cumulative greenhouse gas emissions of industrialised countries, its impacts are disproportionately felt by the poorest and most vulnerable. This dynamic extends to WH properties and their capacities to adapt to climate impacts and will exacerbate an already alarming pattern of WH sites in Danger in less wealthy countries, primarily in Africa and the Arab regions where 71% of WH properties on the WH in Danger list are situated. Table 18.1 outlines the ten most affected countries between 2000 and 2019 as outlined in a

Table 18.1 Long-term climate risk index (10 most affected countries from 2000–2019)

Climate vulnerability	Country	Number of WHS	Lower- and middle-income country (LMIC) status
1	Mozambique	1	Least developed country
2	Zimbabwe	5	Other low-income countries
3	The Bahamas	0	Not LMIC
4	Japan	23	Not LMIC
5	Malawi	2	Least developed country
6	Islamic Republic of Afghanistan	2	Least developed country
7	India	38	Not LMIC
8	South Sudan	0	Least developed country
9	Niger	3	Least developed country
10	Bolivia	7	Lower- and middle-income

Note: After Eckstein et al. (2021)

2021 report by Germanwatch, written in preparation for COP 25 in Madrid (Eckstein et al. 2021), alongside their OECD status, with the number of WH properties in each country. All but four are lower- and middle-income economies, and six are least developed countries. It is imperative that any new tools or techniques be accessible and available to the countries most at risk.

18.5 The Need for a Proactive Values-Based Vulnerability Assessment Tool

So, what is the solution to this dilemma? There is no one answer, but the FooP report stresses that '... conservation management and assessment standards, such as the constructs of authenticity and integrity, will need to be rethought' (ICOMOS, 2019, 16). Khalaf (2020) has recently stressed that the Operational Guidelines must embrace a modality of compatibility if it is to effectively adapt to the challenges posed by climate change. I would suggest here that we must utilise existing mechanisms within the WHC Operational Guidelines to promote explicit, proactive and standardised climate impact assessment at all WH properties. These approaches must be driven by climate science and must include a wide range of values, including the SOUV of properties. They must also be scalable and applicable globally to all property typologies.

Climate impact assessment methodologies already exist for heritage sites. Examples include Perry's (2011, 2019) World Heritage Vulnerability Index (WHVI), which incorporated nine variables but only focused on Natural World Heritage sites, and Guzman et al.'s (2020) landscape-based approaches, which can work in tandem with the conservation and management of OUV through the periodic reporting mechanism. This landscape approach would certainly go some way towards addressing the issues caused by site boundaries. A more recent approach is the climate vulnerability index (CVI), which builds on the existing IPCC assessment framework to identify risks to both the OUV and the socio-economic values of WH properties (Day et al. 2020). Central to the CVI approach is building relationships between heritage professionals and climate scientists to produce downscaled climate models for individual properties, which incorporate both direct and indirect impacts at multiple scales. It then explores the severity of these impacts against key values derived from both the SOUV and workshops involving local stakeholders. Vulnerability is then assessed for individual values (not just OUV) through an assessment of exposure and sensitivity based on adaptive capacity and resilience. The overall vulnerability of a property can be assessed by exploring the cumulative vulnerabilities of individual values. The CVI has been applied at both natural and cultural sites, including Shark Bay in Australia (Heron et al., 2020) and The Heart of Neolithic Orkney (Day et al. 2019), and is currently being undertaken at two African sites as part of the Values-based Climate Change Risk Assessment: Piloting the Climate Vulnerability Index for Cultural Heritage in Africa (CVI Africa Project).

The two sites are the Sukur Cultural Landscape in Nigeria and The Ruins of Kilwa Kisiwani and Ruins of Songo Mnara in the United Republic of Tanzania. This project aims to assess the utility of the CVI technique for sites in the Global South, focusing on distinct and different property typologies and climate threats.

18.6 Case Study: The CVI Africa Project and the Ruins of Kilwa Kisiwani and Ruins of Songo Mnara, United Republic of Tanzania

Climate change has and is already threatening the OUV of WH properties. In some cases, it has been possible to address these impacts within the existing mechanisms of the WHC. One good example is The Ruins of Kilwa Kisiwani and Ruins of Songo Mnara WH site in the United Republic of Tanzania. It was inscribed on the WH list in 1981. The OUV for the site is based on Criterion (iii) and emphasises its architectural, archaeological and documentary values (Chami, 2019; Pollard, 2008). As with so many sites, the statements of integrity and authenticity both emphasise the material completeness of the site as being central to its OUV, while also acknowledging the potential threats to the site from a range of factors, including coastal inundation.

The Ruins of Kilwa Kisiwani and Ruins of Songo Mnara are good examples of how both direct and indirect climate impacts can threaten the OUV of WH properties. In 2004, it was placed on the List of World Heritage in Danger due to inundation by the sea and anthropogenic factors, including encroachment of building and agricultural activities. The site was removed from the list in 2014 following investment and support from the international community, including the construction of barriers to protect structures and the replanting of mangroves (Fig. 18.2); however, there remain ongoing concerns about the lack of engagement with local stakeholders (Chinyele & Lwoga, 2018; Ichumbaki & Mapunda, 2017). Lwoga (2018, 1028) noted that the associated land-use activities impacting the site were largely due to the "limited socio-economic benefits, inconsistent business opportunities, complaints about employment and payment and few feasible alternatives for making a living" for the local community who live in and around the ruins and that the conservation efforts and planning at the site do not properly engage with community needs, and further suggesting that intervention was "relatively limited to the level of tokenism" (Chinyele & Lwoga, 2018, 188; Ichumbaki & Mapunda, 2017). This is worrying as future threats to the property are likely to be considerable. Tanzania is amongst the most vulnerable countries to future impacts of climate change (IPCC, 2015). While existing adaptation efforts protect parts of the site, others remain extremely vulnerable. This disconnect between heritage and community values shows the importance of adopting a values-based approach, which includes local social-economic values alongside threats to the OUV. It is especially important in cases where the SOUV makes limited reference to these wider values beyond their inclusion in a section on the aforementioned conservation issues.

Fig. 18.2 Gabion wall below the Malindi Mosque at Kilwa Kisiwani. (*Note*: Photograph by William Megarry, 2018)

All these factors make the property a good case study for a CVI assessment. Climate impacts at the site have already been substantial, and adaptation efforts have had some success in addressing these; however, the future preservation of the property and its OUV requires greater input from local stakeholders. As part of the CVI Africa Project, an international team including partners from the Tanzania Wildlife Management Authority and ICOMOS is running a CVI workshop. It is working with Tanzanian climate scientists to produce downscaled climate models for the property and with local stakeholders and heritage professionals to identify key values which may be impacted in the future. It is hoped that this assessment will facilitate greater and more inclusive adaptation planning and protect the OUV of the property over the next century.

18.7 Conclusion

This chapter has explored the potential impacts of climate change on WH properties, focusing on key development opportunities within the WHC and its Operational Guidelines. It is proposed here that by adopting a values-based approach to climate impacts and responses within the existing mechanisms of the WHC, which is sympathetic to wider value systems and extends beyond the often narrow geographical extents of properties, it is possible to continue the mission of the World Heritage Convention to 'demonstrate the importance, for all the peoples of the world, of safeguarding this unique and irreplaceable property, to whatever people it may belong' (UNESCO, 1972, 1).

References

Australia ICOMOS. (2013). *The Burra Charter: The Australia ICOMOS Charter for places of cultural significance 2013*. Australia ICOMOS.

Brooks, N., Clarke, J., Ngaruiya, G. W., & Wangui, E. E. (2020). African heritage in a changing climate. *Azania: Archaeological Research in Africa, 55*(3), 297–328.

Chami, F. A. (2019). The archaeology of pre-Islamic Kilwa Kisiwani (Island). *Studies in the African Past, 5*, 119–150. Retrieved June 11, 2021, from http://196.44.162.39/index.php/sap/article/view/2686.

Chinyele, B. J., & Lwoga, N. B. (2018). Participation in decision making regarding the conservation of heritage resources and conservation attitudes in Kilwa Kisiwani, Tanzania. *Journal of Cultural Heritage Management and Sustainable Development, 39*, 88.

Daly, C. (2019). *Built & archaeological heritage climate change sectoral adaptation plan*. Department of Culture Heritage and Gaeltacht. Retrieved July 21, 2020, from http://eprints.lincoln.ac.uk/id/eprint/38705/.

Day, J. C., Heron, S. F., Markham, A., Downes, J., Gibson, J., Hyslop, E., Jones, R., & Lyall, A. (2019). *Climate risk assessment for the heart of Neolithic Orkney World Heritage Site*. Historic Environment Scotland.

Day, J. C., Heron, S. F., & Markham, A. (2020). Assessing the climate vulnerability of the world's natural and cultural heritage. *Parks Stewardship Forum, 36*(1). https://escholarship.org/uc/item/92v9v778.

Eckstein, D., Künzel, V., & Schäfer, L. (2021). *Global Climate Risk Index 2021: Who suffers most extreme weather events? Weather-related loss events in 2019 and 2000–2019*. Germanwatch Nord-Süd Initiative e.V.

Fatorić, S., & Seekamp, E. (2017). Are cultural heritage and resources threatened by climate change? A systematic literature review. *Climatic Change, 142*(1-2), 227–254.

Fluck, H. (2016). *Climate change adaptation report. (Research Report Series 28/2016)*. Historic England.

Guzman, P., Fatorić, S., & Ishizawa, M. (2020). Monitoring climate change in World Heritage properties: Evaluating landscape-based approach in the state of conservation system. *Climate, 8*(39). https://doi.org/10.3390/cli8030039

Heron, S. F., Eakin, C. M., & Douvere, F. (2017). *Impacts of climate change on World Heritage Coral Reefs: a first global scientific assessment*. UNESCO World Heritage Centre.

Heron, S. F., Day, J. C., Cowell, C., Scott, P. R., Walker, D., & Shaw, J. (2020). *Application of the Climate Vulnerability Index for Shark Bay, Western Australia*. Western Australian Marine Science Institution.

Hollesen, J., Callanan, M., Dawson, T., Fenger-Nielsen, R., Max Friesen, T., Jensen, A. M., Markham, A., Martens, V., Pitulko, V., & Rockman, M. (2018). Climate change and the deteriorating archaeological and environmental archives of the Arctic. *Antiquity, 92*.

Ichumbaki, E. B., & Mapunda, B. B. (2017). Challenges to the retention of the integrity of World Heritage sites in Africa: The case of Kilwa Kisiwani and Songo Mnara, Tanzania. *Azania: Archaeological Research in Africa, 52*(4), 518–539.

ICOMOS. (1994). *The Nara document on authenticity*. ICOMOS.

ICOMOS. (2019). *The future of our pasts: engaging cultural heritage in climate action*. ICOMOS.

Intergovernmental Panel on Climate Change. (2015). *Climate Change 2014 – Impacts, adaptation and vulnerability: Part A: Global and Sectoral Aspects: Volume 1, Global and Sectoral Aspects: Working Group II Contribution to the IPCC Fifth Assessment Report*. IPCC. Cambridge University Press.

Khalaf, R. W. (2020). *A proposal to operationalise the concept of compatibility in World Heritage Climate Change Policy* (pp. 1–23). The Historic Environment: Policy & Practice.

Lwoga, N. B. (2018). Dilemma of local socio-economic perspectives in management of historic ruins in Kilwa Kisiwani World Heritage site, Tanzania. *International Journal of Heritage Studies, 24*(10), 1019–1037.

Megarry, W., & Hadick, K. (2021). Lessons from the edge: Assessing the impact and efficacy of digital technologies to stress urgency about climate change and cultural heritage globally. *The Historic Environment: Policy & Practice*. https://doi.org/10.1080/17567505.2021.1944571

Osipova, E., Emslie-Smith, M., Osti, M., Murai, M., Åberg, U., & Shadie, P. (2020). *IUCN World Heritage Outlook 3*. Retrieved January 7, 2021 from https://doi.org/10.2305/iucn.ch.2020.16.en

Perez-Alvaro, E. (2016). Climate change and underwater cultural heritage: Impacts and challenges. *Journal of Cultural Heritage, 21*, 842–848.

Perry, J. (2011). World Heritage hot spots: A global model identifies the 16 natural heritage properties on the World Heritage List most at risk from climate change. *International Journal of Heritage Studies, 17*(5), 426–441.

Perry, J. (2019). Climate change adaptation in natural World Heritage sites: A triage approach. *Climate, 7*(9), 105.

Pollard, E. J. (2008). The maritime landscape of Kilwa Kisiwani and its region, Tanzania, 11th to 15th century AD. *Journal of Anthropological Archaeology, 27*(3), 265–280.

Reimann, L., Vafeidis, A. T., Brown, S., Hinkel, J., & Tol, R. S. J. (2018). Mediterranean UNESCO World Heritage at risk from coastal flooding and erosion due to sea-level rise. *Nature Communications, 9*(1), 4161.

Rockman, M., & Maase, J. (2017). Every place has a climate story: Finding and sharing climate change stories with cultural heritage. In T. Dawson, C. Nimura, E. Lopez-Romero, & M. Yvane Daire (Eds.), *Public archaeology and climate change* (pp. 107–114). Oxbow Books.

Sesana, E., Gagnon, A. S., Ciantelli, C., Cassar, J., & Hughes, J. J. (2021). Climate change impacts on cultural heritage: A literature review. *Wiley Interdisciplinary Reviews: Climate Change*. Retrieved 1/7/21 from https://onlinelibrary.wiley.com/doi/10.1002/wcc.710.

Silva, C. A. V. (2017). Indigenous landscapes Coffee Cultural Landscape of Colombia. *Journal of World Heritage Studies Special Issue: Proceedings of the First Capacity. Building Workshop on Nature-Culture Linkages in Heritage Conservation in Asia and the Pacific*, 44–50.

UNESCO. (1972). *Convention concerning the protection of the world cultural and natural heritage*. UNESCO.

UNESCO. (2008). *Policy document on the impacts of climate change on World Heritage properties*. UNESCO.

UNESCO. (2017). Annex III – declaration of ethical principles in relation to climate change. In *Records of the General Conference, 39th session, Paris, 30 October–14 November 2017* (Vol. 1). Resolutions. UNESCO.

UNESCO World Heritage Centre. (2019). *Operational guidelines for the implementation of the World Heritage Convention*. UNESCO.

Chapter 19
Climate Action and World Heritage: Conflict or Confluence?

Cathy Daly

Abstract In 2007, the *Policy Document on the Impacts of Climate Change on World Heritage Properties* was adopted by the World Heritage (WH) Committee, and a revised policy document, the *Draft Policy Document on Climate Action for World Heritage*, was released in 2021. An English word search on terms related to potential conflicts between WH and climate change was undertaken and utilised as a starting point for an exploration of developments over the 14 intervening years. Four themes were defined and explored, namely, mission, change, loss, and responsibility. In many cases of perceived conflict, professionals and policy makers have been actively working to find solutions. In others, there is the potential for developing new and creative approaches that will ensure the relevance of heritage in an uncertain future.

Keywords Climate change · Climate action · World heritage · Heritage policy

19.1 Introduction

The Paris Agreement of 2015, an agreement within the *United Nations Framework Convention on Climate Change* (UNFCCC), aims to keep global temperature rise below 2 °C through emissions reductions, i.e., the mitigation of greenhouse gasses (GHG) (UN, 1992). Accepting that some degree of climate change (CC) is now

Since this article was accepted for publication the General Assembly of the World Heritage Committee meeting in Paris (24–26 November 2021) failed to adopt the new Policy Document on Climate Action for World Heritage. The policy could not be agreed due to lobbying, by the Australian government in particular, against commitments on Greenhouse Gas Mitigation and use of the List of World Heritage in Danger.

C. Daly (✉)
School of History & Heritage, University of Lincoln, Lincoln, UK
e-mail: cdaly@lincoln.ac.uk

M.-T. Albert et al. (eds.), *50 Years World Heritage Convention: Shared Responsibility – Conflict & Reconciliation*, Heritage Studies,
https://doi.org/10.1007/978-3-031-05660-4_19

inevitable, however, the Paris Agreement has also established the Global Goal on Adaptation, whereby all signatories agree to engage in adaptation planning and the implementation of actions. Adaptation is "*the process of adjustment to actual or expected climate and its effects*…[it] *seeks to moderate or avoid harm or exploit beneficial opportunities*" (IPCC, 2014, p. 1758). The concept of "climate action" encompasses both adaptation and mitigation (UNDP, 2021) and is, therefore, a process with which parties to the Paris Agreement have committed to actively engaging. Although the list of 195 signatories to the Paris Agreement includes every State Party (SP) to the World Heritage (WH) Convention, CC adaptation and mitigation create the potential for disputed approaches for many WH properties. This paper will discuss some of the tensions between climate action and WH governance and what policy efforts have been or could be made to ensure the two coalesce rather than clash.

19.2 Background

Climate change (CC) was first brought to the attention of the WH Committee in 2005 when it received a petition to place four natural heritage sites on the List in Danger due to CC threats (Dannenmaier, 2010; Thorson, 2008). The Committee turned down the proposal, but its resultant decision (05/29.COM/7B.a) made several recommendations that raised the issue of CC as a major concern for both natural and cultural heritage sites for the first time (UNESCO, 2005). The WH Convention (1972) does not specifically mention CC, but SPs are obliged to protect their sites from damaging impacts. In theory, this could be interpreted as an obligation for SPs to the Convention to support the principles of climate action (Gruber, 2008), i.e., mitigating GHGs to prevent increasing CC exposure for WH and implementing adaptation measures to protect against climate impacts.

In 2007, the *Policy Document on the Impacts of Climate Change on World Heritage Properties* was adopted by the WH Committee. Ten years later, they requested the WH Centre and the Advisory Bodies to review and update the policy document "*to make available the most current knowledge and technology on the subject to guide the decisions and actions of the World Heritage community*" (UNESCO, 2016, para. 16). A subsequent meeting of experts tasked with reviewing the 2007 policy gave several recommendations, including that it should be rewritten rather than updated (IANC, 2017). In 2019, two consultants were tasked with drafting this new policy, and one of their first steps was to develop an online questionnaire for WH stakeholders. The response to the questionnaire highlighted the barriers that had inhibited the implementation of the 2007 policy (UNESCO, 2020). These included lack of resources, leadership, knowledge, and political support. The policy document was criticised for being too general and not including site-focused solutions; amendments to the Operational Policy and nomination process as well as Periodic Reporting and Reactive Monitoring processes to include the consideration of CC or make it mandatory were suggested by respondents.

Table 19.1 Potential areas of conflict and related search terms

Conflict area	Search terms			
Mission	Equity	Impact/s	Sustainabl/e/y Sustainability	Justice
Change	Adaptation/ maladaptation	Authenticity	Integrity	Transformation/s/al Transforming
Loss	Loss/es	List of world heritage in danger	Delisting	Operational guidelines
Responsibility	Mitigation	Emissions	Tourism	Renewable/s

Note: Created by the author

The updating of the 2007 policy document on the impacts of CC on WH properties *Draft Policy Document on Climate Action for World Heritage* was released in 2021. Although framed as an update, the document is substantially different from the 2007 policy and departs from it in many significant ways. Interestingly, the draft policy also notes the potential for conflict and tension between CC and WH in a few areas. These include the possible adverse impacts of adaptation or mitigation actions on Outstanding Universal Value (OUV), authenticity and integrity (paras 38 and 94), and the risks of maladaptation and social disruption (para 49). In July 2021, the WH Committee meeting in China, broadcast publicly online, voted to endorse this draft and to send it to the Convention's General Assembly for debate and adoption in the autumn (decision 21/44.COM/7C).

As part of the project *50 Years World Heritage Convention: Shared Responsibility – Conflict & Reconciliation,* points of potential conflict between WH and climate action were discussed and refined at an expert think tank held in March 2021. To explore how policy approaches to these issues have changed from 2007 to 2021, an English word search on both documents was undertaken using related terms (Table 19.1). The word search took the key areas of conflict identified during the think tank discussion as a starting point and, following a period of literature review, these were grouped into sets according to thematic concepts or areas, e.g., "mission". Truncations were used where relevant to ensure all variations of a keyword were found, e.g., sustainab*. There is a degree of overlap between the sets, and the list is iterative; the concept of transform*, for example, was added after its significance in the 2021 policy approach was noted. The relative occurrence of each term in the two texts is utilised in this article as a starting point to discuss developments over the fourteen years since the first policy was written (Figs. 19.1 and 19.2).

19.3 Mission

In 2007, when the UNESCO General Assembly adopted the *Policy Document on the Impacts of Climate Change*, it was accompanied by a report on *Climate Change and World Heritage*, the outcome of an expert advisory group meeting (Cassar

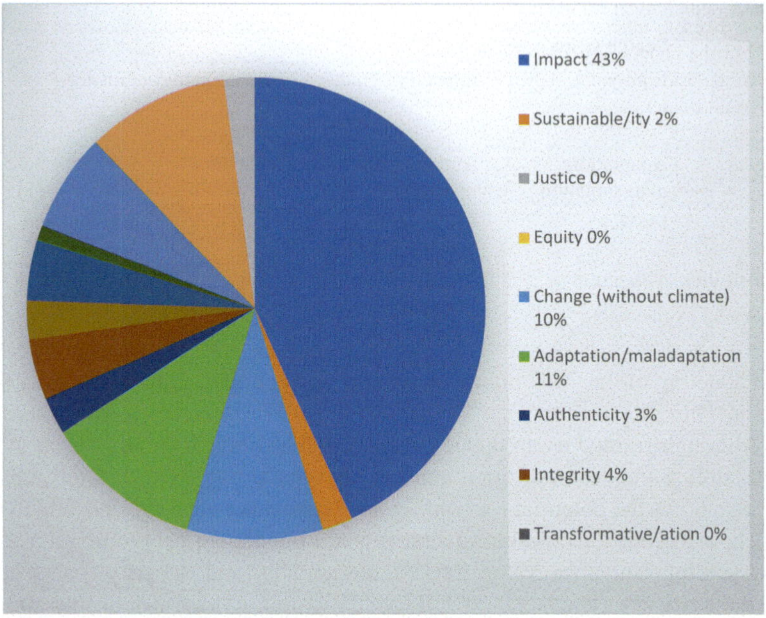

Fig. 19.1 Relative occurrence of search terms in the 2007 policy document. (*Note*: Created by the author)

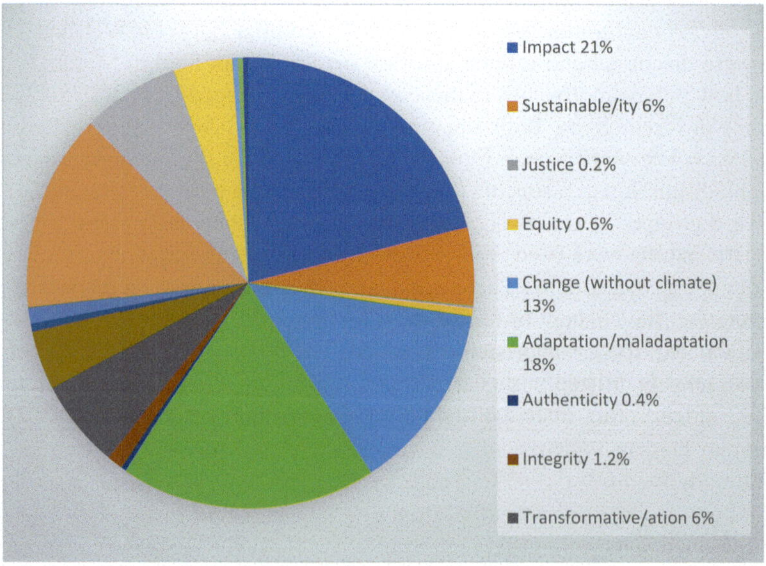

Fig. 19.2 Relative occurrence of search terms in the 2021 draft policy. (*Note*: Created by the author)

et al., 2006; Colette, 2007). The report utilised expert judgement to determine how future CC might impact heritage values worldwide and emphasised the interconnection between the physical and social impacts of CC, suggesting that the way people interact with their heritage and the relevance and value of that heritage to their lives may alter with CC. The WH Committee (07/29.COM/7B.a) also requested that the WH network of sites be used to demonstrate best practice in relation to CC management and raising of public awareness (Cassar et al., 2006). This was accomplished in part by *Case Studies on Climate Change and World Heritage*, a UNESCO publication that used case studies to communicate the issues in an engaging way (Colette, 2009).

In 2005, when UNESCO first called on heritage organisations around the world to embrace the issue of CC, some SPs were vociferous in their opposition to the move for political reasons (Barthel-Bouchier, 2015, p. 157). For example, the Republican US government had not ratified the Kyoto Protocol and was hostile to GHG mitigation at home (Coil, 2007). Barthel-Bouchier (2015) highlights the internal conflict that heritage as an organisational field faced when first engaging with CC, a divisive issue at the time. The danger of perceived 'mission change' for any organisation is that it can alienate public and government support, and it is vital, therefore, that the new direction of the organisation is communicated in a way that makes "*it appear a logical extension of the organisation's charge rather than a betrayal of it*" (Barthel-Bouchier, 2015, p. 151). The construction of a WH mission in relation to CC occurred over the following years and mainly focused on impacts, effectively building consensus by emphasising heritage as a victim of CC (Colette, 2007; ICOMOS, 2008; World Monuments Fund, 2007).

More challenging issues, such as the increase in GHG emissions due to heritage tourism or the need for burden sharing between developed and less developed SPs to the Convention, were largely left out of this early conversation. An exponent of this approach, Terrill (2008) argued that the WH Convention should focus on impacts only and stay out of the politically loaded arena of GHG mitigation as this was the purview of other treaties:

> On grounds of expertise alone, the World Heritage Convention is well advised to avoid moving from describing problems and possible adaptation responses at World Heritage Sites to advocating particular mitigation levels or approaches. (p. 397)

More recently, however, the heritage community have been emphasising issues of climate justice and the potential for heritage to have high ambition and make a more active contribution to climate action (ICOMOS, 2019; IUCN, 2014). The United Nations 2030 Agenda for Sustainable Development (SDG) Goal 13 calls for all countries to "*take urgent action to combat climate change and its impacts*" (UNDP, 2021), and ICOMOS's policy guidance on SDG 13 is to harness heritage "*to enhance the adaptive and transformative capacity of communities and build resilience against climate change*" (Labadi et al., 2021, p. 89). CC is considered a cross-cutting issue and is intrinsic to the other 16 SDG Goals, and ICOMOS presents case studies that illustrate the same is true of heritage (Labadi et al., 2021). The belief that high mitigation ambitions lie outside the remit of WH still persists, however, and was raised during the debate on the CC policy decision 44.COM/7C. Amendments reflecting this view were proposed by Brazil and supported by several other SPs.

The policy word search revealed that the emphasis on "impact(s)" has reduced substantially in 2021, effectively halving in relative occurrence when compared to 2007. This indicates alignment with the wider discourse on the topic. While the 2007 plan does not mention "equity" or "justice" at all, the incidence of these concepts in 2021 is only slightly better (3 and 1, respectively). These occurrences are linked with inclusive and rights-based approaches to the governance of WH (paras 82 and 83) and with sustainable development (para 24). The draft policy points to the 2017 UNESCO *Declaration of Ethical Principles in Relation to Climate Change* (2021, para. 83) for a framework that can address justice and equity and the need for transparent decision making. These ethical principles include the prevention of harm, equity and justice, sustainable development, solidarity and scientific knowledge and integrity in decision making (UNESCO, 2017). The principles could provide an ethical roadmap for the future operation of the WH Convention's climate change policies.

19.4 Change

CC adaptation is likely to require a new approach to change management and the conceptualisation of "acceptable levels" of change (Daly et al., 2020). In the last century, conservation practice had already moved away from the rigidity of arresting change to the flexibility of managing it (Melnick, 2009, p. 41). The Burra Charter, for example, recognises that all places and their components change over time at varying rates (Australia ICOMOS, 2000, p. 2). Given the potential effects of rapid global CC, however, the profession may need to develop a new understanding of what "managing change" means (Melnick, 2009). Harvey and Perry (2015, p. 3) argue that, in a time of rapid change, society needs a new view of heritage, one that "*embraces loss, alternative forms of knowledge and uncertain futures*". They suggest that the heritage field must not only accept transience but should welcome the challenge that this will pose to the current thinking and power structures, using it to transform creatively (Harvey & Perry, 2015, p. 11).

It may indeed be the case that a fundamental shift in ethos is required, and this is likely to be a challenge for the WH system, which relies heavily on concepts of defined values, integrity, and (for cultural sites) authenticity. The idea that heritage managers should facilitate the transformation of the resources in their care to a new state more compatible with a changed climate would seem to be in direct conflict with the notion of maintaining OUV and, therefore, with WH status itself. In the worst-case scenario view, a fear of losing WH designation could inhibit SPs from admitting the scale of CC impact or even undertaking necessary adaptive responses. What Boccardi (2019) calls the "inherent vagueness" of the concept of authenticity may prove sufficiently flexible in interpretation and application to accommodate a response to the challenge of CC. The same does not hold true for the characteristic of integrity, applicable to both natural and cultural properties. Integrity is less mutable than authenticity; relating to "wholeness" and "intactness" of physical elements, it may also include quantifiable indicators such as numbers of species (Stovel, 2007). Paragraph 166 of the Operational Guidelines (UNESCO, 2019) outlines the procedure for modifying the criteria for a property inscribed on the World Heritage List. The initiative for this action must come from the SP concerned and requires a significant burden of paperwork and a minimum of a year and a half preparation.

> Where a State Party wishes to have the property inscribed under additional, fewer or different criteria other than those used for the original inscription, it shall submit this request as if it were a new nomination. (UNESCO, 2019, para. 166)

Statements of value are not static and absolute; they are constructed and defined at a particular time and context (Boccardi, 2019,p. 16). Facilitating the review of statements of OUV, perhaps via the prism of integrity and authenticity, with a less burdensome procedure should be a priority as part of CC adaptation planning. In the word search results of the two policies, the relative incidence of "adaptation" increased only slightly in 2021 when compared to 2007. More significant, however, was the introduction of the term "transformation" in the 2021 plan and the associated concept of transformative change.

Under the heading of *Climate Action*, the 2021 policy states that four key categories are required for WH properties (assessing risks, adaptation, mitigation, and capacity building) but adds an additional fifth concept called "transformative change" (UNESCO, 2021, p. 14. This text is ambitious in the vision it sets out; it stresses the "*urgency and scale of action required by the World Heritage Convention to support bold decisions to transition to a carbon neutral and resilient world that can sustain World Heritage properties for future generations*" (UNESCO, 2021, para. 73). It points to unsustainable cultures of consumption and the call of the Intergovernmental Panel on Climate Change (IPCC) and others for societal reform or "transformative change". The inclusion of this concept in the policy is an acknowledgement that change is not just inevitable but, if embraced as a "*system-wide reorganisation...including paradigms, goals and values*", represents the key to successfully addressing the current climate and ecological crisis both for WH and society as a whole.

> In the context of the World Heritage Convention, transformative change would be exemplified by decisions that contribute towards making World Heritage properties carbon neutral, as much as possible, and more resilient and better adapted to a changing climate, while safeguarding their Outstanding Universal Value...properties can embrace transformative change to become demonstration cases of the change the world needs. (UNESCO, 2021, para. 8)

19.5 Loss

Where CC does result in an irreversible and unmanageable loss of OUV, the WH Committee will face conflict and tensions over how to proceed. Landscape-based approaches offer an approach to exploring environmental change and loss as they are dynamic and capable of navigating time and space, nature and culture (Ferraby, 2015). In a case study of the Jurassic Coast WH site, Ferraby illustrates that loss is part of the process of creating heritage and not necessarily a negative thing.

> It is important to see that these stories and narratives of people and nature can be drawn as much from what is gone as what remains; absent spaces in the environment form and emphasise histories in the living, physical traces. (Ferraby, 2015: 28)

While this may be the case for individual sites, it will be a massive challenge for the WH system to prepare itself for the potential loss of an unknown but likely rapidly escalating number of sites. Entry onto the List of WH in danger, intended to encourage and support governments to take action to protect sites, is often resisted as it is seen as a 'black mark' (Terrill, 2008; Readfearn, 2021).

One of the criteria in the Operational Guidelines for the inscription of a property on the List in Danger is "*threatening impacts of climatic, geological or other environmental factors*" which could have *deleterious effects on its inherent characteristics* (UNESCO, 2019, para. 179–180). The Guidelines go on to specify that such threats "*must be those that are amenable to correction by human action*" (UNESCO, 2019, para. 181). The list of sites in danger is therefore arguably not an appropriate vehicle to address CC, as the causes and solutions are beyond the ability of any one SP to address. The only other mechanism for dealing with sites that have lost their OUV, however, is delisting. According to the Operational Guidelines, properties could be deleted from the WH list "*if the property has deteriorated to the extent that it has lost those characteristics which determined its inscription on the World Heritage List*" (UNESCO, 2019, para. 191). Writing about this issue in 2008, Terrill (2008, p. 395) argued that the credibility of the WH Convention depended on establishing a clear framework for determining loss of OUV to avoid procedural complications when the inevitable difficult decisions must be made.

The UNFCCC established the concept of common but differentiated responsibilities amongst SPs (UN, 1992, Sec. 3.1). This essentially commits developed countries, those most responsible for GHG emissions, to taking the lead in combating CC and its effects. During the WH Committee debate on the policy document, the lack of mechanisms for capacity building assistance and technology transfer (from developed to developing SPs) was noted as an issue that needed to be addressed, as was the lack of reference to the principle of common but differentiated responsibilities. Many early WH losses are likely to be in low-income and small-island-developing nations (Markham et al., 2016). The process of delisting could therefore initiate a political crisis for the WH Convention, and the urgent institution of a process to ensure climate justice and burden sharing is vital for maintaining its reputation. This could be accomplished via UNESCO's sixth ethical principle, "Solidarity", which describes how, on a global scale, assistance should be given to those that are most vulnerable to climate change (UNESCO, 2017, Art. 6). Although there is a much greater occurrence of the word "loss" in the 2021 policy, up from 5 in 2007 to 22 uses, there is a continued lack of concrete proposals for how degradation of OUV will be handled. The draft policy largely avoids confronting the issue. There are no mentions of "delisting" and only 3 of the "List of Heritage in

Danger" in the text, although both were present in 2007 (2 and 8 mentions, respectively). The proposal to place the Great Barrier Reef on the List in Danger during the 2021 WH Committee meeting brought this issue to the fore, with the Australian government successfully lobbying against the action (Readfern, 2021). When it came to the CC policy debate, Australia added an amendment, providing for the formation of an expert group to examine tensions around OUV, delisting, and CC.

The 2007 policy states that existing tools will be utilised to deal with threats to OUV, but, if necessary, the WH Committee will consider taking climate change into account in the revision of these tools (List of World Heritage in Danger, Reactive Monitoring and Periodic Reporting) and of its Operational Guidelines (UNESCO, 2007). The 2021 policy has substantially fewer mentions of the Operational Guidelines (6 vs 13 in 2007) and the other tools of the Convention. It does say that the actions in the plan 'could' be supported at the Committee level by ensuring that basic documents of the World Heritage system, such as the Operational Guidelines and the Resource Manuals, adequately address climate change (UNESCO, 2021, para. 86). Processes and tools were clearly flagged as a priority issue by the stakeholder questionnaire undertaken in advance of the drafting; however, unlike the 2007 policy, which contained several detailed suggestions (never implemented), the 2021 policy takes a strategic approach and requires supplementary implementation guidance.

19.6 Responsibility

The mitigation of GHGs by retrofitting historic buildings or installing renewable energy sources (solar panels on buildings, wind turbines in landscapes) can cause some conflicts (BBC, 2021). This tension is recognised in the 2021 WH policy in paragraph 94, which suggests proactively developing planning processes that avoid or mediate conflicts around proposed renewable projects and other emission reduction actions. More significant was the introduction of several concepts that were either largely (relative incidence <2%) or completely missing in 2007 and which reflect developments over the intervening years. These terms relate to high ambition and the contribution of WH to climate action, namely "sustainability", "emissions", and "transformation" (see Sect. 19.4).

Another problem in respect of emissions is the socio-economic development model that has linked WH status with the generation of international tourism revenue (Rebanks Ltd., 2009). In 2018, the carbon footprint of global tourism was calculated as 8% of worldwide emissions and was projected to grow by 4% every year (Lenzen et al., 2018). What Barthel-Bouchier (2015, p. 161) describes as the "*mythic narrative of sustainable tourism*" continues to be put forward as a solution, yet how the contradiction can be resolved in practice remains unclear. When it comes to climate change adaptation, the tourism imperative will unfortunately also tend to focus resources on sites that are most profitable (e.g., Venice flood defences), also raising the problem of equity and climate justice. Barthel-Bouchier (2015) challenges the heritage community to examine our collective conscience when she writes that, by aligning heritage so closely with economic returns from tourism, we

> …have provided ideological and pragmatic support for a status quo that has made only minimal contributions in the face of the major challenges associated with climate change [and] over time it will become increasingly difficult to…deny evident contradictions between these two missions. (p. 162)

A very notable and welcome development in the 2021 policy, therefore, is the inclusion of references to "tourism". Not referred to at all in 2007, it received 22 mentions in 2021. This inclusion reflects the growing awareness of the contribution heritage tourism makes to economic development but also of the problematic implications this has for GHG emissions reduction (Markham et al., 2016). The emphasis in the main policy text is on the problems of "uncontrolled tourism", the issue of emissions is recognised, but no critical analysis is provided, and the unchallenging solutions are "tourism management" and "sustainable tourism". Experiences during the Covid-19 pandemic point to the importance of heritage for wellbeing, and while international tourism with its high carbon footprint is problematic, the widening of access to a diversity of local visitors should be encouraged. Annex III in the 2021 policy frames key areas for WH properties to reduce emissions such as tourism, and more concrete suggestions are also proposed, including monitoring of GHG emissions from tourism, identifying carbon saving measures, and considering offsetting.

19.7 Conclusion

This paper sought to explore potential conflicts and synergies between the WH system and climate action via a discussion of the 2007 and 2021 policies. As all SPs to the WH Convention are also signatories of the Paris Agreement, it is vital that potential conflicts between commitments under these instruments are recognised and addressed. Heritage professionals and policy makers have been actively working to develop new and creative approaches for dealing with potential conflicts and identifying confluences, ensuring the relevance of heritage in an uncertain future. The new WH policy reframes the debate and clearly represents an attitude shift from the 2007 document, as was seen from the results of a simple word search. The concepts of transformative change, high ambition, and adoption of the UN Ethical Principles all have implications for practice and create clear synergies between WH and other instruments, including the UNFCCC. Changes to the Operational Guidelines, WH tools, and processes are still required, and CC needs to be assessed at all stages of listing and in subsequent reporting. Consideration needs to be given to a range of ideas such as the creation of a specific CC List in Danger, more streamlined methods for SPs to request adjustment of OUV criteria for sites already on the list, mechanisms for dealing with CC in the nomination process, and a means to show solidarity for inevitable losses. Attention is needed to ensure climate justice and burden sharing between nations is part of the process, starting with capacity building and technology transfer between high- and low-income SPs. In the first decade of this century, WH led the heritage sector in highlighting climate change impacts. In this third decade, it could lead in Climate Action, reaffirming the leadership role of WH and ensuring its relevance for generations to come.

References

Australia ICOMOS. (2000). The Burra Charter: The Australia ICOMOS Charter for places of cultural significance 1999, with associated guidelines and code on the ethics of co-existence in conserving significant places. Australia ICOMOS. http://australia.icomos.org/wp-content/uploads/BURRA_CHARTER.pdf

Barthel-Bouchier, D. (2015). Heritage and climate change; organizational conflicts and conundrums. In D. C. Harvey & J. Perry (Eds.), *The future of heritage as climates change loss, adaptation and creativity* (pp. 151–166). Routledge.

BBC. (2021). *Proposed power line would cross Antonine Wall*. British Broadcasting Company. https://www.bbc.com/news/uk-scotland-tayside-central-57061971

Boccardi, G. (2019). Authenticity in the heritage context: A reflection beyond the Nara document. *The Historic Environment: Policy & Practice, 10*(1), 4–18. https://doi.org/10.1080/17567505.2018.1531647

Cassar, M., Young, C., Weighell, T., Sheppard, D., Bomhard, B., & Rosabel, P. (2006). *Predicting and managing the effects of climate change on World Heritage: A joint report from the World Heritage Centre, its advisory bodies, and a broad group of experts to the 30th session of the World Heritage Committee (Vilnius 2006)*. UNESCO World Heritage Centre.

Coil, Z. (2007). How the white house worked to scuttle California's climate law. *San Francisco Chronicle*. https://web.archive.org/web/20130617052823/http://www.commondreams.org/archive/2007/09/25/4099.

Colette, A. (2007). *Climate change and World Heritage: Report on predicting and managing the impacts of climate change on World Heritage and strategy to assist States Parties to implement appropriate management responses*. UNESCO World Heritage Centre.

Colette, A. (2009). *Case studies on climate change and World Heritage*. UNESCO World Heritage Centre.

Daly, C., Engel-Purcell, C. Chan, C., Donnelly, J., MacDonagh, M. & Cox, P. (2020). Climate change adaptation planning, a national scale methodology. *Journal of Cultural Heritage Management and Sustainable Development*. Vol. ahead-of-print No. ahead-of-print. https://doi.org/10.1108/JCHMSD-04-2020-0053

Dannenmaier, E. (2010, September 19). *International tribunals as climate institutions? Strengthening access to climate justice* [conference paper]. Yale/UNITAR Second Conference on Environmental Governance and Democracy. .

Ferraby, R. (2015). Narratives of hange on the Jurassic Coast World Heritage Site. In D. C. Harvey & J. Perry (Eds.), *The future of heritage as climates change loss, adaptation and creativity* (pp. 25–42). Routledge.

Gruber, S. (2008). Environmental law and the threat of global climate change to cultural heritage sites Sydney Law School Research Paper No. 08/117 [conference paper]. *Climate law in developing countries post-2012: North and South perspectives*, IUCN, academy of environmental law, Faculty of Law, University of Ottawa, Canada, pp. 26–28.

Harvey, D. C., & Perry, J. (2015). Heritage and climate change, the future is not the past. In D. C. Harvey & J. Perry (Eds.), *The future of heritage as climates change loss, adaptation and creativity* (pp. 3–21). Routledge.

ICOMOS. (2008). *Heritage at risk: ICOMOS World Report 2006–2007 on monuments and sites in danger*. hendrik Bäßler verlag.

ICOMOS Climate Change and Cultural Heritage Working Group. (2019). *The future of our pasts: Engaging cultural heritage in climate action*. ICOMOS. https://indd.adobe.com/view/a9a551e3-3b23-4127-99fd-a7a80d91a29e

International Academy for Nature Conservation. (2017). *World Heritage and climate change – Towards the update of the policy document on the impacts of climate change on World Heritage properties*. [unpublished report] International Expert Workshop.

IPCC. (2014). Annex II: Glossary [Mach, K.J., S. Planton and C. von Stechow (eds.)]. In R. K. Pachauri & L. A. Meyer (Eds.), *Climate change 2014: Synthesis report. Contribution of working groups I, II and III to the fifth assessment report of the intergovernmental panel on climate change* (pp. 117–130). IPCC.

IUCN. (2014). *The benefits of natural World Heritage: Identifying and assessing ecosystem services and benefits provided by the world's most iconic natural places*. IUCN. https://portals.iucn.org/library/node/44901

Labadi, S., Giliberto, F., Rosetti, I., Shetabi, L., & Yildirim, E. (2021). *Heritage and the sustainable development goals: Policy guidance for heritage and development actors*. ICOMOS.

Lenzen, M., Sun, Y. Y., Faturay, F., Ting, Y. P., Geschke, A., & Malik, A. (2018). The carbon footprint of global tourism. *Nature Clim Change, 8*, 522–528. https://doi.org/10.1038/s41558-018-0141-x

Markham, A., Osipova, E., Lafrenz Samuels, K., & Caldas, A. (2016). *World Heritage and tourism in a changing climate*. United Nations Environment Programme, Nairobi, Kenya and United Nations Educational, Scientific and Cultural Organization. UNESCO.

Melnick, R. Z. (2009). Climate change and landscape preservation: A twenty-first-century conundrum. *APT Bulletin, 40*, 35–42.

Readfern, G. (2021). World Heritage Committee agrees not to place great barrier reef on 'in danger' list. *The Guardian*. https://www.theguardian.com/environment/2021/jul/25/the-great-barrier-reef-is-not-on-the-in-danger-list-why-and-what-happens-next

Rebanks Consulting Ltd. (2009). *World Heritage status, Is there opportunity for economic gain? Research and analysis of the socio-economic impact potential of UNESCO World Heritage site status*. Lake District World Heritage Project. WHS-Lake-District-Economic-Gain-Doct..pdf (engageliverpool.com)

Stovel, H. (2007). Effective use of authenticity and integrity as World Heritage qualifying conditions. *City & Time, 2*(3), 3. http://www.ct.ceci-br.org

Terrill, G. (2008). Climate change: How should the World Heritage Convention respond? *International Journal of Heritage Studies, 14*(5), 388–404. https://doi.org/10.1080/13527250802284388

Thorson, E. J. (2008). On thin ice: The failure of the United States and the World Heritage Committee to take climate change mitigation pursuant to the World Heritage Convention seriously. *Environmental Law, 38*, 139–176.

UNDP. (2021). *Sustainable development goals*. Climate Action. United Nations. https://www.undp.org/sustainable-development-goals#climate-action

UNESCO. (2005). *Decisions of the 29th session of the World Heritage Committee. (05/29.COM/7B.a)*. UNESCO.

UNESCO. (2007). *Policy document on the impact of climate change on World Heritage properties (WHC-07/16.GA/10)*. Sixteenth session of the general assembly of states parties to the convention concerning the protection of the world cultural and natural heritage. UNESCO.

UNESCO. (2016). *Decision 40.COM 7 state of conservation of World Heritage properties*. 40th session of the World Heritage Committee. UNESCO.

UNESCO. (2017). Declaration of ethical principles in relation to climate change Records of the General Conference, 39th session, Paris, 30 October–14 November 2017, v. 1: Resolutions. UNESCO. https://unesdoc.unesco.org/ark:/48223/pf0000260889.page=127

UNESCO. (2019). *Operational guidelines for the implementation of the World Heritage Convention*. Intergovernmental Committee for the Protection of the world cultural and natural heritage. World Heritage Centre, UNESCO. https://whc.unesco.org/en/guidelines

UNESCO. (2020). *Updating of the policy document on the impacts of climate change on World Heritage properties. Summary of the online consultation (30 December 2019–31 January 2020).* [Unpublished].

UNESCO. (2021). *Draft policy document on climate action for World Heritage*. World Heritage Centre, UNESCO.

United Nations. (1992). *United Nations framework convention on climate change,* (A/CONF/151/26). United Nations. https://unfccc.int/files/essential_background/background_publications_htmlpdf/application/pdf/conveng.pdf

World Monuments Fund. (2007). World monuments watch 2008 list of 100 Most endangered sites. https://www.wmf.org/sites/default/files/press_releases/pdfs/2008%20Watch%20List.pdf

Chapter 20
Conflict Areas and Solution Strategies in the Conservation of Ecosystems and Their Services: A Holistic Approach

Esteban Avigliano and Nahuel Schenone

Abstract Natural environments and biodiversity are negatively affected by climate change and non-sustainable human activities around the world. Different management strategies have been implemented to mitigate the loss of habitat and ecosystem functions. Nevertheless, many of these have failed because, in general, they focus on protected areas. The loss of habitat and, thus, biodiversity occurs outside these areas and does not receive attention. Often, the conservation strategies go against the needs of the communities in the surroundings of the protected areas, generating a series of conflicts between the local governments, conservationists, and residents. In this sense, it is necessary to carry out holistic conservation strategies that consider human beings and their socio-cultural complexity within the environment to overcome the effect of climate change on biodiversity loss. This chapter empirically shows how it is possible to apply conservation strategies integrating scientific and political capacities and uniting governmental and non-governmental organizations for the execution of socio-environmental, educational, and research actions. This holistic approach contributes to the restoration of the environment and its services and to the mitigation of climate change in subtropical regions.

Keyword Atlantic forest · Non-protected areas · Holistic model · Policy makers · Biodiversity hotspot

---

E. Avigliano (✉)
Centro de Investigaciones Antonia Ramos (CIAR), Fundación Bosques Nativos Argentinos, Oberá, Misiones, Argentina

Instituto de Investigaciones en Producción Animal (INPA-CONICET-UBA), Universidad de Buenos Aires, Buenos Aires, Argentina
e-mail: estebanaviglian@conicet.gov.ar

N. Schenone
Centro de Investigaciones Antonia Ramos (CIAR), Fundación Bosques Nativos Argentinos, Oberá, Misiones, Argentina
e-mail: nfschenone@bosques.org.ar

M.-T. Albert et al. (eds.), *50 Years World Heritage Convention: Shared Responsibility – Conflict & Reconciliation*, Heritage Studies,
https://doi.org/10.1007/978-3-031-05660-4_20

20.1 Background

Climate change is a phenomenon that has occurred naturally since the beginning of the planet's history; however, during the last century, there is abundant scientific evidence that suggests that certain human activities accelerate change (Mcmichael et al., 2004). These activities are typically related to the emission of greenhouse gases (through the action of industry and agriculture) and the progressive loss of environments, especially forests, which fix these gases such as carbon dioxide (Díaz, 2019; Mcmichael et al., 2004). Forests are transformed into pastures for livestock or are unsustainably exploited, resulting in the degradation and loss of habitats, as well as the ecosystem services that they provide to humanity (Laurance, 2010). Climate change and habitat loss are currently the major causes for the worldwide decline of biodiversity (Brook et al. 2008; Serengil et al., 2011). Habitat fragmentation, conversion, and overexploitation are the leading drivers of biodiversity decline and species extinctions in tropical hotspots (Norman, 2003). Some actions have been implemented in an attempt to mitigate the loss of habitats, such as the creation of specific legally protected areas (PA) (e.g. National Parks) and World Heritage sites (Gray et al., 2016). This concept of creating legally protected areas that preserve samples of specific environments (e.g. vulnerable hotspots) has been used throughout the world (Norman, 2003). There is no doubt that PAs help to conserve environments, which is reflected in the conservation of biodiversity, ensure certain ecosystem services, and contribute to mitigating climate change. However, legally PAs commonly represent a small proportion of the ecosystems they intend to protect (Norman, 2003), and research efforts are often focused only on them (Oldekop et al., 2016), leaving aside legally non-protected areas (NPAs), which are more likely to disappear (Avigliano et al., 2019a). Moreover, many management policies, and even a significant amount of research, are focused on these small PAs (Ferro et al., 2014; Lemes et al., 2014; Loyola et al., 2014; Massara et al., 2015; Paschoal et al., 2016); meanwhile, habitat loss occurs on a large scale outside them. Conservation based solely on Strict Nature Reserves (prohibited access) that often exclude some needs of societies (e.g. refuge, and access to natural resources) generates conflict between different actors such as states, conservationists, and the local population (Lewis, 1996). These conflicts threaten the efficiency of conservation plans because different sectors of society are excluded from them. Currently, we know that conservation based on PAs is not enough to protect the habitats necessary to conserve biodiversity, ensure ecosystem services, and mitigate climate change (Avigliano et al., 2019a). It is necessary to change the paradigm and apply strategies combining legally PAs (e.g. strict nature reserves) with areas of multiple and sustainable use, including society and its needs in conservation and management plans (Oldekop et al., 2016). Considering the wealth of evidence on the importance of NPAs in maintaining ecosystem services and mitigating their exposure to human activities (DeFries et al., 2007; Palomo et al., 2013), greater efforts must be made to conserve and properly manage these areas. The management of NPAs is especially complex because they are typically multipurpose areas and often correspond to a mixture of public and private lots that are subject to different uses and are inhabited by diverse ethnic groups with different relationships with nature (Papalia, 2012). In this sense, conservation actions cannot focus only on inspection, control, research, and, occasionally, environmental education (predominant activities in protected areas) but must go hand in hand with strategies for sustainable use and restoration of biodiversity. To carry out these actions, it is necessary to address different spheres, integrating population, government agencies (ministries, local, provincial, and national governments), scientists, non-governmental organizations, and access to financing from different sources (state, international, private) (Oldekop et al., 2016).

Herein, we present our experience, in which we reveal the importance of an unprotected semi-degraded area from the southern extreme of the Atlantic Rainforest (ARF), which can help provide further tools for sustainable management of biodiversity and environment conservation in NPAs. We have applied a holistic approach, integrating basic research on biodiversity, direct restoration actions, social, educational, and outreach actions, and interactions with different political, scientific, and educational organizations.

20.2 Methods

20.2.1 A Holistic Approach Model

Society uses semi-degraded non-protected areas as part of their economies, culture, and recreation. Consistently, local policy makers should take the community values and characteristics into account in the design and promotion of management plans. This case study transcends the purely conservationist concept by developing a holistic approach (Fig. 20.1) to support ecosystem services, nature conservation, and sustainable development. This approach is the result of the interaction between

Fig. 20.1 Schematic representation of the holistic approach to the management and recovery of ecosystem services, conservation, and sustainable development in the study area. *Note*. All bubbles are closely interrelated through the holistic approach. (*Source*: Prepared by the authors)

different spheres of actors and interactions between national, international, private, and governmental institutions.

20.2.2 *Atlantic Rainforest*

With 1500,000 km^2, the ARF represents one of the main biodiversity hotspots in the world (Norman, 2003) and functions as an important climate shaper (Ledru et al., 2009). It is among the main hotspots with the largest number of endemic plants and vertebrates, amounting to 4.8% of total endemic species on the planet (Norman, 2003). Currently, less than 8% of the original cover remains, principally as small fragments (<50 ha) that are isolated from each other (Ribeiro et al., 2009). More than 110 million people live across the ARF, including colonists of European descent and several Mbyá-Guaraní indigenous groups (~135,000 inhabitants) (Leal et al., 2003; Ribeiro et al., 2011). Forest is commonly replaced by agriculture and livestock pasture, silvopastoral systems, and urban areas at a deforestation rate of 200 km^2/year (INPE, 2015). There are more than 650 PAs in the ARF, with a total surface area of 14,000 km^2, representing only 1% of the original area (Leal et al., 2003).

Therefore, it could be anticipated that semi-degraded non-protected areas from the ARF could be very valuable for the conservation of neotropical habitats, so management efforts should not be concentrated only on PAs. In this context, we have located our holistic management efforts in a semi-degraded non-protected area of the ARF. This area is a 650 ha private parcel where there are no permanent inhabitants. The area borders with farms of 25–50 ha inhabited by single families, which have tobacco, tea, and yerba crops, and a private and uninhabited parcel of semi-degraded forest of around 5000 ha (private property). There are two urbanized centres, with less than 1500 inhabitants each, less than 8 km away.

20.2.3 *Legal Framework*

The Fundacion Bosques Nativos Argentinos para la Biodiversidad (FBNA) is a non-profit, non-governmental organization (NGO) that provides a formal framework for management and conservation actions; it was formalized between 2009 and 2011 by local people and scientists. In terms of a legal framework, the study area (650 ha owned by FBNA) is based on private land without strict restrictions, unlike national parks or other protected natural areas. The main legal instrument concerning land use in the study area is the National Law on Minimum Standards for Environmental Protection of Native Forests (National Forest Law, 26, 331, 2007). This regulation promotes the enrichment, restoration, conservation, and rational and sustainable use of native forests and their ecosystem services.

20.2.4 Identifying Problems That Need To Be Addressed

From 2010, the FBNA brought together a series of specialists from different areas of knowledge, local producers, and people to carry out environmental baselines and identify management problems and priorities (Table 20.1). Scientific research was promoted and facilitated on the NPA by building a scientific station named "Centro de Investigations Antonia Ramos (CIAR)" within the study area. CIAR has accommodation for 16 researchers distributed in two cabins, a laboratory, and terrestrial and aquatic vehicles. Issues about education, communication, and sustainable production were explored through ad hoc interviews and workshops with the local community (Table 20.1).

20.3 Partial Results

20.3.1 Financing

Given the nature of the holistic approach to developing alternatives for preserving natural areas with no legal protection within a climate change context, the financial strategy was open and broad (Table 20.1). The first step was taken by local landlords with high environmental awareness to develop the master plan and the first in-field capacities to obtain clear results in a short time. These results and the master plan were key factors for fundraising. The funding sources were broad and were relevant to all stakeholders to reflect a common goal and common outcomes. Mainstreaming biodiversity and climate change in the project can lead to UNEP-UNDP-GEF funding and other sources of international financing, such as the International Barcode of Life Program (iBOL), national forest funding (National Forest Law), National Science Agency funding, provincial funding, individual donors through the NGO, and private sector.

As an outcome of the financial approach, the GEF-UNDP funding (2016–2018), and multi-stakeholder involvement, an ecosystem services payment scheme was developed to protect the local watershed, encouraging good management practices in local farms.

20.3.2 Scientific Research

Since 2010, we have invited scientists to carry out environmental (bacteriological, pharmaceutical, physicochemical, and agrochemicals studies, among others) and taxonomic (e.g. mammals, birds, arthropods, fish, and fungi) studies in order to generate baselines that facilitate decision-making on the management of the area (Table 20.1). Scientists have carried out their research with the logistical support of

Table 20.1 Summary of the main results obtained through the different activities

Activity	Mechanism	Date	Result
Financing	Constant presentation of financing projects at the local, provincial, national and international levels	2009–present	Acquisition of funds from different sources, for example, International Barcode of Life Program (2013–2014), UNDP-UNEP-GEF 3623; UNDP-GEFARG15/G53: 2016–2018, national forest funding (2011 and 2014), National Science Agency funding (several projects), provincial funding (several projects), and donations
Scientific research	Creation of the CIAR	2010	Biodiversity baseline
	Presentation of financing projects at the local, provincial, national and international levels	2010–present	Pollution baseline
			Peer-reviewed scientific publications
Forest restoration	Workshops and interviews with local inhabitants	2010–2015	Creation and maintenance of a seed bank
	Presentation of financing projects at different scales	2010–present	Creation of plant nurseries (40,000 seedlings/year)
	Identify and differentiate the seedbeds	2010–present	165,000 native trees (plant cover increased 38%)
			Implementation of several river bank recovery projects
Education and communication	Workshops and interviews with local inhabitants and rulers	2010–present	Built a primary level school (2013) involving the local population (around 270 students/year graduated)
	Presentation of financing projects at different scales	2010–present	Implementation of several educative projects (e.g. slingshot project)
	Creation and maintenance of websites and social networks	2011–present	Educational talks from primary to postgraduate
	Creation and maintenance of the biodiversity magazine	2011	Creation of the annual postgraduate course "dynamic biodiversity restoration" (2017)
			Permanent interaction through networks with more than 350,000 followers

(continued)

Table 20.1 (continued)

Activity	Mechanism	Date	Result
Sustainable production	Workshops and interviews with local inhabitants and rulers	2010–present	Creation of sustainable production samples at CIAR
	Presentation of financing projects at different scales	2012–present	Assistance and workshops for local producers and students of different educational levels
			Implementation of several strengthening, training and transfer projects
Policy makers	Permanent work with local and national rulers through workshops and interviews	2011–present	New environmental legislation: e.g. law XVI-116 (2016)
			Support for the creation of the Instituto Misionero de Biodiversidad

Note: Prepared by the authors

CIAR and, in many cases, have published their results in indexed journals (Avigliano et al., 2019a).

20.3.2.1 Biodiversity Baseline

Relatively high specific richness was found in most of the taxa examined, and one new genus (arachnid) and six new species and several putative new species (fish and arthropod) were described (Avigliano et al., 2019a). In addition, six vulnerable species (bird and mammal) and 36 first records for Argentina (fish, arthropod, platyhelminth, and fungi) were reported (Avigliano et al., 2019a). Among the most emblematic and globally vulnerable species, the red brocket deer, *Mazama nana*, the feline *Leopardus tigrinus*, and the helmeted woodpecker, *Celeus galeatus*, were reordered. The DNA of more than 70,000 terrestrial arthropods was studied, and 8,651 different barcode index numbers (which are a close proxy for species) were found. The total number of species of diurnal butterflies found around the CIAR was 500, representing around 70% of the species found in the Iguazú National Park and 25% of the species recorded in the whole ARF, indicating that the CIAR and its surroundings are a butterfly hotspot for the entire ecoregion (Avigliano et al., 2019a).

20.3.2.2 Pollution Baseline

Up to 18 organochlorine pesticides were measured in water, suspended particulate matter, sediment, epiphytic plants, and fish (Avigliano et al., 2019a; Ondarza et al., 2019). Caffeine, norfluoxetine (a metabolite of the antidepressant fluoxetine), benzoylecgonine, and antibiotics were detected in the tissues of three native fish species (Ondarza et al., 2019). The concentration of 20 trace elements, glyphosate, and fecal coliform in surface water have been explored in the CIAR (Avigliano et al., 2019a, b; Avigliano & Schenone, 2015).

20.3.2.3 Forest Restoration

In the last 50 years, selective deforestation (extraction of wood) has affected 85% of the study area, while the remaining 15% was totally deforested (Avigliano et al., 2019a). Tobacco, tea, and yerba mate were cultivated inside the study area. In order to mitigate this scenario, reforestation projects have been implemented as part of the biodiversity restoration master plan. These projects included activities such as the creation and maintenance of a seed bank, construction and maintenance of nurseries, and planting and monitoring (Table 20.1). The first action was a series of workshops and interviews with local inhabitants, many of whom live off the extraction of wood, to identify the highly impacted species. The second action was to identify and differentiate the seedbeds with the support of the local community. Later, plant nurseries of 50 native tree species were built. Finally, annual plantations and monitoring were carried out in the context of different subprojects. FBNA personnel, local residents, and school-age children (see education and communication section) invest in these plantations. Over the last decade, around 165,000 trees have been planted using different methodologies according to the characteristics of the terrain and the degree of deterioration of the forest. Among the main results, we observe that the intermediate stratum (10–20 m) is recovering and, in the most conserved patches of forest, the canopy stratum (20–30 m), the herbaceous and shrub layer was managed to guarantee access to light for seedlings (Avigliano et al., 2019a). After a decade of work, plant cover increased 38%, at an average rate of 3.8% per year (Avigliano et al., 2019a).

Several reforestation projects were associated with the recovery of riverbanks (e.g. Acaraguá River, Ramón, Ramos, and Cascada streams) to mitigate the effect of runoff caused by deforestation and trampling of livestock. These projects need a holistic approach because they must be accompanied by mitigation actions that allow farm animals access to water but not to riverbanks (see financing section). In this sense, electrified fences powered by renewable energy must be built, and reforestation, monitoring of environmental parameters, protection of water springs, and management of pastures and livestock are also required.

20.3.3 Education and Communication

Education and outreach (Table 20.1) were essential to integrate local communities into management projects. The local communities belong to vulnerable settlements, and many people do not have the means to travel to schools. Therefore, the FBNA built a primary level school on the outskirts of the study area, which is attended by 270 students. The school was built by local people, including parents of the students, which created a feeling of belonging. The students frequently attend lectures and environmental activities (actively participate in the plantations) codified by FBNA staff and teachers.

Various educational projects have been implemented and supported, such as the “slingshot project”, in which cycles of talks were held in 16 public schools with more than 100 teachers and attended by more than 1,000 students (2014–2015). The exchange of slingshots for binoculars was used as an opportunity to give educational talks on conservation and environmental management.

At the university level, educational talks and examples of case studies are given in national universities and educational institutes such as museums throughout the country. Since 2017, the postgraduate course “Dynamic Biodiversity Restoration” has been taught, to which more than 400 students are enrolled per edition. Many students carry out their bachelor, master, and doctoral thesis studies on different environmental topics.

The FBNA interacts permanently with its followers through the official website (www.bosques.org.ar n.d.) and social networks such as Instagram and Facebook (www.facebook.com/BosquesNativosArgentinos n.d.). From the beginning, the FBNA has published a free environmental outreach magazine called Biodiversity® (www.bosques.org.ar/publicaciones/ n.d.). Restoration and management actions are also communicated through radio, television, and documentary interviews on a local and national scale.

20.3.4 Sustainable Production

The effectiveness of NPAs as units for the conservation of ecosystem services depends, at least in part, on considering the needs of the local population. In this sense, it is necessary to generate tools that allow the inhabitants to obtain an economic benefit from the services provided by the environment they care for (Table 20.1). CIAR functions as a pilot and experimentation site to generate sustainable production tools and has solar yerba dryers, native bee hives, lots planted with native fruit trees, and yerba mate under the forest, as well as facilities to produce food based on native fruits for educational purposes. Part of the energy used (lighting, irrigation systems, and water distribution for animal drinkers) is based on solar and hydro energy as a demonstration of production based on reduced carbon signatures.

The projects associated with educational and outreach activities seek to produce high-quality organic food within the forest while enriching the forest with native arboreal species at the same time. Other projects aim to replace the afforestation of exotic trees with native fruit trees, which produce food that can be marketed in different ways. For example, from 2020–present, a project (PNUD ARG 15/G53) has been carried out in which a cooperative of women producers were trained to produce native fruit trees and products derived from fruits. A series of field trainings were carried out, four nurseries were built, and private lots were afforested with thousands of native fruit trees.

20.3.5 Policy Makers

The relationship with the local government was part of the strategy to capitalize on the results obtained by the pilot project and improve environmental legislation (Table 20.1). This step was started at the beginning of the process to create a bond between environmental results and political results in a synergic way. The relationship and the fluid information exchange with the local authorities put the environmental concern on the provincial political agenda. New environmental legislation was developed to replicate the pilot model at a provincial scale, and the provincial biodiversity institute (Instituto Misionero de Biodiversidad, IMIBIO) was created for these purposes by law (Law XVI-122) in 2018 (https://imibio.misiones.gob.ar/ n.d). For example, the provincial germplasm law was issued (Law XVI-116) as a result of the holistic model. We have also generated a water quality monitoring manual and a large number of technical documents for specific problems according to the requirements of the provincial managers (Schenone et al., 2014).

20.4 Discussion and Final Remarks

Protecting specific heritage areas and protected areas is crucial, but still not enough to mitigate climate change, conserve biodiversity, and maintain ecosystem services in the long term (Avigliano et al., 2019a). There have to be new ways of understanding the surrounding areas to buffer the effects of climate change in World Heritage natural areas (Gray et al., 2016). The holistic approach based on the interaction between society, science, and environmental policies and applied to the semi-degraded natural pilot site yielded valuable contributions for the integral management of these areas, supporting the restoration of the environment and its services and the mitigation of climate change.

The absence of long-term financing plans and communication and dissemination strategies for management in other natural areas excludes the local community from any possibility of contributing to the management and conservation plans. This is observed in several Multipurpose Protected Areas such as the Yabotí (YBR) and Delta del Paraná (DPBR) Biosphere Reserves. A clear example is the deterioration of both reserve's biological stations, caused by lack of funding and lack of long-term scientific policies. Usually, new knowledge about these areas depends on the will of scientists who carry out specific investigations (Avigliano et al., 2016; Rolón et al., 2021), which are not integrated into a management plan. A large part of the DPBR community does not know what activities are allowed or the benefits of the protected area (Cassini & Túnez, 2019). Therefore, people do not have the tools to integrate with and contribute to management strategies.

In this study, the suitability of the approach used is reflected in the success of reforestation, education and communication activities, research, and economic support to local communities through the execution of projects associated with sustainable production (Table 20.1). The success of this project was also reflected in the creation of a provincial research and management institute (IMIBIO). This institute is based in Iguazú city (located 220 km from CIAR), which has been a biodiversity hotspot and World Heritage property since 1984, and was established to scale up the holistic approach carried out in our pilot site.

The activities carried out during the decade of work resulted in a map of interactions, supported by a series of scientific investigations (Avigliano et al., 2019a), which was essential to generate conservation and management actions (Fig. 20.2). The progress of the projects was communicated at local, national, and international community scales through the scientific community or through environmental education projects. The resulting communications and the scientific evidence contribute to the updating of environmental laws and policies, the action of which requires the interaction between different institutions such as an NGO (FBNA) and local and national governments. Throughout the process, specific funds were required for the different actions (Table 20.1), which were mainstreamed through the FBNA. All these actions added together resulted in the execution of projects for conservation and active restoration of the environment, leading to the implementation of sustainable production models and contributing to regional economies. Finally, the holistic approach was able to provide an important range of ecosystem services and thus contribute to the mitigation of climate change.

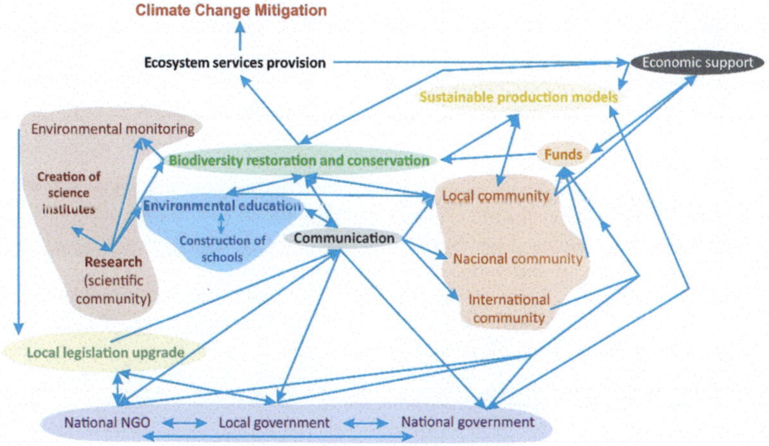

Fig. 20.2 Map of interactions between different actors and work areas carried out to generate holistic conservation and management actions. (*Note*: Prepared by the authors)

Acknowledgements We thank the Centro de Investigación Antonia Ramos (CIAR), Fundación Bosques Nativos Argentinos para la Biodiversidad.

References

Avigliano, E., & Schenone, N. (2015). Human health risk assessment and environmental distribution of trace elements, glyphosate, fecal coliform and total coliform in Atlantic rainforest mountain rivers (South America). *Microchemical Journal, 122*, 149–158.

Avigliano, E., Lozano, C., Plá, R. R., & Volpedo, A. V. (2016). Toxic element determination in fish from Paraná River Delta (Argentina) by neutron activation analysis: Tissue distribution and accumulation and health risk assessment by direct consumption. *Journal of Food Composition and Analysis, 54*, 27–36.

Avigliano, E., Rosso, J. J., Lijtmaer, D., Ondarza, P., Piacentini, P., Izquierdo, M., Cirigliano, A., Romano, G., Nuñez, B. E., Porta, A., Mabragaña, E., Grassi, E., Palermo, J., Bukowski, N., Tubaro, P., & Schenone, N. (2019a). Biodiversity and threats in non-protected areas: A multidisciplinary and multi-taxa approach focused on the Atlantic Forest. *Heliyon, 5*(8), e02292.

Avigliano, E., Clavijo, C., Scarabotti, P., Sánchez, S., Llamazares, V. S., del Rosso, F. R., Caffetti, J. D., Facetti, J. F., Domanico, A., & Volpedo, A. V. (2019b). Exposure to 19 elements via water ingestion and dermal contact in several South American environments (La Plata Basin): From Andes and Atlantic forest to sea front. *Microchemical Journal, 149*.

Brook, B. W., Sodhi, N. S., & BradshawC, J. A. (2008). Synergies among extinction drivers under global change. *Trends in Ecology & Evolution, 23*(8), 453–460.

Cassini, M. H., & Túnez, J. I. (2019). Situación Ambiental Del Bajo Delta Del Paraná: ¿Qué Opinan Los Pobladores? *Ecología Austral, 29*(3), 296–305.

DeFries, R., Hansen, A., Turner, B. L., Reid, R., & Liu, J. (2007). Land use change around protected areas: Management to balance human needs and ecological function. *Ecological Applications, 17*(4), 1031–1038.

Díaz, S., Settele, J., Brondízio, E. S., Ngo, H. T., Guèze, M., Agard, J., … Zayas, C. (2019). *Summary for policymakers of the global assessment report on biodiversity and ecosystem services of the intergovernmental science-policy platform on biodiversity and ecosystem services*. Intergovernmental Science-Policy Platform on Biodiversity and Ecosystem Services (PBES).

Ferro, V. G., Lemes, P., Melo, A. S., & Loyola, R. (2014). The reduced effectiveness of protected areas under climate change threatens Atlantic Forest Tiger Moths. *PLoS One, I*(9), e107792.

Gray, C. L., Hill, S. L. L., Newbold, T., Hudson, L. N., Boïrger, L., Contu, S., Hoskins, A. J., Ferrier, S., Purvis, A., & Scharlemann, J. P. W. (2016). Local biodiversity is higher inside than outside terrestrial protected areas worldwide. *Nature Communications, 7*(1), 1–7.

INPE. (2015). *Atlas Dos Remanescentes Florestais Da Mata Atlântica – Período 2013–2014*. Fundação SOS Mata Atlântica.

Laurance, W. F. (2010). Habitat destruction: Death by a thousand cuts. In N. S. Sodhi & P. R. Ehrlich (Eds.), *Conservation biology for all*. Oxford University Press.

Leal, C. G., de Gusmão Câmara, I., & Sayre, D. (2003). *The Atlantic forest of South America: Biodiversity status, threats, and outlook* (1st ed.). Island Press.

Ledru, M. P., Mourguiart, P., & Riccomini, C. (2009). Related changes in biodiversity, insolation and climate in the Atlantic rainforest since the last interglacial. *Palaeogeography, Palaeoclimatology, Palaeoecology, 271*(1&2), 140–152.

Lemes, P., Sanches, M. A., & Dias, L. R. (2014). Climate change threatens protected areas of the Atlantic forest. *Biodiversity and Conservation, 23*(2), 357–368.

Lewis, C. (1996). *Managing conflicts in protected areas*. IUCN.

Loyola, R. D., Lemes, P., Brum, F. T., Provete, D. B., & Duarte, L. D. S. (2014). Clade-specific consequences of climate change to amphibians in Atlantic Forest protected areas. *Ecography, 37*(1), 65–72.

Massara, R. L., De Oliveira, P. A. M., Doherty, P. F., Hirsch, A., & Garcia, C. A. (2015). Ocelot population status in protected Brazilian Atlantic Forest. *PLoS One, 10*(11), e0141333.

Mcmichael, A. J., Campbell-lendrum, D., Kovats, S., Edwards, S., Wilkinson, P., Wilson, T., Nicholls, R., Hales, S., Tanser, F., Le Sueur, D., & Schlesinger, M. (2004). Global climate change. In M. Ezzati, A. D. Lopez, A. Rodgers, & C. J. Murray (Eds.), *Comparative quantification of health risks* (2nd ed.). World Health Organization.

Norman, M. (2003). Biodiversity hotspots revisited. *Bioscience, 53*(10), 916–917.

Oldekop, J. A., Holmes, G., Harris, W. E., & Evans, K. L. (2016). A global assessment of the social and conservation outcomes of protected areas. *Conservation Biology, 30*(1), 133–141.

Ondarza, P. M., Haddad, S. P., Avigliano, E., Miglioranza, B. W., & Brooks, K. S. B. (2019). Pharmaceuticals, illicit drugs and their metabolites in fish from Argentina: Implications for protected areas influenced by urbanization. *Science of the Total Environment, 649*, 1029–1037.

Palomo, I., Martín-López, B., Potschin, M., Haines-Young, R., & Montes, C. (2013). National parks, buffer zones and surrounding lands: Mapping ecosystem service flows. *Ecosystem Services, 4*, 104–116.

Papalia, M. (2012). Construcción de Demandas Políticas de Comunidades Mbyá Guaraníes En Contextos de Conservaciónde La Naturaleza. *Cuadernos de Antropología Social, 36*, 119–150.

Paschoal, A. M. O., Massara, R. L., Bailey, L. L., Kendall, W. L., Doherty, P. F., Hirsch, A., Chiarello, A. G., & Paglia, A. P. (2016). Use of Atlantic forest protected areas by free-ranging dogs: Estimating abundance and persistence of use. *Ecosphere, 7*(10), e01480.

Ribeiro, M. C., Metzger, J. M., Camargo, M. A., Ponzoni, F. J., & Makiko, H. M. (2009). The Brazilian Atlantic Forest: How much is left, and how i the remaining forest distributed? Implications for conservation. *Biological Conservation, 142*(6), 1141–1153.

Ribeiro, M. C., Al, C. M., Metzger, J. P., Tabarelli, M., Scarano, F., & Fortin, M. J. (2011). The Brazilian Atlantic Forest: A shrinking biodiversity hotspot. In F. Zachos & J. Habel (Eds.), *Biodiversity Hotspots* (pp. 405–434). Springer.

Rolón, E., Ondarza, P. M., Miglioranza, K. S., Rosso, J. J., Mabragaña, E., Volpedo, A. V., & Avigliano, E. (2021). Multi-matrix approach reveals the distribution of pesticides in a multipurpose protected area from the Atlantic rainforest: Potential risk for aquatic biota and human health? *Environmental Science and Pollution Research, 28*, 34386–34399.

Schenone, N. F., Moscuzza, H., Avigliano, E., Rosso, J. J., & Mabragaña, E. (2014). *Plan Estandarizado de Muestreos de Calidad de Agua Superficial*. Fundación Bosques Nativos Argentinos para la Biodiversidad.

Serengil, Y., Augustaitis, A., Bytnerowicz, A., Grulke, N., Kozovitz, A. R., Matyssek, R., Müller-Starck, G., Schaub, M., Wieser, G., Coskun, A. A., & Paoletti, E. (2011). Adaptation of forest ecosystems to air pollution and climate change: A global assessment on research priorities. *IForest Viterbo, 4*, 44–48.

Chapter 21
Historic Gardens as a Cultural Task: Climate Adaptation Strategies and Understanding of Nature

Michael Rohde

Abstract The cultivation and preservation of gardens, parks and cultural landscapes as fine art have been expressions of culture for millennia and are becoming essential tasks of cultural property protection in times of climate change. This is because the visible effects of climate change are increasingly threatening the historical aesthetics and current uses of historic gardens. Strategies for climate adaptation require not only thorough and networked experiential knowledge in the field of conservation and restoration sciences but also specific and interdisciplinary research expertise. Gardens as cultural assets must become scientific model laboratories to understand cultivation and conservation as essential cultural tasks of our societies. These challenges must lead to a new understanding of nature that initiates and perpetuates a responsible, humane sense of life through the gardens.

Keywords Climate change · Gardens · Garden art · Garden preservation · Cultural task · Nature

21.1 Gardens as an Expression of Cultural History

Carl Friedrich von Weizsäcker emphasized the importance of cultural landscapes and the role of the natural sciences as mediators between culture and nature in his historical anthropology more than 40 years ago, in 1977, a year after the Federal Republic of Germany ratified the *Convention Concerning the Protection of the World Cultural and Natural Heritage* (UNESCO, 1972). Man had "discovered too late how his silent hostility to nature has undermined the natural foundations on which his culture also rests" (Weizsäcker, 1977, p. 68). At the time, he meant "environmental damage, famine catastrophes, loss of freedom, war" (Weizsäcker, 1977, p. 69), to which the effects of increasing climate change must be added today.

M. Rohde (✉)
Gardens Direktor SPSG/Prof. h.c. TU Berlin, ILAUP, Berlin, Germany
e-mail: m.rohde@spsg.de

M.-T. Albert et al. (eds.), *50 Years World Heritage Convention: Shared Responsibility – Conflict & Reconciliation*, Heritage Studies,
https://doi.org/10.1007/978-3-031-05660-4_21

Weizsäcker countered this with the perception of the beautiful "as a mode of appearance of the good" (1977, p. 105). If man in the technical age perceives ecological balance – sensually represented in the example of the meadow with flowering plants and bees – as beautiful, he perceives harmony, without which he cannot live: "How magnificent are the old cultural landscapes [...], where for centuries every tree and every house stood where people with a sense of beauty wanted it to be" (Weizsäcker 1977, p. 105).

In 1992, the World Heritage Committee reviewed cultural criteria to ensure recognition of "combined works of nature and man" of "outstanding universal value" (UNESCO, 1992). With the associated recognition of cultural landscapes as protected properties and the newly formulated categories, "castles and gardens" were also defined as cultural heritage. In the Directive for the Implementation of the World Heritage Convention at that time, Article 1 stated, "The most readily identifiable is the clearly defined landscape, intentionally designed and created by man. This includes garden and park landscapes produced for aesthetic reasons and often (but not always) associated with religious or other monumental buildings and ensembles" (UNESCO, 1972). Individual historic gardens and smaller ensembles have now been included within cultural landscapes (Rössler, 2003, 220–227; UNESCO 2019; Dornbusch, 2017, 196–200) (Fig. 21.1).

The palaces and parks of Potsdam and Berlin belong to these very cultural properties, which are recognized "as a masterpiece of human creativity" and "for a period of time or in a cultural area of the earth, demonstrates a significant intersection of human values in relation to the development [...] of landscape design" (ICOMOS, 1990; Giersberg, 2000, p. 17). Currently, in the context of Agenda 2030s holistic approach to the protection of cultural and natural heritage for sustainable

Fig. 21.1 Drought in Sanssouci Park as an outstanding example of architectural creations and landscape designs, UNESCO World Heritage Site since 1990. (*Note*: Photo by Michael Rohde, June 2020)

development, it is being discussed whether the pre-1992 World Heritage site "Potsdam-Berlin Cultural Landscape" should also be classified as a cultural landscape (Ringbeck, 2020, p. 143–145). Examples of cultural landscapes include Sintra, Lednice-Valtice (Eisgrub-Feldsberg) and Aranjuez, as well as the Garden Kingdom of Dessau-Wörlitz and Muskauer Park (Park Muzakowski).

Culture, derived from the term *colere* of the Latin *cultura*, encompasses the creative and preserving power of man. It means *nurturing, cultivating and educating*, broadly defined, everything that man himself has brought forth in a formative way in contrast to the surrounding, unchanged nature, including spiritual goods.

Since the differentiation of early societies around 5500 years ago and the process of human civilization that began with the first advanced civilizations in cities, the formal design and diverse use of gardens have been part of the cultural world of man. With the handing down of garden art, including newer inventions such as irrigation or planting techniques, knowledge and culture or cultural skills unfolded. Traditions were handed down through the active preservation of gardens as well as orally and pictorially, from the discovery of writing to digitalization. Tradition formation is a social expression and can be understood "as a cultural construction of identity set in perpetuity" (Assmann, 1999, p. 60).

Garden art generally refers to the artistic shaping of limited green spaces into specific functions through woody plants, flowers, meadows, orangeries or other plant furnishings, paths, stones and water installations, modelling or ground reliefs, architectural elements, and pictorial works that bear witness to a respective historical relationship between man and nature. As with other works of art, green spaces can be understood as carriers of meaning and thus have not only aesthetic but also historical, allegorical and symbolic significance.

21.2 Climate Impacts Threaten Historic Gardens

Published in August 2021, the first part of the Sixth Assessment Report of the Intergovernmental Panel on Climate Change (IPCC, Intergovernmental Panel on Climate Change) is alarming. Global warming could already exceed 1.5 degrees in ten years. The consequences of climate change – extreme heatwaves, floods and hurricanes – are becoming more intense and more frequent. As living, dynamic cultural monuments, historic gardens and parks around the world are more affected by these increasing weather extremes of human-induced climate change than any other art form.

The *summer dry periods with high temperatures* increasingly cause water shortages with simultaneously reduced groundwater recharge. These conditions not only cause woody plants to die in a relatively short time but also weaken the entire flora through further phenomena. The vegetation period is lengthened, leaf development and the beginning of flowering start earlier. In connection with late frosts, this leads to an additional weakening of the woody plants. Furthermore, parameters such as vegetation duration, temperature minima and maxima, and precipitation amounts affect oak dieback or beech bark disease via new pests (Kehr & Schumacher, 2014, pp. 64–69).

Storms and hurricanes occur more frequently as weather becomes more extreme, damaging and destroying not only structural assets but also trees in historic gardens. For example, at the beginning of October 2017, Storm Xavier recorded gale-force winds of around 150 km per hour in Berlin-Brandenburg, the second-highest since records began; the last similar event was in November 1972 (Rohde, 2018). However, since the trees were still in full leaf and soaked, they offered particularly large attack surfaces. In the "Prussian Gardens" (SPSG), in addition to soil erosion, there was uprooting and a total loss of around 1000 large trees and, thus, sensitive disruption of the historic park image. The task of tree control and traffic safety (removal of dead branches, fencing off park areas, etc.) in the historical gardens has increased in the meantime.

Heavy rainfall events are also becoming more frequent. The devastating flood disaster in the Ahr Valley in July 2021 demonstrated some of the most serious effects of these events. The Potsdam Institute for Climate Impact Research (PIK) predicts that storms and extreme rainfall events will intensify by 2040 and not just in the USA, parts of India, Africa and Indonesia (Willner et al., 2018). Heavy rainfall events will also increase sevenfold in Central Europe, especially in Berlin-Brandenburg. For historic gardens, the risk of destruction by flooding will increase, as was seen in Thuringia (Hagner & Seidel, 2014, pp. 80–85) and the castle gardens of Het Loo, Netherlands in 2009 (Dulk, 2014, pp. 110–111). In the future, there will be an increase in flooding in the winter and spring as soils become oversaturated, while there will be more extreme heat storms in the summer months (Gerstengarbe, 2014, pp. 48–51).

21.3 The Tasks of Garden Conservation Science

As predominantly living monuments, historic gardens and parks represent a unique cultural asset with a special monument character, in contrast to works of art made of solid matter such as buildings, castles, sculptures, furniture or paintings. They are particularly dependent on the environment and, due to competing public interests, are more easily subject to misuse and damage that is alien to their nature. In Germany, historical gardens represent cultural monuments according to the respective monument protection laws of the federal states due to special characteristics.

The artistic criterion can be traced back to an artistic personality as well as stylistic direction due to the design according to form and function. The historical criterion identifies, protects and documents sources and evidence of human developments. This includes diverse fields of historical knowledge, including art or social history, intellectual history, religious or economic history and technological history. Scientific and urban planning criteria also play a major role, with the latter meaning urban architecture, ground plan architecture or the evolved structure as an expression of intellectual creation or significance for the townscape or landscape.

For more than 100 years, the specific discipline of garden monument preservation has stood for the methodical effort to research historic green spaces, gardens, parks and plant remnants that represent cultural monuments in the public interest because of their historical, artistic, scientific and urban planning significance, and to preserve and, if necessary, repair them through administrative, planning, and gardening measures (Rohde, 2008).

Monument status requires a thorough analysis of the respective historic gardens and parks, ideally, based on a monument concept (Rohde, 2010, pp. 221–227). This includes a) the analysis of the historical design phases up to the present, b) the presentation of the current inventory including the overlaps or comparisons with the formerly completed design phases, c) a monument evaluation addressing the formerly and currently existing inventory, and d) a concept to be carefully justified in the future handling of the garden monument.

There are basically two options for the monument methodology, which must be professionally implemented, documented and communicated to the public:

- Conservation or preservation and maintenance: gardens must be professionally maintained on a permanent basis. This also applies to neglected, converted, altered or even partially destroyed gardens.
- Restoration: as complete a return as possible to earlier design conditions, i.e. structures that are demonstrably still (partially) present. This requires a careful monument evaluation beforehand.

21.4 Climate Adaptation Strategies for Gardens in the Context of Conservation Science

In Germany, the state and municipal garden administrations have traditional, extremely diverse empirical knowledge. However, responding to climate damage involves not only promoting technical and personal equipment for garden maintenance but also incorporating science and research, which has to be more networked. Additionally, interdisciplinary knowledge must be integrated into the maintenance and repair measures and scientifically accompanied. This requires a greater understanding in politics, administration and science

- for garden monument preservation as an academic and craft discipline of the "conservation and restoration sciences" of landscape architecture, as is now common practice in building monument preservation and restoration;
- for gardens as "living monuments" – the only category of valuable cultural assets that can be regenerated in their authentic formal artistic expression through replanting and constant expert care; and
- for the need to generate the necessary knowledge from special sciences to sustainably preserve the historical gardens as cultural assets in terms of experiencing testimonial, historical and artistic values – also with regard to climate adaptations.

The management of climate impacts for historic gardens and parks is now developing into a core task within the framework of the aforementioned monument methodology. In Europe, different analyses and strategies are being discussed on the topic of water scarcity or woody plant replacement, which are in the area of tension between possible adaptation and preservation of the historical and testimonial value. For example, in the United Kingdom, the University of Sheffield, in cooperation with the National Trust, has developed examples of optimized irrigation or composting methods for climate adaptation in "climate adaptation plans (for garden and plant collection)," among others (Dix, 2019; Woudstra, 2019). In Italy, the example of archaeological parks in Rome shows how experience from history combined with the possibilities of modern techniques can lead to optimized water use or drought-resistant replacement plantings can be used (Fallani, 2019) (Fig. 21.2).

Among the most important climate adaptation strategies for gardens are "water management plans" that aim to retain precipitation to minimize stormwater runoff. The goals of these plans include the necessary provision of surface water (service water) or the use of groundwater, as well as the use of new irrigation options such as drip or night irrigation (Sellinger, 2014, pp. 168–173; Schröder, 2014, pp. 186–191). The dike reinforcements and mobile protective dikes made of plastic tubes of the Dessau-Wörlitz Cultural Foundation represent exemplary protective measures against flood hazards (Trauzettel, 2014, pp. 158–161).

Soil management, including the handling of biomasses in the context of optimized composting, will (re)gain importance in the future. Soil sites for replanting and, in some cases, existing tree sites need attention in terms of fertilization and

Fig. 21.2 The Royal Botanic Gardens, Kew (Kew Gardens), UNESCO World Heritage Site since July 2003, formerly pleasure gardens, since 1759 as botanical gardens. Pictured is the Chinese Pagoda designed by Sir William Chambers in 1762. Professionals of the "Capability Brown" congress led by Jonathan Finch and Jan Woudstra. (*Note*: Photo Michael Rohde, June 2016)

irrigation options. To improve the water and nutrient balance of the soil, mulching offers a variety of advantages (Schneider & Hüttl, 2014, pp. 140–143).

Trees and shrubs create the actual spatial images of historic gardens. The existing, formerly artistically used plant material at respective historic sites should not only be respected but preserved as far as possible. Vegetation will build up a stronger resilience through plant associations, i.e. woody plant lots as opposed to solitary woody plants, and will also promote biodiversity. As early as the 1990s, the Eberswalde State Competence Centre developed a practicable indicator system for forest monitoring that records the physiological performance of trees and evaluates it in terms of stress tolerance, adaptation and vitality. This so-called biomarker concept can also be applied to historic gardens (Kätzel & Löffler, 2014, pp. 152–157).

Plant selection for future replanting needs to be carefully reviewed to counteract susceptibility to drought stress. For example, the DBU-funded TU Berlin research project on the future management of woody plants using the example of the SPSG's World Heritage Gardens (2015–2017) provides recommendations on woody plant selection for replanting: Self-sown woody plants or those from park stands of genetically adapted natural regeneration are more resistant than additional purchases (Kühn et al., 2017; Butenschön, 2014, pp. 210–213).

The gardens, parks and cultural landscapes of state and municipal garden administrations increasingly represent scientific model laboratories in the context of climate adaptation. At the end of 2020, for example, the "Protecting cultural assets from extreme climate events and increasing resilience (KERES 2020–2023)" research project was launched under the leadership of the Fraunhofer-Gesellschaft (FhG), in cooperation with the Climate Service Center Germany (HZG-GERICS) and the SPSG – funded by the German Federal Ministry of Education and Research (BMBF). Using the example of the "Prussian Gardens" (and buildings), the stability and branch breakage risk of trees (vitality) or risk potentials of park paths (traffic safety) are being investigated. A variety of measures are planned to improve the understanding of the acute threat scenarios caused by extreme weather events among decision makers and all stakeholders in the cultural heritage sector. Examples of such measures include detailed climate forecasts for selected cultural heritage sites in different climate zones in Germany, using high-resolution regional climate models for IPCC climate scenario RCP 8.5 ("Climate Fact Sheets"). Sustainable, i.e. long-term and resource-saving, water supplies and sufficient nutrient supplies for woody plants will also play an important role. Model measures of these adaptation strategies will be published on a knowledge platform by the end of 2023.

21.5 The Preservation of Historic Gardens Promotes a New Understanding of Nature by Humans

"The preservation of historic garden artworks is a societal task by which our willingness not to accept cultural losses due to anthropogenic climate change can also be measured," according to Roland Bernecker (2014, p. 14), on behalf of UNESCO (also Sanssouci Declaration, 2014).

In 1983, Hermann Lübbe also linked social issues with the preservation of cultural assets from the perspective of cultural philosophy. What matters, he argued, is not constantly new goals for action within the process of civilization, but rather the "relatedness of these goals to the humane sense of life of our civilization" (138), which includes moral action and capacities such as initiative, criticism, and self-criticism (Lübbe, 1983, p. 138, 145).

"We need garden thinking," demanded biologist Hubert Markl (1938–2015). As a consequence of our mental mastery of nature, Markl (1986) argued that our species would, for the first time in the history of life, acquire the ability "to change and *disturb* the whole of living nature on this earth in such a lasting way that its destruction-and thus also that of man-becomes a real possibility" (Markl, 1986, p. 9). Thus, for humans, as "beings *out of* nature, as beings *in* nature, and as beings *against* nature" (Markl, 1986, p. 10), nature would become "inescapably more and more a mission for the fulfilment of which he bears responsibility, and since all of man's activity is an expression of his capacity for culture - which constitutes his being - nature becomes for him a threefold cultural task" (Markl, 1986, p. 7): the exploration of nature, the care of nature, and the preservation of nature, or "more precisely: for the preservation of its capacity to bear and endure human culture" (Markl, 1986, p. 9).

On the question of the development and future of culture and civilization, Immanuel Kant (1724–1804) combined ethical perspectives. The determination of man as a culture-creating being takes place in relation to nature because "man can only be a final purpose of creation as a moral being" (Kant, 1790; after Markl, 1986, p. 341). In the sense of the categorical imperative, the "idea of morality still belongs to culture" (Kant, 1784, p. 26; Elias, 1989, p. 8 f.). Without this guiding principle, man is only able to develop in a purely technical way.

In his main work, *The Principle of Responsibility - An Attempt at Ethics for Technological Civilization*, the philosopher Hans Jonas (1903–1993), following Kant, among others, extends the scope of human responsibility to the whole of animate nature and to the dimension of the future. His imperative is "Act in such a way that the effects of your action are compatible with the permanence of genuine human life on earth" (Jonas, 1979, p. 36). Thus, his ethic of preservation and conservation of nature encompasses not only the survival of the biosphere but also the integrity of its essence and respect for its dignity. It exhorts us to reverence human dignity in accordance with nature's "moral proper" (Jonas, 1979, p. 29; Wetz, 2005, p. 115–116, 120).

21.5.1 Changes in the Perception of Nature

Gardens in cultural history bear witness to the formation of tradition and orientation as social expression (Renn & Rohde 2020, pp. 23–37). The cultural world of man includes the design and use of gardens, which acquired aesthetic, religious, scientific, ideological, political and even functional significance through symbolic meaning. Designed nature and artistic gardens existed in the advanced civilizations of Mesopotamia or Egypt, later in Roman and Greek antiquity, and from there were finally adopted in northern Europe. The princely, public and private gardens were always characterized by political structures, sciences and arts.

But man's relationship with nature has changed again and again in the course of history. If the garden was considered a metaphor for earthly and heavenly paradise until the Renaissance, the mastery of nature as a means of enhancing one's sense of life followed with a new sense of power and order. The elaborately architecturally designed gardens stood in contrast to "wild nature." They served as a representative expression of an absolutist and hierarchized state and world order. Currents of the Enlightenment, and later Romanticism, again shaped a changed sense of nature from the eighteenth century onward. Philosophers and painters as garden artists derived the creation of picturesque landscape gardens from nature via a socially new idea of freedom, now also via a view of the ethical and aesthetic. Since the twentieth century, this has given rise to an ethic of preservation and conservation of nature, which should enable survival on the planet into the future while also respecting its dignity.

In his *History of German Garden Art*, Dieter Hennebo (1965) took a holistic view of the development of garden ideals: "They move in a common direction of development determined by social changes, because the respective culture-shaping society also determines the ostensible garden image of its time through its relationship to nature and the garden, through its conception of art, and through the demands it makes on the garden" (Hennebo, 1965, p. 9).

21.5.2 Gardens and Parks for a New Understanding of Nature

How do we understand gardens, parks, and historic cultural landscapes as cultural monuments and World Heritage sites, and how do they help us in the Anthropocene of the twenty-first century? (cf. Crutzen, 2002; Renn, 2020). This question was recently explored at the DBU-sponsored international SPSG congress "Historical Gardens and Society" (Rohde & Schmidt, 2020, pp. 13–16).

The rich potential of historic gardens, parks and cultural landscapes as cultural monuments can counter the trends of the unprecedented dynamics of our civilization as well as environmental and natural destruction. They have the capacity to initiate and perpetuate a responsible, humane sense and style of life. Their formal artistic design and their centuries-long preservation encompass cultural and ecological as well as economic and social contexts of impact. Gardens are transcultural places of understanding that connect people, and, in this capacity, also promote the development of a sense of global citizenship.

Historical gardens can create significant social identities of culture and education and have enormous potential biodiversity (Agenda 2030, SDG 15 Life on Land). These gardens can also achieve, in addition to the added value of tourism and job security, hardly quantifiable economic values and welfare effects, which are gaining importance in times of evident transformations: enhancing air quality, strengthening health resources, reducing noise pollution, promoting relaxation and stress reduction, and improving physical and psychological quality of life through movement, art events, creativity and perceptions of nature.

Ultimately, the focus is on people, as "stewards" of gardens and cultural landscapes and their development, to intervene much more strongly than before and to achieve a (re-)renewed, sustainable relationship between people and nature (Schmidt, 2020). Educational processes must always be updated for this purpose, explicitly in connection with the cultivation of international gardens (reference Agenda 2030, SDG 4 High-quality education worldwide): This requires "insights into central problem contexts that determine future viability - such as nutrition, climate change, biodiversity conservation, and cultural diversity" (Stoltenberg, 2010, p. 294; Bernecker & Grätz, 2017). The essential goals of preserving, expanding and disseminating knowledge should be achieved, among other things, "by preserving and protecting the world's heritage of books, works of art, and monuments of history and science," according to Article 1, 2c of the Constitution of the United Nations Educational, Scientific, Cultural, and Communication Organization (UNESCO), founded in 1945 (Constitution UNESCO, 1945). A variety of examples, including international cases, of the appreciation and benefits of gardens as cultural monuments for society, as well as sustainable future strategies for education, can lend themselves to achieving these goals.

21.6 Epilogue

Gardens, parks and cultural landscapes have been under World Heritage status as cultural monuments for decades. As early as November 1972, the UNESCO World Heritage Convention stated that "cultural and natural heritage are increasingly threatened with destruction, not only by the traditional causes of deterioration, but also by the changes in social and economic conditions which aggravate the situation by even more pernicious forms of damage or destruction."

In times of complex and dramatic changes in societies and environments, man's relationship with nature is changing. We, humans, are part of nature on this earth and thus, in a sense, stewards of "gardens." In this context, we act on all levels of society: politics, economy and administration, science, technology and culture. Historical gardens are not just cultural constants: by newly recognizing and implementing sustainable action goals and ways of life, cultural landscapes can and must contribute to an expanded, ethical understanding of nature.

References

Assmann, A. (1999). *Zeit und tradition*. Böhlau.

Bernecker, R. (2014). Foreword of the Secretary General of the German Commission for UNESCO. In S. P. Schlösser & G. Berlin-Brandenburg (Eds.), *Historic gardens and climate change – Recommendations for preservation* (p. 14). Edition Leipzig.

Bernecker, R., & Grätz, R. (2017). *Global Citizenship – Perspektiven einer Weltgemeinschaft*. Steidl.

Butenschön, S. (2014). Tree nurseries in parks – Centres of regionally adapted woodland in the past and future. In S. P. Schlösser & G. Berlin-Brandenburg (Eds.), *Historic gardens and climate change – Recommendations for preservation* (pp. 210–213). Edition Leipzig.

Crutzen, P. J. (2002). Geology of Mankind. *Nature, 415*, 23.

den Dulk, L. (2014). Climate change and its implications for historic parks and gardens in the Netherlands. In S. P. Schlösser & G. Berlin-Brandenburg (Eds.), *Historic gardens and climate change – Recommendations for preservation* (pp. 110–111). Edition Leipzig.

Dix, B. (2019). The reconstruction of historic parks and gardens in the context of climate change. In R. Hüttl, K. David, & B. U. Schneider (Eds.), *Historische Gärten und Klimawandel – Eine Aufgabe für Gartendenkmalpflege, Wissenschaft und Gesellschaft* (pp. 349–355). de Gruyter.

Dornbusch, R. S. (2017). Historische Kulturlandschaft als Kulturgut. In D. Martin & M. Krautzberger (Eds.), *Handbuch Denkmalschutz und Denkmalpflege* (4th ed., pp. 196–200). C.H. Beck.

Elias, N. (1989). Über den Prozeß der Zivilisation. Erster Band, Kapitel: Über den Entwicklungsgang des Gegensatzpaares: „Zivilisation“ und „Kultur“. Berlin: Suhrkamp.

Fallani, M. de V. (2019). Klimaanpassung. Anmerkungen zu den archäologischen Parks in Rom. In R. Hüttl, K. David, & B. U. Schneider (Eds.), *Historische Gärten und Klimawandel – Eine Aufgabe für Gartendenkmalpflege, Wissenschaft und Gesellschaft* (pp. 356–368). de Gruyter.

Gerstengarbe, F. W. (2014). Climate change and the possible consequences for the Potsdam region. In S. P. Schlösser & G. Berlin-Brandenburg (Eds.), *Historic gardens and climate change – Recommendations for preservation* (pp. 48–51). Edition Leipzig.

Giersberg, H.-J. (2000). Zur Pflege und Erhaltung des UNESCO-Welterbes der Potsdam-Berliner Kulturlandschaft in den zehn zurückliegenden zehn Jahren. In S. P. Schlösser & G. Berlin-Brandenburg (Eds.), *Zehn Jahre UNESCO-Welterbe der Potsdam-Berliner Kulturlandschaft* (pp. 17–22). Stiftung Preussische Schlösser und Gärten Berlin-Brandenburg.

Hagner, D., & Seidel, C. (2014). Consequences of weather events in Thuringia's historical gardens. In S. P. Schlösser & G. Berlin-Brandenburg (Eds.), *Historic gardens and climate change – Recommendations for preservation* (pp. 80–85). Edition Leipzig.

Hennebo, D. (1965). Geschichte der deutschen Gartenkunst. Band II. Der architektonische Garten Renaissance und Barock. Broschek.

ICOMOS (1990). Eintragungsvorschlag der Schlösser und Parks von Potsdam-Sanssouci. https://www.icomos.de/admin/ckeditor/plugins/alphamanager/uploads/pdf/HefteIII.pdf

Jonas, H. (1979). *Das Prinzip Verantwortung – Versuch einer Ethik für die technologische Zivilisation*. Insel Verlag.

Kant, I. (1784). Idee zu einer allgemeinen Geschichte in weltbürgerlicher Absicht. *Akademie-Ausgabe Band, 8*.

Kant, I. (1790). *Kritik der Urteilskraft*. Reclams Universal-Bibliothek.

Kätzel, R., & Löffler, S. (2014). Physiological indicators for assessment of water-deficiency stress of trees. In S. P. Schlösser & G. Berlin-Brandenburg (Eds.), *Historic gardens and climate change – Recommendations for preservation* (pp. 152–157). Edition Leipzig.

Kehr, R., & Schumacher, J. (2014). Diseases and pests of woody plants – What role does climate change play? In S. P. Schlösser & G. Berlin-Brandenburg (Eds.), *Historic gardens and climate change – Recommendations for preservation* (pp. 64–69). Edition Leipzig.

Kühn, N., Gillner, S. & Schmidt-Wiegand, A. (2017). Gehölze in historischen Gärten im Klimawandel – Transdisziplinäre Ansätze zur Erhaltung eines Kulturguts. Universitätsverlag der TU Berlin.

Lübbe, H. (1983). Zeit-Verhältnisse. Zur Kulturphilosophie des Fortschritts. Styria.

Markl, H. (1986). Natur als Kulturaufgabe. Über die Beziehungen des Menschen zur lebendigen Kultur. Deutsche Verlags-Anstalt.

Renn, J. (2020). *The evolution of knowledge*. Rethinking science for the Anthropocene.

Renn, J., & Rohde, M. (2020). Gardens should engage in dialogue with society. In S. P. Schlösser & Gärten (Eds.), *Historic gardens and society. Culture – Nature – Responsibility* (pp. 23–37). Schnell und Steiner.

Ringbeck, B. (2020). Historic garden and park landscapes and the appreciation of the "green heritage" from the perspective of the World Heritage Convention. In S. P. Schlösser & Gärten (Eds.), *Historic gardens and society. Culture – Nature – Responsibility* (pp. 143–145). Schnell und Steiner.

Rohde, M. (Ed.). (2008). *Pflege historischer Gärten – Theorie und Praxis*. Edition Leipzig.

Rohde, M. (2010). Denkmalkonzepte, Denkmalmethodik und ständige Pflege der preußischen Stiftungsgärten. In T. Drachenberg, A. Klausmeier, R. Paschke, & M. Rohde (Eds.), *Denkmalpflege und Gesellschaft* (pp. 221–227). Hinstorff.

Rohde, M. (2018). Grünes Welterbe in Gefahr. Stürme und Sturzregen in den königlichen Gärten: Zwei Jahrhundertereignisse und die Folgen. In *Magazin der Stiftung Preußische Schlösser und Gärten Berlin-Brandenburg* 2017 (04) (p. 12–13).

Rohde, M., & Schmidt, F. (2020). Introduction. In S. P. Schlösser & Gärten (Eds.), *Historic gardens and society. Culture – Nature – Responsibility* (pp. 13–16). Schnell und Steiner.

Rössler, M. (2003). Die Verknüpfung von Kultur und Natur – Der Schutz von historischen Gärten und Kulturlandschaften nach der UNESCO-Welterbekonvention. In M. M. Rohde & R. R. Schomann (Eds.), *Historische Gärten heute* (pp. 220–227). Edition Leipzig.

Sanssouci Declaration. (2014). Erklärung von Sanssouci zum Erhalt von historischen Gärten und Kulturlandschaften. http://www.spsg.de/presse-foto-film/2014-09-05-erklaerung-von-sanssouci/

Schmidt, F. (2020). Sustainability and responsibility. The (historic) garden, singular and planetary. In S. P. Schlösser & Gärten (Eds.), *Historic gardens and society. Culture – nature – responsibility* (pp. 229–231). Schnell und Steiner.

Schneider, B. U., & Hüttl, R. (2014). Die Bedeutung des Mulchens für den Wasser- und Nährstoffhaushalt des Bodens. In S. P. Schlösser & G. Berlin-Brandenburg (Eds.), *Historic gardens and climate change – Recommendations for preservation* (pp. 140–143). Edition Leipzig.

Schröder, K. (2014). Complex: The water system in the Babelsberg Park. In S. P. Schlösser & G. Berlin-Brandenburg (Eds.), *Historic gardens and climate change – Recommendations for preservation* (pp. 186–191). Edition Leipzig.

Sellinger, W. (2014). Water as an influencing factor: Historical irrigation systems, effective irrigation, dealing with severe weather. In S. P. Schlösser & G. Berlin-Brandenburg (Eds.), *Historic gardens and climate change – Recommendations for preservation* (pp. 168–173). Edition Leipzig.

Stoltenberg, U. (2010). Kultur als Dimension eines Bildungskonzepts für eine nachhaltige Entwicklung. In O. Parodi, G. Banse, & A. Schaffer (Eds.), *Wechselspiele: Kultur und Nachhaltigkeit* (pp. 293–311). Edition Sigma.

Trauzettel, L. (2014). Flood defences in the Gartenreich Dessau-Wörlitz UNESCO World Heritage Site. In S. P. Schlösser & G. Berlin-Brandenburg (Eds.), *Historic gardens and climate change – Recommendations for preservation* (pp. 158–161). Edition Leipzig.

UNESCO. (1945). United Nations Educational, Scientific and Cultural Organization Constitution. UNESCO.

UNESCO. (1972). Convention concerning the protection of the world cultural and natural heritage. UNESCO.

UNESCO World Heritage Centre. (1992). *Operational guidelines for the implementation of the World Heritage Convention*. UNESCO.

UNESCO World Heritage Centre. (2019). *Operational guidelines for the implementation of the World Heritage Convention*. UNESCO.

Weizsäcker, C. F. von (1977). *Der Garten des Menschlichen. Beiträge zur geschichtlichen Anthropologie. Fischer-Verlag.*

Wetz, F.J. (2005). *Hans Jonas. Eine Einführung* (orig. 1994). Panorama Verlag.

Willner, S. N., Levermann, A., Zhao, F., & Frieler, K. (2018). Adaptation required to preserve future high-end river flood risk at present levels. *Science Advances, 4*, 1–8.

Woudstra, J. (2019). Dealing with the consequences of climate change in historic parks and gardens in the United Kingdom. In R. Hüttl, K. David, & B. U. Schneider (Eds.), *Historische Gärten und Klimawandel – Eine Aufgabe für Gartendenkmalpflege, Wissenschaft und Gesellschaft* (pp. 337–348). de Gruyter.

Chapter 22
The Highest Mountain in the Shadow of Climate Change: Managing Tourism and Conservation in a World Heritage Site: Sagarmatha National Park, Nepal

Sushma Bhatta, Robin Boustead, and Kurt Luger

Abstract The unique diversity of the world's highest flora and fauna is a sacred landscape for the resident Sherpa communities, but climate change has been casting a threatening shadow over World Heritage Sagarmatha (Mt Everest) National Park (SNP) for years, causing rapid and pronounced impacts. Tourism is a key driver to the local economy but is exacerbating the impacts of climate change. Through extensive community and individual surveys in major villages, combined with a wealth of data from other studies, we focus on how the impacts of climate change and tourism development can be countered by measures at the local level. We identified two sources of conflict that need to be addressed: (1) conflict between tourism businesses and park management and (2) a lack of awareness of the need for an overarching conservation strategy among residents and stakeholders. To solve these issues, site management needs considerable enforcement and support from the State Party.

Keywords Sagarmatha · Everest · Tourism · Climate change

S. Bhatta
Dignity Without Danger, Kathmandu, Nepal
e-mail: sushmabhatta29@gmail.com

R. Boustead
Great Himalaya Trail, Kathmandu, Nepal
e-mail: robin@greathimalayatrail.com

K. Luger (✉)
UNESCO Chair Cultural Heritage and Tourism, University of Salzburg, Salzburg, Austria
e-mail: Kurt.luger@plus.ac.at; http://www.kurt-luger.at

M.-T. Albert et al. (eds.), *50 Years World Heritage Convention: Shared Responsibility – Conflict & Reconciliation*, Heritage Studies,
https://doi.org/10.1007/978-3-031-05660-4_22

22.1 Introduction

Mount Everest, the highest point on earth, 8848 m above sea level, is known as Sagarmatha by the Nepalese. The magnificence of this matchless peak, an incomparable icon of nature's splendour, has inspired human longings and attracted mountaineers from all over the globe. Since the first successful summit expedition in 1953, the stream of visitors to the surrounding Khumbu region has grown exponentially. In 1976, the Government of Nepal established Sagarmatha National Park (SNP) to conserve its unique diversity of flora, fauna and culture. In 1979, UNESCO's justifications for including SNP on the World Heritage Site (WHS) List were geological, biological, aesthetic and based on humanity's interaction with its environment alongside the peculiar evolutionary relationship of the indigenous Sherpa people with their own natural environment (UNESCO, 2009).

Settled for centuries by the Sherpa ethnic group, about 7000 people now live in 20 villages within SNP's 1148 km^2 and adjacent buffer zone (SNPBZ). As Buddhists, Sherpas interpret the Khumbu as a sacred valley (Sherpa, 2008), and their indigenous natural resource management has been a major contributing factor to conserving the Khumbu region (Daconto & Sherpa, 2010).

Nepal is a developing country with weak social and technical infrastructure and is among the poorest countries in the world with a per capita GDP of US$3417 (United Nations Development Program, n.d.) Over the last 40 years, tourism has become a key driver in the local Sherpa economy through increased house construction and infrastructure with indirect improvements to the standard of health and education systems but has deprecated traditional agropastoral practices (NPC, 2017).

SNP's extraordinary Outstanding Universal Value (OUV) lies in the world's highest ecologically characteristic flora and fauna, intricately blended with rich Sherpa culture. This inter-relationship is the foundation of the sustainable protection and management of the park for the benefit of local communities (UNESCO, n.d.). SNP management, which is also responsible for WHS management, relies on state measures and local support through the SNP and Wildlife Conservation Office (Department of National Parks and Wildlife Conservation, Ministry of Forests). Additionally, about 4.7% of the park is covered by glaciers that provide freshwater benefits to millions of people downstream (Salerno et al., 2017). Therefore, in addition to conservation, monitoring the impacts of global warming and climate change on the hydrological regime is another priority for the park.

The addition of a buffer zone to form the SNPBZ in 2002 was deemed necessary to reduce biotic pressure on SNP and introduce a revenue plough-back system to improve the socio-economic status of local communities (DNPWC, n.d.). As in other countries, inhabited protected areas require a preservation policy that must be supported by the population and inherently contain the potential for conflict. Without reconciling conflicts and mitigating impacts, sustainability-oriented development in the sense of the 2030 Agenda is impossible.

22.2 The Challenge of Climate Change

Climate change, which is becoming increasingly evident, has been casting a threatening shadow over the Khumbu region for years. Reports from the Intergovernmental Panel on Climate Change (Intergovernmental Panel on Climate Change, n.d.) and the Hindukush Himalaya Assessment (ICIMOD, 2019) document rapid and pronounced climate change in the Himalaya since the mid-1970s. Progressive warming with elevation is resulting in substantial glacial melt and reduced snowfall, turning the 'abode of snow' into bare, grey, rocky mountains. The most devastating impacts concern the hydrological regime, with melting glaciers increasing the magnitude and frequency of catastrophic glacial lake outburst floods (GLOFs). From 1962 to 2003, the number and area of glacial lakes in SNP increased (ICIMOD, 2019), and three major GLOF events have occurred. In August 1985, the Digtso Lake GLOF completely destroyed the previous Namche Hydropower Station, infrastructure, farming land and livestock and killed at least 20 people (UNESCO, 2009). The most recent GLOF occurred in September 1998. Changes in atmospheric temperature and rainfall patterns are affecting the equilibrium between the amount of precipitation stored in the winter and summer thaw. These impacts are reducing the Himalaya's ability to serve as water towers for those living downstream, thus causing water scarcity that will impoverish lives and may breed conflicts at local and regional scales.

WHSs are obligated to implement the SDGs and their respective educational programs (World Heritage Centre, n.d.). A major target of SDG13 is to integrate measures that strengthen resilience and adaptive capacities to climate-related hazards and natural disasters. The need for climate adaptation approaches is recognised by SNP in the SNP State of Conservation (SoC) Reports (SNP, 2017, 2019), which confirm that climate change is seriously challenging biodiversity conservation. One of the most likely impacts of climate change is a shift in spatial and temporal patterns to the availability of suitable habitats for terrestrial species (UNESCO, 2009). The risks are considered so significant that the potential impact of climate change might create devastating losses to SNP and its OUV if the appropriate measures are not taken.

Issues like the lack of a comprehensive tourism management plan, site-specific tourism problems and inequitable benefits, unregulated and concentrated tourism, poaching of wildlife, poor sanitation capacity, waste management, helicopter overuse, inappropriate or inadequate zonation within SNP and poor goods transportation capacity were raised in studies and advisory missions (SNP, 2016; WNF, 2013; UNESCO Advisory Mission, n.d.). Directly or indirectly, these issues impact a range of SDG goals within SNP, particularly poverty alleviation, good health and well-being, gender and social equity, access to affordable and clean energy, responsible consumption and production, and partnerships to achieve SDGs.

22.3 Problem Identification and Research Question

Even if the Sherpa people in the Khumbu region cannot be held responsible for climate change in general, it must be their concern to get negative impacts under control as much as possible. These impacts are exacerbated by tourism, which contributes to the environmental burden. Therefore, we concentrate on the impacts of climate change and tourism that can be countered by measures at the local level. Our research asks, are the multi-stakeholder linkages between SNP management and Sherpa residents sufficient protection against current challenges, and could they be used more effectively?

The 'Pentagon of Sustainable Tourism' (Luger & Ripp, 2021) aligns with UNESCO's Sustainable Tourism Programme imposed in 2011 (UNESCO, STP, n.d.) and introduces sustainability principles into the mechanism of WHS management SNP WHS management is committed to the overarching goal of sustainability (SNP, 2016), and a comprehensive sustainable tourism development strategy is, therefore, necessary. Tourism is on a sustainable path when it is

(1) possible in the long term through enhanced stakeholder responsibility and community resilience;
(2) culturally compatible and focused on reconciliation, such that local residents are fully involved in negotiations and decisions and accountable for their behaviour;
(3) socially balanced by spreading benefits and disadvantages equally and regional disparities are avoided;
(4) ecologically viable by placing the lowest possible pressure on the environment, preventing biodiversity damage and promoting environmental awareness; and
(5) economically sensible and productive because it is profitable for the local and national economy.

We, therefore, take tourism as our central starting point because almost the entire population live directly or indirectly from it, and it has contributed significantly to changes in lifestyles and culture over the years. In particular, we focus on aspects of tourism activities, house building culture, transport, ecology, conservation and cultural adaptation. Our findings incorporate empirical studies that contribute to the overall understanding and assessment of the situation, as well as our own field research on how the impacts of climate change and tourism development are perceived by resident Sherpa people conducted in 2020 (Boustead & Bhatta, 2021).

We consider SNP development over a 10-year period in which there have been considerable improvements to residents' living conditions. For example, a basic solid waste management system was implemented in SNP (EcoHimal SAME, n.d.), and the Thame hydropower plant was upgraded to one Megawatt (EcoHimal Hydro, n.d.). A number of micro-hydro projects now serve villages along the main trekking route to Everest Base Camp (EBC) (André-Lamat & Sacareau, 2019) and, with the expansion of telecommunication infrastructure through FM radio, mobile phones and internet, there is now an extensive communication system. However, tourism poses increasing threats to vulnerable landscapes and reduces livelihood-adaptation and risk-mitigation options and cultural identity and traditional practices (Sherpa, 2021).

22.4 Methodology

Our study consists of two components. Firstly, voluntary-participation community (group meetings) and individual surveys in 8 villages and with 49 residents along the main EBC trekking route (Fig. 22.1). Participants included community leaders, business owners and individuals associated with the tourism value chain. Referring to the Sustainable Livelihoods Approach, the survey was developed by the MyGHTi Project (established by the Great Himalaya Trail program) to collect data on resident Sherpa knowledge and perceptions of socio-cultural and climate change indicators relating to quality of life and liveable environment (Boustead & Bhatta,

Fig. 22.1 Survey route (following the main EBC trail). (Source: Boustead R. & Bhatta S. (2021). Assessment of the Social Impacts of Community Based Tourism. https://doi.org/10.1515/tw-2021-0010)

2020). The survey used both closed and open-ended questions combined with semi-structured interviews and community discussions. Muccione et al. (2019) consider this a co-production of knowledge for sustainability, where researchers, policy makers and civil society actors cooperate in the production and dissemination of knowledge that is based on mutual recognition and learning. Secondly, we integrated data from other studies, including SNP State of Conservation Reports (SNP, 2019;2017) and the World Nature Forum (WNF, 2013) SNP benchmarking analysis based on research data collected in 2011 and ideally suited for comparison.

According to local municipal records, the current resident populations within the SNPBZ totals 7161 (comprising Khumjung, Namche, and Chaurikarka VDCs; Sagarmatha Next, 2018), many of whom reside elsewhere during the off-season periods (winter and monsoon months). The district administration office estimates that annually about 120,000 tourist guides and porters visit the SNPBZ as well as 4000 local immigrants working in hotels and in restaurants. For the purposes of the study, tourism support staff, immigrant and seasonal labour were not included in community meetings nor individual interviews.

In the year 2018–2019 (July to June), SNP received 57,289 tourists. Tourists visit SNP during two significant peak seasons, April–May (32%) and October–November (42%), for mountaineering, trekking and learning about Sherpa communities (MoTCA, 2020). Visitors are not evenly distributed throughout SNP, with the majority focused on the trail to and from the airport at Lukla to EBC (Fig. 22.2).

The survey was conducted during October and November 2020, which is normally peak season. However, due to Covid-19, tourist numbers were less than 5% compared to the same period in 2019, which gave communities and individuals the time and the will to voice their concerns.

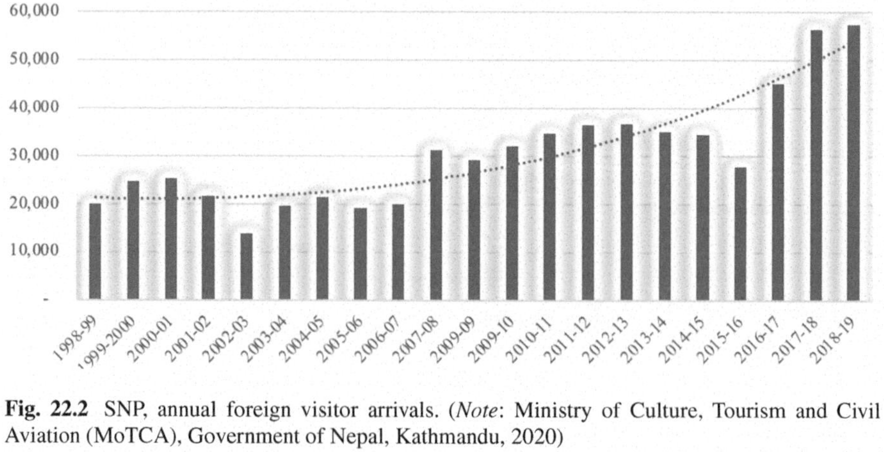

Fig. 22.2 SNP, annual foreign visitor arrivals. (*Note*: Ministry of Culture, Tourism and Civil Aviation (MoTCA), Government of Nepal, Kathmandu, 2020)

All interviews were conducted using the offline KoBoConnect (KoBoConnect n.d.) survey tool. Communities included in the survey represent 2850 Sherpa residents within the SNPBZ (approximately 40% of the population).

22.5 Results and Comparative Analysis

Survey responses for a range of criteria were assessed against the baseline study (WNF, 2013), including Sherpa resident awareness of WHS status, awareness of climate change, satisfaction with tourism, cultural erosion, waste and litter management and illegal activities. One criterion that was not identified by residents but frequently observed by the research team and mentioned in the SoC reports was the use of helicopters in SNP and is therefore included in our findings. These selected criteria correspond with site-specific issues identified by park management (SNP, 2016) relating to inequitable tourism benefits, unregulated and concentrated tourism, poor capacity for sanitation, waste and garbage management, poaching of wildlife, overuse of helicopters, inappropriate or inadequate zonation within the park and buffer zone and poor goods transportation capacity.

- *Awareness of World Heritage Status*

During community and individual surveys, the WHS status of SNP was considered to be of little or no benefit and thought of as a form of restriction or control in local management committees. A female Sherpa from Dingboche succinctly summarises a common sentiment in all communities surveyed: *"Park management and WHS do nothing for us, they only make more rules to stop us from doing business."* (respondent 15, Field Study, 2021, p. 12).

This indicates that site management is failing to communicate the importance of WHS status to conservation efforts and tourism development. This finding concurs with Sherpa (2021) and the baseline assessment of WNF (2013), which both noted little or no understanding of the significance of SNP's WHS status among locals.

- *Awareness of Climate Change*

Climate change was considered a threat and resource constraint in all community surveys and 30% of individual surveys. This opinion of a Sherpa lady was shared by many villagers: *"Farming is changing, more insects, less harvest, rain at different times, there are many problems."* (respondent 20, Field Study, 2021, p. 14).

Specific issues identified included reduced irrigation water flows, especially during dry seasons; the increase in forest fires; irregular rainfall and unpredictable seasonality (impacting sowing and harvesting); decreasing grazing land; reduced harvest volumes and quality; and a reduction in the mountain aesthetic (reduced snowfall and receding glaciers). This finding concurs with both Sherpa (2021) and Nyaupane et al. (2014), who reported that residents perceive climate change as the single greatest threat to the region and livelihoods.

- *Satisfaction With Tourism*

Community and individual surveys indicated an overwhelmingly positive satisfaction level with tourist flows and a general desire for even more tourists. Potential overcrowding on trails during the high season was not mentioned. All survey respondents ranked supplemental income as the greatest perceived benefit derived from tourism, followed by the ability to interact with tourists, a process often considered a source of learning and knowledge sharing. Sacareau (2009) also notes the positive development impacts of tourism in the SNPBZ, which has brought better houses as well as improved basic services. A Sherpa lady from Namche succinctly summarises a common sentiment: *"People were better before, but life is better now."* (respondent 35, Field Study, 2021, p. 17).

Research into tourist satisfaction with their experience in SNP found similar responses, with 91% satisfied (Posch, 2013) and 87% of tourists recording positive reviews of their experience (ATTA, 2018). However, the rapid and extensive development of tourism in SNP does raise questions of carrying capacity in what is a delicate alpine environment.

Tourism in SNP engages more than 85% of the resident Sherpa population (Rai, 2017). Running lodges is considered especially lucrative along the main tourist trail from Lukla to EBC (Rai, 2017), where income from the main season is sufficient for running a household and providing child education for an entire year. However, the distribution of economic benefits rapidly reduces away from the main trekking route.

The heavy reliance on seasonal tourism has also contributed to increasing vulnerabilities in SNP as tourism demand is, in general, a volatile industry, being extremely susceptible to external developments over which communities have little control, such as natural disasters, exchange rate fluctuations, political unrest and pandemics (Kruk & Banskota, 2007).

- *Cultural Erosion*

UNESCO recognises the intricate contribution of Sherpa culture to SNP's unique OUV, and it is considered a major tourist highlight. Most survey responses identified tourism as a positive source of cultural exchange and learning about the 'outside world'. Respondents did not feel that Sherpa culture was generally weakening, thus, confirming what has been found in other studies (Nyaupane et al., 2014). A female Sherpa from Dingboche summarises her benefits by saying, *"I learn English from tourists and how to do different dances!"* (respondent 14, Field Study, 2021, p. 11).

The changing nature of buildings and the landscape due to tourism was not mentioned by respondents but has received attention from researchers and was identified in the SoC reports (SNP, 2017, 2019) and baseline study (WNF, 2013). In all communities, there is a marked increase in construction without any evident planning criteria and an associated loss of traditional architecture (Nyaupane et al., 2014), challenging the authenticity and integrity of the site.

- *Waste and Litter Management*

Waste management was identified as a very strong negative impact in all community and individual surveys, SoC reports (SNP, 2017, 2019) and in the baseline study (WNF, 2013). All community respondents strongly emphasised the need for a more comprehensive waste management solution. "*Rubbish problem is growing every year.*" A male Sherpa from Chheplung succinctly summarises this negative impact of tourism (respondent 7, Field Study, 2021, p. 8).

That tourism in SNP is the largest direct and indirect contributor to waste generation has been thoroughly established (Manfredi et al., 2010; Sagarmatha Next, 2018; Zuser et al., 2011). Waste is categorised as kitchen/organic (fed to livestock or composted), burnable (22% paper, 14% plastic and 8% PET bottles) and non-burnable (5% metal, 5% glass, 4% textiles and 2% aluminium). A Kathmandu University study (Sagarmatha Next, 2018) calculated the average total waste generation in SNP to be 787 kg/day, of which non-degradable categories are often burnt in one of 75 dumping pits during low tourist seasons or disposed of in open areas (Sagarmatha Next, 2018; SPCC, 2018). This contributes significantly to local and regional air pollution and poses a significant hazard to human and animal health (Byers et al., 2021; ICIMOD, 2019). Additionally, black carbon soot emitted from burn-pits and fuelwood contributes directly to increased glacial melt (Mani, 2021). Most dump sites are located close to rivers that are prone to regular flooding during monsoon, thereby directly contaminating river water (SNP, 2016). There is currently no comprehensive system to collect and dispose of hospital waste or other toxic substances (Lichtberger, 2015). The most visible form of waste is litter, for which the Sagarmatha Pollution Control Committee (SPCC) has built 106 bins (SPCC, 2018) as part of the ongoing waste management plan. SNP management has introduced a number of rules to ban glass beer bottles and plastic bags; however, inadequate implementation means they remain commonplace. Tightening regulations for expeditions (SPCC, 2018) is a further step towards more effective waste management but is being offset by increasing tourism flows.

- *Illegal Activities*

The SoC reports (SNP, 2017, 2019) and baseline study (WNF, 2013) identified illegal activities (tree felling without permission and poaching) as a threat to SNP. Strong community support combined with rigorous patrolling of the park by authorities over the last decade has seen these activities reduced to near-zero levels (SNP, 2016). Accordingly, this issue was not identified by any survey respondents.

- *Air Traffic*

The SoC reports (SNP, 2017, 2019) and baseline study (WNF, 2013) identify air traffic as an issue, but it was not mentioned by any community or individual respondents. However, during the survey process, the team noticed frequent, regular helicopter flights in SNP.

During the high season (approximately 20 weeks of the year), there are 40–100 inbound fixed-wing flights to Lukla. Over 95% of tourists who visit SNP fly to and from Lukla from Kathmandu, so tourists contributed at least 22,000 tons of CO_2 to atmospheric emissions in 2019. Helicopter traffic is focused on flights beyond Lukla into the park area, with an average of 70–100 flights per day. Flights are roughly equally split between cargo (sheet metal and construction wood) and passengers (mountaineers and equipment, sightseers and medical evacuations). The estimated total contribution of helicopter flights is 4000 tons of CO_2 to atmospheric emissions (both calculations based on *facteurs mobitool*; mobitool, n.d.).

Without no monitoring or evaluation of CO_2 emissions, nor offsetting programs within Nepal, these results suggest that tourism is a significant contributor to greenhouse gas emissions. This is especially concerning given that the SNP WHS is embedded in the SDGs.

22.6 Threats Comparison

A comparison of threats from the 2013 benchmark study (WNF, 2013) with those from the 2020 survey can be made with reservations. SoC reports and the benchmark study are primarily based on expert interviews (decision makers, stakeholders, scientists, journalists) who were asked about an indicator set. The benchmark analysis showed that SNP performed significantly worse than the other protected mountain regions in the study; Kilimanjaro, Mt. Kenya, Swiss World Heritage Jungfrau-Aletsch and Sardona tectonic region. Although the methodology of the two studies is different – condensed expert interviews versus participatory methods like group surveys and interviews of residents – the results are relevant (Table 22.1).

The comparison concurs with the SNP Management Plan (SNP, 2016) that the threats posed by climate change, environmental pollution and vulnerabilities posed

Table 22.1 Summary comparison of threats to the SNP WHS

	Benchmark study (WNF 2013)				Survey 2020				
Issues of concern	++	+	–	–	++	+	–	–	Threat trend
Awareness of WHS status			X				X		Remains poor
Climate change impacts		X			X				Worsening
Economic activities: Tourism		X					X		Improving
Cultural decay/erosion			X				X		Considered small
Environmental pollution (waste and litter)		X			X				Worsening
Illegal lumbering/poaching		X					X		Improving
Air traffic				X		X			Worsening

Note: World Nature Forum (2013). *Benchmarking World Heritage & Tourism*. WNF; Boustead R. & Bhatta S. (2021). Assessment of the Social Impacts of Community Based Tourism. *Zeitschrift für Tourismuswissenschaft*. https://doi.org/10.1515/tw-2021-0010; ++ very large threat; – very small threat by uncontrolled tourism development have continued to increase. Although these issues are identified in the current management plan, the control or measures to combat them do not appear to be sufficient. After 40 years of WHS status and numerous international donor-driven development projects, effective implementation of protective measures only appears possible through more stringent management and stronger involvement of responsible local committees and user groups.

22.7 Saving Mount Everest – Towards an Integrated Preservation Strategy

To reduce pressures related to tourism and climate change, the SNP-cum-World Heritage Management Plan must be stringently implemented within a multi-stakeholder governance system and engage residents in sustainable development, as suggested by Luger and Ripp (2021).

The SNPBZ Management Plan vision is to be a "*representative example of biodiversity in the highest point of the world which is managed to enhance the unique biodiversity and maintain OUV of the area with active community participation that eventually supports the welfare of human being in perpetuity*" (SNP, 2016, p. 15). Even though the conservation of SNPBZ's biodiversity is central to the success of tourism, individual businesses are often in conflict with park interests, such that park rules are only followed by locals if they do not contradict their own interests. This is most obvious in the lack of appropriate measures to manage waste and litter and air traffic. This untenable state of affairs can only be remedied by park administration taking a much stricter approach through stronger management, but there is a lack of capacity, staff, budget and willingness to deal with conflicts of this kind with the leading stakeholders.

The second source of potential conflict is a lack of awareness of the need for an overarching WHS management strategy among residents and stakeholders. Park management has the objective '*to promote sustainable tourism and regulate it for maintaining ecological integrity and cultural heritage*' (SNP, 2016, p. 13) and unregulated tourism and land encroachment are considered a threat. However, there is no significant input from SNP management to establish a suitable tourism development policy as recommended by UNESCO.

During the past five years of the SNPBZ plan (to be revised in 2021/22), the number of tourists increased considerably, and tourism's leading role in the local economy will no doubt continue. However, uncontrolled development of multi-storey hotels, waste generation and air traffic are changing SNP's integrity and authenticity and, therefore, threatening OUV. Reconciliation is possible through sustainable tourism that integrates local cultural values, especially those that support SNP conservation and preservation agendas. Such an approach would appear to be an obvious pathway for greater community participation and to mitigate cultural erosion pressures. Reducing the impact of illegal activities is a case in point where locals and park management have collaborated effectively. Conversely, the Ministry of Tourism and Civil Aviation will not take strict measures voluntarily because licences create revenue for the state. Therefore, the UNESCO World Heritage Centre should urge the State Party to impose restrictions on air traffic and to introduce a carbon emission offset scheme through carbon sequestration projects in SNP.

Waste management requires a comprehensive review with a focus on the extraction of recyclable materials and appropriate disposal of non-recyclables. This requires resident involvement and greater responsibility by incorporating the 'preserving character' of Sherpa culture more strongly into the self-responsible conservation goals of the WHS. To do so means educating local communities on the importance of WHS status and benefits to SNP, on climate-smart practices and on increasing community engagement with day-to-day park management issues. Sagarmatha Next (2018) proposes a new waste management structure with a solid financial basis for SPCC, staff training and upgrading technical equipment (Byers et al., 2021). Open incineration of waste in dumping sites should be strictly prohibited. Undoubtedly, the best solution would be to not bring any goods into SNP and to pursue waste avoidance more consistently. The proposal for a goods-only ropeway between Namche and Lukla, to a site near the soon to be completed overland road in the buffer zone, should be seriously considered, as this would appear to offer an almost ideal low-cost solution by reducing impacts of porterage and helicopter use. To achieve a higher level of community engagement and SNP management effectiveness, more staff, more training, more budget and a stronger right of intervention to enable the staff to fulfil their monitoring and control tasks are required.

SNP WHS receives worldwide attention, and successful site management could be an exemplary example for other mountain areas. Academic institutions have described and analysed the challenges and problems over the years, but too little of this knowledge has reached the local population or been integrated into the body of knowledge by park management and then implemented. To solve this problem, a well-staffed and skilled department is needed to coordinate and enact research through SNP's multi-stakeholder, community-engaged preservation strategy.

The challenges in the SNP WHS are great, as both conflicts and threats are increasing. Climate change is challenging the OUV of the site, and the negative aspects of tourism are exacerbating the problem. International backing through science and development projects that support the multi-stakeholder environment of residents and national and local authorities are therefore welcome.

References

Adventure Travel Trade Association. (2018). *The 2018 report: An adventure travel scorecard*. ATTA.

André-Lamat, V., & Sacareau, I. (2019). Dispositifs techniques de la micro-hydroélectricité et reconfiguration d'un territoire touristique de haute montagne (region de l'Everest, Népal). *Dévelopement durable et terrioires, 10*(3). https://doi.org/10.4000/developpementdurable.16190

Boustead, R. & Bhatta, S. (2020). Field study: Unpublished final report, assessment of the social impacts of community based tourism in the Solu-Khumbu.

Boustead, R., & Bhatta, S. (2021). *Assessment of the social impacts of community based tourism.* Zeitschrift für Tourismuswissenschaft. https://doi.org/10.1515/tw-2021-0010

Byers, A. C., Gustafsson, T., Shrestha, M., & Chhetri, N. (2021). A sustainable solid waste management plan for Sagarmatha (Mt Everest) National Park and Buffer Zone, Nepal. *Mountain Research and Development, 40*(3). https://doi.org/10.1659/MRD-JOURNAL-D-20-00018.1

Daconto, G., & Sherpa, L. N. (2010). Applying scenario planning to park and tourism management in Sagarmatha National Park, Khumbu, Nepal. *Mountain Research and Development, 30*(2), 103–112. https://doi.org/10.1659/MRD-JOURNAL-D-09-00047.1

DNPWC. (n.d.). http://www.dnpwc.gov.np/en/conservation-area-detail/72/

EcoHimal Hydro. (n.d.). www.ecohimal.org

EcoHimal SAME. (n.d.). www.savingmounteverest.org

Intergovernmental Panel on Climate Change. (n.d.). https://www.ipcc.ch/reports/

International Centre for Integrated Mountain Development. (2019). The Hindu Kush Himalaya Assessment Mountains, climate change. In *Sustainability and People*. ICIMOD. https://doi.org/10.1007/978-3-319-92288-1

KoBoConnect. (n.d.). https://www.kobotool-box.org/

Kruk, E., & Banskota, K. (2007). *Mountain tourism in Nepal: From impacts to sustainability.* International Centre for Integrated Mountain Development.

Lichtberger, P. (2015). *Hazardous waste management in a natural heritage site: A case study from the Sagarmatha National Park and Buffer Zone in Nepal.* Masterthesis, Universität für Bodenkultur. http://www.ecohimal.org/fileadmin/user_upload/lichtberger_wastemanagement.pdf

Luger, K., & Ripp, M. (2021). Heritage management – Committing to preservation and facilitation. Theoretical and practical considerations. In K. Luger & M. Ripp (Eds.), *World Heritage, place making and sustainable tourism towards integrative approaches in heritage management* (pp. 15–38). Studien Verlag.

Manfredi, E. C., Flury, B., Vivano, G., Thakuri, S., Khanal, S. N., Jha, P. K., Maskey, R. K., Kayastha, R. B., Kafle, K. R., Bhochhibhoya, S., Ghimire, N. P., Shrestha, B., Chaudhary, G., Giannino, F., Carteni, F., Mazzoleni, S., & Salerno, F. (2010). Solid waste and water quality management models for Sagarmatha National Park and Buffer Zone, Nepal. *Mountain Research and Development, 30*(2), 127–142. https://doi.org/10.1659/MRD-JOURNAL-D-10-00028.1

Mani, M. (Ed.). (2021). *Glaciers of the Himalayas: Climate change, black carbon, and regional resilience*. South Asia Development Forum. World Bank. https://doi.org/10.1596/978-1-4648-0099-3

Mobitool. (n.d.). www.mobitool.ch

MoTCA. (2020). *Nepal tourism statistics 2019*. Ministry of Culture, Tourism and Civil Aviation (MoTCA), Government of Nepal.

Muccione, V., Huggel, C., Bresch, D., Jurt, C., Wallimann-Helmer, I., Mehra, M., & Caicedo, J. (2019). Joint knowledge production in climate change adaptation networks. *Current Opinion in Environmental Sustainability, 39*, 147–152.

National Planning Commission. (2017). *Nepal's sustainable development goals: Baseline report.* NPC Government of Nepal.

Nyaupane, G. P., Lew, A., & Tatsugawa, K. (2014). Perceptions of trekking tourism and social and environmental change in Nepal's Himalayas. *Tourism Geographies, 16*(3), 415–437. https://doi.org/10.1080/14616688.2014.942233

Posch, E. (2013). *Solid waste management in Sagarmatha National Park: Understanding tourists' perceptions, attitudes and behaviours*. [Master thesis, University of Vienna].

Rai, D. B. (2017). Tourism development and economic and socio-cultural consequences in Everest Region. *The Geographical Journal of Nepal, 10*(89–104), 2017.

Sacareau, I. (2009). Changes in environmental policy and mountain tourism in Nepal. *Journal of Alpine Research, 97*(3) http://journals.openedition.org/rga/1031

Sagarmatha National Park. (2016). *Management plan: Sagarmatha National Park and its Buffer Zone 2016–2020*. SNP, Ministry of Forests and Soil Conservation, Department of National Parks and Wildlife Conservation, Government of Nepal.

Sagarmatha National Park. (2017). *State of conservation report Sagarmatha National Park*. SNP Department of National Parks and Wildlife Conservation, Ministry of Forests and Soil Conservation, Government of Nepal.

Sagarmatha National Park. (2019). *State of conservation report Sagarmatha National Park, Nepal*. SNP. Department of National Parks and Wildlife Conservation, Ministry of Forests and Soil Conservation, Government of Nepal.

Sagarmatha Next. (2018). *From mountain waste to art and inspiration*. Sagarmatha Next. www.sagarmathanext.com

Sagarmatha Pollution Control Committee. (2018). *Annual report 2018, Sagarmatha pollution control Committee (SPCC)*. SPCC.

Salerno, F., Buraschi, E., Bruccoleri, G., Tartari, G., & Smiraglia, C. (2017). Glacier surface-area changes in Sagarmatha National Park, Nepal. *Journal of Glaciology, 54*(187), 738–752. https://doi.org/10.3189/002214308786570926

Sherpa, L. N. (2008). *Through a Sherpa window*. Vajra Publications.

Sherpa, A. R. (2021). *Climate change in the Himalayas: A case from Solu Khumbu*. The Partners Nepal.

UNESCO. (2009). *Case studies on climate change and World Heritage*. UNESCO World Heritage Centre.

UNESCO. (n.d.). https://whc.unesco.org/en/list/120

UNESCO Advisory Mission. (n.d.). https://whc.unesco.org/en/soc/3663

UNESCO, STP. (n.d.). https://whc.unesco.org/en/tourism

UNESCO World Heritage Centre. (n.d.). http://whc.unesco.org/en/sustainabledevlopment/

United Nations Development Program. (n.d.). http://hdr.undp.org/en/countries/profiles/NPL

World Nature Forum. (2013). *Benchmarking World Heritage & tourism*. WNF.

Zuser, A., Fellner, J., & Lederer, J. (2011). *Managing the solid waste of Sagarmatha National Park and its Buffer Zone*. Vienna University of Technology.

Chapter 23
Technological Change – Risk or Opportunity for UNESCO World Heritage?

Alexander Siegmund and Anca Claudia Prodan

Abstract This chapter provides reflections on the consequences of technological change in relation to World Heritage properties. While technological change is a core means of human adaptation and survival, it becomes a risk if the pace is too fast. This has increasingly affected societies worldwide since the industrial revolution, resulting in many negative consequences for people and the environment. Technological change is also associated with positive developments, such as those brought about by digital technology. Insights into both risks and opportunities are given in this chapter, and they are illustrated with examples, such as mining and digital geomedia. Technological change appears as a double-edged sword, but there is currently no methodology for assessing its consequences for World Heritage properties. Therefore, the chapter turns to lessons learnt from the Historic Urban Landscape approach, the UNESCO Man and the Biosphere Programme, and from impact assessment methods. While these provide useful inspiration and a basis for further reflection, the chapter concludes by emphasizing the necessity of a methodology for assessing the impacts of technological change on World Heritage properties against the background of the Sustainable Development Goals.

Keywords Geotechnology · Sustainable development · Impact assessment · Technological innovation

---

A. Siegmund (✉)
Heidelberg University of Education, Department of Geography – Research Group for Earth Observation (rgeo), UNESCO Chair on Observation and Education of World Heritage and Biosphere Reserve, Heidelberg University, Heidelberg Centre for the Environment (HCE) & Institute of Geography, Heidelberg, Germany
e-mail: siegmund@ph-heidelberg.de

A. C. Prodan
Institute Heritage Studies at the Internationale Akademie Berlin (INA) gGmbH, Berlin, Germany
e-mail: prodan@inaberlin.org

M.-T. Albert et al. (eds.), *50 Years World Heritage Convention: Shared Responsibility – Conflict & Reconciliation*, Heritage Studies,
https://doi.org/10.1007/978-3-031-05660-4_23

23.1 Introduction and Problem

Our time is characterized by technological change – there is hardly any area of personal or professional life, which has not been affected by increasing mechanization and digitization. This triggers profound changes in working and living conditions, and it has diverse ecological, economic, social, cultural and political consequences. The consumption of resources has increased significantly because of increasing industrialization and technologization. These are often associated with a wide range of negative impacts for the environment and people, resulting from the extraction, processing and use of energy raw materials such as coal and oil and metallic raw materials such as iron and copper or rare earths. These developments have been greatly influenced by rapid population growth, as the need for food, energy and urbanization have been increasing – over 50% of the world's population now live in cities, with the associated land consumption. As a result of these processes, humans, through activities involving technological change, leave behind a clear "human footprint".[1] We consume significantly more resources than the Earth can regenerate within 1 year, which makes the aims of sustainability impossible. The so-called "Earth Overshoot Day", which marks the day when the needs of people exceed the capacity of the Earth, was on July 29, 2021 (https://www.overshootday.org/).

These developments have extensive direct and indirect effects on those 1154 cultural and natural properties inscribed on the World Heritage List. Many properties are directly influenced by the expansion of cities and urban infrastructure and the associated increase in land and resource consumption. This includes deforestation to obtain raw materials and arable land, which leads to further changes in use. In addition, there are factors such as global tourism and climate change, indirectly associated with technological change, which threaten the survival of World Heritage properties. It is worth noting that the preamble of the World Heritage Convention opens with the acknowledgement that heritage is increasingly threatened "also by changing social and economic conditions" (UNESCO, 1972, Preamble). This threat has not diminished since the adoption of the Convention in 1972. If anything, it has increased, being partly facilitated by technological change. Some of the 52 properties currently inscribed on the "List of World Heritage in Danger"[2] (UNESCO, n.d.-d), such as the Historic Centre of Vienna or the delisted sites of the Dresden Elbe Valley and Liverpool, provide an illustration.

Despite potential negative consequences, technological change is also linked to a wide range of opportunities for the protection, preservation and sustainable development of World Heritage properties. For example, modern digital (geo)technologies such as satellite and drone data, digital applications for processing spatial data,

[1] Human footprint is a quantitative analysis measuring the relationship between the consumption of resources by humans and the number of resources the Earth can produce.

[2] The List of World Heritage in Danger is defined in Article 11(4) of the World Heritage Convention, and it foresees the adoption of special financial and other support measures for highly endangered properties.

geographic information systems (GIS) or GPS-supported surveying techniques can help to provide documentation about World Heritage properties, to record and analyse their state of conservation, thus contributing to their long-term preservation. In addition, in line with Article 5 of the World Heritage Convention, which names measures for States Parties to take, including for the presentation of World Heritage properties, digital media offer many opportunities. For example, they help create 3D animations and other forms of visualization for a larger audience and promote various uses, as exemplified further in this paper.

Processes that are linked directly or indirectly to technological change affect World Heritage properties for better or worse. If we consider the negative consequences, the question may arise whether the rapid pace of technological change and the associated consequences such as resource consumption and urbanization are compatible with the protection principles of the World Heritage Convention. If we consider the positive aspects, we cannot but notice the opportunities brought about by digital technology to present and experience World Heritage properties in new ways. However, there has been no comprehensive analysis of the impacts associated with technological change for World Heritage properties. Against this background, the aim of this article is to reflect on the risks and opportunities of technological change for World Heritage protection and on ways to mitigate the risks. The reflection is based primarily on insights from geography and examples of World Heritage properties, and it includes both positive and negative developments.

23.2 Signs of Technological Change and Their Consequences

Technological change has always accompanied human development. At the beginning of human history, the dynamics of these processes were still low. Nonetheless, even in earlier times, technological change sometimes led to extensive ecological, social and environmental upheavals. This began with the settling down of people during the Neolithic Cultural Revolution about 10,000 years ago, and the associated transition from hunters and gatherers to agriculture and animal husbandry, as well as the emergence of permanent settlements. (Haviland et al., 2016, 226). In particular, industrialization, starting in the second half of the eighteenth century, was accompanied by profound changes in economic and social conditions, which have since led to a worldwide increase in population, (over)consumption of resources and associated environmental pollution.

The manifold impacts of technology and technological innovations have been studied in a variety of fields, and they have become an important aspect of Science and Technology Studies (Hackett et al., 2008). For the purpose of this article, it is worth highlighting that the dynamics of technological change have continued to accelerate since the beginning of industrialization (Haviland et al., 2016, 607). This is evident in the number and spread of innovations, such as the invention of the steam engine, the railroad, electrical engineering, the automobile and, more recently, renewable energies and digital technology, the latter having increased the pace of change even more. The broader consequences, also resulting from global population growth, which has multiplied over the past 200 years, with more than half now living in urban areas, is evident in many statistics. They show socio-economic trends since 1750 of various indicators on the relationship between population growth and other variables such as land use, transportation or global tourism (Steffen et al., 2016). The deeper impacts may not be readily obvious in statistics, but technological change often goes along with environmental damage and the disruption of human settlements. Mining offers a good example.

In order to extract raw materials, large amounts of land are destroyed by the associated opencast mines and their production and transport facilities. These areas of land are not only lost for other uses, but the associated changes in ecological cycles between soil, plants and atmosphere also affect the immediate vicinity of these mining areas, with effects reaching even beyond. Technological change is also associated with new means of transportation and working conditions and with increased mobility, which makes people use their time differently, including their leisure time, as reflected in the increased numbers in global tourism. Due to the great influence of humans on the environment, it is now often spoken of as the era of the Anthropocene (Crutzen, 2006). From this, we can also infer that humans have not only an impact but also a special responsibility for the future of the planet in terms of sustainable development and the sustainability of the Earth.

The insights provided may create the impression that technological change is always negative, but technology has been crucial to human adaptation and survival. The adoption of technological innovations can lead to either disruption and abandonment of existing practices and tools or to adaptation, depending on how they are used (Haviland et al., 2016). In the next section, we give selected examples of both aspects as they relate to World Heritage.

23.3 Risks for UNESCO World Heritage Through Technological Change

Some world cultural and natural heritage properties reflect changes caused by natural processes or cultural-historical developments, including technological ones. For example, the Ancient Ferrous Metallurgy Sites of Burkina Faso illustrate the first phase of iron production development in Africa along with traditional iron ore smelting techniques (UNESCO, n.d.-a). Another example is the major mining sites of Wallonia in Belgium, considered to represent a testimony to the early dissemination of the technical, social and urban innovations of the industrial revolution (UNESCO, n.d.-c). Ironically, while World Heritage properties are valued for reflecting technological change, they do not remain unaffected by its consequences, like those described in the previous section. Plenty of cases can be found in reports on the state of conservation of World Heritage properties.

Mining and other extractive industries – to continue the example given previously – are often mentioned as factors affecting World Heritage properties. The City of Potosí in Bolivia is one such example, which has been inscribed on the List of World Heritage in Danger since 2014; one threat is mining, which leads to the degradation of the historic site (UNESCO World Heritage Centre, 2020). Mining was also the factor that led to the first removal of a property from the World Heritage List, Oman's Arabian Oryx Sanctuary, in 2007 (UNESCO, n.d.-b). The goal of the World Heritage Convention is to protect the properties, for which it has dedicated mechanisms, such as the "List of World Heritage in Danger". However, Oman wished to reduce the property to 90% in order to proceed with hydrocarbon prospection (UNESCO, 2007). In fact, if one looks at the 14 primary factors listed by the World Heritage Centre as affecting World Heritage properties, 4 of them are directly related to technological change: buildings and development (47%), transportation infrastructure (33%), service infrastructure (17%) and physical resource extraction (17%). Other factors, such as pollution (16%), are often indirectly related to technological change (UNESCO, n.d.-e; UNESCO World Heritage Centre, 2014) (See Fig. 23.1).

Technological innovation and change, as well as the social and ecological transformation processes that are associated with them (Veuve, 2020), often result from human striving for (economic) prosperity, greater efficiency in work processes and the production of goods and services to ensure subsistence or increase capital and

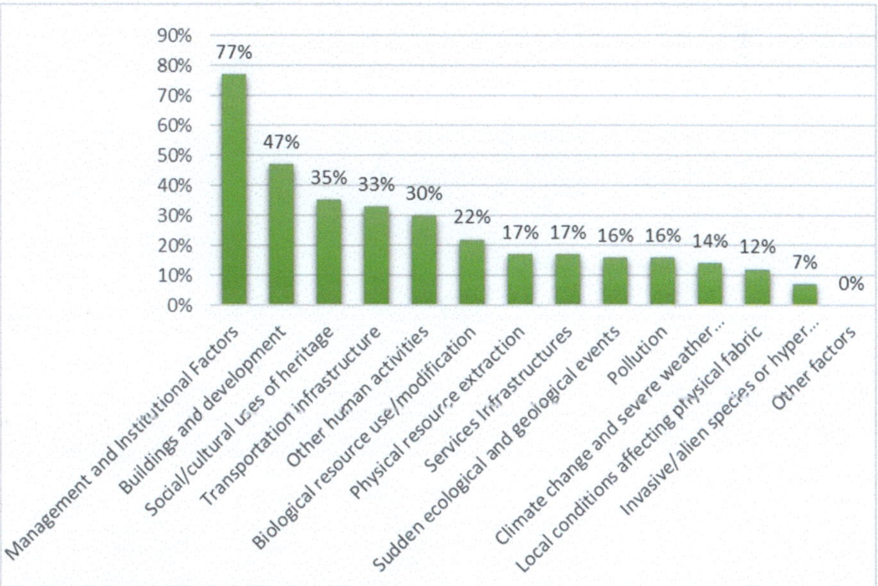

Fig. 23.1 The main threats affecting World Heritage properties. (*Note.* The statistical analysis covers the period 1979–2013 and includes 13 factors. The threats have not changed since then, but a new category entitled "other factors" has been added. However, to date, there is no statistical information for this threat. [Graph A. Siegmund 2021])

productivity (Jischa, 2007). Technological change has always been part of human life and is often a necessary component of human adaptation and survival. Thus, technological change per se is not the problem, but, as already indicated in Sect. 23.2, the scale and pace of change certainly are. World Heritage historic cities or other properties in areas experiencing rapid growth and infrastructure development illustrate this problem (UNESCO World Heritage Centre, 2010). Sometimes World Heritage properties are virtually "enveloped" by settlements and economic land, as the example of the Pyramid of Cheops in Egypt shows (see Fig. 23.2) (Hemeda & Sonbol, 2020).

While technological change may affect all properties directly or indirectly, through influences on climate, carbon dioxide emissions and other forms of environmental pollution, the extent of the threat depends on the local conditions and characteristics of the property. Nevertheless, it seems to have a greater impact on properties located in urban growth regions, in opencast mining areas or areas rich in natural resources. We presented illustrations regarding mining and urbanization. An example concerning impacts on natural areas is the tropical forests of Sumatra in Indonesia. The property has been on the List of World Heritage in Danger since 2011, and it is highly affected by deforestation, illegal logging and agricultural encroachment (Fig. 23.3) (UNESCO, 2011a).

As already mentioned in passing, and as illustrated by examples such as Venice, the Great Wall of China, or Machu Picchu (UNESCO World Heritage Centre, 2021), World Heritage properties can also be endangered by mass tourism. This is accompanied by environmental damage associated with visitor transport, accommodation and supply, and a lack of appropriate infrastructure such as waste disposal. Even the causes associated with armed conflicts could go hand in hand with technological

Fig. 23.2 Endangerment through urbanization Cheops-Pyramid/Egypt. (*Note.* Sentinel-2 (ESA) image courtesy of the U.S. Geological Survey)

Fig. 23.3 Endangerment through deforestation, Kerinci-Seblat National Park, Indonesia. (*Note.* Sentinel-2 (ESA) image courtesy of the U.S. Geological Survey)

change, for example, by fostering disputes over resources (Ferguson, 2001). Many other examples could be added, but those already given illustrate the risks of technological change to World Heritage properties. They also reveal that most of the risks stem from one fundamental problem, namely the need to strike a balance between conservation on the one hand and use, development and change on the other. In other words, there is a need to approach conservation as sustainable change.

23.4 Potentials of (Geo-) Technologies for the Sustainable Development of World Heritage Properties

Technological change does not have to lead to disruptions. It may also lead to adaptation and bring about opportunities for World Heritage properties. Digital technologies are perhaps the best example in this regard, and many believe these technologies greatly contribute to the sustainable development of World Heritage properties. This is not to say that digital technology may not have unwanted consequences. Each technology can be a curse or a blessing. Research shows that there are direct environmental effects from the production, use and disposal of digital technology, such as global warming and e-waste, and indirect effects from changes in patterns of consumption and production (Bieser & Hilty, 2018; Bedford et al., 2021). Yet, in many regards, digital information and communication technologies can present numerous opportunities (Xiao et al., 2018).

The rapidly growing range of digital technologies is just as extensive as the diverse potential uses in the context of World Heritage – a comprehensive overview is hardly possible. Yet, the potential can be illustrated with the example of modern geotechnology, although even the spectrum of such digital geotechnologies is extremely large. It ranges from the use of remote sensing methods based on satellite and aerial image data and the digital processing of spatial data using geographic information systems (GIS) to laser and GPS-supported surveying methods. On this basis, geotechnologies can make an important contribution to the recording, analysis and monitoring, reconstruction, restoration and conservation and sustainable planning and management of World Heritage properties (Xiao et al., 2018). This has been well illustrated by several authors in the Technological Change section of this book, who present applications ranging from digital maps to interactive tools. However, in order to present their potential compactly, we can use a principle known to geoinformatics, namely the IMAP principle (abbreviated from Input, Management, Analysis and Presentation) associated with the use of digital geomedia. This is presented briefly below.

Input data can be generated through remote sensing methods using satellite and aerial image data. This can serve to record the state of a property without any physical contact. Through the additional use of drones and the associated high spatial resolution of the aerial image data, this is possible even with small-scale structures down to the size of a centimetre. With the help of aircraft and drone-assisted laser scanning, the structures of World Heritage properties can be recorded in a higher resolution and even in three dimensions, without the sites themselves being accessed and damaged. It is often only through the use of remote sensing data that the extent of a World Heritage property becomes visible (Xiao et al., 2018). Furthermore, the use of satellite, aerial photo or drone data can be used to map and explore, at different scales, areas that are otherwise inaccessible or difficult to access due to a lack of transport infrastructure or for security reasons. Such methods are becoming increasingly important as non-contact and thus "non-destructive" methods in the context of World Heritage, and they have been applied to properties such as the Old Town of Ávila, Spain, Kathmandu Valley, Nepal, or My Son Sanctuary, Vietnam (Xiao et al., 2018, 397–402).

Beginning with the satellite Landsat in the 1970s, a variety of Earth observation satellites are now available, with data of varying characteristics (e.g. spatial, temporal, spectral resolution) available free of charge, such as the satellite data and derived data products under the European Union's Copernicus program. They offer a wide range of possibilities to promote the protection, preservation, management and sustainable planning as well as communication of the universal values of World Heritage properties. This is evident in the increasing number of specialist conferences and calls for tenders and special issues in journals such as "Earth Observation for Heritage Documentation", in preparation under the "International Journal of Applied Earth Observation and Geoinformation".

Following the IMAP principle, with the help of GIS, spatial data of World Heritage properties not only can be generated but also managed and analysed. Datasets with different scales, underlying coordinate systems and properties (vector and pixel-based data) from different sources can be integrated into a kind of digital spatial database. The resulting different data layers can be further analysed with GIS by combining or blending different datasets to generate new information. Finally,

GIS serves to visualize and thus present the corresponding data in the form of (interactive) maps, animations or three-dimensional representations. Thus, modern geotechnologies are of particular importance for the documentation, management and presentation of complex structures, as is the case with many World Heritage properties. In combination with historical data and maps, which in turn can be digitized, the comprehensive development of the properties can be traced (Nicu, 2017).

In addition to facilitating more efficient management, geotechnologies may contribute to research, knowledge and appreciation of the universal values of these properties by enabling accessibility for a broader audience as well as participation. According to a study, 71% of the population in the USA in 2015 already used digital media to access UNESCO cultural and artistic assets instead of visiting them on site (Nicu, 2017). Thus, digital technology may reduce the environmental impact of World Heritage tourism, such as carbon dioxide emissions and resource consumption associated with the transport, accommodation and supply of visitors. (Xiao et al., 2018). The data obtained with geotechnologies can also be combined with Augmented Reality (AR) and Virtual Reality (VR) technology, providing novel means of knowledge transfer and interaction with World Heritage (Kenderdine et al., 2008, 275). Geotechnology presents opportunities not simply for World Heritage but for its use in a way that responds to the Sustainable Development Goals related to the protection and safeguarding of cultural heritage (SDG 11.4) and the promotion of sustainable tourism (SDG 8.9) (Xiao et al., 2018).

23.5 The Way Forward – Reflections on Risks and Opportunities

As we have seen above, technological change may be a risk or an opportunity. However, the question of whether technological change in the balance sheet tends to favour or hinder World Heritage conservation cannot be answered conclusively. Not only is "technological change" too broad a concept, including as many technologies as humans have created, but its impacts also depend too much on the particular conditions of individual World Heritage properties. Thus, the question is whether and how it is possible to ensure that the properties can be protected and used sustainably despite or precisely because of technological change.

To tackle this question, inspiration can be taken from related activities regarding World Heritage and other programmes relevant for heritage conservation, such as the Historic Urban Landscape approach (HUL) or the UNESCO Man and the Biosphere Programme (MAB). HUL is an approach to the management of heritage properties promoted through the Recommendation on the Historic Urban Landscape (UNESCO, 2011b). It was developed because the previous conservation paradigm, based on a separation of the property, with its core components expressing the Outstanding Universal Value (OUV), from the surrounding area, was no longer appropriate. Today, sustainable conservation requires perceiving the site in context as part of a region in which people live and work (Kloos, 2014). The HUL initiative was specifically launched for World Heritage properties in urban areas, hoping to achieve a stronger integration of urban World Heritage protection within the respective socio-economic context (Kloos, 2014). In a similar vein, a shift in perspective is needed for a broader view of the impacts of technological change on World Heritage properties, not limited to the boundaries of a property and its buffer zones but in relation to its use and consequences locally and regionally.

The MAB Programme may also offer some insights. MAB was launched by UNESCO in 1971 with a focus on the sustainable use and conservation of the resources of the biosphere, and it aims to establish a scientific basis for the relationship between people and their environment. This programme's strategy is specifically adapted to support the 2030 Agenda for Sustainable Development, the Sustainable Development Goals (SDGs) and the Paris Climate Agreement (UNESCO, 2017). Such measures have also been taken in the context of World Heritage, in particular since the adoption of the *Policy Document for the Integration of a Sustainable Development Perspective into the Processes of the World Heritage Convention* by the General Assembly of States Parties to the World Heritage Convention in 2015 (UNESCO World Heritage Centre, 2015). However, the biosphere reserves protected under MAB serve as models for national or regional demonstration of sustainable development (UNESCO, 2017, 22). While a similar idea exists as a modest suggestion in the *Policy Document* (paragraph 5), the emphasis is much stronger in the Lima Declaration on the UNESCO Man and the Biosphere (MAB) Programme and its World Network of Biosphere Reserves (WNBR) as well as in its action plan adopted in 2016 (UNESCO, 2017). World Heritage properties reflect technological change, as exemplified above, but they could more strongly serve as models to illustrate sustainable adaptation strategies to technological change.

Furthermore, it is worth considering the potential of impact assessment methods. They are available and have been used in the context of World Heritage for about a decade (ICOMOS, 2010; Pereira Roders & van Oers, 2012). Environmental Impact Assessment (EIA) and Heritage Impact Assessment (HIA) are cases in point. As authors who have assessed these methods explain, EIA focuses on "major development projects such as roads, industrial plants or airports" and their potential impacts on cultural heritage, including larger areas, while HIA focuses on proposals for change and "the analysis is confined to the impacts on cultural significance" (Pereira Roders & van Oers, 2012, 105). Both EIA and HIA incorporate the impacts of technological change, but it would be worth considering the potential of an assessment tool with technological change at its core. Such methods have been used since the 1970s. They are known as Technology Assessment (TA) and continue to be used in adapted forms, based on the lessons learnt over time (Grunwald, 2018). It would be worth considering how such methods can be tailored to World Heritage. They can be enhanced by the potential of digital technology in building future scenarios to capture and evaluate the risks associated with technological change and its potential for World Heritage (Weyer, 2017; Xiao et al., 2018). To align fully with the Sustainable Development Goals, they could even include the negative impacts associated with the use of digital technology, not only environmental, as noted above, but also those resulting from digital obsolescence and the need to consider the preservation of World Heritage-related digital data.

23.6 Conclusion & Outlook

Technological change is a key characteristic of our time. While it has always accompanied human development as a necessary means of adaptation and survival, the pace and scale of change have intensified greatly, making technology one of the main factors influencing the dynamics of human societies today. As the examples provided show, technological change has resulted in a series of negative consequences for people and the environment. World Heritage properties, with all their typological diversity, have not remained unaffected. The tensions between development needs and conservation requirements appeared to be one of the main factors negatively affecting many properties, whether cultural or natural. At the same time, not all forms of technological change are negative. There are positive examples, and one of them, chosen for purposes of illustration in this chapter, was digital geomedia. It has not only proven beneficial for World Heritage properties but also contributes to achieving the Sustainable Development Goals (Xiao et al., 2018).

Many other examples of risks and opportunities could have extended our presentation; in fact, so many so that a comprehensive overview is hardly possible. Technological change is a very broad concept. Furthermore, its impacts depend heavily on local and regional contexts, and they are manifold. Yet, how can we foresee the impacts of change in the absence of a methodology for assessing the consequences of our actions today? How can we proceed efficiently in the absence of guidelines, which capture the complexities of the problems we are facing? How can we use the opportunities of technological change while avoiding or at least minimizing the risks it brings? No answers can be given today, but answers must be given in the future if our aim is the sustainable conservation of World Heritage properties. Thus, when envisioning the way forward, a methodology for assessing the impacts of technological change on World Heritage properties, developed against the background of the Sustainable Development Goals, as well as a policy instrument with technological change at its core, emerge as indispensable tools.

References

Bedford, L., Mann, M., Walters, R., & Foth, M. (2021). A post-capitalocentric critique of digital technology and environmental harm: New directions at the intersection of digital and green criminology. *International Journal for Crime, Justice and Social Democracy*. [In Press]. https://eprints.qut.edu.au/213773/

Bieser, J. C. T., & Hilty, L. M. (2018). Assessing indirect environmental effects of information and communication technology (ICT): A systematic literature review. *Sustainability, 10*(8), 2662. https://doi.org/10.3390/su10082662

Crutzen, P. J. (2006). The "Anthropocene". In E. Ehlers & T. Krafft (Eds.), *Earth system science in the Anthropocene*. Springer. https://doi.org/10.1007/3-540-26590-2_3
Ferguson, R. B. (2001). Materialist, cultural and biological theories on why Yanomami make war. *Anthropological Theory, 1*(1), 99–116. https://doi.org/10.1177/14634990122228647
Grunwald, A. (2018). *Technology assessment: Practice and theory*. Routledge.
Hackett, E. J., Amsterdamska, O., & Lynch, M. (2008). In J. Wajcman (Ed.), *The handbook of science and technology studies*. The MIT Press.
Haviland, W. A., Prince, H. E. L., Walrath, D., & McBdride, B. (2016). *Anthropology: The human challenge* (15th ed.). Cengage Learning. (Original work published 1974).
Hemeda, S., & Sonbol, A. (2020). Sustainability problems of the Giza pyramids. *Heritage Science, 8*(8), 1–28. https://doi.org/10.1186/s40494-020-0356-9
ICOMOS. (2010). *Guidance on heritage impact assessment for cultural world heritage properties*. ICOMOS.
Jischa, M. F. (2007). Herausforderung Zukunft. Technischer Fortschritt und Globalisierung. *Chemie Ingenieur Technik, 79*(1–2), 29–41. https://doi.org/10.1002/CITE.200600142
Kenderdine, S., Shaw, J., Del Favero, D., & Brown; N. (2008). PLACE-HAMPI Co-evolutionary narrative and augmented stereographic panoramas, Vijayanagara, India. In Y. E. Kalay, T. Kvan, & J. Affleck (Eds.), *New heritage: New media and cultural heritage* (pp. 275–293). Routledge.
Kloos, M. (2014). *Landscape 4. Landschaftsideen Nordeuropas und die visuelle Integrität von Stadt und Kulturlandschaften im UNESCO-Welterbe*. [Doctoral Dissertation, Rheinisch-Westfälisch Technische Hochschule Aachen]. RWTH Publications. https://publications.rwth-aachen.de/record/444768/files/5078.pdf
Nicu, I. C. (2017). Tracking natural and anthropic risks from historical maps as a tool for cultural heritage assessment: A case study. *Environmental Earth Sciences, 76*(330). https://doi.org/10.1007/s12665-017-6656-z
Pereira Roders, A., & van Oers, R. (2012). Guidance on heritage impact assessments: Learning from its application on world heritage site management. *Journal of Cultural Heritage Management and Sustainable Development, 2*(2), 104–114. https://doi.org/10.1108/20441261211273671
Steffen, W., Broadgate, W., Deutsch, L., Gaffney, O., & Ludwig, C. (2016, September 29). Zum Entwicklungsverlauf des Anthropozäns: ‚Die Große Beschleunigung'. *Bundeszentrale für Politische Bildung*. https://www.bpb.de/gesellschaft/umwelt/anthropozaen/234831/entwicklungsverlauf-des-anthropozaens
UNESCO. (1972). *Convention concerning the protection of the world cultural and natural heritage, Paris 1972*. UNESCO.
UNESCO. (2007). *Oman's Arabian Oryx Sanctuary: First site ever to be deleted from UNESCO's World Heritage List.* https://whc.unesco.org/en/news/362/
UNESCO. (2011a, June 22). *Danger listing for Indonesia's Tropical Rainforest Heritage of Sumatra*. https://whc.unesco.org/en/news/764/
UNESCO. (2011b). *Recommendation on the historic urban landscape*. https://whc.unesco.org/uploads/activities/documents/activity-638-98.pdf
UNESCO. (2017). A new roadmap for the man and the biosphere (MAB) programme and its world network of biosphere reserves. .
UNESCO. (n.d.-a). *Ancient Ferrous Metallurgy Sites of Burkina Faso*. https://whc.unesco.org/en/list/1602
UNESCO. (n.d.-b). *Arabian Oryx Sanctuary*. https://whc.unesco.org/en/list/654
UNESCO. (n.d.-c). *Major Mining Sites of Wallonia in Belgium*. https://whc.unesco.org/en/list/1344
UNESCO. (n.d.-d). *List of world heritage in danger*. http://whc.unesco.org/en/danger
UNESCO. (n.d.-e). *State of conservation information system*. https://whc.unesco.org/en/soc/
UNESCO World Heritage Centre. (2010). *Managing historic cities* (World heritage paper series 26) (p. UNESCO). https://whc.unesco.org/documents/publi_wh_papers_27_en.pdf
UNESCO World Heritage Centre. (2014). *State of conservation of World Heritage properties: A statistical analysis (1979–2013)*. UNESCO World Heritage Centre.

UNESCO World Heritage Centre. (2015). *Policy for the integration of a sustainable development perspective into the processes of the World Heritage Convention*. UNESCO.

UNESCO World Heritage Centre. (2020). *State of conservation report COM 44, Plurinational state of Bolivia, City of Potosí (1987, ref. 420)*. Gobierno del Estado Plurinacional de Bolivia; Ministerio de Culturas y Turismo; UNESCO World Heritage Centre & ICOMOS.

UNESCO World Heritage Centre. (2021). *The World Heritage List*. https://whc.unesco.org/en/list/

Veuve, A. (2020, April 20). 12 Thesen für das Zeitalter des immer schneller werdenden technologischen Wandels. *AV Digital Transformation Blog*. https://www.alainveuve.ch/12-thesen-fuer-das-zeitalter-des-immer-schneller-werdenden-technologischen-wandels/

Weyer, J. (2017, March 8). Technischer Fortschritt – Fluch oder Segen? *Bundeszentrale für Politische Bildung*. https://www.bpb.de/dialog/netzdebatte/243905/technischer-fortschrittfluch-oder-segen

Xiao, W., Mills, J., Guidi, G., Rodríguez-Gonzálvez, P., Barsanti, S. G., & González-Aguilera, D. (2018). Geoinformatics for the conservation and promotion of cultural heritage in support of the UN sustainable development goals. *ISPRS Journal of Photogrammetry and Remote Sensing, 142*, 389–406.

Chapter 24
Change in Water Technology in Anatolia: From Use to Energy, Conflicts to Climate Action

Yonca Erkan

Abstract UNESCO World Heritage properties in Anatolia show a great variety in their land use morphologies as a response to their environmental differences, developing different ways of managing water for daily use and agricultural irrigation. History testifies to the many conflicts and wars that occurred in defending and/or accessing these important water resources. The remnants of this infrastructure form part of invaluable cultural heritage and present opportunities for the embodied traditional knowledge to mitigate the impacts of climate change. However, ruthless water regimes (i.e. hydroelectric plants), which disregard the importance of water for communities, have prioritized water as a source of energy over its value for daily and agricultural use and have impacted the environment and climate, which directly affects both tangible and intangible cultural heritage. Dried creeks leave water-related infrastructure and equipment without a purpose and people deprived of water. In the age of Anthropocene, such an approach victimizes people through the idea of taking over nature while at the same time making people the victims of nature's response. In return, new regional conflicts are instigated, and migration becomes inevitable, diminishing neighbourly peace and also aggravating climate change, causing negative impacts on cultural heritage and jeopardising many of the UN Sustainable Development Goals.

Keywords Water heritage · Dams · Hydroelectric plants · Climate action · World heritage

---

Y. Erkan (✉)
UNESCO Chair, Management and Promotion of World Heritage Sites: New Media and Community Involvement at Kadir Has University, Istanbul, Turkey

ARCHES, University of Antwerp, Antwerp, Belgium
e-mail: yonca.erkan@khas.edu.tr; yonca.erkan@uantwerpen.be

M.-T. Albert et al. (eds.), *50 Years World Heritage Convention: Shared Responsibility – Conflict & Reconciliation*, Heritage Studies,
https://doi.org/10.1007/978-3-031-05660-4_24

24.1 Introduction

Civilization has evolved thanks to humanity's ability to control water. The creative genius of humankind in relation to water manifests itself in water structures for collecting (wells, boreholes), storage (dams, open and closed cisterns), distribution (qanat, aqueducts, water pipes, reservoirs, water towers), presentation (fountains, ponds), sanitary and bathing facilities (pools and bathhouses), use of water for transportation (channels), water sanctuaries and water treatment systems, among others. The discovery of the water level alone, as an instrument that made it possible to build irrigational channels as well as exemplary structures such as pyramids and temples, is worthy of recognition. Thus, as water is essential to all living creatures, it is equally important to understand the role it plays in the development of civilization (Graff et al., 2019). Although humankind mastered water technology a long time ago, issues related to water are at the heart of global discussions (Sanyanga et al., 2020; Wrong Climate for Big Dams, 2011). As our relationship to water changes through increased consumption water assumes new roles in energy production. We witness drying lakes and rivers, declining groundwater resources, land subsidence, the establishment of sinkholes, water contamination, water supply shortages, forced migration, agricultural losses, salinization of the soil and sand storms and resulting conflicts and ecosystem damage, which are the contemporary water-related issues (Madani, 2014).

Within the World Heritage context, water structures can be viewed in two ways. The first is to look at the World Heritage List as an exemplary register of outstanding water structures, reflecting various creative technologies of different civilizations in response to geographical circumstances (Douet, 2018). Secondly, the properties on the World Heritage List can be seen as a useful resource to study how water technologies have contributed to the creation of these outstanding places. Understanding the traditional relationship between the site and its water use/technologies is also important for site conservation and dealing with global challenges where water plays a major part (Wittenberg & Schachner, 2013; Sevimli & Uzel, 2003; Willems & Schaik, 2015; Angelakis, 2012).

24.2 Change in Water Technology, Dams and Hydroelectric Power Plants (HEPPs)

In the shift of conceptualization of water from being a free public good to a rare and costly commodity, the most dramatic change occurred at the beginning of the twentieth century (İlhan, 2017). Access to clean water has always been a valid concern, especially in the face of pollution and climate change due to the industrialization, population increase and urbanization of the last 200 hundred years. In the present day, the need for higher amounts of water and energy jeopardises the limited resources of the world and makes the UN 2030 Agenda more difficult to achieve.

Specifically, water technology has become a source of conflict since it evolved from being an essential element of daily life and agricultural use to something that is expected to generate energy for masses greater than its immediate surrounding population. This tension is not only between different nations sharing the water resources but also within nations between communities that have direct access to water and their governments that disregard the social, cultural and human rights of those communities. Likewise, the ruthless water regimes (i.e. dams and hydroelectric plants), which disregard the importance of water for communities in terms of social and environmental impacts, prioritized its potential for energy as the sole criterion.

The changing water technology, through the construction of the large dams of the modern era, has an impact on the environment and regional climatic conditions and directly affects cultural heritage (both tangible and intangible), leaving some of the tangible heritage submerged in the water collection basin while displacing populations (such as Hasankeyf). Dried creeks leave water-related infrastructure and equipment without a purpose (i.e. wooden and masonry bridges, waterwheels, mills), and people become deprived of water. Such interventions with the intent of taking over nature for development result in people becoming a victim of nature's response (i.e. droughts, heatwaves, flush rains, floods). Dams were originally constructed to secure water; however, in the face of long-lasting droughts, they have become obsolete, especially when the short lifespan of the dams are considered. Through these extreme environmental conditions and climate change, new regional conflicts have been instigated, and migrations have become inevitable, diminishing neighbourly peace and also aggravating climate change, causing negative impacts on cultural heritage and jeopardising many of the Sustainable Development Goals set out in the UN 2030 Agenda (Al-Muqdadi, 2019).

For a while, it was believed that hydroelectric power plants (HEPPs) were a form of green energy, but this belief is now being heavily challenged due to their various negative impacts on the environment and socio-cultural life (Wrong Climate for Big Dams, 2011; İlhan, 2013; World Bank, 2017). A Joint Statement by Civil Society Organizations on the occasion of the 2019 World Hydropower Congress in Paris reviewed the threats that arise from hydroelectric dams, that they have already forced the displacement of people, estimated to be around 40–80 million worldwide. The building material of the dams disrupts the natural flow of water and sediments, deteriorating water quality, eliminating unique habitats (flora, fauna, human) and undermining biodiversity. Dam constructions are accompanied by new access routes, escalating illegal land-grabbing, deforestation and mining and cause urban violence (Jensen-Cormier, 2019).

According to one assertion, dams reduce vulnerabilities to fluctuating weather events and prevent floods (Berga, 2016), but there are other concerns about how they might aggravate climate change through evaporation of water from their large surface areas and, furthermore, through carbon dioxide (CO_2), methane (CH_4) and nitrous oxide (N_2O) emissions from reservoirs (World Bank, 2017). Groundwater depletion in dam basins, for example, in the Euphrates–Tigris River Basin, with its many dams and HEPPs, has led to alarmingly low groundwater levels over the last two decades (Voss et al., 2013; Thaman, 2020).

It is believed that at least 20% of natural World Heritage sites are affected or threatened by dams, and this percentage has tended to increase over the last 5 years (The False Promises of Hydropower, 2019). Concerns over the impacts of dams on the natural World Heritage sites were raised in a 2015 IUCN Report (IUCN, 2015).

Tim Badman, the Director of IUCN's World Heritage Program, suggests that.

> Dams can have a huge impact on World Heritage sites, reducing precious natural wetland areas, changing river flows and impacting local communities. It is essential to consider better alternatives that avoid such constructions where possible and to properly assess how dams will affect our World Heritage before they are built. Nature transcends national borders, and efforts to preserve what we recognise as our collective heritage must also look beyond national borders. (IUCN, 2015).

Similar concerns were expressed at the Extended 44th Session of the World Heritage Committee (2021) by a coalition of relevant NGOs (Rivers without Boundaries International Coalition et al., 2021).

24.3 Ancient Water Technology in Anatolia

Anatolia, historically known as Asia Minor, is part of transcontinental Turkey between Asia and Europe. Geographically, the Anatolian peninsula is defined by the Black Sea, Aegean Sea and the Mediterranean Sea, while the south-eastern borders meet the Fertile Crescent, where civilization is accepted to have been born. The water culture of Anatolia, especially the nineteen properties on the UNESCO World Heritage List, among other cultural sites (Öziş, 1996; Tanyeli & İkiz, 2017), reflect cultural exchanges throughout centuries and between civilizations, and unique structures attest to specific typologies of water structures and land uses that are still a source of inspiration for present and future generations.

Recent archaeological evidence from the Göbekli Tepe (~9500–8000 calBC), as a prehistoric site (inscribed on the WH List in 2018), demonstrates the transition from hunter-gathering to farming economies and the mastering of water collection for human use (Clare, 2020). Initial beliefs based on the first excavations in the region positioned Göbekli Tepe as only a ritual site, looking at the elaborately carved monolithic T-shaped pillars together with the absence of domestic structures and lack of reliable water sources near the site. However, recent discoveries of domestic buildings and structures for rainwater harvesting and distribution rectified this view (Clare, 2020). Carved channels and cisterns on the natural bedrock are considered evidence of human beings' creative ingenuity to address a lack of freshwater sources through onsite water acquisition mechanisms (Clare, 2020).

Another important Neolithic site is Çatalhöyük (inscribed on the WH List in 2012), which contains eighteen levels of Neolithic occupation between 7400 BC and 6200 BC. This site reveals the use of water channels dug by hand for irrigational purposes and is considered to be a pioneering example (Bildirici, 2020). Boğazköy, the Hattusha Hittite Capital (WH List, 1986) holds a spring collection chamber of masonry (Öziş et al., 2010). Furthermore, Pergamon and its multi-layered cultural landscape (inscribed on the WH List in 2014) has a 20 km Hellenistic water supply system composed of clay pipe systems of a masonry-vaulted gallery and a long aqueduct (Tanyeli & İkiz, 2017) that was used and developed through consecutive Roman, Byzantine and Ottoman periods (Öziş et al., 2010). The Kings of Pergamon also created Hierapolis, a Graeco-Roman thermal spa near Pamukkale that utilised the natural calcite-laden water pools (Hierapolis-Pamukkale, WH List, 1988; Tanyeli & İkiz, 2017). Troy (inscribed on the WH List in 1998), known as the Ilium from *The Iliad*, enjoyed water being carried from the slopes of Mount Ida with aqueducts and clay pipes.

Safranbolu (inscribed on the WH List in 1994) exemplifies Ottoman rural settlement and its water management mechanisms perfectly. Water is brought into the city through aqueducts, and the general layout of the settlement was planned according to water resources where the preindustrial tannery workshops, which required abundant water resources, were placed at the juncture of two streams (Gümüş and Akçasu) keeping the wastewater away from the city centre through vaulted underground systems (Editor, 2019). On the other hand, the summer residential quarter demonstrates the creative use of indoor water pools as a solution for climatic and acoustic relief. Additionally, Safranbolu has earlier traces of a deep underground water well dug into the rock formations from 2500 years ago (Doğa araştırmacısı iki arkadaş …, 2021).

Göreme National Park and the Rock Sites of Cappadocia (inscribed on the WH List in 1985) presents interesting responses to water collection. Recent research identified complex underground hydrological mechanisms carved into the rocks (Bixio et al., 2020). This study presents creative water management solutions developed in an extreme natural environment, where fresh surface water resources are scarce. The rectangular wells were dug vertically, while the aquifer has footholes dug on the sides to provide ventilation and maintenance. Also, there are underground cisterns fed by springs, rainwater or meltwater. A unique form of water collection is made through funnels that collect water on the surface at the edge of cliffs and convey it to the underground reservoir. Additionally, one can find underground aqueducts that carry water from one point to another (e.g. the Uçhisar channel system is 3600 m long). Drainage tunnels and tunnel cisterns (linear cisterns) are found in the vicinity of the Göreme Open Air Museum. Drainage tunnels remove the flash-flood water, enabling agricultural activity at the bottom of deep valleys. This further enhanced the control of erosion and sediment accumulation at the base of the valley. The tunnel cisterns can be observed in Kılıçlar Valley, which are mechanisms to provide irrigation water for agricultural use. Water collection in these tunnel cisterns was achieved through trenches filled with rubble that enabled proper drainage, which can be accepted as a unique solution. This complex mechanism showcases a year-long approach that enabled water collection in short periods of time in winter seasons, which was then used over the course of dry summer months. This system ensured maximum water collection and minimum water loss, as there was no surface evaporation (Bixio et al., 2020).

The water systems of İstanbul (inscribed on the WH List in 1985) deserve a dedicated research paper. Constantinople, a metropolis throughout the centuries, was the centre of attention due to its strategic location along the land and sea trade ways. Although there are no substantial freshwater resources in the city, the integrated water systems developed in the consecutive periods of Roman, Byzantine and Ottoman periods have ensured that a remarkable water heritage survives to the present day. The water supply system composed of aqueducts, open pools, large cisterns, water pipes, dams, water towers, bathhouses, public fountains and drainage channels put Istanbul in a unique position (Çeçen, 1996; Crow, 2012). The intangible cultural heritage woven around water developed in the Ottoman times further makes this aspect more interesting (Reyhan & Yazıcı, 2020).

Xanthos-Letoon (WH List, 1988) sanctuary consists of three temples dedicated to Leto, Artemis and Apollo· as well as the ruins of a nymphaeum from the times of Hadrian, built on the water source and considered to be sacred, which was the reason for the creation of the sanctuary (Çulcuoğlu et al., 2013). The continuous use of the sacred site from antiquity until the seventh century AD is a testament to the importance of water for the society, which sheds light on present-day cultural practices exercised in the region.

Complimentary early examples can be found on the World Heritage Tentative List of Turkey. Tanyeli and İkiz (Tanyeli & İkiz, 2017) shed light on archaeological evidence from Eflatun Pınarı, the Hittite Spring Sanctuary, a late Bronze Age (roughly 1400–1175 BCE) that signifies the importance attributed to water and its sanctity (Harmanşah, 2018). The Archaeological Site of Perge, Allianoi, located in close proximity to the ancient city of Pergamon, is a well-known example of hydrotherapy. However, the 2014 Periodic Report highlights that WH properties in Turkey are facing certain threats that are associated with the changing use of water in the region, such as sudden ecological events, pollution, climate change and water extraction.

24.4 Changing Water Technology in Anatolia Through Dams and HEPPs

Starting in 1989, mainly with the GAP Project (South-east Anatolian Project), Turkey prioritized the construction of dams and hydroelectric plants in relation to the development of the country. The objectives of the GAP Project were to increase the income level and living standards of the local people using the resources of the South-eastern Anatolia Region and to increase the rural efficiency and employment opportunities in the area while enhancing the national economic development and social stability of the country. In this regard, 22 dams and 19 HEPPs were planned in the region, of which 14 HEPPs are already completed as of 2019, generating 22.8 Billion kW of energy. This achievement encouraged extensive dam and HEPP constructions in other regions of the country.

In Turkey, there are 25 river basins and 320 lakes. The total water potential of the country is 112 billion m^3. The first hydroelectric plant in Turkey was installed in Tarsus in 1902 (Öztürk & Mutlu Öztürk, 2019). As of 2019, the overall number of dams is 860 and HEPPs is 683, of which 616 were built by the private sector (Saygılı, 2018). From the developmental point of view, dams and hydroelectric plants are presented as vehicles of progress (Altinbilek, 2002). However, some believe that they have negative impacts on the environment and socio-cultural life, including tangible and intangible cultural heritage (Yeşil & İnal, 2019; Thaman, 2020; Eken et al., 2016; Wrong Climate for Big Dams, 2011; Başkan, 2018; IUCN, 2015; The False Promises of Hydropower, 2019).

Cultural heritage affected by dam and hydroelectric plant construction can be analysed under cultural heritage (tangible, intangible) components and in relation to flora and fauna (Öztürk & Mutlu Öztürk, 2019). To start with the environmental side, Key Biodiversity Areas are poorly protected in Turkey, with less than 14% protected (Eken et al., 2016). In a 2016 study, the construction of dams and hydroelectric plants emerged as the number one threat to the biodiversity areas of Turkey (Eken et al., 2016). This study identified irrigation and drainage, tourism development, urban development, roads and infrastructure and mining followed by dams and hydroelectric plants (Eken et al., 2016) as major threats to Key Biodiversity Areas, which have lasting irreversible impacts. Although Turkey does not have a natural heritage site on the World Heritage List, these Key Biodiversity Areas form the most valuable natural resources, where we observe a close geographical match with those properties on the World Heritage List. Due to this reason, we can easily say that dams and hydroelectric plants do change the landscape and therefore play an important role in the challenges that World Heritage properties in Turkey face.

The most emblematic site that is affected by dam construction in Anatolia is the Hasankeyf settlement, which contains traces of 12,000 years of human presence in the region. Civil society, national and international experts have tried extensively to highlight the importance of the site. However, due to the construction on Ilısu Dam in 2006, the historic centre of the city was submerged in 2020. Over the years, through rescue excavations, only a minimal part of this rich cultural heritage could be salvaged (Figs. 24.1 and 24.2). Some examples of the losses in the region include the following: 300 historical mounds were affected, several historical monuments (Zeynel Bey Tomb, fifteenth century, Er-Rızk Mosque, fifteenth century, and Artukid Bath, fourteenth century, among others) in the city centre were relocated, and five villages were moved to a distant place. The waters of the Tigris River have shaped the landscape of Hasankeyf through millions of years of erosion at the cliffs of the valley. Ethnobotanical studies reveal either a dramatic or gradual loss of traditional knowledge and practice in the case of Hasankeyf. The flora of Ilısu (Hasankeyf) and its surroundings embodies 472 taxa belonging to 279 genera and 64 families, of which twenty taxa are threatened by the dam project, as they are endemic to this region (Yesil & Inal, 2019).

Another example is the Yortanlı Dam that affected the Allianoi (second century AD) Roman archaeological site near Bergama. Allianoi is the extraordinary site of a health resort from antiquity, where thermal baths, pools and surgical equipment

Fig. 24.1 Hasankeyf before dam construction. (*Note*. Photograph by B. Gün, 2017)

Fig. 24.2 Hasankeyf during dam construction. (*Note*. Photograph by B. Gün, 2018)

used in operations were identified in the region along with famous mosaics. Against extensive civil society appeals made to stop the dam construction to save the site, this important archaeological site has been submerged under since 2011. Rescue excavations in the region continued between 1994 and 2006 and managed to dig only 30% of the entire site (Hamamcioglu-Turan et al., 2013).

These examples reveal that dam constructions may have direct and indirect impacts on cultural and natural heritage. The prioritization of energy over the socio-cultural components (including the cultural heritage), such as in Hasankeyf and Yortanlı Dams, have led to irreversible damage and caused the loss of archaeological strata in submerged areas. Translocation of cultural heritage has raised questions regarding the relationship between cultural property and its setting, as well as issues related to the cultural memory of the area. Biological loss in the region is another layer added to the costs of dams to the region. Furthermore, the bond between endemic plants and intangible cultural heritage practices alters the gastronomic practices and rituals of the region.

Additionally, increased humidity due to extensive surface water harms building materials and alters their reaction to climatic conditions. The changing groundwater level is another potential risk for the existing cultural heritage that may cause structural problems due to ground subsidence. Especially considering that water consumption per person has increased dramatically in recent years, the lack of sewage systems and the poor maintenance of existing systems bear significant risks. Change in land use is another factor that severely affects groundwater levels. Unsuitable crop choices and irrigation systems are contributing to this issue. Multiple factors contribute to other hazards, including sinkholes that have been more frequently seen in Anatolia in recent years (Demir & Keskin, 2020). Droughts and extreme climatic events are expected due to the desertification of the region. From October to December 2020, precipitation across the country measured 48% lower than the average for 1981–2010.

24.5 Conclusions: Learning from the Water Technology of the World Heritage Properties of Turkey

Historical and archaeological evidence shows that direct access to water was not of primary concern for the selection of the sites for World Heritage properties in Anatolia. In the absence of immediate freshwater sources, human creative genius stepped in, harvesting rainwater and transporting it from distant places as water was always scarce and therefore treated with care. In order to overcome these challenges, various inspiring forms of water technology were developed to provide sustenance for the communities. Furthermore, ancient systems diversified water resources, which accommodated all forms of water (springs, rivers, lakes, rainwater, transported water etc.). Ancient communities lived and consumed proportionately and appropriate to environmental givens (certain crop types, building materials and typologies) were preferred above others. To make the most of the existing resources, systems that did not expose large surfaces of water were favoured (cisterns, channels, wells), keeping water underground or undercover. Mechanisms for clean and equal sharing of water were developed. Water was considered sacred, and intangible practices were developed as part of healing and religious purposes. Knowledge of underground water (groundwater, aquifers etc.) was incorporated into design solutions.

Therefore, lessons from traditional water management systems in Anatolia can be key guiding principles for mitigating impacts of climate change and reaching the 2030 Agenda. Specifically, how we use and manage water will affect our ability to achieve Goals 11: Sustainable Cities and Communities, 12: Responsible Consumption and Production, 13: Climate Action and 16: Peace. Since the notion of sustainability relies on a delicate balance between what we take from nature and what we give in return, human society and culture that is bound to water stand in a vulnerable position.

The existing and planned new interventions in the natural flow of rivers in Anatolia clearly puts communities within Turkey and those downstream in an isolated position in their fight against climate- and water-related challenges. Innovative new technologies available to communities can help bridge the gap between actual realities created and developed at the state level and the enhancement of the resilience of communities. Evidence-based new technologies such as meteorological data on droughts, flash floods and extreme weather events are becoming more easily available and could be of great assistance. The existing and future struggle for climate mitigation at the community level can benefit from awareness-raising on traditional knowledge and water technology practices. Developing integrated design-based solutions for water harvesting (from rain, air and seawater), cheap low-power irrigation and smart local use of water locally seem to be the only existing strategies that communities can develop in the absence of governmental commitment at the global level.

References

Al-Muqdadi, S. W. H. (2019). Developing strategy for water conflict management and transformation at Euphrates-Tigris basin. *Water, 11*(10). https://doi.org/10.3390/w11102037

Altınbilek, D. (2002). The role of dams in development. *Water Science and Technology, 45*(8), 169–180. https://doi.org/10.1080/07900620220121620

Angelakis, A. N. (2012). Evolution of water supply through the millennia. *IWA Publishing*. https://doi.org/10.2166/9781780401041

Başkan, S. (2018). Artvin Baraj Rezervuarları Altında Kalan Maddi Tarihi Kültürel Mirasın Değerlendirmesi. *ZfWT, 10*(1), 199–217.

Berga, L. (2016). The role of hydropower in climate change mitigation and adaptation: A review. *Engineering, 2*(3), 313–318. https://doi.org/10.1016/J.ENG.2016.03.004

Bildirici, M. (2020). *Tarihi Kültürel ve Çevre Değerleri ile Konya Kapaşı Havzasına Su Aktaran Konya-Çumra 2.ve 3.* Merhale (ADIM) Sulama Projesi, Devlet Su İşleri Genel Müdürlüğü.

Bixio, A., Bixio, R., & Pascale, A. D. (2020). Underground hydraulic systems and anthropogenic cavities in Cappadocia. In P. Boschiero & L. Latini (Eds.), *Güllüdere and Kızılçukur: Rose Valley and Red Valley in Cappadocia. Premio Internazionale Carlo Scarpa per Il Giardino* (31st ed.). 2020–2021 Fondazione Benetton Studi Ricerche-Antiga.

Çeçen, K. (1996). *The longest Roman water supply line*. Istanbul Sınai Kalkınma Bankası.

Clare, L. (2020). Göbekli Tepe, Turkey. A brief summary of research at a new world heritage site (2015–2019). *e FORSCHUNGSBERICHTE, 2*, 81–88.

Crow, J. (2012). Ruling the waters: Managing the water supply of Constantinople, AD 330–1204. *Water History, 4*(1), 35–55. https://doi.org/10.1007/s12685-012-0054-y

Çulcuoğlu, G., Erkan, Y. K., & Karakul, Ö. (Eds.). (2013). *UNESCO World Heritage in Turkey*. Turkish National Comission for UNESCO.

Demir, V., & Keskin, A. Ü. (2020). Water level change of lakes and sinkholes in Central Turkey under anthropogenic effects. *Theoretical and Applied Climatology, 142*(3–4), 929–943. https://doi.org/10.1007/s00704-020-03347-5

Doğa araştırmacısı iki arkadaş 2 bin 500 yıllık gizli geçidi ortaya çıkardı: Safranbolu tarihinde yeni bir … (2021, February 27). *Yeni Şafak*. Retrieved October 30, 2021, from https://www.yenisafak.com/hayat/doga-arastirmacisi-iki-arkadas-2-bin-500-yillik-gizli-gecidi-ortaya-cikardi-safranbolu-tarihinde-yeni-bir-sayfa-acilabilir-3601256

Douet, J. (2018). *The water industry as World Heritage thematic study*. TICCIH.

Editor. (2019, October 7). *Safranbolu'nun Gün Yüzü Görmemiş 400 Yıllık Gizli Tünelleri*. Nereye. https://nereye.com.tr/safranbolunun-gun-yuzu-gormemis-400-yillik-gizli-tunelleri/

Eken, G., Isfendiyaroğlu, S., Yeniyurt, C., Erkol, I. L., Karataş, A., & Ataol, M. (2016). Identifying key biodiversity areas in Turkey: A multi-taxon approach. *International Journal of Biodiversity Science, Ecosystem Services and Management, 12*(3), 181–190. https://doi.org/10.1080/21513732.2016.1182949

Graff, S. R., Branting, S., & Marston, J. M. (2019). Production requires water: Material remains of the hydrosocial cycle in an ancient Anatolian city. *Economic Anthropology, 6*(2), 234–249. https://doi.org/10.1002/sea2.12147

Hamamcıoğlu-Turan, M., Arısoy, Y., Nuhoğlu, A., & Erturan, Y. P. (2013). Protection of archaeological remains in the Yortanli dam reservoir in Turkey. *International Journal of Architectural Heritage, 7*(6), 653–672. https://doi.org/10.1080/15583058.2012.655404

Harmanşah, Ö. (2018). Geologies of belonging: Political ecology of water in Central Anatolia. In E. Holt (Ed.), *Water and power in past societies* (pp. 259–279). State University of New York Press.

İlhan, A. (2013). Rivers for life. *Review, 3*(6), 666–667. https://doi.org/10.1111/an.2004.45.3.22

İlhan, A. (2017). Suya Erişim Hakkından Suyun Akma Hakkına. In A. İlhan, E. Erdoğan, N. Yüce, & Ö. Özbay (Eds.), *Türkiye' de ve Dünyada Su Krizi ve Su Hakkı Mücadeleleri* (pp. 29–45) [Brochure] https://www.suhakki.org/docs/kitap_su-krizi-su-hakki-mucadeleleri-web.pdf). Sivil ve Ekolojik Haklar Derneği.

IUCN. (2015). *Climate Change and dams threaten natural World Heritage, warns IUCN New report on Impacts of Mae Wong Dam in Thailand underlines dangers*. Retrieved October 30, 2021, from https://www.iucn.org/sites/dev/files/import/downloads/cc_and_dams_threaten_natural_world_heritage__warns_iucn.pdf

Jensen-Cormier, C. (2019, November). *Watered down: How big hydropower companies adhere to social and environmental policies and best practices*. International Rivers. Retrieved October 30, 2021, from https://3waryu2g9363hdvii1ci666p-wpengine.netdna-ssl.com/wp-content/uploads/sites/86/2020/10/watered-down-full-report-english-compressed.pdf

Madani, K. (2014). Water management in Iran: What is causing the looming crisis? *Journal of Environmental Studies and Sciences, 4*(4), 315–328. https://doi.org/10.1007/s13412-014-0182-z

Öziş, Ü. (1996). Historical water schemes in Turkey. *International Journal of Water Resources Development, 12*(3), 347–384. https://doi.org/10.1080/07900629650213

Öziş, Ü., Arisoy, Y., & Alkan, A. (2010). Ancient waterworks in Anatolia. In V. Mladenov & Z. Bojkovic (Eds.), *Latest trends on cultural heritage and tourism* (pp. 63–71). WSEAS.

Öztürk, H. K., & Mutlu Öztürk, H. (2019). Hidrolik Santrallerin Turizm ve Kültürel Mirası Üzerine Etkisi (the impact of hydroelectric power plants on tourism and cultural heritage). *Journal of Tourism and Gastronomy Studies, 7*(3), 2321–2335. https://doi.org/10.21325/jotags.2019.474

Reyhan, K., & Yazıcı, M. (2020). Su kültürü, tarihsel, toplumsal gelişme sürecinde su ile ilişkili oluşturulan bütün maddi ve manevi değerler ile bu değerlerin oluşturulmasında ve sonraki nesillere iletmede kullanılan bilgi, özgün düşünce, yaşam ve sanat eserlerinin bütünlüğüne işar. *Avrasya Uluslararası Araştırmalar Dergisi, 8*(23), 45–57.

Rivers without Boundaries International Coalition, Wilderness Society Tasmania, Ohrid SOS, Yayasan HAkA, SOSOrinoco, World Heritage Watch, & International Rivers. (2021). *Oral statement on rivers and increasing dam damage (item 7, closing of the item)*. World Heritage

Watch. Retrieved October 30, 2021, from https://world-heritage-watch.org/content/wp-content/uploads/2021/07/WHW-Statement-on-44COM-7Rivers-and-Dam-Damage.pdf

Sanyanga, R., Simon, M., & Hill, C. (2020). *Transforming power: A gender guide for organisations campaigning on dams and for rivers*. International Rivers. Retrieved October 30, 2021, from https://www.internationalrivers.org/wp-content/uploads/sites/86/2020/09/Gender-Guide-English-single-pages.pdf

Saygılı, R. (2018). *Turkiye hidroelektrik santralleri haritası*. Cografya Harita. Retrieved October 30, 2021, from http://cografyaharita.com/haritalarim/4eturkiye-2017-hidroelektrik-santralleri-haritasib2.png

Sevimli, Ş., & Uzel, İ. (2003). Water technology and cleanliness on the Hittite B. C. 1700–1200. In *Proceedings of the first international exergy, energy and environment symposium, 13–17 July 2003 İzmir Turkey, July* (pp. 264–271).

Tanyeli, G., & İkiz, D. (2017). The cultural heritages of water in Turkey. In M. Cotte (Ed.), *Cultural heritages of water les patrimoines culturels de l ' eau* (pp. 221–240). ICOMOS.

Thaman, R. M. (2020). The Ilisu dam and its impact on the Mesopotamian marshes of Iraq: Implications for the future directions of international water law. *International Water Law, 5*(4), 3–92. https://doi.org/10.1163/23529369-12340020

The False Promises of Hydropower. (2019). *The false promises of hydropower*. https://archive.internationalrivers.org/sites/default/files/attached-files/declaration_iha2019_21x30_english_for_printing.pdf

Voss, K. A., Famiglietti, J. S., Lo, M., De Linage, C., Rodell, M., & Swenson, S. C. (2013). Groundwater depletion in the Middle East from GRACE with implications for transboundary water management in the Tigris-Euphrates-Western Iran region. *Water Resources Research, 49*(2), 904–914. https://doi.org/10.1002/wrcr.20078

Willems, W. J. H., & Van Schaik, H. P. J. (Eds.). (2015). *Water & heritage*. Sidestone Press. https://www.ifugao-archaeological-project.org/uploads/6/4/4/7/6447606/acabado_&_martin_2015_-_between_pragmatism_and_cultural_context.pdf

Wittenberg, H., & Schachner, A. (2013). The ponds of Hattuša – Early groundwater management in the Hittite kingdom. *Water Science and Technology: Water Supply, 13*(3), 692–698. https://doi.org/10.2166/ws.2013.025

World Bank. (2017). Greenhouse gases from reservoirs caused by biogeochemical processes. *World Bank*. https://doi.org/10.1596/29151

Wrong Climate for Big Dams. (2011, October). *International rivers*. Retrieved October 30, 2021, from https://www.internationalrivers.org/wp-content/uploads/sites/86/2020/05/intlrivers_wrongclimate_4.pdf

Yeşil, Y., & İnal, I. (2019). Traditional knowledge of wild edible plants in Hasankeyf (Batman Province, Turkey). *Acta Societatis Botanicorum Poloniae, 88*(3), 1–22. https://doi.org/10.5586/asbp.3633

Chapter 25
Mineral Extractive Industries in the Context of European World Heritage Cultural Landscape Conservation and Management: The Case Study of the Erzgebirge/Krušnohoří Mining Region

Friederike Hansell

Abstract Current prospection activities lead not only to the opening of new mines but also to a revival of activity in historic mining areas that are partly in or close to protected areas. Consequently, the issue of mining and its potential negative impact on protected areas, including natural World Heritage sites in particular, has increased over the past two decades. Considering that attributes and values assigned to natural World Heritage sites differ from those assigned to cultural World Heritage sites, the paper focuses on the evaluation of the potential impact of mining activities on cultural World Heritage sites and outlines management and conservation strategies as well as recommendations for the assessment of potential negative and positive impacts of mining activities on the OUV.

Keywords Mining heritage · Historic mining regions · Cultural landscapes · World Heritage sites · Mining resumption · Mining impacts

25.1 Introduction

Mineral extraction plays a crucial role in the global economy. Current prospection activities lead not only to the opening of new mines but also to a revival of activity in historic mining areas that are partly in or close to protected areas. Consequently, the issue of mining and its potential negative impact on World Heritage sites has increased

F. Hansell (✉)
World Heritage Desk Officer, Saxon State Office for the Conservation of Monuments, Dresden, Germany
e-mail: friederike.hansell@lfd.sachsen.de

M.-T. Albert et al. (eds.), *50 Years World Heritage Convention: Shared Responsibility – Conflict & Reconciliation*, Heritage Studies,
https://doi.org/10.1007/978-3-031-05660-4_25

over the past two decades. The most severe impacts were primarily concerned with natural World Heritage sites, and the development of guidelines and principles were guided by experiences derived from such sites. Considering that attributes and values assigned to natural World Heritage sites differ from those assigned to cultural World Heritage sites, and, accordingly, the experiences are not commonly transferrable to cultural sites, the paper focuses on the evaluation of the potential impact of mining on cultural World Heritage sites. The geo-cultural region considered is Europe due to the relatively high number of mining-related World Heritage sites, similarities in technical heritage and a shared European mining identity. The Erzgebirge/Krušnohoří Mining Region, inscribed in 2019 as a prominent example of a classic European mining landscape, serves as a case study. The paper examines three key aspects for the conservation and management of World Heritage sites in Europe: (1) The role of past and new mining in the statement of Outstanding Universal Value; (2) Conflict-solving and management strategies in place to manage future mining activities; and (3) the potential contribution of the resumption of mining.

25.2 Mining, and the Resumption of Mining, in Cultural World Heritage Sites: A Controversial Topic?

The significance of mining for the history of humankind has been recognised in recent years, leading to the inscription of a number of mining-related sites on the World Heritage List. To date, of the current 1154 World Heritage sites, 33 mining-related sites are inscribed on the World Heritage List, out of which 19 are located in Europe (UNESCO, 2021a). Eight European mining areas were inscribed under the organically evolved cultural landscape category: Hallstein Dachstein/Salzkammergut Cultural Landscape (1997, (iii) (iv), Austria), Blaenavon Industrial Landscape (2000, (iii) (iv), UK), Cornwall and West Devon Mining Landscape (2006, (ii), (iii) (iv), UK), Nord-Pas de Calais Mining Basin (2012, (ii) (iv) (vi), France), Erzgebirge/Krušnohoří Mining Region (2019, (ii), (iii), (iv), Germany), Krzemionki Prehistoric Striped Flint Mining Region (2019, (iii) (iv), Poland), Roşia Montană Mining Landscape (2021, (ii), (iii), (iv), Romania) and The Slate Landscape of Northwest Wales (2021, (ii), (iv), UK) (UNESCO, 2021b). In terms of values and attributes, each mining landscape in a World Heritage context has its own specific character. However, most landscapes commonly combine relict and living features. The application of criterion (iii) emphasises the strong relationship of a historic mining region with a cultural tradition or civilization that may have disappeared in historic areas. In more recent mining areas, the landscapes are considered as "continuing", encompassing intangible values in which mining retains an active social role in contemporary society. The structure and pattern of mining landscapes are characterised by an inseparable connection of mineral extraction and mineral practices to the location of mineral deposits (Tost et al., 2021, 7). This close relationship to potentially viable ore resources in mining-related World Heritage sites makes the resumption of mining likely.

The German-Czech World Heritage Site Erzgebirge/Krušnohoří Mining Region can be considered a prominent example of a classical European mining landscape. The site was inscribed on the World Heritage List in 2019, under criteria (ii), (iii) and (iv). Ore mining and its influence on the landscape and the people were at the heart of the prestigious designation. The property itself is a substantially relict mining landscape with a strong mining tradition, albeit no mining activities have taken place since the cessation of the last operational mine in 1990. Located in a world-class polymetallogenic province, the evolutionary process is still in progress. Saxony is a "mining country", and prospection activities in the Ore Mountains region continue – a fact that was diligently considered during the nomination process and addressed in the conservation and management planning of the site. Referring to Decision 43 COM 8B.26 of the World Heritage inscription in 2019 reveals an apparent contradiction. Based on the recommendation of ICOMOS, the Advisory Body to UNESCO for cultural heritage sites, the World Heritage Committee have recommended the State Parties "to formally committing that no mining activities or processing will be allowed in the future within the boundaries of the component parts of the serial property" (UNESCO, 2019a). The question arises, what is the basis of such a recommendation? Can mining developments be considered under all circumstances to represent a threat to Outstanding Universal Value (OUV)? Or can there be exceptional cases or a suitable protocol whereby mining developments can take place in and/or adjacent to World Heritage sites without the fear of negatively impacting OUV? Affolder (2007, 24), points out that there is "no express prohibition on all mining within World Heritage sites in the text of the Convention nor is it clear that such a sweeping prohibition could be implied from the Convention text". The Operational Guidelines do not foresee such a formal blanket commitment.

It is undisputed that certain mining activities either in or adjacent to a World Heritage site can threaten its OUV. Being aware of the potential threats caused by mining, in particular to natural World Heritage sites, a number of international meetings have been held on the subject since 1999 (UNESCO, 2021c). In 2003, the members of the International Council on Mining and Metals (ICMM) endorsed a "No-go commitment" to "respect legally designated protected areas" and "not explore or mine in World Heritage properties" (ICMM, 2003, 3; Affolder, 2007, 25). An independent study commissioned in 2012 by the International Union for Conservation of Nature (IUCN) in cooperation with the World Heritage Centre and ICMM and Shell sums up the issues, challenges and opportunities related to extraction and natural World Heritage sites (Turner, 2012).

> The World Heritage Committee has expressed many times its clear position that mineral, oil and gas exploration or exploitation is incompatible with World Heritage status and that such activities should not be undertaken within World Heritage properties. It is, in any case, essential that the Outstanding Universal Value of a World Heritage property is not impacted. (UNESCO, 2021c)

There is no doubt about the potential negative impacts of mining activities and developments on natural World Heritage sites. However, the values and attributes of natural World Heritage sites that are the subject of protection and management are different from the values and attributes of cultural World Heritage sites. Experiences with natural World Heritage sites, therefore, are not commonly transferrable to cultural sites. In conformity with paragraph 172 of the Operational Guidelines, the statement of OUV determines the assessment of individual development projects (UNESCO, 2019b).

25.3 What Is the Current Situation in the Context of Cultural World Heritage Sites?

Analysing state of conservation reports reveals the reasons for the position of the World Heritage Committee. From 1984 to 2021, UNESCO reports that pressure from extractive industries is rising, resulting in 530 reports on 87 properties in 52 States Parties concerning extractive practices. 384 reports on 59 properties in 40 State Parties particularly address the issue of mining. 76% of reports concern threats by mining development activities in natural World Heritage sites, 5% in mixed sites and 19% in cultural sites. In recent years, the number of concerns related to cultural heritage sites increased. 42 reports on 11 properties in 9 State Parties concern World Heritage sites inscribed on cultural criteria (UNESCO, 2021d).

A screening of mining sites on the World Heritage List in which mining activity occurs demonstrates that mining does not necessarily have a negative impact on the cultural heritage site. In the case of the World Heritage site "City of Potosi", Bolivia, certain recent mining activities have a clearly defined negative impact on the cultural heritage. The potential degradation of the historic site by continued and uncontrolled mining operations in Cerro Rico Mountain is one of the threats to OUV that contributes to the property being placed on the List of World Heritage in Danger (UNESCO, 2019c). The collapse of the summit of the Cerro Rico remains a concern and, apart from geotechnical stabilisation, miners working above a certain elevation have now been relocated. In contrast, mining activities also take place in the World Heritage site "Historic Town of Guanajuato and Adjacent Mines", Mexico. Here a miners' cooperative established in the 1930s continues to mine silver ore in La Valenciana mine at considerable depth, activities which, coupled with concentrating the ore outside the World Heritage site, do not negatively impact OUV. Moving to Europe, a recent screening showed that, to date, no World Heritage sites are imminently threatened by mineral extraction (Tost et al., 2021, 3). In the case of Hallstatt-Dachstein/Salzkammergut Cultural Landscape, active salt mining is still taking place within the boundaries of the World Heritage cultural landscape. The heritage designation is based on salt mining in the region, and the active salt mines were pre-existing and part of the selection process in 1997 (Tost et al., 2021, 13).

One of three properties in Europe, and the most recent that was raised for examination by the World Heritage Committee, is the case of Cornwall and West Devon Mining Landscape. Criterion (iii) refers to the transformation of the urban and rural landscapes by copper and tin mining that presents a vivid and legible testimony to the success of Cornish and West Devon industrialised mining (UNESCO, 2021e). The issue of the potential resumption of mining was addressed in the nomination file and the management plan stating that "proposals for the resumption of mining will be supported where they do not adversely affect the Outstanding Universal Value of the Site" (Gamble, 2005, 197&151). ICOMOS (2006, 312), highlighted the following in its evaluation report: "Although the re-opening of mines in the nominated areas could be considered as re-invigorating the cultural landscape, great care would be needed with any such proposals to ensure that the values associated with early steam technologies are not harmed, especially as manifest by engine houses. It is therefore recommended that any such proposals, within the nominated areas, or their setting, are forwarded to the World Heritage Committee for debate and scrutiny." In 2012, the State Party submitted a state of conservation report notifying the World Heritage Centre about plans to resume mining at South Crofty, located within the World Heritage property, which has been inactive since 1998 (UNESCO, 2012). From 2012 onwards, state of conservation reports dealt with the potential resumption of mining at South Crofty (UNESCO, 2021d). The position was that the resumption of mining was justifiable and is achievable without adverse impact on OUV. Mining has not yet resumed, and no plant has been constructed.

The case of Roşia Montană Mining Landscape in Romania shows that the proposed large-scale open-pit mining would adversely affect its OUV. In fact, it would almost totally obliterate the Roman gold mining heritage for which it was inscribed in 2021. The justification of criterion (iii) is based on "cultural traditions of one of the oldest documented mining communities in Europe, anciently founded by the Romans" and a heritage that provided "an authentic picture of daily life and cultural practice in this ancient frontier mining community" (ICOMOS, 2021, 264). The property has also been inscribed on the List of World Heritage in Danger because of the still pending potential of the resumption of large-scale opencast mining that could lead to significant damage to the cultural heritage. The mining company has not yet received a licence, and the arbitration process is ongoing, yet there still remains a risk of activation of the licence that extends until 2024 under existing Romanian law (ICOMOS, 2021, 269).

25.4 The Erzgebirge/Krušnohoří Mining Region Case Study

25.4.1 The World Heritage Site

The Erzgebirge/Krušnohoří Mining is located in a mountainous region that spans the German-Czech border. The landscape has been profoundly and irreversibly shaped by 800 years of almost continuous mining, from the twelfth to the twentieth centuries.

The value of the cultural landscape is based on the interaction between people and their environment and is tangibly manifested by a diverse cultural heritage. Mining activities have always been central to the cultural, social and economic heritage of the community. The Erzgebirge/Krušnohoří Mining Region is anchored as a substantially relict landscape, in which an evolutionary process came to an end at some time in the past, but is also partly relevant as a continuing landscape in that parts of the landscape retain an active social role. The inscription is associated with strong mining traditions based on criterion (iii), that is, justified by the exceptional testimony to technological, scientific, administrative, educational, managerial and social aspects that underpin the intangible dimension of living traditions, ideas and beliefs of the people associated with the Ore Mountains' culture. Mining in the region is considered a core part of identity, and the World Heritage inscription reflects the strong relationship between people in the region and their tangible and intangible heritage (Fig. 25.1).

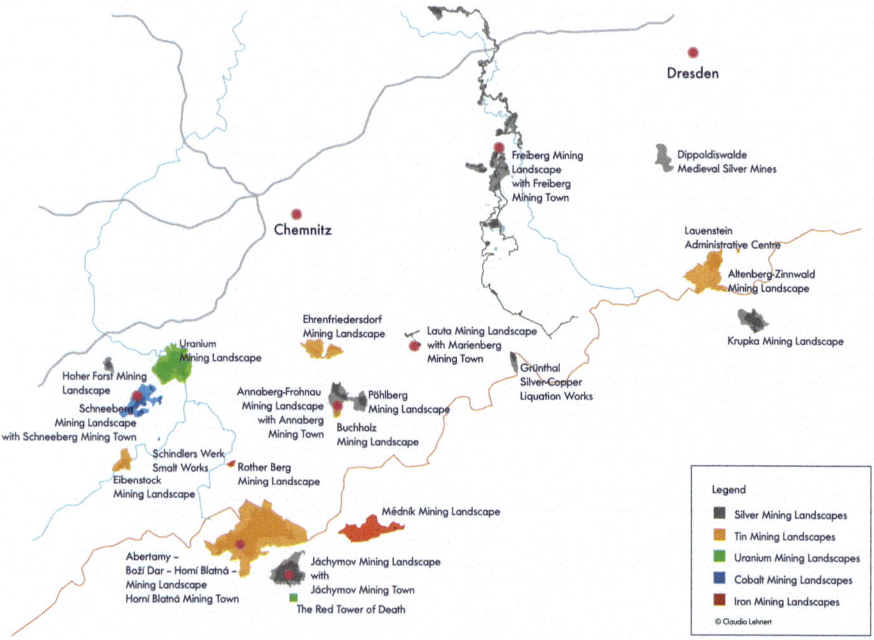

Fig. 25.1 Twenty-two component parts reflect the unevenly distributed locations of the most important raw material deposits that dictated land use and are characterised by specific and formative contributions made by the exploitation of different metals at different times. (*Note*. Source: C. Lehnert, 2020, © IWTG)

25.4.2 The Potential Resumption of Mining

The Ore Mountains region is considered as a living landscape, in which the new mining activity constitutes a continuation of traditional mining, including its controlled influence on the landscape. Mining continues to play an important role in the life and economy, and the resumption of mining is most likely. On a global scale, most of Saxony's deposits may be considered rather small to medium-sized, but they can play a new economic role as demand grows and world market prices rise. Tin, zinc, copper and tungsten are of particular interest, as well as fluorspar, barite and other metallic raw materials such as lithium, of which economically viable reserves are available. The handling of raw material reserves is reflected in the raw material strategy of the Free State of Saxony, which integrates the potential into an overall economic concept for a sustainable raw material economy. In the strategy, it is stated that "As a region rich in raw materials, Saxony advocates placing an additional focus on ensuring and developing the local supply of raw materials" (Saxon State Ministry of Economic Affairs, Labour and Transport, 2012, 11). The strategy outlines the guidelines and objectives of Saxon raw material policy and practical tasks for their implementation. The need for a continuous supply of diverse mineral resources leads to continuous prospection activities to safeguard a sufficient level of known mineral resources to meet the needs of future generations. In recent years, a number of exploration licences in the Ore Mountains region were approved. The validity of exploration licences is limited, and the processes of granting are dynamic. As of 2019, 18 licences were assessed in relation to their potential impacts on the property. Only one mining licence needs attention as it is located in and adjacent to the World Heritage site, and possible impacts on the present landscape cannot be precluded. To assess the potential impact of the proposed lithium mining, a preliminary Heritage Impact Assessment was prepared in close cooperation with the mining company, mining authority, monument protection authority and World Heritage management. This was specifically undertaken to inform any final decision and concludes with the position, pending more detailed design, that there could be negligible adverse effects and that the resumption of mining in this protected landscape may be justifiable. The awareness of the concept of a mining landscape as an assemblage of surface and subsurface spatial patterns, landscape features and elements are of particular importance when it comes to heritage impact assessments, evaluating the possible impact of mineral practices in a protected landscape or those from possible mining operations located outside the site (Tost et al., 2021, 8).

25.4.3 Conservation and Management Strategies

A key to the holistic approach of conservation and management was the early involvement of all concerned stakeholders at all levels and intensive cooperation. The whole cultural landscape is significant and requires clear definition and understanding in order to secure its protection, including mineral resource assessments. The preparation of the World Heritage nomination was designed as a collaborative process of recognising and understanding the OUV of one's own heritage from a global perspective. Both the technical preparation of the nomination and the drafting of the application document took place within the framework of broad-based consultation. All relevant stakeholders such as municipalities, districts, building and planning authorities, monument owners, property owners, associations and interested citizens as well as external experts were involved. The process allowed for complete transparency.

During the nomination process, potential sources of conflict, including mining activities, were identified and addressed. Conflict-solving strategies were developed as well as measures to integrate the issue of responsible mining in the future management of the site. The World Heritage Centre, IUCN and ICOMOS were informed regarding mining-related activities prior to designation. In addition, a procedure has been developed to identify potential conflicts at an early stage and to jointly find solutions for the development of the region in line with the protection and preservation of the World Heritage site. The statement of the OUV and the definition of the contribution of each component part to the OUV are key to the conservation and future sustainable management of the property. The World Heritage attributes and values were identified as crucial both for the assessment of authenticity and integrity and for the determination of the boundaries of the proposed property, and thus also for the future management, protection and conservation of the World Heritage site. Such clear determination provides a robust basis to assess the impact of mining activities and other developments on the component parts and to mitigate negative impacts on the OUV of the property, its integrity and authenticity. In addition to defining the attributes and values in terms of the rationale for the criteria, the tangible elements that convey the proposed OUV have also been clearly identified and described. This characterisation enables World Heritage values to be defined in more detail and also considers the landscape context that may not be recorded within the existing conservation framework and which should be considered in possible future development projects. Specific actions to ensure the effective management of potential new mining activities included:

1. A management system was developed that involves the responsible ministries on the national level and the local management institutions and defines responsibilities to secure a constant flow of information regarding potential new developments in the region (Albrecht et al., 2017, 479ff.) Regular meetings between the institutions secure the flow of information. Moreover, there is continuous monitoring of the component parts that include reporting on the state of conservation as well as on development projects.
2. Issues of monument protection, including proposed or acknowledged World Heritage attributes and values, are considered in the approval procedure. A combination of legal mechanisms is in place within the component parts that enable the protection of the World Heritage site and the effective operation of its buffer zone.

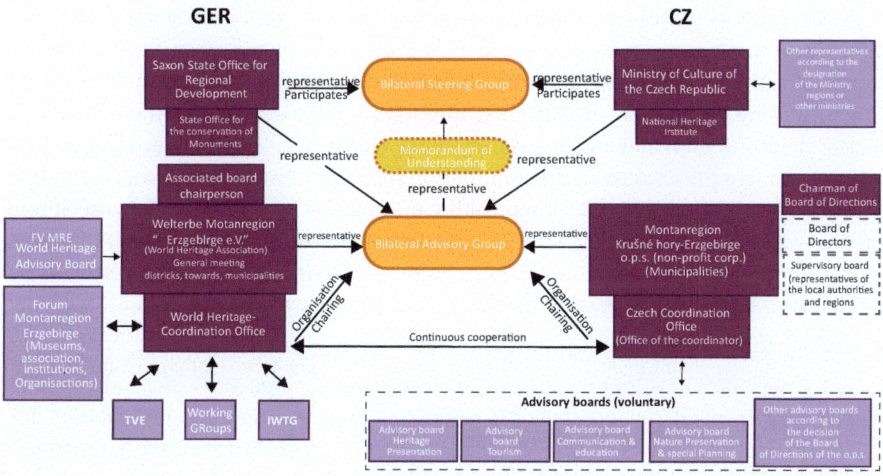

Fig. 25.2 The management system includes all relevant stakeholders. Experts on various themes such as monument protection and regional development are involved by working groups. (*Note*. Source: Albrecht et al. (2017). *Erzgebirge/Krušnohoří Mining Region Nomination for Inscription on the UNESCO World Heritage List*. Online Source. UNESCO/World Heritage Centre. URL: https://whc.unesco.org/document/166295)

3. An agreement was made between the Saxon local management institution and the Saxon mining authority stipulating that mining enterprises will be informed about the World Heritage site and its potential of overlapping future mining interests. The agreement allows early consideration of potential impacts as well as facilitating the consultation process with mining enterprises.
4. All planned and proposed mining activities are continuously monitored by the managing institution, supported by the technical agency on the State level. The proposed resumption of mining within the World Heritage site and its buffer zone will be informed by a Heritage Impact Assessment that focuses on OUV. Any resumption of mining in or close to the site will be subject to a paragraph 172 notification and respective impact assessments (Fig. 25.2).

25.5 World Heritage-Compatible Mining? Recommendations for Assessment in Future

The examples illustrate that there is not a single approach for considering the impact of mining activities on World Heritage sites. Considering that mining-related World Heritage sites are partly bound to still-viable mineral deposits and especially given the recent developments in the demand for lithium to supply the renewable energy battery storage market and the wider demand for metals including tin, tungsten, and copper increasingly used in low-carbon technologies, there can be no intention to completely deny access to mineral resources for the future. In each case, site-specific circumstances, as well as close reference to the justification for inscription, must guide the assessment of potential impacts of mining activities and the resumption of mining. Therefore, of particular importance for the assessment of impacts is a clear determination and understanding of the World Heritage attributes that should be outlined from the outset in the nomination file; a requirement that was taken into account in the revision of the format for the nomination of properties for inscription on the World Heritage List in 2021. It is recommended that the assessment of potential negative and/or positive impacts of mining activities on the OUV should be guided further by three key aspects:

1. Relationship to mining traditions: A key aspect for the assessment of mining activities and developments, and their potential impact, is the relationship between mineral resources as an important land use and the justification for World Heritage inscription. In certain cases, e.g. the Erzgebirge/Krušnohoří Mining Region and the Cornwall and Devon Mining Landscape, the resumption of responsible mining can be considered strongly in accordance with the intangible values of the sites that are also inscribed for their mining tradition, with mining still playing an active role in today's society. Mining continues to be considered as a strong factor for regional identity, and the potential resumption of mining activities exists and is widely supported among the community. Accordingly, the resumption of mining has the potential to support intangible heritage values.
2. Technological change in mining: It has to be noted that licence areas do not necessarily imply mining or, if mining does proceed, corresponding potential landscape impact. In cases of the discovery of economic deposits, the underground mining fields are usually much smaller, commonly located beneath historic mines and do not allow any conclusions regarding the location of surface installations. Moreover, today's underground mining technology coupled with modern high-tech surface facilities mean that extraction can be more efficient and use less space and fewer resources. Methods may involve micro-invasive mining with no surface waste dumps, minimal impact on the landscape through new surface structures, and secondary processing remote from any protected landscapes. These changes in mining technologies have the capacity, at least, for neutral impacts on the values and attributes and the visual integrity of a cultural World Heritage site. The ability to devise a compatible operation is encouraging.
3. Contribution of mining to Sustainable Development Goals: As mining-related cultural landscapes may be located in globally significant polymetallogenic provinces, the potential contribution to SDGs should be considered. In 2016, an atlas was initiated and published by the World Economic Forum, the Columbia Centre on Sustainable Investment, the United Nations Development Program and the Sustainable Development Solutions Network. This maps the relationship between mining and the SDGs (World Economic Forum, 2016, 3). The atlas presents a broad overview of opportunities and challenges to demonstrate the

actual and potential contributions of the mining sector to the achievement of the SDGs.

25.6 Conclusion

Concerning the key aspects for the conservation and management of World Heritage sites in Europe that this paper examined, the following conclusions could be drawn:

1. The role of past and new mining in the statement of OUV: There is a need to carefully differentiate between types of cultural sites and the potential that mining has to impact upon these sites. In certain cases, mineral exploration and extraction seem to be deemed compatible within the boundaries of World Heritage sites, at least to some degree. This might be applied in particular to evolving, living landscapes for which criterion (iii) was applied and in which mining retains an active role in contemporary society.
2. Conflict-solving and management strategies in place to manage future mining activities: Inscription confirms that the statement of OUV is the key to the conservation and future sustainable management of the property, together with a clear definition of the contribution made by each component part. World Heritage attributes and values are implicit in the authenticity and integrity of properties, and for the justification of property boundaries, and thus for the future management, protection and conservation of the World Heritage site. The management plan should carefully consider potential impacts – positive and/or negative – as well as provide strategies for mining activities within and adjacent to the site. Mineral planning policies of the respective country should be outlined, and the authorities concerned involved in management processes.
3. Potential contribution of the resumption of mining: The potential of recent or new mining activity being supported in the future management of World Heritage sites was indicated by measures to integrate the issue of responsible mining in property management plans that were submitted along with nomination dossiers prior to inscription. This tells us that recent or new mining activity within cultural World Heritage sites is not necessarily or automatically considered as incompatible. On the contrary, new mining might have the potential to contribute to the values of a specific type of site, in contrast to the common experiences with natural World Heritage sites. Management and conservation strategies should guide the effective management of potential new mining activities within or adjacent to World Heritage sites and must be considered in management planning.

In the framework of the 50th anniversary of the World Heritage Convention, particular attention should be drawn to the potential contribution of mining to Sustainable Development Goals, an opportunity that in the context of mining-related World Heritage sites is not yet sufficiently addressed. Protecting UNESCO World Heritage sites for all people of the world and for future generations requires sustainability in the preservation and use of the sites. Responsible mining at World Heritage sites could serve as an example that demonstrates how a potentially conflicting use could contribute to the wellbeing of local communities – as long as it is well managed and the attributes of OUV are fully protected. A first step was undertaken by the Cornwall and West Devon Mining Landscape. The newly updated management plan, for the period 2020–2025, comprehensively addresses the issue of mining and the resumption of mining and its importance for sustainable future-orientated development. Climate resilience and biodiversity, renewable energy and carbon-neutral agendas, environment and culture as an enabler of sustainable development and economic prosperity: these are all considered in this forward-looking plan (Cornish Mining World Heritage Site, 2020, 102). This approach could provide a useful case study as a way ahead for dialogue between other mining-related World Heritage sites in Europe and worldwide.

References

Affolder, N. (2007). Mining and the World Heritage Convention: Democratic legitimacy and treaty compliance. *Pace Environmental Law Review, 24*(1), 35–66. https://digitalcommons.pace.edu/cgi/viewcontent.cgi?article=1053&context=pelr

Albrecht, H., Hansell, F., & Urban, M. (2017). *Erzgebirge/Krušnohoří Mining Region Nomination for Inscription on the UNESCO World Heritage List*. Free State of Saxony/Germany and Czech Republic. Online Source. UNESCO World Heritage Centre. https://whc.unesco.org/document/166295

Cornish Mining World Heritage Site. (2020). *The Cornwall and West Devon mining landscape World Heritage site. Management plan 2020–2025*. Cornish Mining World Heritage Site. https://www.cornishmining.org.uk/media/Conservation/Management%20Plan/PDFs/CM_WHS_Management_Plan_page_90-148.pdf

Gamble, B. (2005). *Cornwall and West Devon mining landscape. Nomination for inscription on the UNESCO World Heritage List*. Cornish Mining World Heritage Site. https://whc.unesco.org/uploads/nominations/1215.pdf

ICMM. (2003). *Mining and protected areas: Position statement*. International Council on Mining and Metals. http://www.icmm.com/website/publications/pdfs/mining-principles/position-statements_protected-areas.pdf

ICOMOS. (2006). *Cornwall and West Devon mining landscape (United Kingdom)*. Advisory Body Evaluation. UNESCO/World Heritage Centre. https://whc.unesco.org/en/list/1215/documents/. Accessed 5 Aug 2021.

ICOMOS. (2021). Roşia Montană Mining Landscape (Romania) No 1552rev. In *Evaluations of nominations of cultural and mixed properties. ICOMOS report for the World Heritage Committee. 44th extended session, Fuzouh (China)/Online – Meeting, 16–31 July 2021* (pp. 269–275). ICOMOS. https://whc.unesco.org/archive/2021/whc21-44com-8Binf1-en.pdf

Saxon State Ministry of Economic Affairs, Labour and Transport. (Ed.). (2012). *Saxon Raw Material Strategy 2012. The raw material economy – An opportunity for the Free State of Saxony*, 2. Auflage, Staatsministerium für Wirtschaft, Arbeit und Verkehr.

Tost, M., Ammerer, G., Kot-Niewiadomska, A., & Gugerell, K. (2021). Mining and Europe's world heritage cultural landscapes. *Resources, 10*(2), 8. https://doi.org/10.3390/resources10020018

Turner, S. (2012). *World Heritage sites and the extractive industries*. IUCN. https://portals.iucn.org/library/sites/library/files/documents/Rep-2012-013.pdf

UNESCO. (2012). *State of Conservation Cornwall and West Devon Mining Landscape (United Kingdom of Great Britain and Northern Ireland).* https://whc.unesco.org/en/soc/141/
UNESCO. (2019a). *Decision: 43COM 8B.26 – Erzgebirge/Krušnohoří Mining Region (Czechia and Germany).* UNESCO. https://whc.unesco.org/en/decisions/7387
UNESCO. (2019b). *Operational guidelines for the implementation of the World Heritage Convention.* UNESCO. https://whc.unesco.org/en/guidelines/
UNESCO. (2019c). *State of conservation City of Potosí (Bolivia (Plurinational State of). Factors.* UNESCO. https://whc.unesco.org/en/soc/3882/
UNESCO. (2021a). *World Heritage List.* UNESCO. https://whc.unesco.org/en/list/
UNESCO. (2021b). *Cultural Landscapes.* UNESCO. https://whc.unesco.org/en/culturallandscape/#2
UNESCO. (2021c). *World Heritage and extractive industries.* UNESCO. https://whc.unesco.org/en/extractive-industries/. Accessed 31 Aug 2021.
UNESCO. (2021d). *World Heritage state of conservation information system.* Online Source. https://whc.unesco.org/en/soc. Accessed 20 July 2021.
UNESCO. (2021e). *Cornwall and West Devon mining landscape.* UNESCO. https://whc.unesco.org/en/list/1215. Accessed 30 July 2021.
World Economic Forum. (2016). *Mapping mining to the sustainable development goals: An atlas.* World Economic Forum. https://www.undp.org/sites/g/files/zskgke326/files/publications/Mapping_Mining_SDGs_An_Atlas_Executive_Summary_FINAL.pdf

Chapter 26
Cultural Landscape Compatibility Study Upper Middle Rhine Valley – A Proactive Tool for Preventive Monitoring of Complex World Heritage Landscapes

Michael Kloos

Abstract In recent years, many "living" cultural and urban landscapes on a large scale were inscribed on the World Heritage List. However, such complex World Heritage properties generate frequently challenges concerning their management due to transformations caused by to pressure to change. As a result, there is a need for new proactive systematic approaches to assess such transformations combining innovative technical solutions with a systematic approach to using attributes and values conveying their Outstanding Universal Value. Taking the World Heritage cultural landscape *Upper Middle Rhine Valley* as a case study, this paper investigates such a systematic instrument to monitor transformations and to assess their impact on the OUV and integrity of UNESCO World Heritage properties. It is concluded that such systematic technical instruments can be helpful to support strategies for an integrated management combining preservation and sustainable development. However, an in-depth theoretical knowledge of sites' OUVs and attributes and values related thereby, as well as a sound integration in existing legislative frameworks and the participation of stakeholders on various levels is indispensable to guarantee their full effectiveness.

Keywords UNESCO · World Heritage · Sustainability · Sustainable development · Heritage Impact Assessment

M. Kloos (✉)
UNESCO Chair for Historic Urban Landscapes and Heritage Impact Assessments, Professorship of Preservation and Sustainable Development of Cultural and Historic Urban Landscapes, Faculty of Architecture and Engineering, RheinMain University of Applied Sciences, Wiesbaden, Germany
e-mail: michael.kloos@hs-rm.de

M.-T. Albert et al. (eds.), *50 Years World Heritage Convention: Shared Responsibility – Conflict & Reconciliation*, Heritage Studies,
https://doi.org/10.1007/978-3-031-05660-4_26

26.1 Introduction

Since its adoption in 1972, the World Heritage Convention has faced continuous evolution. The first inscribed cultural properties had an iconic monumental character, were supposed to be conserved in a certain historic state, and were mainly managed without the participation of local communities. The decision of the UNESCO World Heritage Committee in 1992 to include *cultural landscapes* as a separate category of cultural sites representing the "combined works of nature and of man" designated in Article 1 of the Convention contributed significantly to altering this situation since it influenced several other strategic decisions of the World Heritage Committee. This resulted in a major paradigm shift with regard to the Convention's implementation. In the beginning, cultural World Heritage properties were considered mainly as isolated "islands" regarding their conservation. Meanwhile, due to the Convention's evolution and due to the introduction of cultural landscapes, many "living" World Heritage properties were inscribed on the World Heritage List that are supposed to be managed with the integration of local communities. However, such properties on a large scale mostly have a high degree of complexity and are vulnerable to pressure to change, thus requiring integrated management strategies. Using the World Heritage cultural landscape *Upper Middle Rhine Valley* as a case study, it will be argued in this paper that both the abovementioned paradigm shift and the increasing challenges to manage complex World Heritage properties cause the need for new proactive instruments to assess transformations. Such new instruments should combine innovative technical solutions with a systematic approach to using attributes and values conveying their Outstanding Universal Value and must be thoroughly embedded in local social, political and legislative frameworks.

26.2 World Heritage Landscapes and Pressure to Change

Acknowledging cultural landscapes as cultural heritage had a decisive influence on the evolution of the World Heritage Convention. This step enabled underrepresented State Parties to suggest cultural sites without a monumental character for inscription in the World Heritage List and can be seen as a precursor of both the *Global Strategy* adopted in 1994 by the World Heritage Committee and the so-called *Gap Report* published by ICOMOS in 2004 (ICOMOS, 2004). It was also an important step towards the Committee's decision to approve the *2011 Recommendation on the Historic Urban Landscape* as an "holistic and interdisciplinary" approach addressing the "inclusive management of heritage resources" (WHITRAP, 2016, 11), thus leading to an integrated approach to urban management, which should support the integration of urban conservation in broader urban development considerations (Bandarin & van Oers, 2015). Meanwhile, many cultural and urban landscapes on a large scale, conveying both tangible and intangible values, have been inscribed on the World Heritage List. Mostly, such properties are embedded in or related to "living" urban agglomerations, and a considerable number of them are transboundary or serial transnational properties.

In contradiction to properties with a monumental character, such complex landscapes require strategies for integrated management combining preservation and sustainable development (Rössler, 2012; Kloos, 2017). Besides, after the turn of the millennium, due to many discussions with regard to their visual, structural and functional integrity, it became obvious that such complex World Heritage cultural and urban landscapes easily can get affected by pressure to change. Common questions are related to management problems and pressure due to development, inter alia, caused by planned high-rise tower blocks, traffic and service infrastructure, as well as socio-economic transformations (Veillon, 2014; van Oers, 2010; Bandarin & van Oers, 2015). At the latest, after the Committee's decision to withdraw the cultural landscape *Dresden Elbe Valley* from the World Heritage List due to the realisation of the so-called *Waldschlößchen Bridge* in 2009, it became obvious that the increasing size and complexity of cultural properties inscribed on the World Heritage List cause new challenges concerning their management (Ringbeck & Rössler, 2011). As a consequence, the World Heritage Committee successively adapted the regulations to manage World Heritage properties. The Operational Guidelines state, since 2005, that nomination proposals for the World Heritage List have to contain *management plans* explaining how the OUV and attributes conveying tangible and intangible values of potential World Heritage properties can be maintained, how this can be combined with their sustainable development and how such strategies will be organised and coordinated with the integration of relevant stakeholders, especially local communities. Besides, it was decided that properties inscribed earlier on the World Heritage List should also be provided with management plans as soon as possible.

In parallel, new monitoring mechanisms were adopted by the Committee. The so-called *Periodic Reporting* serves as a regular monitoring system for World Heritage properties; State Parties have to submit *State of Conservation (SOC) Reports* every 6 years. In parallel, *Reactive Monitoring* was introduced as a mechanism of: "reporting by the Secretariat, other sectors of UNESCO and the Advisory Bodies to the Committee on the state of conservation of specific World Heritage properties that are under threat." (UNESCO World Heritage Centre, 2019, para 169) Meanwhile, the Operational Guidelines also state that: "to this end, the States Parties shall submit *specific reports and impact studies* each time exceptional circumstances occur, or work is undertaken which may have an impact on the Outstanding Universal Value (OUV) of the property or its state of conservation." (UNESCO World Heritage Centre, 2019, para 169) Hence, the introduction of Reactive Monitoring is closely related to a third monitoring instrument, the aforementioned *impact studies*. As a reaction to a growing number of properties affected by pressure to change, an increasing number of *Heritage Impact Assessments (HIAs)* has been requested by the Committee in recent years. Due to the publication of the so-called *ICOMOS Guidance on Heritage Impact Assessments for Cultural World Heritage Properties 2011*, which is planned to be updated soon, HIAs meanwhile can be considered as a standardised instrument to identify and assess positive and negative impacts of planned or cumulative transformations with regard to the OUV and integrity of World Heritage properties (ICOMOS, 2011; Kloos, 2017).

Additionally, the Convention's evolution during the last 50 years resulted in a second major change concerning its implementation. While the first iconic monuments were chosen as the "best of the best"(Cameron, 2008, 71–79) and entered on the World Heritage List as their "global significance was beyond question"(Ringbeck, 2021, 117–130) the inscription of the "new generation" of more complex World Heritage properties caused a need for a more systematic approach to justify their OUV. As such sites consist of a large range of different attributes conveying tangible and intangible values, a new systematic approach to identify the potential OUV during nomination processes was defined in 2005. Since then, so-called *Statements of Outstanding Universal Value (SOUV)* serve "as the central reference document for justifying inscription and assessing developments, risks and threats following recognition as a World Heritage property" (Ringbeck, 2021, 117–130) · As an essential tool for describing the attributes and values, SOUVs define "the thinking at the time of inscription on the basis of the criteria in force at the time" so as to provide "a clear, shared understanding of the reasons for inscription". Besides, requirements for the management to sustain the Outstanding Universal Value for the long-term should be mentioned. (Ringbeck, 2021, 117–130). Formally, SOUVs consist of a Brief Synthesis, the Justification of Criteria, a Statement of Integrity and Authenticity (only cultural sites), as well as Requirements for Protection and Management (UNESCO World Heritage Centre, 2019, Annex 10). For earlier inscribed properties, so-called *Retrospective Statements of Outstanding Universal Value (RSOUV)* have to be compiled, which also must be adopted by the Committee.

In conclusion, the Convention's evolution during the last 50 years resulted in four major consequences with regard to its implementation:

(a) In general, the shift to larger and more complex World Heritage properties increased the risk of pressure to change.
(b) World Heritage management became far more challenging than it used to be as management strategies have to address, inter alia, preservation and sustainable development of inhabited areas on a wider scale with a high degree of complexity, as well as a wide range of different stakeholders.
(c) Such management strategies also have to respond to more refined monitoring mechanisms, including Periodic Reporting and Reactive Monitoring, as well as HIAs as additional assessment instruments.
(d) Nominations, management and monitoring of World Heritage properties have to be handled in line with (R)SOUVs, and the attributes and values described thereby. Hence, the identification and listing of attributes and values became far more important than it used to be in the beginning of the Convention's implementation.

In the following, the World Heritage property *Upper Middle Rhine Valley* will be used as an example to demonstrate that these consequences of the Convention's evolution result in a need for more systematic instruments to monitor transformations and to assess their impact on the OUV and integrity of UNESCO World

Heritage properties. Thereby, such instruments should address and activate a clear understanding of the OUV, as well as attributes and values of such complex World Heritage properties, so as to support UNESCO's overall strategy to use cultural heritage as a pillar of sustainable development.

26.3 Pressure to Change and Present Monitoring Activities in the World Heritage Property Upper Middle Rhine Valley

The Upper Middle Rhine Valley was inscribed on the World Heritage List in 2002. The property, located in southwest Germany, was inscribed according to the Operational Guidelines as a so-called "organically evolved landscape", sub-category "continuing landscape" (UNESCO World Heritage Centre, 2019, Annex 3). It stretches 67 kilometres from the two towns Bingen and Rüdesheim in the south to the city of Koblenz in the north. The cultural landscape of the Upper Middle Rhine Valley is characterised by both its spectacular scenery characterised by River Rhine, which formed a canyon in the Rhenish Slate Mountains, and by its rich cultural heritage. Moreover, its particular geomorphological setting consisting of about 60 towns and villages, as well as numberless castles and steep vineyards terraced by dry stone walls, characterise the valley. Besides its size and these various attributes, the political landscape of the area is also complex. The site is located in the two federal states Rhineland-Palatinate and Hesse, which, according to the legislation of the Federal Republic of Germany, have full and sovereign rights concerning cultural affairs. Hence, both federal states are jointly responsible for the upkeep of the OUV of the site. Additionally, the World Heritage cultural landscape consists of 59 municipalities. To coordinate and manage the property, the so-called *Upper Middle Rhine Valley World Heritage Association* (Zweckverband Oberes Mittelrheintal) was established. This administrative institution comprises representatives from all the local and "county" authorities in the World Heritage area and its buffer zone, as well as the federal states of Hesse and Rhineland-Palatinate.

The OUV of the property was justified as follows (UNESCO World Heritage Centre, 2021):

- *Criterion ii: As one of the most important transport routes in Europe, the Middle Rhine Valley has for two millennia facilitated the exchange of culture between the Mediterranean region and the north;*
- *Criterion iv: As an outstanding organic cultural landscape, the present-day character of which is determined both by its geomorphological and geological setting and by the human interventions, such as settlements, transport infrastructure, and land use, that it has undergone over two thousand years;*
- *Criterion v: As an outstanding example of an evolving traditional way of life and means of communication in a narrow river valley. The terracing of its steep slopes in particular has shaped the landscape in many ways for more than two*

millennia. However, this form of land use is under threat from the socio-economic pressures of the present day.

Justification criterion (v), where it is stated that the characteristic vineyards terraced by dry stone walls as the traditional way of land use are under threat, shows that the property faces pressure to change. Many of these steep vineyards were neglected in recent years and, therefore, have been transformed into a forested landscape (Ministerium für Wirtschaft, Klimaschutz, Energie und Landesplanung Rheinland-Pfalz, 2013). However, other socio-economic factors also affect the property. Currently, the valley serves as the main transport corridor for freight traffic between main ports located in Northern Italy and Northern Europe. Approximately 130,000 trains passed through the valley in the year 2019 (Ministerium für Wirtschaft, Klimaschutz, Energie und Landesplanung Rheinland-Pfalz, 2013), thus, leading to an enormous amount of noise and considerable air pollution. Younger inhabitants also frequently move to the higher parts or leave the valley, which leads to demographical imbalances of the population (Ministerium für Wirtschaft, Klimaschutz, Energie und Landesplanung Rheinland-Pfalz, 2013). Besides, numberless planned infrastructure projects document the current pressure to change in the valley. Issues during recent years included, inter alia, plans for new wind turbines, alternative railway tracks due to the need to replace existing tunnels (Goedkoop et al., 2014) and a new railroad crossing at the town of Rüdesheim (Kloos et.al., 2021). Additionally, to facilitate the crossing of the River Rhine, a new bridge between the towns of St Goar and St Goarshausen is currently conceived (Kloos et al., 2009), and a new ropeway was installed in Koblenz.

Most of the abovementioned projects were mentioned in the various SOC Reports submitted by the State Party of Germany. Due to the large number of such issues, the Committee has started a Reactive-Monitoring process and requested the State Party to submit SOC Reports every 2 years (instead of 6 years normally). Frequently, the planned projects led to lengthy and partly controversial discussions with the UNESCO World Committee and its Advisory Body, ICOMOS. Consequently, when the Committee requested that the State Party update the property's management plan in 2018, the idea arose to conceive a more systematic approach to submit information about planned projects, which could possibly affect the OUV of the property. Up until now, the various planned projects were separately submitted to the Committee. Frequently, the Committee recommended compiling additional studies such as Heritage Impact Assessments to assess these projects. With the goal to accelerate this process and to avoid unnecessary communication, the new system should provide a more unified and proactive assessment methodology, which should allow the evaluation of planned projects prior to the information of the Committee concerning their compatibility with the OUV of the World Heritage property.

26.4 Cultural Landscape Compatibility Study (CLCS) as a New Proactive Approach to Monitor Change

Against this background, it was decided by the Upper Middle Rhine Valley World Heritage Association and the responsible ministries in Rhineland-Palatinate and Hesse to set up both the new management plan and this new systematic approach to assess planned projects in parallel. The general starting point of this new approach, called *Cultural Landscape Compatibility Study (CLCS)*, is a simple idea. Visualisations of planned projects, generated on the basis of superimpositions of a 3D computer model and GPS-related digital photographs, should serve as a means to assess the potential impact of planned projects on the OUV and integrity of the World Heritage property. Such a methodology has already proven to be useful in numberless HIAs both in the Upper Middle Rhine Valley and other World Heritage properties, as it can show and evaluate planned transformations transparently from an independent point of view (Kloos, 2015, 2017).

However, generating such visualisations can be complex because they are based on 3D models of the planned projects, which have to be shown in their environmental surroundings. Normally, such models have to be built up from scratch by combining and elaborating LiDAR laser measurements (so-called point clouds) with high-resolution aerial photographs, which is a time-consuming process. An additional requirement is that according to the ICOMOS Guidance 2011, the OUV of World Heritage properties and the attributes and values conveying this OUV should serve as a starting point for the assessments (ICOMOS, 2011). However, as SOUVs mostly provide only relatively general criteria, which cannot be used directly for assessments, consultants frequently have to set up separate analytic studies during the assessment process to identify relevant attributes and values. Additionally, HIAs, other than *Strategic Impact Assessments (SEAs)* and *Environmental Impact Assessments (EIAs)*, are anchored neither in German nor in EU legislation at present. Frequently, this leads to long political decision processes about which institutions are responsible for conducting and financing HIAs. Due to these time-consuming factors, HIAs are often only commissioned in a late stage of planning processes or even when projects have already been realised, and they are seldomly compiled as a proactive and process-related instrument to generate a basis for bottom-up, cultural heritage-led planning processes (Kloos, 2017).

To avoid these problems, the systematic approach of the new instrument *CLCS* is based on the following three interrelated elements so as to accelerate assessment processes of planned projects in the Upper Middle Rhine Valley:

1. First, a *3D model of the entire World Heritage property* should serve as the basis of the assessment of planned projects in order to save time concerning the generation of visualisations. This integrated 3D computer model of the entire Upper Middle Rhine Valley was generated from LiDAR laser measurement datasets, high-resolution aerial photographs (DOP 20) and City Geography Markup Language datasets (City GML);

Fig. 26.1 Generation of the Integrated 3D Computer Model for CLCS. (*Note.* 3D models for CLCS from V-Cube, by A. Walther, [3D models], 2020)

2. Second, a *clear and thorough understanding of OUV* should be provided by a systematic and unified identification of the property's various attributes and values;
3. Third, an independent *Monitoring Advisory Body (MAB)* consisting of several experts was established to evaluate planned projects. The MAB should also provide recommendations for which of the planned projects would probably be compatible with the World Heritage status and which ones would need a more in-depth investigation. The second task of this advisory body was to support the development of the *CLCS* (Fig. 26.1).

26.5 Experiences During the CLCS Development Process

During the elaboration of the CLCS, several presentations for various stakeholders revealed that many parties were interested in the 3D computer model for different reasons. Planning authorities and local municipalities wanted to use the model as a basis for planning processes, e.g. the planned garden exhibition *BUGA 29* and architectural competitions. Additionally, questions appeared whether the 3D computer model could also be used to assess smaller projects conceived by private parties concerning their compatibility with the OUV of the property. As the model contained a large amount of data due to the size of the World Heritage property, it was decided to design it as user-friendly as possible in order to respond to these various requests. It is planned now to rent out separate parts of the 3D model to third parties, e.g. architectural offices, as a basis for their planning activities. After the finalisation of such planning processes, these supplemented parts can be re-integrated into the 3D model (Fig. 26.2).

In parallel to the generation of the 3D computer model, a participation process was started to provide a clear and thorough understanding of the OUV of the World Heritage property and the attributes and values conveying this OUV. In so doing, the elaboration process of the management plan was used as a unique opportunity to organise several workshops with various relevant stakeholders. A crucial idea of these workshops was to build up a broad basis for the understanding of the OUV, attributes and values, especially on the level of representatives of the various administrative institutions in the property.

Fig. 26.2 Separated parts of the integrated 3D computer model. (*Note.* 3D computer models from Michael Kloos Planning and Heritage Consultancy and V-Cube, by P. Tebart and A. Walther [3D Models], 2020)

Nevertheless, during the elaboration process of the CLCS, it turned out to be complicated to identify these relevant attributes. Particularly in the beginning of the process, it appeared to be an obstacle that "except for authenticity, attributes have so far only been defined outside the Operational Guidelines in the questionnaire for the third cycle of periodic reporting" (Ringbeck, 2021). In this phase, the suggestion of the Monitoring Advisory Body to carry out an in-depth landscape study covering the entire Upper Middle Rhine Valley turned out to be helpful. A second helpful element during the process was the requirement of ICOMOS within the 3rd cycle of Periodic Reporting to restrict the number of attributes. Generally, no more than 15 attributes should be mapped (Ringbeck, 2021). This restricted approach led to a compact table where both key attributes and attributes can be shown in one compressed overview. Additionally, to support clear and transparent information for all stakeholders, it was decided to also show these attributes on a set of maps (Fig. 26.3).

These new tools of the CLCS approach – the 3D computer model and the unified system to map attributes – were later tested concerning their efficiency with regard to planned projects in the Upper Middle Rhine Valley. During these first test assessments, plans of various projects were inserted in the 3D computer model in order to visualise their potential impact on the OUV of the World Heritage property. In the second step, relevant attributes in the investigation areas were derived from the unified attribute table and displayed on panoramic photographs. This combined approach of visualisations and visual analysis of key attributes turned out to be very useful. As it could clearly be assessed and graded how attributes would be affected, it could also be stated which of the planned alternative projects would be compatible with the OUV and which ones not. Consequently, incompatible versions can now already be ruled out prior to the information of from the World Heritage Committee (Fig. 26.4).

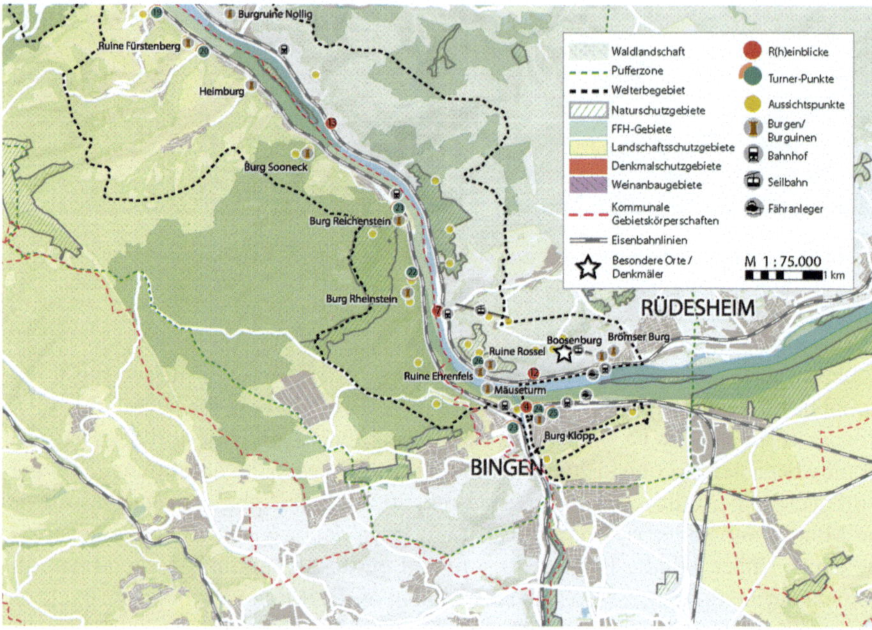

Fig. 26.3 Extracted map with identified attributes conveying the OUV of the World Heritage property Upper Middle Rhine Valley. (*Note.* Map of Upper Middle Rhine Valley from Michael Kloos Planning and Heritage Consultancy and v-cube, by P. Tebart and A. Walther [Map], 2021)

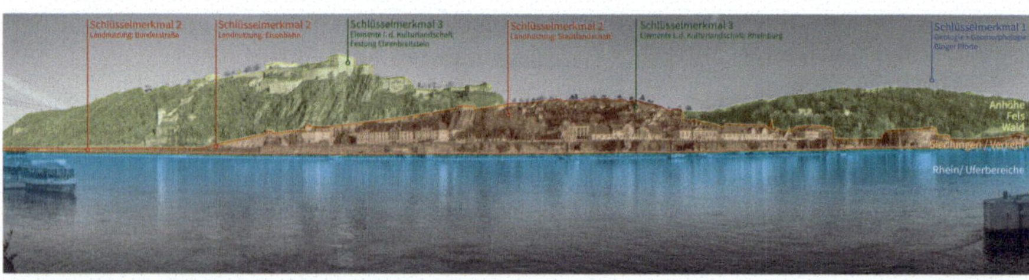

Fig. 26.4 Visual analysis of attributes of an inner investigation area in the World Heritage property Upper Middle Rhine Valley. (*Note.* Colored digital photograph of an inner investigation area from Michael Kloos Planning and Heritage Consultancy, by M. Kloos and P. Tebart [Digital Photograph], 2021)

26.6 Conclusion: Recommendations for Future Research Activities

The case study in the Upper Middle Rhine Valley reveals that it is possible to provide efficient tools to monitor and assess transformations caused by planned projects in complex World Heritage properties. At present, further test assessments are being carried out to elaborate the CLCS approach as a new systematic and proactive evaluating instrument. However, the development process of CLCS in the Upper Middle Rhine Valley also shows that various questions remain to be solved, both on a theoretical and practical level.

It is a crucial theoretical issue that both OUV and attributes play a very important role concerning both the assessment of changes in World Heritage properties and the identification of the OUV during nomination and management processes (Ringbeck, 2021). However, up until now, the Operational Guidelines only define how to identify attributes with regard to the authenticity of World Heritage properties. In this context, it turned out to be helpful to follow the abovementioned statement of ICOMOS within the 3rd cycle of Periodic Reporting that recommended using a limited number of approximately 5 key attributes and 10 attributes. However, further research should be carried out to determine whether this approach could serve as a general starting point to establish a unified system to identify attributes and values in World Heritage properties and their surroundings.

A second more practical issue concerning the implementation of CLCS is related to juridical and political levels. Up until now, it is unclear how initiatives of private building owners and project developers in the Upper Middle Rhine Valley will be affected. Consequently, it must be clarified how planned projects of private parties can be assessed with this new tool and who has to cover the costs for that. These questions are particularly relevant if such preliminary assessments reveal that an in-depth assessment will be necessary since HIAs are also not yet embedded into German legislation. Even though it turned out to be possible to develop a new systematic approach to evaluate transformations on a technical level, it is obvious that future research activities also have to cover such practical juridical and political questions.

Fifty years ago, the theoretical idea of sustainability of the World Heritage Convention was to safeguard the most iconic sites of Outstanding Universal Value for future generations. Today, it appears that UNESCO's approach to considering cultural and urban landscapes as important elements to provide identity for local communities is highly dependent on efficient systematic tools that can combine the preservation of their OUV with their sustainable development on a practical level. Systematic technical approaches and instruments such as CLCS can be helpful to support the sustainable preservation of complex World Heritage properties. However, such instruments also require an in-depth theoretical knowledge of sites' OUVs and attributes and values related thereby, as well as a sound integration in existing legislative frameworks. It should also be noted that such strategic instruments can only be helpful if they are developed with the participation of stakeholders on various levels because this is an indispensable step for their broad acceptance on local and regional levels. In other words, innovative technical instruments such as CLCS can be considered valuable to support strategies combining preservation and sustainable development, but they should not be considered as stand-alone instruments. To guarantee their full effectiveness, they should be embedded in a multidimensional management strategy.

References

Bandarin, F. and van Oers, R. (2015). Reconnecting the city. The historic urban landscape approach and the future of urban heritage. .

Cameron, C. (2008). *Evolution of the application of the "Outstanding Universal Value" for cultural and natural heritage*, Keynote Paper in Kazan (2005), published in: *What is OUV? Defining the Outstanding Universal Value of Cultural World Heritage Properties*. An ICOMOS study compiled by Jukka Jokilehto, with contributions from Christina Cameron, Michel Parent and Michael Petzet, Monuments and Sites XVI. ICOMOS.

Goedkoop, M., Kloos, M., & Wachten, K. (2014). *Visualisierungen von Tunnelein- und -ausgängen im Rahmen der Variantenuntersuchung zur Erneuerung von Bank-, Bett- und Kammerecktunnel zur Beurteilung von Auswirkungen auf das UNESCO Welterbe Oberes Mittelrheintal*. Institute for Urban Design and Regional Planning. Unpublished.

ICOMOS. (2004). *The World Heritage List: Filling the gaps – An action plan for the future an analysis by ICOMOS*. ICOMOS.

ICOMOS. (2011). *Guidance on heritage impact assessments for cultural World Heritage properties*. ICOMOS.

Kloos, M. (2015). Heritage impact assessments as a tool to open up perspectives for sustainability: Three case studies related to discussions concerning the visual integrity of World Heritage cultural and urban landscapes. In M. T. Albert (Ed.), *Perceptions of sustainability in heritage studies* (pp. 215–230). De Gruyter.

Kloos, M. (2017). Heritage impact assessments as an advanced tool for a sustainable management of cultural UNESCO World Heritage sites – From theory to practice. In M.-T. Albert (Ed.), *Going beyond – Perceptions of sustainability in heritage* (Vol. 2, pp. 335–359). Springer.

Kloos, M., Korus, C., Tebart, P., & Wachten, K. (2009). *Independent assessment of the visual impact of the planned Rhine bridge between Wellmich and Zu Fellen on the integrity of the World Heritage property "Upper Middle Rhine Valley"*. Institute for Urban Design and Regional Planning. https://www.staedtebau.rwth-aachen.de/go/id/guva/lidx/1. Accessed 5 Nov 2021.

Kloos, M., Tebart, P., & Tebart-Tirai, D. (2021). *Heritage Impact Assessment of the planned removal of the railway crossing in the location of Rüdesheim concerning the UNESCO World Heritage property "Upper Middle Rhine Valley"*. [Unpublished manuscript].

Ministerium für Wirtschaft, Klimaschutz, Energie und Landesplanung Rheinland-Pfalz (Ed.). (2013). *Masterplan Welterbe Oberes Mittelrheintal Herausforderungen und Visionen für die zukünftige Entwicklung*. Landesregierung Rheinland-Pfalz.

Ringbeck, B. (2021). OUV & Attributes. In Japan Center for International Cooperation in Conservation (Ed.), *Attributes – A way of understanding OUV* (pp. 117–130). Tokyo National Research Institute for Cultural Properties.

Ringbeck, B., & Rössler, M. (2011). Between international obligations and local politics: The case of the Dresden Elbe Valley under the 1972 World Heritage Convention. *Informationen zur Raumentwicklung Heft, 3*(4), 205–211.

Rössler, M. (2012). Partners in site management. A shift in focus: Heritage and community development. *World Heritage Papers, 31*, 27–31.

UNESCO. (1972). *Convention concerning the protection of the world cultural and natural heritage*. UNESCO.

UNESCO World Heritage Centre. (2019). *Operational guidelines for the implementation of the World Heritage Convention*. UNESCO.

UNESCO World Heritage Centre. (2021). https://whc.unesco.org/en/list/1066/

van Oers, R. (2010). Managing cities and the historic urban landscape initiative – An introduction. *World Heritage Papers, 27*, 7–17.

Veillon, R. (2014). *State of conservation of World Heritage properties. A statistical analysis (1979–2013)*. UNESCO.

WHITRAP. (2016). *The HUL guidebook: Managing heritage in dynamic and constantly changing urban environments; a practical guide to UNESCO's recommendation on the historic urban landscape*. UNESCO. WHITRAP, City of Ballarat.

Chapter 27
Geoheritage to Support Heritage Authorities: Research Case Studies on Maya Archaeological Sites

Mario Hernández, Philippe De Maeyer, Luc Zwartjes, and Antonio Benavides Castillo

Abstract Since the adoption of the World Heritage Convention (1972), modern technologies have significantly changed the way our society behaves and operates, with an increased demand for energy, fast and reliable communications, etc. Some modern technologies might contribute to negative impacts on heritage sites, e.g. through climate change and/or excessive tourism; however, modern digital technologies can also be extremely beneficial for heritage activities. In this paper, we focus on how modern digital geo-science and geo-technology can support heritage authorities' daily work. We introduce herein the concept of digital Geoheritage, which can help heritage authorities to discover and understand the enormous benefits that geomatics can provide for their daily heritage activities. This research case, implemented through an interdisciplinary scientific approach, originally aimed to support the preservation, restoration and management of a cultural heritage site; however, it was later expanded to also support archaeological research, stability risk assessment, planning, design, education, dissemination and promotion. The use of digital geo-sciences for the benefit of the local Maya communities living around a heritage site is also illustrated. Our objective, within the current book, was to present a paper that is oriented toward heritage authorities, and, therefore, technical language has been avoided.

Keywords Cultural heritage · Modern technologies · Applied geomatics · Citizen science · Geosciences · Community participation

---

M. Hernández (✉)
International Society for Digital Earth (ISDE), Mexico City, Mexico
e-mail: mariohernandezvaldes@gmail.com

P. De Maeyer · L. Zwartjes
Department of Geography, Ghent University, Ghent, Belgium
e-mail: philippe.demaeyer@ugent.be; Luc.Zwartjes@ugent.be

A. B. Castillo
Instituto Nacional de Antropología e Historia (INAH), Campeche, Mexico
e-mail: abc999@prodigy.net.mx

M.-T. Albert et al. (eds.), *50 Years World Heritage Convention: Shared Responsibility – Conflict & Reconciliation*, Heritage Studies,
https://doi.org/10.1007/978-3-031-05660-4_27

27.1 Heritage and Modern Technologies: An Introduction

Some modern technologies might be a threat to heritage sites, while other modern digital technologies can be beneficial for heritage activities. Recent advances in digital technologies have enabled a new emergent heritage assistance methodology, herein referred to as digital Geoheritage. The use of Geoheritage can have many advantages for heritage sites; the digitalisation of cultural heritage sites is important for the protection, conservation, restoration, research, dissemination and promotion of tangible and intangible cultural assets. For heritage sites, the possibilities created by advances in digital technologies are impressive and ever-growing, e.g., three-dimensional modelling or virtual/augmented reality that enables the concept of a "virtual heritage site" in such a way that if the heritage authorities cannot go daily to the heritage site, then the heritage site can come to them virtually. These non-invasive digital technologies are providing significant support in all aspects of heritage-related activities. The following research case, implemented through an interdisciplinary scientific approach, originally aimed to support the preservation, restoration and management of a cultural heritage site; however, it was later expanded to also support archaeological research, stability risk assessment, planning, design, education, dissemination and promotion. The paper further illustrates the use of digital geo-sciences for the benefit of the local Maya communities living around a heritage site. As our objective was to present a paper that is oriented toward heritage authorities, technical language has been avoided.

27.1.1 Modern Technologies, a Threat to Heritage Sites

More than half of the world's population now live in urban areas. Modern technologies, e.g., remote-controlled building cranes, are facilitating the growth of urban areas, and this sometimes affects World Heritage sites. The World Heritage Committee (WHC) identified a list of factors that affect World Heritage properties (WHC, 2008) in a report that specifically mentions "Buildings and Development". Urban growth, facilitated by modern technologies, is invading the historical landscape[1] of certain World Heritage sites, e.g., the *Giza Pyramids* (Egypt) (Vaz, 2011) or *Teotihuacan* (Mexico) (WHC, 2007b).

In 2015, the United Nations adopted Resolution 70/1, "Transforming our World: the 2030 Agenda for Sustainable Development" (UN, 2015). The document lays out the 17 Sustainable Development Goals (SDGs), which aim to end poverty and hunger, protect human rights and human dignity, protect the planet from degradation, and foster peace. Within SDG 11, "*Make cities and human settlements inclusive, safe, resilient and sustainable*", Target 11.4 makes a vague reference to heritage

[1] Historical landscape is used herein as the geographical area around a heritage site, having remains of the human activities of the habitants that used to live in the heritage site.

(UN, 2015, 21–22). Such a reference is not nearly enough to safeguard heritage sites, especially from the uncontrolled urban growth facilitated by modern construction technologies.

Digital technologies, e.g., artificial intelligence and robotics, have revolutionized travelling by air, making it affordable and easier. Before the current worldwide pandemic, the airline industry was carrying 3.6 billion passengers yearly. Consequently, heritage sites are receiving a larger number of visitors. Overcrowding at a heritage site leads to degradation of the site and affects the quality of life of the local population.

Modern digital communication technologies have transformed our society and the way we communicate. There are many advantages of these incredible methods of communication; however, there are also disadvantages, for example, social media has created an anxious tendency to take a "selfie"[2] exactly as posted by others on social media. This is causing significant concentrations of social media tourists and damaging overcrowding at specific heritage sites, and it is a form of tourism that is uninterested in the various heritage values of such sites. European social media data shows that Paris, Istanbul and Rome have the largest number of posts for cultural World Heritage cities (between 50 and 110 million posts). The Acropolis in Athens is the most popular archaeological site (Thomas, 2021). Social media will continue showing users' preferred places and travel experiences, contributing to severe congestion and associated damages to certain heritage sites.

27.1.2 Digital Modern Technologies: Non-invasive Tools Supporting Heritage

Digital techniques that support heritage documentation, preservation, protection and presentation cover a wide range of technologies, mainly grouped in three major areas: data capture, virtual reconstruction and visual communication. Contemporary advances in science and technology facilitate the elaboration of accurate 3D models of heritage sites' features. Heritage 3D models can range from using high-resolution satellite images to model entire historical landscapes with a resolution of meters (m) to the modelling of small archaeological features (e.g., small pottery artefacts) with a resolution less than 1 mm.

Today, thanks to powerful laptops and digital cameras, digital Geoheritage can be used at heritage sites to capture data and upload such data to the internet for processing. In fact, during the last decade, digital applications have become part of the archaeological toolbox. Together with archaeological sciences, digital databases

[2] A "selfie" is a self-portrait photograph, typically taken with a digital camera or smartphone and showing the person taking the picture with a well-known element in the background. Selfies are often shared on social media to inform friends that the author has been visiting such a place.

and other computer methods, modern digital technologies are now present in every respectable archaeological investigation (Nicollucci, 2020).

27.2 The Heritage Site and Project Partners

Our research case study focuses on an outstanding Maya archaeological site that represents a masterpiece of human creative genius; it exhibits an important interchange of human values over a span of time within the Maya culture. The archaeological Maya site of Edzná, located in the State of Campeche, Mexico, was the ancient urban area that hosted the *"Itzaes"* (in Maya *Ytzná/Edzná*) (Fig. 27.1). *Edzná has an urban extension of 25 km*^{2}. The main elements of the heritage site buildings correspond to the Maya *Petén* architectural style (Benavidez, 1997, 2014). *Edzná* was abandoned around 1450 A.C., and then the exuberant vegetation of the tropical forest grew between the stones in such a way that *Edzná* became a series of hills covered by vegetation. In modern times, *Edzná* was well known by the local Mayas living in its surroundings. Being very humble people, they never thought about letting others know about this heritage site. The discovery of *Edzná* is then erroneously attributed to an Austrian explorer who published about it in a European journal. In 1943, Mexican heritage authorities began to remove the vegetation and initiated the associated restoration of *Edzná* (Benavidez, 2014). Today, heritage management, archaeological research, restoration, safeguarding and dissemination

Fig. 27.1 Heritage site of Edzná. (*Note*. Source: Photograph, by Hernandez M., 2018)

related to the site is under the authority of the Mexican Institute of Anthropology and History (INAH).

The type of research described herein, originated in 2000, when the European Space Agency (ESA) and UNESCO launched the "Open initiative on the use of space technologies to support World Heritage sites, From Space to Place" (ESA, 2003). This was a call to all geo-space actors to join in supporting heritage sites. Over eighty space partners joined the initiative, and Ghent University (UGent), Belgium, was among these. Financed by the Belgian Science Policy Office (BELSPO), UGent, through its Department of Geography, provided invaluable contributions. UGent, jointly with ESA, produced the very first full and accurate cartography for the five World Heritage sites of the Democratic Republic of Congo and Rwanda. In 2005, UGent significantly assisted with the inscription of *Calakmul*, Mexico, as a mixed site (Belspo, 2003). Since then, UGent and Dr. M. Hernandez have been working jointly with INAH Campeche supporting Maya archaeological heritage sites. For this research, the team is composed of INAH Campeche, UGent and Dr. Mario Hernandez.

27.3 Heritage-Related Issues and Objectives

We identified various challenges with respect to *Edzná*. As with many other heritage sites, common issues related to insufficient funding, lack of staff, inappropriate or non-sufficient tools for the cleaning of vegetation, etc., were mentioned by INAH heritage authorities. We decided to focus on the following challenges where Geoheritage could be of assistance: 3D digital models to enable office research, avoiding frequent field visits to the heritage site, as well as facilitating the measurement of different archaeological components and significantly reducing the complexity of on-site research; temporal digital 3D models (and 4D, where time is the fourth component) to monitor the vegetation cleaning and restoration work; accurate digital architectural plans to speed up the process of vegetation cleaning; assessment of eventual ground subsidence causing structural damage to the main archaeological buildings; use of 3D models for education, dissemination and promotion.

It is also important to understand that the main goal of our research work was for the results to support the *Edzná* heritage authorities in the following: developing a methodology that can be used locally in the long term, eventually with the support of a local university (know-how and technological transfer); selecting equipment for data capture that is affordable for the heritage authorities; implementing all digital processing methods in such a way that they become open access and accessible as web services. On the other hand, our research also aimed to identify high-level research topics that enable the involvement of Masters or PhD degree students at UGent.

27.4 Geoheritage Assisting Site Managers

Human eyes work together to gauge distance (called depth perception); each eye sees a 2D image, which is similar to photographs. These two images are then processed internally by our brain to extrapolate depth (stereoscopic vision). Conversely, photogrammetry uses a sensor (e.g., photographic camera) to capture a surface. The line between the camera and the centre of the object is called the "line of sight" (sometimes called "ray" or "beam"). However, if we capture a second photograph of the same object from a different location, being careful to cover almost the same surface as the first photograph (overlap), we can then obtain a different 2D photograph. The different "lines of sight" of each photograph can then be mathematically intersected to produce the 3D coordinates of the various points located on the common surface covered by the two 2D photographs.

Modern digital technologies are facilitating the acquisition of 2D digital photographs. Using the Global Positioning System (GPS), extremely accurate digital measurements are obtained for the position of the camera and the distances between the archaeological object and the camera. In our research, to obtain high-cartographic accuracy, we used ground control points (GCP) in a geographic network. Thanks to digital cameras that can autofocus and hold thousands of photos on their internal memory card, an enormous number of digital images of an archaeological object can be captured. Big Data methodologies (Pence, 2014) and artificial intelligence can then be used to process all digital images, including the GPS points in the process. The results are extremely accurate 2D architectural plans of the archaeological monument or, even better, highly accurate 3D models.

Digital technologies have also enabled the development of digital devices that can capture the 3D coordinates of points located on archaeological monuments, e.g., hand-held laser scanners make it possible to obtain 3D models with higher than 1 mm precision. An airborne LiDAR scanner (LiDAR stands for Light Detection and Ranging) can also be used to model a historical heritage landscape. LiDAR on an aeroplane can map terrain at 30 cm resolution.

Appropriate sensors (to capture digital images and/or digital points) can be chosen depending on the needs of the site manager. A series of new data capture sensors as well as tools have been developed in the last 15 years, including high-resolution satellite images, laser scanning, rapid prototyping, red-green-blue-depth (RGB-D) sensors, high dynamic range imaging, spherical and infrared imaging, mobile mapping systems, unmanned aircraft systems (UAS), based imaging, augmented and virtual reality, etc. In summary, Geoheritage comprises a wide range of digital scientific and technological methodologies (Fig. 27.2).

Structure from motion (SfM) photogrammetry is a method of deriving a three-dimensional structure by using two-dimensional images. For our research case study, we used the structure from motion (SfM) photogrammetry methodology, which can derive hyper-scale three-dimensional (3D) landform models. This methodology is based on the use of overlapping geo-referenced images acquired from different perspectives. As stated, our objective was to use techniques that would be

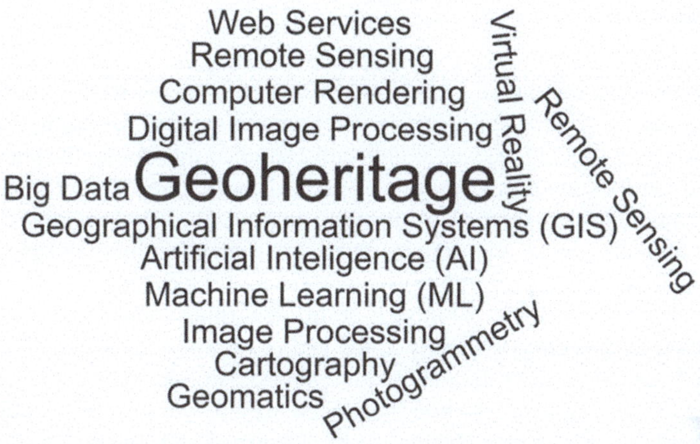

Fig. 27.2 Scientific disciplines under the "umbrella" of Geoheritage. (*Note*. Source: diagram created by Hernandez M. 2021 ©M.Hernandez)

affordable within the framework of the heritage authorities of *Edzná*. SfM is widely used in geo-sciences applications as it is a low-cost topographic survey technique that can produce dense 3D point clouds, digital elevation models and topographic maps (Carrivick et al., 2016). As it is a well-structured repetitive methodology, one advantage is that AI can be applied to optimize and speed up the processing of digital images.

A point cloud dataset is a large collection of points that are placed on a three-dimensional coordinate system. Point cloud files greatly speed the design process by providing real-world context where you can re-create the referenced objects or insert additional models. When deriving a point cloud dataset, it is important to eliminate erroneous points that are due to artefacts in the original images and/or erroneous recording of geo-referenced parameters. The automatic detection and elimination of noise in point cloud datasets is an important area of research (Cheng & Lau, 2017).

Digital data sensors and the associated processing software are well known for geomatics experts, but they are not typically within the knowledge of the heritage site managers. Therefore, the main challenge for us has been determining how the produced accurate archaeological 3D models can be provided to heritage site managers who do not have the complex software or educational background to process 3D geomatics data. We have, therefore, developed 3D services based on geographical information systems (GIS) on the web (WebGIS). Such services can then be easily accessed by heritage site managers to search and query, in real-time, segments of the 3D model at different resolutions.

For heritage sites and their associated historical landscape, digital geometric documentation has many advantages. The principal advantage of this new science and technology is, according to our heritage expert partners, that they enable a non-invasive methodology to interact with the different heritage objects.

27.5 Results

Thanks to the accurate digital models that were obtained, the work of the heritage authorities related to archaeological research and associated restoration (cleaning from encroached vegetation) was significantly speeded up. Due to the current pandemic and the lockdown in Campeche, Mexico, the heritage staff were unable to undertake any fieldwork on the site but could continue their archaeological restoration measurements using the digital 3D model from the office and home (tele-archaeology-restoration preparation).

The analysis between the 3D models at different times indicated terrain subsidence of some mm occurring mainly at the end of the rainy season. This was confirmed in 2020 when heavy rain caused a collapse onto part of the main building. The use of infrared cameras on drones has been extremely useful to identify archaeological vestiges. The water channels were identified and mapped, and the wall surrounding the site in *Uxmal* was clearly detected and mapped. A larger replica of our 3D model is being printed with a 3D plotter. Such a model will be shown at the Museum in the city of Campeche and used for education, promotion and dissemination of the importance of this heritage site.

In summary, as the *Edzná* site manager stated, "the digital 3D model, seen from various angles and with different angles of illumination brings per se a completely new set of ideas for further research as well as questions to be addressed".

The more we work jointly with the heritage authorities, the more they identify additional applications. Therefore, we continue with data acquisition and processing to further implement solutions to support the heritage authorities in their daily tasks. The most attractive results can be visualized in 3D on the associated website for this research project (UGent, 2013) (Figs. 27.3 and 27.4).

Fig. 27.3 Digital 3D model of the "Edificio de cinco pisos". Multitemporal 3D models were used to assist the cleaning of vegetation. (*Note*. Source: Screenshot from http://cartogis.ugent.be/edzna/ 2021 ©UGent)

Fig. 27.4 Digital "mesh" for the building Hochna of Edzná. This result was used to restore the main stairs. (*Note*. Source: Screenshot from http://cartogis.ugent.be/edzna/ 2021 ©UGent)

27.6 Geoheritage Science and Technology, Supporting Communities of the Surroundings of a World Heritage Site

While undertaking field research studies in *Edzná,* we met Archaeologist José Huchim, the site manager of the World Heritage *Uxmal* (WHC, 1996). Mr. Huchim was having a series of meetings with the Maya communities of all villages surrounding *Uxmal*, one of the main concerns being that the heritage site was not perceived by the locals as a benefit but rather as a nuisance to their daily lives. According to the 2015 census, over 70% of people in all the municipalities surrounding *Uxmal* live in poverty.

The WHC has been encouraging the participation of the local communities in heritage-related activities. At the 31st WHC session, a fifth "C" was adopted, with the five "Cs" being the following: strengthening Credibility, ensuring effective Conservation, promoting effective Capacity-building measures, increasing Communication and enhancing the role of Communities in the implementation of the WH Convention (WHC, 2007a). In the case of *Uxmal,* the Maya communities are not asking to be involved in the implementation of the Convention but rather to benefit from the large number of visitors who come to *Uxmal*. Such an expectation is extremely valid, and if the heritage site can help improve the poverty situation of the Mayas, then the local communities will appreciate the site and become interested in its associated safeguarding. However, if the heritage site only causes inconvenience for the local community, it is understandable that they would not want to participate in any related activity.

The *Uxmal* local communities have identified offering eco-tourism services as potential activities that might encourage visitors to *Uxmal* to stay longer in the area and enjoy the richness of the habitat of the Maya communities and unique intangible heritage, including traditional food, handicrafts, etc. The area has significant potential for the proper development of tourism services. Modern Mayas live within an outstanding richness of natural heritage and tangible and intangible cultural heritage. Mexico is among the world's largest megadiverse countries, with the Yucatán area being home to the country's largest remaining swath of tropical forests (Varns et al., 2018). The three peninsular states (Yucatán, Campeche, and Quintana Roo) have long recognized that they share one ecosystem in the great Mayan Forest, as well as a common cultural heritage.

Therefore, the Maya communities aim to locally design, implement, manage and operate tourism services. Such locally designed tourism services would then bring necessary financial income to alleviate the poverty of the Maya population. These activities will contribute towards a successful implementation of the UN Sustainable Development Goals: SDG 1 no poverty; SDG 2 no hunger; SDG 3 good health and wellbeing; SDG 5 gender equality; SDG 6 clean water and sanitation; SDG 8 decent work and economic growth; SDG 10 reduce inequalities; SDG 15 life on land (by promoting the protection of nature).

After a full overall assessment, we identified that we could use a series of geo-packages to provide the local Maya population with a simple end-user interface (geo-tools) to collect main points of interest (POI) for the automated elaboration of attractive and accurate eco-tourism cartography. We define POIs as being geo-referenced points related to, e.g., Maya hut where local food can be eaten; view-point for biodiversity; points on a path for bike tours inside the tropical forest; *cenote* for swimming etc. This participatory activity of collecting ground data can then be used to automatically elaborate tourist maps. The resulting eco-tourism maps would be a first step in initiating the promotion of associated eco-tourism services. Based on UGent's long tradition of developing geo-educational packages (Zwartjes, 2016), we made use of complex GIS services that can combine with an easy-to-use front-end smartphone application (app software) and facilitate the Maya communities collecting of geo-referenced POIs. The Mayas can further enrich this information by adding pictures and descriptive text. The collected POIs can also be uploaded to the world wide web through the app, using the internet facilities at the *Uxmal* heritage authority office. The POIs are then automatically downloaded into a complex web Geographical Information System (WebGIS) developed by UGent. Once in UGent's server, the cartography is built through WebGIS tools and artificial intelligence. Such cartography is enriched with the automatic addition of satellite imagery as backgrounds, as well as GIS administrative boundaries from the Mexican Institute for Geography and Statistics (INEGI). The resulting cartography is sent, through the web, back to *Uxmal* headquarters in the form of a printable file (Fig. 27.5).

At this stage, it is important to understand that our research has developed the app for the collection of POIs, but the current pandemic has prevented us from returning to *Uxmal* to work jointly with the local communities.

Our solution of automatically elaborating tourist cartography resolves one major challenge of eco-tourism services. However, the Mayas are now facing the

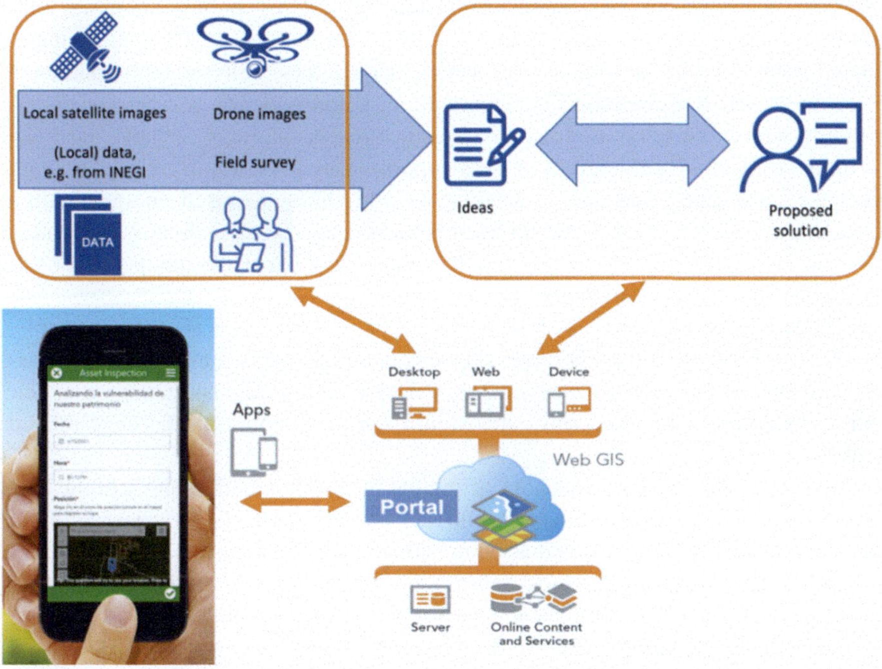

Fig. 27.5 The local Maya population can use the app to capture POIs. (*Note*. Source: Diagram created by Zwartjes L. 2021 ©UGent)

challenge of not being able to promote their eco-tourism services because there is a strong and well-established tourism monopoly, where everything is predefined in advance, including the time that tourists spend in *Uxmal*, souvenir shops where the bus will stop, places where they will have lunch, etc. This makes it impossible for the Mayas to offer their eco-tourist services due to the strong dominance of existing tourism operators. This issue illustrates the complexities of providing support for sustainable development using science and technology. This situation clearly reflects that the involvement of the local population in heritage-associated benefits and activities is not as simple as it is stated in the decisions of the WHC.

27.7 Conclusions

We have described an applied research study with the main objective of assisting and supporting heritage authorities in their daily work and how a second opportunity emerged to assist the local Maya communities using Geoheritage.

We are thankful to INAH Campeche for their enormous support. The main success of our research study is that we were able to put together a multidisciplinary team of scientists with a wide range of expertise, covering satellite remote sensing, cartography, geography, geographical information systems, photogrammetry, education, computer sciences, archaeology, heritage restoration, etc. Clearly, heritage-related issues require assistance from various disciplines, with the main challenge being to establish a common language and, as a consequence, a common understanding within the scientific team. The science and technology we have used can be transferred to a local team of experts supported by a local university. Although we have colleagues that are archaeologists who have acquired an outstanding knowledge of Geoheritage, we recommend that heritage experts, already overloaded by their daily work, remain focused on their domain of expertise and that other sciences should be used to assist them.

There are many advantages of using Geoheritage for heritage sites; for example, the digitalization of cultural heritage sites offers unique opportunities to share heritage information among heritage experts, heritage authorities, and many other scientific disciplines. All this contributes significantly to supporting heritage-related activities.

Working in *Edzná*, questions concerning the inscription methodology of the WH List emerged, as *Edzná* could easily fulfil all the requirements. However, without the WH label, *Edzná* is perceived as being a third-class heritage site. Tourists prefer to visit already overcrowded sites, causing severe damage. A better distribution of tourism crowds would be more beneficial for all the heritage sites. A serial nomination of Maya Heritage sites from El Mundo Maya should be considered.

Advances in digital technologies are continuously improving what we have referred to as Geoheritage science and technology. This is creating new opportunities for the cultural heritage sector, offering innovative non-invasive methodologies for heritage research, presentation, dissemination, education and enjoyment. Digital Geoheritage also makes it easier for other scientific disciplines to participate in heritage research activities, constituting a unique and powerful multidisciplinary scientific platform for supporting heritage.

There are also new challenges that heritage experts alone may not be able to solve, including the tremendous amount of digital data, the complexity of algorithms and computer processes involved, as well as the expertise required to produce accurate 3D heritage models. Therefore, the main challenge is providing heritage experts with easy-to-use end results that do not need to be installed or require particular expertise for dealing with sophisticated software packages. All complex data processing can be completed in "the cloud"[3] using artificial intelligence. With the support of WebGIS services, we have been able to make the final results available to heritage experts. Heritage experts can then visualize the 3D model on the web, observing virtual tours, and, when necessary, selecting different sections at different scales of 3D models. This enables them to have accurate digital replicas of their selected sections in the office. The number of emerging applications is infinite: heritage authorities can virtually undertake a digital restoration to

[3] "The cloud" refers to computer servers, software and databases that are accessed over the Internet. By using cloud computing, users do not have to manage physical servers themselves or run software applications on their own machines.

see the results and decide then if such a methodology should be applied or not to the real heritage site. Heritage authorities can upload their results on private websites to exchange these results with other cultural experts working on similar issues elsewhere, providing a fast and unique exchange of results and expertise. The main success of our research case study is how much the heritage site managers and heritage authorities appreciated the support provided by Geoheritage.

With respect to the WH Convention, UNESCO's Science Sector could have a significant role to play in supporting heritage (Cultural Sector). Unfortunately, UNESCO carries out its programmes through individual sectors, with each sector remaining a "closed silo", struggling to obtain visibility for survival. As an example, *Calakmul* has dual labelling as a UNESCO Biosphere Reserve and a mixed World Heritage site. Issues related to sustainability have to be addressed within the Sciences Sector, while issues related to the Outstanding Universal Value have to be addressed within the Cultural Sector. However, both of these issues cannot be separated in the field.

The Educational Sector also has significant potential to contribute to heritage. If the different sectors were focused on supporting heritage, it would give UNESCO a unique role within the UN system, in contrast to its current widely dispersed roles. Outside UNESCO, in pursuit of the UN SDGs, scientists are setting up multidisciplinary scientific teams as the best and only option to address the complexity of sustainable development (Hernandez, 2017), and the time has come for UNESCO to consider this approach.

References

Belspo. (2003). *Calakmul 4D GIS:* http://www.belspo.be/belspo/fedra/UN/CALAKMUL_en.pdf

Benavidez, C. A. (1997). *Edzná: Ciudad Prehispánica De Campeche*. Instituto Nacional De Antropología e Historia. ISBN-10: 9701800214.

Benavidez, C. A. (2014). *Arquitectura precolombina de Edzná*. Ayuntamiento de Campeche. ISBN: 978-607-96394-0-2.

Carrivick, J. L., Smith, M. W., & Quincey, D. J. (2016). *Structure from motion in geosciences*. Wiley.

Cheng, S.-W., & Lau, M.-K. (2017, November). Department of Computer Science and Engineering, Denoising point cloud for surface reconstruction. *ArXiv*, arXiv:1704.04038.

ESA. (2003). www.esa.int, European Space Agency http://www.esa.int/Applications/Observing_the_Earth/Satellites_to_focus_on_UNESCO_World_Heritage_sites

Hernandez, M. (2017). A digital earth platform for sustainability. *International Journal of Digital Earth, 10*(4), 342–355. https://doi.org/10.1080/17538947.2016.1272646. Taylor and Francis.

Nicolucci, F. (2020, June). From digital archaeology to data-centric archaeological research. *Magazén, 1*(1). Edizione Ca'Foscari.

Pence, H. (2014, December 1). What is big data and why is it important? *Journal of Educational Technology Systems, 43*(2), 159–171. https://doi.org/10.2190/ET.43.2.d. Sage Publications.

Thomas, M. (2021). Visitor flows to World Heritage sites. *Journal of Sustainable Tourism, 29*(10), 1547–1564.

UGent. (2013). *Edzna 3D models*. Ghent University. http://cartogis.ugent.be/edzna/

UN. (2015). *Resolution adopted by the general assembly on 25 September 2015*. United Nations. https://www.un.org/en/development/desa/population/migration/generalassembly/docs/globalcompact/A_RES_70_1_E.pdf

Varns, T., Cortez, R., Hovani, L., & Kingsbury, P. (2018). *Yucatán peninsula, Mexico: A jurisdictional approach to conserving the Maya forest*. The Nature Conservancy.

Vaz, E. (2011, May). A multi-level spatial urban pressure analysis of the Giza pyramid plateau in Egypt. *Journal of Heritage Tourism, 6*(2), 99–108. https://doi.org/10.1080/1743873X

WHC. (1996). *Decision: CONF 201 VIII.C 'Prehispanic Town of Uxmal, (Mexico)*. https://whc.unesco.org/en/decisions/3000

WHC. (2007a). *Decision 31COM/13B Proposal for a "Fifth C" to be added to the Strategic Objectives*. UNESCO World Heritage Centre. https://whc.unesco.org/archive/2007/whc07-31com-13be.pdf

WHC. (2007b). *Decision: 31 COM 7B.127 Pre-Hispanic City of Teotihuacan (Mexico)*. UNESCO World Heritage Centre. https://whc.unesco.org/en/decisions/4922/

WHC. (2008). *List of factors affecting the properties*. UNESCO World Heritage Centre. https://whc.unesco.org/en/factors/

Zwartjes, L. (2016). *Creating a learning line on spatial thinking*. Report from the EU sponsored project: How to teach GI science in education. 2021 GI Learner – GI Pedagogy. https://www.gilearner.ugent.be/wp-content/uploads/2016/05/GI-Learner-SpatialThinkingReview.pdf

Chapter 28
Adopting Digital Tools & Technology to Evolve Sustainable Tourism at World Heritage Sites: Case Studies from India and Greece

George N. Zaimes, Valasia Iakovoglou, Fergus T. Maclaren, and Pankaj Manchanda

Abstract There is a strong interconnection between tourism and World Heritage Sites. This interconnection, on the one hand, can have a positive impact by presenting these sites to the public and helping generate conservation funds, but, on the other hand, if done unsustainably, can lead to their degradation. The adoption of digital technologies in tourism has made travel and visitations, even in remote areas, relatively easy. The adoption of new technologies at World Heritage Sites can also prove to be beneficial and help evolve a more sustainable tourism model at these venues. Furthermore, the new conditions due to the COVID-19 pandemic, while having a detrimental impact on global tourism, provide an opportunity to remotely promote and generate revenue to preserve a regions' tangible and intangible heritage. A technology-based intervention, if adopted correctly, can help to develop sustainable visitation capacity and management at World Heritage Sites while also enhancing and enticing visitation at lesser-known sites. Case studies from Greece and India are presented to demonstrate how to increase visitation to lesser-known sites and enhance the overall tourism experience at these sites. A variety of digital

G. N. Zaimes (✉) · V. Iakovoglou
UNESCO Chair Con-E-Ect, Conservation and Ecotourism of Riparian and Deltaic Ecosystems, International Hellenic University, Thermi, Greece
e-mail: zaimesg@for.ihu.gr; viakovoglou@yahoo.com

F. T. Maclaren
MAC-DUFF Tourism | Heritage | Planning, Macduff, UK
e-mail: tourman@mac-duff.ca

P. Manchanda
Augtraveler, ICOMOS International Cultural Tourism Committee, ICOMOS SDG Working Group (Sustainable Development Group), New Delhi, India
e-mail: pankaj.manchanda@augtraveler.com

M.-T. Albert et al. (eds.), *50 Years World Heritage Convention: Shared Responsibility – Conflict & Reconciliation*, Heritage Studies,
https://doi.org/10.1007/978-3-031-05660-4_28

tools are presented, from simplistic to technologically advanced ones. These digital tools could be adopted and used globally for other World Heritage Sites to enhance visibility and sustainability.

Keywords Ecotourism · Virtual interactive maps · Eco-routes · Augmented reality · Intangible cultural heritage · Geo-location

28.1 UNESCO World Heritage Sites (WHSs)

There are currently 1154 cultural, natural and mixed (cultural and natural) World Heritage Sites (WHSs) located in 167 countries (UNESCO, 2021). Many of these are national symbols, such as the Acropolis in Greece, Taj Mahal in India, Machu Pichu in Peru, and the Pyramids in Egypt. A substantial visitor economy exists alongside WHSs. UNESCO, the responsible governing body within the United Nations (UN), acknowledges that there is an interdependent relationship between WHSs and tourism (UNESCO, 2018, 2019a):

1. WHSs are often major attractions for the tourism sector.
2. Tourism offers World Heritage stakeholders the ability to meet the requirements of the Convention to "present" WHSs to the public and to generate funds for their conservation while also realizing community and economic benefits through sustainable use.

The advent of the COVID-19 pandemic, however, has impacted national tourism sectors worldwide. In WHSs worldwide, visitation has dropped by almost 90% at some of the more popular urban venues. On a positive note, this decrease provides the local tourism ecosystem with an opportunity to *Build Back Better* (OECD, 2020). The opportunity to adopt digital technology-based interventions can help build a sustainable and resilient tourism model in alignment with the UN Sustainable Development Goals (SDGs) (UN, 2021).

"Better," when defined in terms of the SDGs, is when technology models can offer platforms that collect data from all stakeholders of different regions and utilize data analytics to design sustainable policies and governance models. Digital technologies can align with the direction of SDG 11 Sustainable Cities and Communities – *Make cities and human settlements inclusive, safe, resilient and sustainable* and its specific Target 11.4 – *Strengthen efforts to protect and safeguard the world's cultural and natural heritage*.

Curated content and stories of host communities can align with SDG Targets 8.7 – "*Take immediate and effective measures to eradicate forced labour*" and 12.b – "*Develop and implement tools to monitor sustainable development impacts for sustainable tourism which creates jobs, promotes local culture and products*", and help onboard local communities and provide them with an opportunity to highlight and preserve their local tangible and intangible cultural heritage via their unique crafts, art, cuisine, music, dance, etc. (UNWTO, 2021). This value proposition's intention is to help host communities develop sustainable livelihoods in the vicinity of WHSs by *promoting* and *preserving* the tangible and intangible heritage of regions worldwide.

As part of SDG 4 – *Quality Education*, younger generations at schools and universities can be reached out to and onboarded to help better appreciate the world's rich history and culture in an engaging way. Digital technology can also be used to train them as contributors in documenting and curating the content for WHSs. The UNESCO World Heritage Education Programme encourages young people to get actively involved in the protection of cultural and natural heritage. Its goal is to enable future decision makers in the conservation and protection of WHSs from the threats they face.

This study assessed the implication that digital technologies can have on WHSs. Specifically, it briefly presents case studies from Greece and India. These case studies are from simple to more elaborate uses of digital technologies to enhance WHS visibility and/or visitation experiences that could be adopted by managers of other WHSs.

28.2 World Heritage Sites and Technological Advances

Technological advances in transportation allow the visitation of even remote areas in a less expensive and time-consuming way. From 2010 to 2017, the number of international tourists increased by 50% (Dinu, 2018). While this might be economically favourable, many of the locations were not ready for this increase. Some researchers have found a positive relationship between WHSs and tourist numbers that can lead to long-term GDP growth (Farid, 2015). Unfortunately, in most cases, emphasis has not been placed on the negative impacts of mass tourism. Tourism pressures along with increased urbanization continue to negatively impact WHSs. There is often no planning for the infrastructure and other types of development and no appropriate regulation or implementation of regulations that directly impact the cultural values of WHSs (Wu, 2010).

In the WHS "Venice and its Lagoon," after the enforcement of the lockdown conditions due to COVID-19, major changes occurred. A sharp decrease in different pollutants such as plastic marine litter and microplastic leachable contaminants were recorded as the result of the lack of mass tourism to this site (Cecchi, 2021). In another study, pre-COVID mass tourism increased greenhouse gas emissions that required mitigation measures (Cavallaro et al., 2017). Similar problems are occurring in many WHSs, even in remote ones. These new urban and tourism conditions require the implementation of strategies and policies to protect WHSs. Accurate and prompt tracking of the effectiveness of conservation measures should be a priority in WHSs with extensive urban development and where mass tourism is expected to continue; otherwise, the damages that may occur will be irrevocable (Pham et al., 2021).

Technological advances, if utilized properly, can benefit WHSs. Earth observation satellites and detection methods are recent examples of such advances. Through these methods, WHSs can be mapped and monitored more accurately with minimal impact on the actual sites, thus preserving them better. In the European Union, the Copernicus program provides satellite imagery, models and field measurements as free and open data (European Commission, 2018). One of the objectives is to use this data for the better protection of WHSs. GIS technologies (e.g., kernel maps, viewshed analyses) and a radial intensive archaeological survey model were combined to develop a new tool to more diligently assess archaeological surface assemblages (Mingo et al., 2021). This was implemented in the WHS of Rock Art of the Mediterranean Basin on the Iberian Peninsula in Spain. In the Cahokia Mounds WHS in Illinois, USA, Unmanned Air Vehicle photogrammetry provided a very accurate, fast, cost-effective and relatively large-scale surveying tool of archaeological sites with low-lying vegetation (Vilbig et al., 2021).

In modern archaeology, remote sensing tools and geospatial data can help in the preservation of sites and the discovery of new ones. Interferometric synthetic aperture radar (InSAR) methods are being used to assess the potential impacts of geohazards on cultural heritage sites (Elliott et al., 2016). Tapete and Cigna (2017) found that InSAR provides data on many WHSs in Italy, Netherlands, Germany, Spain, Greece and the UK. The existing InSAR geoinformation covers 36% of overall WHSs in Europe. It was also found that InSAR covers a similar percentage of "urban" (40%) and "rural" (34%) WHSs.

Virtual experiences for visitors are another example of the positive use of digital technologies. Avebury is a Neolithic heritage site in the UK. This area is part of the Stonehenge, Avebury, and Associated Sites WHS. Visitors can get a better sense of the place and presence by the virtual simulations of the Avebury area (Falconer et al., 2020). It was also found that such simulations have wide appeal for heritage and museum visitors, regardless of age, gender or familiarity with technology. For the old fortress site of A Famosa in Melaka, Malaysia, a mobile application was developed to assist visitors in the walkthrough of this WHS (Izani et al., 2020). Overall, this application was an effective method to promote cultural heritage because it makes the experience more interesting and engaging. In Indonesia, researchers developed a digital heritage knowledge platform that showcases its national WHSs online (Permatasari et al., 2020). This digital platform provides associated websites and mobile apps for these sites. Overall, new technologies can help in the discovery, protection, conservation, enhanced experience and increased visibility of WHSs.

The potential usage of new digital technologies is described in detail in the next sections. Initially, simple and applied technologies used in Greece are presented, followed by complex technologies in the case of India. Specifically, the Greek case studies showcase the experiences and the role that UNESCO Chairs can have in supporting and enhancing the World Heritage programme. In addition, the Greek case studies highlight how digital technology contributes to tourism diversification in order to meet the SDGs. In India, Augtraveler has implemented a model that links tourism, heritage and sustainable development that has already been recognized for its potential within UNESCO and ICOMOS.

28.3 Case Studies from Greece

Greece is a touristic hotspot, where this sector is a major part of its economy. Before the COVID-19 pandemic, particularly in the summer, certain cities and areas were extremely crowded. The new COVID conditions are leading to a shift from traditional mass tourism. Despite being a relatively small country, Greece has 18 WHSs (16 cultural and 2 mixed) and 14 more sites on its Tentative List (sites that are intended to be considered for WHS nomination) (UNESCO, 2021). Some of the sites (e.g., the Acropolis) are very heavily visited, while others (e.g., the Archaeological Site of Philippi) have substantially fewer visitors. The new COVID conditions indicate that that the number of visitors per site can be better balanced to achieve sustainable management. In addition, many of the visitors in Greece do not visit some of the WHSs because they are unaware of their existence, even though they might be vacationing nearby. Promoting WHS awareness should help diversify tourism in Greece and elongate the touristic period. This diversification of tourism will help meet SDG 8 – *Promote sustained, inclusive and sustainable economic growth, full and productive employment and decent work for all* (United Nations, 2021).

The mission of the UNESCO Chair Con-E-Ect (Conservation and Ecotourism of Riparian and Deltaic Ecosystems) is to promote sustainable tourism in unique ecosystems (Emmanouloudis et al., 2017). A university or a higher education or research institution group partners with UNESCO to establish a chair of practices and knowledge in areas that both the institution and UNESCO focus on. Greece has many such unique ecosystems, including the Gorge of Samaria and Mount Olympus, which are part of the UNESCO Man and the Biosphere Programme. Ecotourism is the best way to promote the conservation and protection of natural ecosystems, as long as it is done in a responsible, environmentally friendly way (Iakovoglou & Zaimes, 2018). Currently, in the WHSs of Greece, there are only two mixed sites (Meteora and Mount Athos), while Mount Olympus is still on the Tentative List. Ecotourism can be used as a tool to promote awareness and education among the general public, which would enhance the protection of sites to meet SDG 13 – *Climate Action*, SDG 14 – *Life Below Water* and SDG 15 – *Life on Land* (United Nations, 2021).

The diversification of tourism can be achieved with the utilization of new technologies. An example is the development of an interactive webmap. Specifically, the UNESCO Chair Con-E-Ect, in collaboration with the Municipality of Avdyra, Greece, developed an interactive map with the most important cultural and natural sites of the area (Fig. 28.1). For example, the Greek philosopher Democritus, who is considered the father of the "atom theory", was born in Avdyra, but this has not

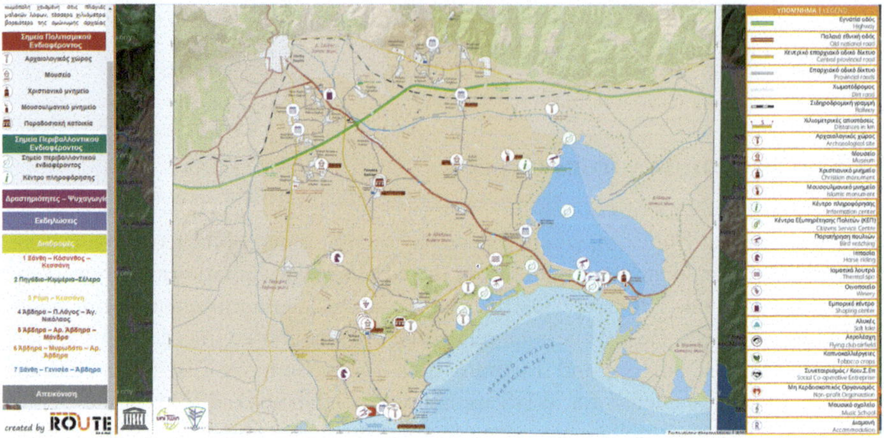

Fig. 28.1 The interactive webmap developed for the Municipality of Avdyra, Greece. (*Note.* Webmap from ROUTE MAPS. G. N. Zaimes 2020. Source: www.routemaps.gr)

been well advertised to visitors to the area. The "interactive webmap" informs visitors about Democritus and the potential sites of interest to visit. In addition, it provides information about the roads, railroads, services and amenities, as well as cultural sites: Christian and Muslim monuments, traditional-maintained houses, restaurants where traditional meals can be tasted or shops where traditional artefacts or products can be purchased. This was the first interactive webmap developed by a municipality in the region. Similar approaches could be adopted for lesser-known WHSs in Greece.

In another project with the Municipality of Drama, eco-routes are being developed, such as the "Water-Riparian route," where tourists can visit the urban springs and riparian areas of the city (Fig. 28.2) (Gkiatas et al., 2021). Drama has the unique riparian-wetland ecosystem of Agia Varvara within its urban area. The other is the "Forest-Urban Route," where people can visit the suburban forest and important cultural areas (e.g., Archaeological Museum, Muslim Mosque, Byzantine walls, Macedonian Tomb) of the city of Drama (Fig. 28.2). The next step is to develop an interactive webmap for the city with the two routes. In addition, an easily downloaded app will be developed that will provide navigation of the route that the visitor chooses. Signs have been installed depicting the "connections points" of the city's history with water and the riparian areas. The final activity will be to develop barcodes for the signs that will allow the visitor to get additional information on these areas. These are relatively inexpensive and easily implemented activities for lesser-developed WHSs that can really promote interest and visitation, thus, reinforcing their sustainable management (Argyropoulou et al., 2011).

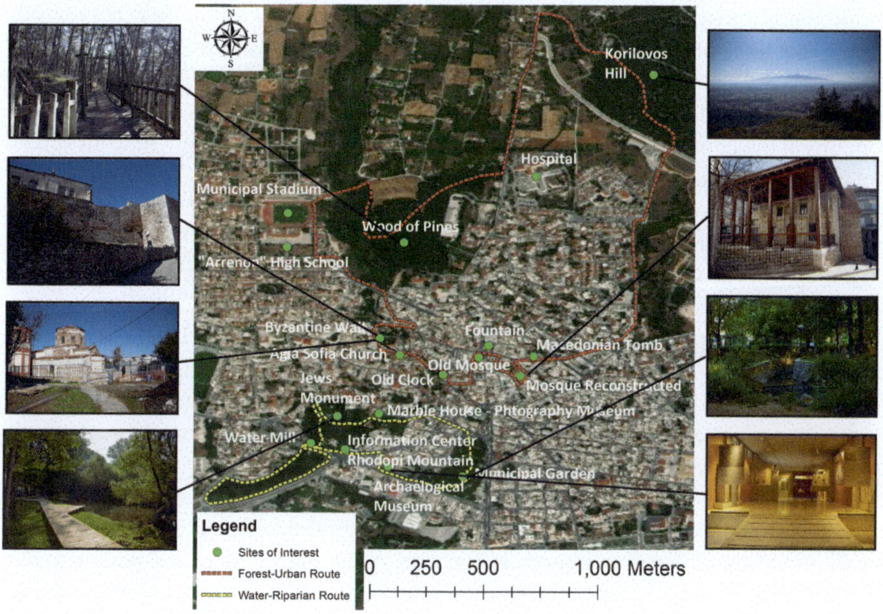

Fig. 28.2 The suggested two eco-routes (Forest-Urban and Water-Riparian) for ecotourism activities. (*Note*. Prepared by I. Kasapidis and G. Gkiatas, 2021)

28.4 Case Studies from India

India was one of the first countries to ratify the World Heritage Convention in 1977. Today, India has 40 nominated WHSs: 32 cultural, 7 natural and 1 mixed and 48 on its Tentative List (UNESCO, 2021). The Indian Government spends millions of rupees in conserving WHSs and maintaining their appeal as destinations. The "Incredible India" tourism marketing campaign, initiated by the Government in 2002, projected the country as an attractive destination by showcasing different aspects of Indian tangible and intangible culture heritage (NITI Aayog, Government of India, 2020).

However, the problems of inadequate visitor interpretation within the WHSs have been arguably below expectations. The limited and qualified guide services are an issue, while hardware-based audio guides in the current scenario of COVID, site visitations might spread infection and, therefore, are not feasible.

The Augtraveler app-based platform was subsequently conceived in 2017 as a concept and adopted a digital technology stack of Augmented Reality (AR), Geo-location, Multimedia and E-commerce, which is implicitly curated in a storytelling narrative of the WHSs and the communities that live around them. These experiences are consumed by users on their mobile phones and, therefore, are accessible to all age groups and user backgrounds in their vernacular languages. It is worth noting that 91% of respondents identified their mobiles as preferred devices in daily use across age groups (Falconer et al., 2020).

The project's design approach and methodology ensure that the host community remains at the centre to promote heritage-based tourism, highlighting how all stakeholders across the value chain can benefit from cultural tourism. The Augtraveler platform helps host communities highlight their local and tribal crafts, cuisines, handlooms, theatre, art, meditation, experiences and homestays, through a value-added approach, which reflects, extends and builds on the heritage and culture of any region, but also helps promote a "viable economic model."

The goal is to promote sustainable tourism at WHSs where the local community connect with the international and domestic tourists directly and, therefore, may receive a greater share of the visitor economy through the development and promotion of experiential and immersive travel. This Augtraveler sustainability approach has been documented as a case study in the ICOMOS SDG Policy guidance document and aligns with UN SDG 8 (Fig. 28.3) (ICOMOS, 2021).

The global pandemic has also seen digital interventions such as Augtraveler take on an important destination interpretation role. As an independent and personalized user-led experience, the app extends its value as a self-contained, COVID-ready innovation that facilitates independent, socially distanced travel.

Another mandate for Augtraveler has been to reach out to students at schools and encourage them to build a deeper appreciation and constituencies of support for these WHSs and their associated cultural heritage. The WHS experiences on the app

Fig. 28.3 The Augtraveler Case Study listed in the ICOMOS SDGs policy guidance document. (*Note.* Augtraveler platform proposition. Pankaj Manchanda 2021. Source: Augtraveler Concept Deck)

are complimented by multi-disciplinary activity books, which are aligned to the grade-level school curriculum and are focused on Science Technology Engineering and Maths (STEM)-based learning. The intent is that the monuments and sites should evolve from being merely Instagram photo opportunities or picnic spots to knowledge dissemination and learning zones. The vision is to develop this module to align with the intents of SDG Target 4.7 – *Ensure all learners acquire knowledge and skills needed to promote sustainable development* (Fig. 28.4) (UNESCO, 2019b).

One of Augtraveler's recent projects was in the "Pink City" of Jaipur, where the firm was invited by its knowledge partner DRONAH (www.dronah.org) to co-create a project to curate the tangible and intangible heritage of the historic neighbourhood of "Chowkri Modi Khana" within the UNESCO WHS-prescribed area of the Old Town.

Jaipur was founded in 1727 and originally envisaged as a trade capital in the state of Rajasthan, where the main avenues of the urban ensemble were designed as markets, which remain characteristic bazaars of the city to date. Due to this extensive tangible and intangible cultural heritage, Jaipur was named as a UNESCO Creative City of Craft and Folk Art in 2015 and as a WHS in 2019.

A mandate of the municipal government is to elaborate a mechanism to preserve and protect its 710-hectare World Heritage-designated urban landscape. A detailed

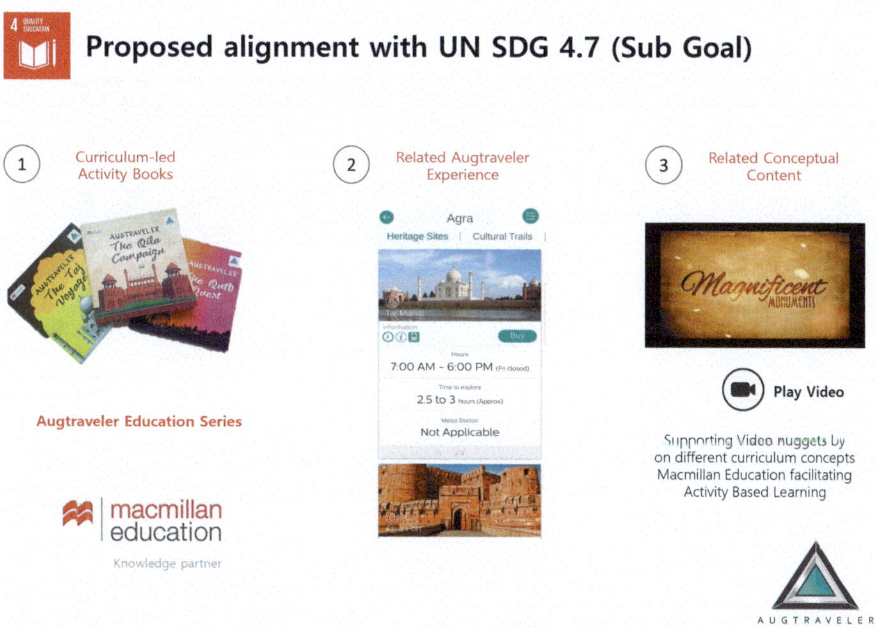

Fig. 28.4 Augtraveler's School Outreach program aims to evolve WHSs as Knowledge Dissemination Zones whilst building on the Framework of Systemic Social and Emotional Learning. (*Note*. Augtraveler Education Series. Pankaj Manchanda 2021. Source: Augtraveler Concept Deck)

inventory of every building within the site's perimeter and its various elements, including the socio-economic relations with the local population, will be prepared to understand the distinctive tangible and intangible cultural heritage features of each street and district.

Through this approach, Augtraveler and DRONAH developed the Chowkri Modi Khana heritage trail to provide visitors with an opportunity to experience the city's many different facets, including the beautiful architecture of the old city (Fig. 28.5), from Rajput/Mughal style to Colonial and Art Deco-themed structures. This trail also extends to AR tours of high-profile venues such as the Amer Palace Fort, Jantar Mantar historic observatory and the iconic Hawa Mahal.

Augtraveler's digital trail allows visitors to directly experience Jaipur's living intangible cultural heritage traditions at a COVID-safe distance through the many havelis (traditional courtyard houses) and temples, along with the work of traditional crafts communities. This includes being informed about the living practice of brass utensil making.

Through a multimedia approach combined with AR, the Augtraveler platform offers historical narratives and production techniques of local arts and crafts. Additionally, in line with the vision to promote sustainable livelihoods of host communities, the platform makes them discoverable on trail maps by geo-tagging local community businesses in the historic city that craft and sell traditional brassware authentic in Jaipur. The methodology provides the visitor with an opportunity to

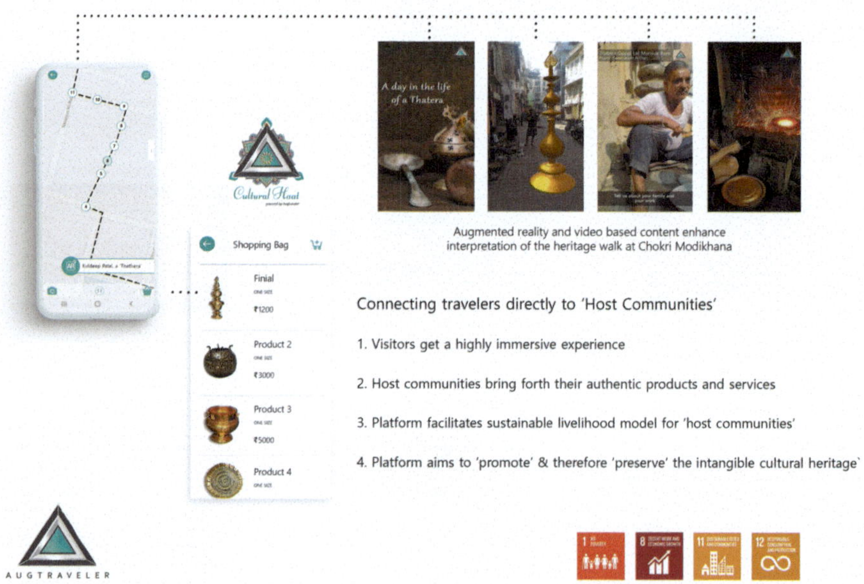

Fig. 28.5 Chowkri Modi Khana Walk heritage trail segment and example of traditional brassware with an online community marketplace for local products and services. (*Note*. Augtraveler Culture Haat – Promoting and Preserving Livelihoods and Cultural Heritage Pankaj Manchanda 2021. Source: Augtraveler Concept Deck)

explore community-based cultural tourism opportunities while developing a resilient livelihood model for host communities through a unique intervention (UNWTO, 2021). The Chowkri Modi Khana walk on Augtraveler has been listed in the UNESCO Creative Cities report as a use case of creative economies that can benefit from new-age digital technology (UNESCO, 2019b).

The Augtraveler methodology and its platform, including its use of AR/VR reality interfaces accessible on mobile phone devices, are well recognized as a first-in-class comprehensive digital interpretive, educational and community interaction platform. The platform has also received several commendations, including being named the Best COVID-ready Innovation at the India Responsible Tourism Awards 2021, a Finalist at the World Tourism Forum Lucerne's Indian Startup Innovation Camp 2019 and a top innovative Startup in Heritage Travel by the Ministry of Tourism and Invest India in 2017.

28.5 Conclusions

Technological advances and traditional tourism practices, in some cases, may have detrimental impacts on WHSs. However, the adoption and implementation of new digital technologies can also provide an opportunity to establish sustainable tourism and enhance public awareness while educating decision makers and visitors about their importance and wide-ranging benefits. Technology can also onboard host communities, which are often marginalized in traditional tourism circuits, and their active participation can eventually promote the conservation, protection and sustainable management of WHSs. By applying such approaches, as exemplified by the case studies of Greece and India, digital platforms can serve as vital instruments to instil resiliency and also opportunities to pivot when dealing with unforeseen calamities like the COVID-19 global pandemic.

References

Argyropoulou, A., Dionyssopoulou, P., & Miaoulis, G. (2011). Tourist destination marketing and management using advanced ICTs technologies. *Advances on Information Processing and Management, 1*, 365–369.

Cavallaro, F., Irranca Galati, O., & Nocera, S. (2017). Policy strategies for the mitigation of GHG emissions caused by the mass-tourism mobility in coastal areas. *Transportation Research Procedia, 27*, 317–324.

Cecchi, T. (2021). Analysis of volatiles organic compounds in Venice lagoon water reveals COVID 19 lockdown impact on microplastics and mass tourism related pollutants. *Science of the Total Environment, 783*, 146951.

Dinu, A. (2018). The importance of transportation to tourism development. *Academic Journal of Economic Studies, 4*, 183–187.

Elliott, J. R., Walters, R. J., & Wright, T. J. (2016). The role of space-based observation in understanding and responding to active tectonics and earthquakes. *Nature Communications, 7*, 13844. https://www.nature.com/articles/ncomms13844

Emmanouloudis, D., Zaimes, G. N., & Iakovoglou, V. (2017). UNESCO Chair Con-e-ect: Promoting conservation in riparian and deltaic ecosystems. *Procedia Environmental Science, Engineering and Management, 4*(4), 235–243.

European Commission. (2018). *Copernicus services in support to cultural heritage*. European Commission.

Falconer, L., Burden, D., Cleal, R., Hoyte, R., Phelps, P., Slawson, N., Snashall, N., & Kate Welham, K. (2020). Virtual Avebury: Exploring sense of place in a virtual archaeology simulation. *Virtual Archaeology Review, 11*, 50–62.

Farid, S. (2015). Tourism in world heritage sites and its impact on economic development: Some African countries case studies. In *Proceedings of the II international conference on best practices in World Heritage* (pp. 721–737). ISBN: 978-84-606-9264-5.

Gkiatas, G., Kasapidis, I., Koutalakis, P., Iakovoglou, V., Savvopoulou, V., Germantzidis, I., & Zaimes, G. N. (2021). Enhancing urban and sub-urban riparian areas through ecosystem services and ecotourism activities. *Water Science and Technology: Water Supply, 21*, 2974–2988.

Iakovoglou, V., & Zaimes, G. N. (2018). Enhancing rural areas while safeguarding ecosystems through sustainable practice of ecosystem based approaches (EBA) with emphasis on ecotourism. *International Journal of Bio-Resource and Stress Management, 9*(1), 129–131.

ICOMOS. (2021). *ICOMOS releases Sustainable Development Goals policy guidance for heritage and development actors*. International Council of Monuments and Sites. https://www.icomos.org/en/focus/un-sustainable-development-goals/91455-icomos-releases-sustainable-development-goals-policy-guidance-for-heritage-and-development-actors

Izani, M., Samad, A., & Razak, A. (2020). Augmented reality application based navigating the Famosa fortress site. Augmented reality application based navigating the a Famosa fortress site. In *Proceedings of the third international conference on intelligent sustainable systems [ICISS 2020]* (pp. 285–290).

Mingo, A., Ramos, F., Izquierdo, L., Francesa, V., Armengol, A., & Lozano-Lopez, N. (2021). The cultural ascription and evaluation of archaeological surface assemblages: Development and application of the Archaeological Surveying Consistency Index (ASCI) in the radial archaeological survey of Minateda rock art site (SE Spain). *Journal of Archaeological Science: Reports, 36*, 102843.

NITI Aayog Government of India. (2020). *Working Group Report on improving heritage management of India*. NITI Aayog. http://www.niti.gov.in/sites/default/files/2020-06/Improving-HeritageManagement-in-India.pdf

OECD. (2020). *Building Back Better: A sustainable, resilient recovery after COVID-19. OECD policy responses to coronavirus (COVID-19)*. OECD. https://www.oecd.org/coronavirus/policy-responses/building-back-better-a-sustainable-resilient-recovery-after-covid-19-52b869f5/#contactinfo-d7e2074

Permatasari, P. A., Qohar, A. A., & Rachman, A. F. (2020). From web 1.0 to web 4.0: The digital heritage platforms for UNESCO's heritage properties in Indonesia. *Virtual Archaeology Review, 11*, 75–93.

Pham, V.-M., Nghiem, S. V., Pham, C. V., Thi Luu, M. P., & Bui, Q.-T. (2021). Urbanization impact on landscape patterns in cultural heritage preservation sites: A case study of the complex of Huế Monuments, Vietnam. *Landscape Ecology, 36*, 1235–1260.

Tapete, D., & Cigna, F. (2017). InSAR data for geohazard assessment in UNESCO World Heritage sites: State of-the-art and perspectives in the Copernicus era. *International Journal of Applied Earth Observation and Geoinformation, 63*, 24–32.

UN. (2021). *Sustainable development – The 17 goals*. United Nations. https://sdgs.un.org/goals

UNESCO. (2018). Tourism management at UNESCO World Heritage Sites.

UNESCO. (2019a). *Operational guidelines for the implementation of the World Heritage Convention*. UNESCO.

UNESCO. (2019b). *UNESCO Creative City of Crafts & Folk Art Network Monitoring Report 2019*. UNESCO. https://en.unesco.org/creative-cities/sites/default/files/jaipur_uccn_monitoring_report_2019.pdf

UNESCO. (2021). *World Heritage List*. UNESCO. https://whc.unesco.org/en/list/

UNWTO. (2021). *Inclusive Recovery Guide. Sociocultural impacts of COVID-19 issue 2: Cultural tourism*. UNWTO. https://www.e-unwto.org/doi/epdf/10.18111/9789284422579

Vilbig, J. M., Sagan, V., & Bodine, C. (2021). Archaeological surveying with airborne LiDAR and UAV photogrammetry: A comparative analysis at Cahokia Mounds. *Journal of Archaeological Science, 33*, 102509.

Wu, J. (2010). Landscape of culture and culture of landscape: Does landscape ecology need culture? *Landscape Ecology, 25*, 1147–1150.

Chapter 29
The Commodification of World Heritage: A Marxist Introduction

Thomas M. Schmitt

Abstract The commodification of World Heritage potentially takes place in various contexts or "markets", such as tourism markets, media markets, but also in the sessions of the World Heritage Committee as an inscription market. Loosely following Marxian categories, but based on a broader range of scholars, for example, from philosophical anthropology, several problem areas in the commodification of World Heritage can be distinguished: first, exploitation (e.g. of a World Heritage title, heritage values or of the environment of a site), second, alienation (of residents and visitors towards a site, or between residents of a site and its visitors) and, third, a possible "fetishism" around the title. The article offers a systematic conceptual approach for the analysis of commodification phenomena related to heritage and especially the World Heritage system.

Keywords World Heritage · Commodification · Critical theory · Alienation · Heritage tourism · Heritage studies

29.1 Introduction

The World Heritage List is devoted to the preservation of outstanding cultural artefacts, natural features and wildlife and also to the mutual understanding of humanity. According to these idealistic ascriptions, there should not be any place for the commodification of World Heritage. However, this might be perceived as a problem, for example, in the context of tourism at World Heritage sites, but also regarding negotiations within the World Heritage Committee which are not only driven by scientifically based or universalist ethical arguments. The aim of this paper is (1) to offer fundamental concepts for an understanding of commodification processes, (2) to discuss the appropriateness of their application in World Heritage and (3) to discuss possible solution approaches and their limits. The systematic presentation of this nexus of heritage and commodification is preceded by a brief literature review

T. M. Schmitt (✉)
Heidelberg Center for Cultural Heritage (HCCH), University of Heidelberg, Heidelberg, Germany
e-mail: Thomas.Schmitt@hcch.uni-heidelberg.de

M.-T. Albert et al. (eds.), *50 Years World Heritage Convention: Shared Responsibility – Conflict & Reconciliation*, Heritage Studies,
https://doi.org/10.1007/978-3-031-05660-4_29

on this topic. For reasons of logical coherence, the broader spectrum of heritage studies is considered here. A first hypothesis, which must always be empirically tested, is that corresponding problems are potentially aggravated at World Heritage sites.

Within heritage studies, phenomena that can be associated with the term commodification are discussed in the context of tourism valorisation and branding, e.g. for sites, in particular. The majority of the relevant literature does not use the term commodification but related terms, such as commercialisation, valorisation, marketing/marketisation or branding, which are not necessarily identical. Since the 1970s at the latest, the change of cultural traditions – interpretable as intangible heritage – and of local settings, such as around historical monuments, occurring as a result of adaptations to the interests of the tourism industry and the supposed needs of tourists, have been critically discussed (cf. MacCannell, 1973; UNESCO, 1975; Tangi, 1977; Vorlaufer, 1999). In 1977, Tangi distinguished three major areas that may be affected by tourism, namely the "natural environment", the "man-made environment" (i.e. the built environment or cultural landscapes, including the preservation of historic monuments and sites) and the "socio-cultural environment", including the commercialisation and banalisation of socio-cultural traditions.

This simple systematisation is still helpful today in order to structure the now unmanageably extensive literature on the nexus of heritage and tourism from a factual point of view. The Advisory Bodies to the World Heritage Convention (UNESCO, 1972) have repeatedly analysed problems of tourism at World Heritage sites and tried to address standards for sustainable, environmentally and socially compatible tourism (ICOMOS, 1999; IUCN, 2011).

As traced by Dicks (2003), the formation of British heritage studies in the 1980s can also be understood as a reaction to the increasing marketing of historic sites and local traditions by the "heritage industry", whose products were seen as "fantasies of a world that never was" (Hewison, 1987, 10). The widespread "marketisation" of heritage in the U.K. since the Thatcher era has been interpreted as a neoliberal strategy of restructuring the national economy; the educational mission of heritage institutions is being undermined in new kinds of "heritage centres" in favour of a market-like representation of the past (Lumley, 1988; Walsh, 1992; Dicks, 2003). In her influential essay *Theorizing Heritage*, the U.S. anthropologist B. Kirshenblatt-Gimblett (1995, 369) even called heritage as "a 'value added' industry": "Heritage produces", according to Kirshenblatt-Gimblett, "the local for export". Undoubtedly, this essay had a stimulating effect within the emerging field of heritage studies. However, the author of this chapter explicitly does not share the equation of heritagisation and economic valorisation, i.e. commodification, suggested by the essay's formulations.

In numerous publications, the World Heritage List and comparable institutions are interpreted as brands, for example, in applied publications with a tendency towards affirmative and other publications with neutral to critical connotations (Hall & Piggin, 2003; Quack & Wachowiak, 2013); a tourist (mis)understanding of the World Heritage List thus prevails here. In a critical reflection, the question of how this brand understanding of the World Heritage List affects our perception of World Heritage sites arises, as discussed below in Sect. 29.4.

There is a vast body of literature on the nexus of heritage, on the one hand, and tourism, marketisation, commercialisation and branding, on the other (Bendix, 2018), the review of which exceeds the scope of this short contribution. The term *commodification,* however, is rarely explicitly used in heritage studies. A noticeable part of the contributions that use this term – by no means all of them; Bui and Lee (2015) – explicitly draws on theoretical concepts from the Marxist theoretical tradition or Critical Theory (Walsh, 1992; Henning, 2006; Smith, 2007; Baillie et al., 2010; Aggenbach, 2017; Su, 2015). Between these contributions and the present introduction, which aims to develop the topic of the commodification of (World) heritage systematically, by drawing on these theoretical traditions, there are thus recognisable content-related affinities.[1]

29.2 Relations of Commodification and Heritagisation: A Conceptual Framework

The word *commodification* is derived from Latin and contains the noun *commoditas* (commodity) and the verb *facere* (to make), thus expressing the concept that something is made into a commodity. Commodification as a social phenomenon has accompanied civilisation for several thousand years. In a more specific sense, the term commodification is used when objects that were previously not treated as goods, or only to a small extent, are now (also) traded as goods according to market principles. A core idea of critical observation of commodification processes is that the objects – e.g. World Heritage sites in our case – change significantly in their social perception, functioning and associated social practices due to commodification.

Several "markets" can be distinguished, in which World Heritage or at least specific goods relating to World Heritage are negotiated. These include the following:

- *Tourism markets*: Cities or regions compete for tourists and the money they spend. The World Heritage title is used in marketing a city or region as a destination or in advertising a tourism product associated with the site.
- *Location markets*: Cities or regions compete for the attention of investors, skilled labour or state subsidies, where a World Heritage title can be used for location branding.
- *Media markets*: Sites on the UNESCO World Heritage List or the list as a whole are the focus of numerous media, such as books, films or photo calendars.

[1] The author of the contribution would like to clarify that he does not see himself as a Marxist; in particular, he does not share ontological positions of Marxism. However, for the discussion of commodification, as for numerous social phenomena, Marxism and especially Critical Theory offer substantial starting points.

- *Markets for movable goods*: Movable cultural goods, for example, wild animals, ivory or fossils, are extracted (as a rule illicitly) from World Heritage sites and traded.
- *The inscription market*: The World Heritage Committee is the central body that decides on the inscription or non-inscription of sites on the World Heritage List, and, from a certain perspective, the sessions of the Committee might be conceptualised as inscription markets, as discussed below in Sect. 29.3.3.

The concept of commodification is theorised in more depth in the next section. What has been said so far is sufficient to develop a formal analytical framework that explicates possible relationships between heritagisation and commodification (See Fig. 29.1). Heritagisation is understood here as the signifying practice whereby social institutions (such as the World Heritage Committee) or collectives recognise an object or phenomenon as "heritage", whether in a formal or informal way. Such an object could be, for instance, a single building like Notre Dame Cathedral or a cultural landscape or a nature reserve like the Serengeti. From a formal point of view, heritagisation and commodification can be understood as comparable processes in which additional meanings are attributed to objects, and additional practices are assigned to them.

Both are analytically separable phenomena that can run parallel but do not necessarily have to do so (cf. the discussion above on Kirshenblatt-Gimblett 1995). The protection of historical monuments or natural areas, for example, is not automatically economically motivated, but can be – this would be the ideal systemic case – due to "intrinsic" motivations of monument and nature conservation or other "extrinsic" but not economic motives, such as nationalist objectives of stabilising identities and power by recourse to a selected past.

With the official designation of such an object as heritage, defined significances or heritage values are attributed to it; in the case of World Heritage, this is done

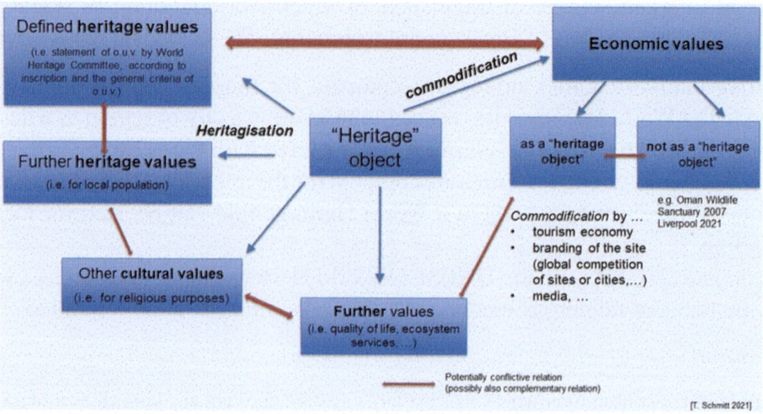

Fig. 29.1 Heritagisation and commodification – a conceptual framework. (*Note.* [Chart] prepared by the author, T. Schmitt 2021)

through the statement of outstanding universal value (OUV). In addition, this heritage object may be associated with other heritage values or significances, such as personal or collective memory values. These two types of heritage values/significances are not coherent in all cases and at least occasionally conflictive. This also applies to other cultural significances of the property. A listed church or synagogue is, for example, often perceived not only as a monument but also, in its original primary functions, as a place of prayer. The heritagisation of the monument may often be a *conditio sine qua non* for its long-term preservation, but, at the same time, heritagisation changes its primary perception and the way it is dealt with. Heritage objects can also have other values ascribed to them, such as protected rainforests which have the function of CO_2 storage; in economised terms: forests provide ecosystem services.

Heritage objects can also be *commodified* in the sense that they can be assigned *economic values*. This can be done (1) by abstracting or negating their heritage significances. The most striking example within the history of the World Heritage Convention is the Oman Wildlife Sanctuary, which the Omani government dedicated to oil production in the 2000s rather than maintaining it as a nature reserve and World Heritage site. On the other hand (2), it is precisely the (World) Heritage title that can be economically valorised, for example, for the tourism economy or location branding, and here we are dealing with the commodification of (World) Heritage in the narrower sense.

It remains to be said that all these different value attributions can stand in a potentially conflicting relationship in this framework, but they could potentially also complement each other. Which relationships – conflicting or complementary – are realised must be analysed separately for each individual case.

29.3 Theoretical Aspects of Commodification and Their Transfer to (World) Heritage

In this section, the concept of commodification and the attempt to transfer it to World Heritage will be elaborated in more detail. It is hard to talk about commodification meaningfully without looking at Karl Marx's philosophy and political economy, which have significantly influenced subsequent thinking on this subject (see Ibe & Lohmann, 2005; Watts, 2009), including non-Marxist thinkers like Karl Polanyi. Three key terms used by Marx are to be discussed in our context: exploitation, alienation and commodity fetishism. They are first briefly introduced in the following in a Marxian sense and discussed with reference also to non-Marxist thinkers from the social sciences and humanities. The sub-sections each conclude with a discussion of the extent to which these concepts can be usefully transferred to the field of heritage and specifically to the implementation of the UNESCO World Heritage Convention.

29.3.1 Exploitation

Exploitation can be understood in a general sense as a condition in which, for the benefit of one person (or institution, collective, organisation), other persons (or institutions, collectives, organisations, animals) are unfairly harmed (Zwolinski & Wertheimer, 2017). In Marx's analysis of capitalism, the capitalist exploits the worker by paying him a wage that allows him to live only at a subsistence minimum, while the capitalist siphons off the surplus value of labour for himself. Labour itself becomes a commodity, which is traded on labour markets where the capitalist is undoubtedly de facto in the dominant position; exploitation is not primarily the result of individual malice but is a structural phenomenon. The concept of exploitation was later transferred to other areas, such as the exploitation of women by men or the exploitation of nature (Zwolinski & Wertheimer, 2017). In this respect, it seems legitimate to transfer the concept of exploitation tentatively to the fields of heritage and World Heritage. The common denominator of different understandings of exploitation is the idea of a parasitic relationship or a harmful, instrumental utilisation of a person, an animal, a resource or a system to one's own advantage (Zwolinski & Wertheimer, 2017).

In such a sense, one possible form of exploitation of a heritage site or the World Heritage idea would be persons, companies, organisations or governmental bodies trying to make financial, social or symbolic profit from the title or the prominence of the site without paying attention to its adequate protection or other legitimate interests, such as those of the local population. This could be done by (1) not providing sufficient financial, human and material resources to protect the site or (2) deliberately allowing, seeking or encouraging overuse of the site, for example, through tourism, at the risk of damaging its material substance, socio-culture or environment. (3) On the global level, the World Heritage system could be exploited for national prestige or personal careers (diplomats, perhaps also scholars researching on heritage). The "profit" of the social actors at the expense of World Heritage would be financial income (especially tourism-generated) or, e.g. in the increase of personal, regional or national symbolic capital.

In the introduction, the thesis was put forward that problems of the commodification of heritage are particularly evident at World Heritage sites. With surveillance by the international community and specific instruments such as the List of World Heritage in Danger, the World Heritage system has, on the other hand, a strong potential to respond appropriately to such local undesirable developments. It would be the task of a site-specific assessment to judge to what extent World Heritage is affected by local forms of exploitation. However, the instrument of the List of World Heritage in Danger is not applied consistently, as national delegations often pull out all the stops to prevent an entry on the Danger List (Schmitt, 2009, 117–118).

29.3.2 Alienation

The second essential phenomenon of commodification for Marx is the socio-cultural phenomenon of the alienation of the worker. This initially manifests itself as the alienation of the worker from the object, the product of his labour and the labour process. This forces the worker not to perform the work in a self-determined way but only by fulfilling given norms. According to Marx's analysis, this first alienation from the product of labour is immediately followed by (1) alienation of the worker from nature (as the material basis of commodity production), (2) alienation from oneself, (3) alienation from the human species and (4) alienation from concrete fellow human beings, who are only regarded as a means of securing life and satisfying needs (Marx 1844/2018a, 183). Marx borrowed the concept of alienation from G.F. W. Hegel and applied it to the realm of economics; previously, Jean-Jaques Rousseau had asserted a self-alienation of human beings through the artificiality of culture (Barth, 1959, 21). After Marx, the concept was taken up in philosophical anthropology and within the Christian theologies (Schrey, 1975; Zima, 2014). The core element of all meaningful concepts of alienation is a separation between a subject and an object, whereby this object can also extend to the entire environment or refer to itself. This separation does not have to be subjectively conscious but can also be stated from the outside (Leopold, 2018; Zima, 2014, 3). Alienation is usually assessed as extremely disadvantageous. Many authors no longer (exclusively) attribute alienation to capitalism but interpret it, for example, as an effect of rationalisation and modernisation processes, postmodern constellations and civilisational mechanisms in general (Schrey, 1975; Zima, 2014).

To what extent can the concept of alienation be meaningfully applied to the field of heritage/World Heritage? Let us start by looking at tourist visitors as well as residents at a World Heritage site. The latter, in its cultural meanings or its "natural" aspects, can certainly represent something "foreign" for visitors, perhaps also for residents of the surrounding area, and this experience of foreignness should not necessarily be equated with a negatively evaluated alienation according to the above explanations.[2] This sense of foreignness is potentially productive; it also prevents a hasty nostrification, for example, of the remnants of a past cultural epoch or even of a "wild" nature.

While newly awarded World Heritage titles meet with an extraordinarily positive response in many countries, at least in the published opinion of regional media, cases have been documented in which an award of a title meets with indifference or even rejection (Schmitt, 2011, 306). Representatives of indigenous groups, in particular, describe the World Heritage designation as a form of expropriation of their own cultural traditions by national stakeholders and the international community (Disko & Tugendthat, 2013, 16); this experience can be understood as a form of alienation from their own cultural resources by the World Heritage system. In these contexts, it is not commodification that creates a potential alienation situation, but

[2] See Landmann (1975) on the general relation between "the foreign" and "alienation".

the official heritagisation or the confrontation of locally and globally different cultural patterns.

It is now conceivable that the commodification of World Heritage sites could trigger even more serious alienation processes. Such alienation can be loosely coupled with exploitative economic structures, for example, of employees in the tourism economy, but it is analytically separable from them as a socio-cultural and existential-psychological problem and potentially affects a larger group of people than, for example, precariously employed people. Alienation in this sense can concern (1) the relation of local residents to the site and (2) of external visitors/tourists to the site but also (3) between and among residents and visitors and (4) self-alienation effects. Alienation in this context can mean that people are not able to "access" a site – in a phenomenological sense rather than in the sense of physical accessibility; they do not understand its cultural and natural features and significances (anymore). Alienation can further affect the relationship between visitors and local people or between different members of the local population, as potentially all social relations at the site are subordinated to the dictate of its commodification. This reflects Marx's spectrum of meaning, in which alienation can refer to things, nature, other people and oneself.

Black and white images should be avoided in such analyses: A certain degree of commodification is often a positive prerequisite for generating income that enables regional development and the adequate protection of a heritage site. Moreover, not every market interaction necessarily poisons the social relations between the participants. Many readers will probably be able to cite experiences of positive human exchange associated with market interactions from their travels to (World) Heritage sites. It seems to be an elementary prerequisite that tourist visitors do not ignore the economic disparity that usually exists between them and a large part of the employees, e.g. in the service sector; it is part of socially good relations that this disparity is alleviated. Conversely, it is known from experimental economics that simulated market situations, such as those emulated by stock exchanges, lead to an erosion of ethical behaviour, which can be understood as a form of self-alienation (Falk & Szech, 2013). Thus, if the economy at a World Heritage site and therefore the social life is largely based on its commercialisation, this will have potentially serious negative social and psychological effects. In principle, corresponding problems are now recognised in the World Heritage system: "World Heritage conservation and management strategies that incorporate a sustainable development perspective embrace not only the protection of the OUV, but also the wellbeing of present and future generations" (UNESCO, 2015, 2). This gives rise to a responsibility of the World Heritage Committee that goes beyond conservation issues.

29.3.3 Commodity Fetishism

In addition to exploitation and alienation, another Marxian concept relating to commodification became famous in the critical social sciences, namely the *fetish character of the commodity* (Marx, 1867/2018a). This occurs when a product no longer appears to people as the result of human labour but as a thing whose properties are presented as external and natural. The *exchange value*, the price, detaches itself from the *use value*. Contemporary readers can exemplify this by looking at branding: a brand is symbolically charged and filled with emotion, detached from the concrete usefulness of the objects. Marx (1867/2018b, 332, 337) makes a comparison here with religious categories: a commodity is only "at first sight a trivial thing", but "full of metaphysical subtleties and theological capers" surrounded by "magic and phantoms". Referring to these arguments, Michael Watts (2009, 99) states, "It is as if our entire cosmos, the way we experience and understand our realities and lived existence in the world, is mediated through the base realities of sale and purchase. Virtually *everything* in modem society *is* a commodity". Thus, one may assume that (partially) unconscious hegemonic patterns of understanding at least partially influence our perception of World Heritage. The inscription, the World Heritage designation, would then be the equivalent of the fetish around which all activities and efforts revolve, be it acquiring the title (through appropriate nomination dossiers), maintaining it (through preservation measures) or communicating about World Heritage. The heritage values that art historians or, for example, nature conservationists appreciated about a site, and whose reception had initially drawn attention to it, fade then into the background – analogously to the *use value* in relation to the *exchange value* in Marx's theory. Whereas Marx developed the idea of fetishism from a consideration of cultural phenomena and transferred it to the economy, the preceding considerations again transfer it back to the realm of culture.

The previous remarks had shown the theoretical possibility for a "commodity fetishism" around the World Heritage title. In an ethnographic study of the World Heritage Committee in the 2000s, the author described "[r]eciprocal expectations and claims of national states" (Schmitt, 2009, p. 117; Schmitt, 2011). Meskell (2015, 3) speaks for the 2010s of, marked as a euphemism, "gifts and exchanges on a global stage". The World Heritage Committee appears here as a

> global marketplace where the inscription of heritage properties is prized more for its capillary transaction potentials than its conservation values. World Heritage Committee debates (…) are becoming largely irrelevant in substance, yet highly valued in state-to-state negotiations and exchanges of social capital. (Meskell, 2015, 3)

This and similar assessments,[3] insofar as they are accurate, can be interpreted as an expression of a partial "commodity fetishism" within the World Heritage system. In the 2000s, the author had recognised corresponding tendencies but had seen them limited by the desire for the hegemony of – however justified – scientific positions

[3] See Brumann, 2011, and Brumann & Meskell, 2015 for further readings.

in Committee decisions (Schmitt, 2009, 117). Some delegations, such as the Algerian delegation, deliberately evaded the expectation to increase the number of their own World Heritage sites for reasons of national prestige and decided to ensure better protection of existing sites before increasing the quantity (Schmitt, 2011, 228). If one leaves the scientific observer's perspective and asks for practical solutions, the attitude of the Algerian delegation at that time reveals probably the most difficult *remedy* for such commodity fetishism: self-restraint. The global public cannot rely on this alone: Academia, media, NGOs, advisory bodies and the UNESCO administration have the task of critically reflecting the work of the Committee in this regard.

29.4 Summary

The article attempted to present a systematic outline of the nexus of commodification and (World) Heritage, as far as this is possible in the limited scope of a book chapter (see Table 29.1). Table 29.1 also takes into account "markets" such as the tourism market and media markets (or the media presentation of World Heritage), which were dealt with in greater detail, based on approaches of Critical Theory, in earlier versions of this paper.[4]

This contribution took a conceptual starting point in central ideas of Marxian thinking on commodities, namely exploitation, alienation and commodity fetishism, which were also taken up outside the Marxist tradition, for example, in philosophical anthropology, and applied to social and cultural phenomena beyond economic production. In this respect, it seems permissible and promising to use

Table 29.1 Possible aspects of commodification in heritage-related markets – a heuristic assessment

Heritage related markets Aspects of commodification	**Inscription** market (the World Heritage Committee)	**Tourism** markets	**Media** markets	Markets for (illegal) **movable goods**
Socio-economic *exploitation inequalities*	Possibly favouring or tolerating	XX	(X)	X
Socio-cultural *alienation misunderstanding commodity fetishism*	Possibly favouring or tolerating XX	XX	X	X
Damaging of **physical-material** features of sites	Possibly favouring or tolerating	X		XX
Damaging of **environmental** features	Possibly favouring or tolerating	XX		X

Note. (X), X, XX: minor … major estimated relevance. [Table] prepared by the author, T. Schmitt, 2021

[4]This book chapter is based on a presentation held at the online conference "50 Years World Heritage Convention" of the Institute Heritage Studies (June 2021).

these concepts tentatively as a lens for analysing the social embeddedness of the World Heritage system. The aspects of commodification were applied to different "markets", the "inscription market" of the World Heritage Committee on the global scale, tourism markets on the local/regional scale, as well as – drawing on approaches of Critical Theory – the question of representation of World Heritage in media. The problem of illegal trade in artefacts, animals and plants with a connection to World Heritage sites had to be left out of this article for reasons of space. The concepts used are suitable for naming and classifying the consequences of existing practices related to World Heritage – be it in the local tourism sector or, for example, in the global decision-making arenas.

Marxian concepts and partly also those of Critical Theory make a claim to totality, which is not adopted by the author. For the author, for example – until empirically proven to the contrary – the World Heritage Committee is not per se a pure bazaar for titles and national prestige but also the possible place of reasonable or engaged debates; in what mix this happens is a question for empirical research. Furthermore, tourism at World Heritage sites does not automatically lead to the exploitation of people and the environment and cultural content of the site but potentially to positive encounters in the sense of "sharing heritage", income for the local population and cultural understanding. The concepts used here are thus not intended to provide an inappropriately one-sided explanation of the empirical world; rather, these conceptual lenses can sensitise us to relevant problem areas. At the same time, it became clear that there are no one-dimensional solutions for the problem areas under consideration, but that possible solution strategies have to apply different levers, usually with a combination of structural changes, changed awareness and individual practices.

References

Aggenbach, A. (2017). *The effects of commodification on cultural significance: Two African fortifications*. [MA-dissertation, University of Cape Town]. Retrieved from https://open.uct.ac.za/handle/11427/24986

Baillie, B., Chatzoglou, A., & Taha, S. (2010). Packaging the past. The commodification of heritage. *Heritage Management, 3*(1), 51–71.

Barth, H. (1959). Über die Idee der Selbstentfremdung des Menschen bei Rousseau. *Zeitschrift für Philosophische Forschung, 13*(1), 16–35.

Bendix, R. (2018). *Culture and value: Tourism, heritage, and property*. Indiana University Press.

Brumann, C. (2011). Unser aller Kulturgut. *Sociologus, 61*(1), 19–43.

Brumann, C., & Meskell, L. (2015). UNESCO and new world orders. In L. Meskell (Ed.), *Global heritage: A reader* (pp. 22–42). Wiley-Blackwell.

Bui, H. T., & Lee, T. J. (2015). Commodification and politicization of heritage: Implications for heritage tourism at the Imperial Citadel of Thang Long, Hanoi (Vietnam). *ASEAS – Austrian Journal of South-East Asian Studies, 8*(2), 187–202.

Dicks, B. (2003). Heritage, governance and marketization: A case-study from Wales. *Museum and Society, 1*(1), 30–44.

Disko, S., & Tugendhat, H. (2013). *Report. International expert workshop on the World Heritage Convention and indigenous peoples. 20–21 September 2012 – Copenhagen. Denmark*. International Work Group for Indigenous Affairs.

Falk, A., & Szech, N. (2013). Morals and markets. *Science, 340*, 707–711.

Hall, C. M., & Piggin, R. (2003). World Heritage sites: Managing the brand. In A. Fyall, B. Garrod, & A. Leask (Eds.), *Managing visitor attractions* (pp. 203–219). Butterworth Heinemann.

Henning, M. (2006). *Museums, media and cultural theory*. Open University Press.

Hewison, R. (1987). *The heritage industry*. Methuen.

Ibe, C., & Lohmann, G. (2005). Ware; Warencharakter; Warenfetischismus (engl. commodity, commodity fetishism). In J. Ritter, K. Gründer, & G. Gabriel (Eds.), *Historisches Wörterbuch der Philosophie* (Vol. 12, pp. 319–325). Schwabe.

ICOMOS. (1999). *International cultural tourism charter. Managing tourism at places of heritage significance*. ICOMOS.

IUCN. (2011). *Sustainable tourism and natural World Heritage*. IUCN.

Kirshenblatt-Gimblett, B. (1995). Theorizing heritage. *Ethnomusicology, 39*(3), 367–380.

Landmann, M. (1975). Das Fremde und die Entfremdung. In H.-H. Schrey (Ed.), *Entfremdung* (pp. 180–232). Wissenschaftliche Buchgesellschaft.

Leopold, D. (2018). Alienation. In *The Stanford Encyclopedia of Philosophy*. https://plato.stanford.edu/archives/fall2018/entries/alienation/

Lumley, R. (1988). Introduction. In R. Lumley (Ed.), *The museum time machine: Putting cultures on display* (pp. 1–24). Routledge.

Smith, L. (Ed) (2007). *Cultural heritage. Volume III. Heritage as an industry. Routledge.*

MacCannell, D. (1973). Staged authenticity: Arrangements of social space in tourist settings. *American Journal of Sociology, 79*(3), 589–603.

Marx, K. (2018a). Die entfremdete Arbeit (= MEW vol. 40: p. 510-522). In F. Butollo & O. Nachtwey (Eds.), *Karl Marx. Kritik des Kapitalismus* (pp. 177–191). Suhrkamp. Original work published 1844.

Marx, K. (2018b). Der Fetischcharakter der Ware und sein Geheimnis (= MEW vol. 23, pp. 85-98). In F. Butollo & O. Nachtwey (Eds.), *Karl Marx. Kritik des Kapitalismus* (pp. 332–343). Suhrkamp. Original work published 1867.

Meskell, L. (2015). Transacting UNESCO World Heritage: Gifts and exchanges on a global stage. *Social Anthropology, 23*(1), 3–21.

Quack, H.-D., & Wachowiak, H. (2013). Welterbe und Tourismus: ausgewählte Forschungsergebnisse. In H.-D. Quack (Ed.), *Kulturtourismus zu Beginn des 21. Jahrhunderts* (pp. 279–295). Oldenbourg.

Schmitt, T. (2009). Global cultural governance: Decision making about world heritage between politics and sciences. *Erdkunde, 63*(2), 103–121.

Schmitt, T. (2011). *Cultural governance. Zur Kulturgeographie des UNESCO-Welterberegimes*. Franz-Steiner-Verlag.

Schrey, H.-H. (Ed.). (1975). *Entfremdung*. Wissenschaftliche Buchgesellschaft.

Su, X. (2015). Urban entrepreneurialism and the commodification of heritage in China. *Urban Studies, 52*(15), 2874–2889.

Tangi, M. (1977). Tourism and the environment. *Ambio, 6*(6), 336–341.

UNESCO. (1972). *Convention concerning the protection of the world cultural and natural heritage*. UNESCO.

UNESCO. (1975). *Les effets du Tourisme sur les valeurs socio-culturelles (/SHC/OPS/TST/100)*. UNESCO.

UNESCO. (2015). *Policy document for the integration of a sustainable development perspective into the processes of the World Heritage Convention* (WHC-15/20.GA/INF.13). UNESCO. whc.unesco.org/document/139146

Vorlaufer, K. (1999). Tourismus und Kulturwandel auf Bali. *Geographische Zeitschrift, 87*(1), 29–45.

Walsh, K. (1992). *The representation of the past: Museums and heritage in the post-modern world*. Routledge.
Watts, M. (2009). Commodity. In D. Gregory, R. Johnston, G. Pratt, M. Watts, & S. Whatmore (Eds.), *The dictionary of human geography* (5th ed., pp. 99–101). Wiley-Blackwell.
Zima, P. V. (2014). *Entfremdung: Pathologien der postmodernen Gesellschaft*. A. Francke.
Zwolinski, M., & Wertheimer, A. (2017). Exploitation. *The Stanford Encyclopedia of Philosophy*. https://plato.stanford.edu/entries/exploitation/ (2021-09-28).

Chapter 30
Tourism Without Commodification at a Hungarian World Heritage Site

Lia Bassa

Abstract This study – based on a concrete example of a Hungarian World Heritage site, the Millenary Benedictine Abbey of Pannonhalma and its Natural Environment – focuses on the topic from the point of view of heritage value, providing attraction and a special experience for visitors without harming local traditions, and contributing to the improvement of the local economic situation, where the outcome is the safeguarding and raising public awareness of the Outstanding Universal Value of a unique site.

Keywords Heritage tourism · Benedictine traditions · Presentation · Meaningful visit · Safeguarding · OUV

30.1 Introduction

The Millenary Benedictine Archabbey of Pannonhalma (Fig. 30.1) and its natural environment was inscribed on the World Heritage List in 1996. It is a great example of how a religious site can positively take part in the development of a monument and its tourism.

The monks manage their own operation, economic life by selling goods and receiving tourists, which shows clearly that these activities do not necessarily mean the commodification of heritage; on the contrary, they can reconcile tourism with their own religious interests by using the income of their activities for the purpose of preservation and the maintenance of their traditions. This is not an innovation, but the Archabbey of Pannonhalma could make its measures sustainable, as they have always been functioning, developing and – at the same time – preserving their values while following their Benedictine conviction.

L. Bassa (✉)
Budapest Metropolitan University; Foundation for Information Society, Budapest, Hungary
e-mail: dr.bassa.lia@gmail.com

M.-T. Albert et al. (eds.), *50 Years World Heritage Convention: Shared Responsibility – Conflict & Reconciliation*, Heritage Studies,
https://doi.org/10.1007/978-3-031-05660-4_30

Fig. 30.1 The Porta Speciosa of Pannonhalma. (Note: Photo by Lia Bassa, 2016)

30.2 Economy and Heritage

The Benedictine Archabbey of Pannonhalma was founded as the first Hungarian Benedictine monastery in 996, and it was inscribed on the UNESCO World Heritage List exactly 1000 years later. The first Benedictine monks went on to convert the Hungarians, found the country's first school and, in 1055, write the first document in Hungarian. From the time of its founding, this monastic community has promoted culture throughout central Europe. Its 1000-year history can be seen in the succession of architectural styles of the monastic buildings (the oldest dating from 1224), which still house a secondary school and the monastic community. Their archbishop says that they make ventures not only for profit but as part of their "ars poetica" or creed, as it belongs to their mission, and they represent value in themselves.

The economic life in the Archabbey is also a moral issue, originating from spirituality. According to the monasterial tradition, Saint Benedict says that the monks must live from the achievements of their own work (*ora et labora*). Therefore, step by step, they have built up independent and ecclesiastic ventures to supplement budgetary resources and donations. The core message of their exceptional system is that the preservation of the World Heritage site and the consideration of sociocultural concerns related to tourism can go hand in hand in the so-called *commodification process*.

The Archabbey manages its life mainly from the income of enterprises that include sacral, charity, cultural and social institutions. The ventures are based on local – Benedictine – characteristics: education, culture, visitors and the products of the Abbey (liquors, teas, lavender, publications and wine cellar).

30.3 Tourism

In such a religious institution, tourism basically means handling three types of visitors: religious visitors such as monks and pilgrims, tourists interested in experiencing the Abbey's religious activities and cultural tourists who are interested in the institution's outstanding attractions. Furthermore, the touristic activities must be well-done and not disturb the everyday life of the monks or the local community, as detailed in the ICOMOS Charter of International Tourism (1999). In each case, the arriving guests – who may also have different cultural backgrounds – must be prepared for the content and purpose of the religious site visit and may also need clothing and behavioural instructions.

The visitors have access to the following attractions:

1. an archaeological site and the history of the construction of the Archabbey, the architectural monument of the main church and the other edifices of the Abbey and the "Our Lady Church";
2. the uniquely rich library and the ecclesiastic museum with relics and outstanding historic documents (including the first Hungarian written text);
3. the past and present life of the Benedictine order (on film);
4. the continuation of educational traditions: the secondary school (its operation can be seen on a film), as well as the transfer of their knowledge for the next generation and equally for the visitors;
5. the plantation of herbs (https://pannonhalmifoapatsag.hu/arboretum-es-gyogynovenykert/), their traditional utilisations in the processing plant including a "smelling museum" and a Tea shop with a store for cosmetic, healing products, liquors, chocolates;
6. the surrounding flora and fauna of the natural environment (also belonging to the WHS);
7. the local traditions of the vineyard (https://bortkostolunk.hu/pinceszet/pannonhalmi-apatsag-pinceszet/): winemaking from the grapes to bottling, including tasting, product exhibition and selling;
8. the orchard of the Hospodár garden with an open-air theatre;
9. Saint Jacob Pilgrim's House and Forest Chapel;
10. their own restaurant, "Viator"; and
11. other enterprises related to the Abbey where they invest workforce, work or money.

The sights are all connected to the basic activities of the historic ecclesiastic commitment of the Benedictine monks to conduct honest and ethical economic activities. The enterprises are owned and operated by them with the intention of continuing the traditional occupations of the order like winemaking, processing of herbs into healthcare products as well as education in their schools including information provision for tourists. Although they obviously earn money by these activities, the selling of goods cannot be called commodification, as the construction of both the production and business segment are strictly serving the traditional occupations of the Benedictine monks.

The vineyard has been revived where it used to be, but wine production is implemented in the most modern way. Nevertheless, the production of traditional types of quality wine is based on the existing descriptions by the Benedictine predecessors. There is a guided tour in the wine cellar, and the products can be tasted either there or in the Abbey's own restaurant, "Viator", next to the reception building. Even interesting legends survive: a person is authorised to drink one hemina of wine per day. How much is a hemina? As much as you are gifted to drink with pleasure.

The Hospodár Garden is named after a former chief cook of the Abbey, who started to cultivate the garden and later returned it to the order as a gift. It has a special atmosphere and fine collection of ancient Hungarian fruit varieties and offers the refreshment of a taste of fruit to those who have come here in deep contemplation. In the last decade, the garden has hosted the open-air theatre of the Abbey, as well.

The Saint Jacob Pilgrim's House and Forest Chapel was built in this century on Kosaras Hill that previously housed the steam laundry of the Abbey. The quiet forest area provides an ideal opportunity for welcoming and hosting guests (pilgrims, families, youth groups) arriving at the Abbey. The chapel cannot be separated from the pilgrim house, as they form an integral spiritual and infrastructural unit. The two buildings offer rest and revitalisation for both the body and the soul of their visitors.

The organisation of visits (partly for controlling and avoiding mass tourism) is also the Abbey's own venture. The life of the monks and that of the secondary students must not be disturbed; therefore, they cannot be visited. Instead, there are two short films made about them and projected in the reception building after arrival. Then – in order not to disturb the valuable forest but to learn about the flora and the fauna of the World Heritage site – there is a special walking path built for the tourists to reach the entrance of the monument itself.

The World Heritage site management plan of the Abbey maintains their basic cultural (ecclesiastic, touristic) and educational (school) undertakings in line with Strategic Development Goal 4, as well as the environmental protective (park) activities according to SDG 15. They reinvest the significant profit from their entrepreneurial accomplishments for further developments. It is an excellent example of how the spiritual, natural, built and tangible heritage of an ecclesiastic institution, which is a World Heritage site, can be properly organised (Fig. 30.2).

30.4 Conclusion

The World Heritage Convention of 1972 aims to protect unique natural and cultural heritage having an Outstanding Universal Value. However, this value is not only a financially inestimable aesthetic value but must also be made part of the local economic system because the site must be preserved, maintained and made public for the transmission of cultural knowledge for the next generation, as well as for members (visitors) of other cultures. The process of sharing a unique site benefits everyone if knowledge is effectively communicated and visitors respect the presented heritage. In this case study, it can be clearly seen how these values are separated with no populist devaluation of built, tangible or intangible heritage happening due to commercial interest. The result is more than 100,000 visitors/year, guided by the students of the Abbey. In addition to the Sustainable Development Goals mentioned above, they have planned tourism development, a new energy program, advancement of product supply and extension of the herbal program. Moreover, all related activities contribute to the safeguarding of all heritage branches of the 1000-year-old ecclesiastic institution.

Fig. 30.2 Pannonhalma library. (Note: Image by Lia Bassa, 2016)

References

ICOMOS. (1999). *International cultural tourism charter*. ICOMOS.
UNESCO. (1972). *Convention concerning the protection of the world cultural and natural heritage*. UNESCO.

Chapter 31
Natural Heritage in Danger. Native Forests, New and Old Forms of Extractive Activities and Sustainability from the Perspective of a New Generation of Scientists, Activists and Entrepreneurs in Argentina

Claudia Lozano

Abstract This article aims to address conflictive projects and their pathways of resolution in the field of land use and territorial and natural resource management in Latin America in relation to one of the protected areas of Argentinean Patagonia, Los Alerces National Park (PNLA), inscribed by UNESCO as a Natural World Heritage Site in 2017. I argue that the changes driven by the commodity boom (2000–2014), i.e., the development of extractive activities, deforestation and the expansion of grain and mineral exports, have had and continue to have a high environmental impact. Since the 2000s, the continent has experienced a series of protests that brought to light the unease caused by projects based on the development of large-scale extractive activities, which impacted legislation and reinforced the implementation of restrictive regulations for territorial management and land use, as seen from the nomination of the PNLA as a Natural World Heritage Site. The article shows that the legislative changes, as well as requirements of the PNLA inscription, provide fundamental legal support to the formulation, management and technical implementation of a new agro-silvo-pastoral culture capable of reconciling the conservation of scenic value (vii), biodiversity (x) and sustainability in protected areas. This proposal opens up the possibility of expanding the protected areas within the framework of the Andean-North Patagonian Biosphere Reserve.

Keywords Natural world heritage · Extractive activities · Commodification · Parque Nacional Los Alerces

C. Lozano (✉)
Independent Scholar, Member of the Advisory Board for Humanities and Social Sciences, Network of Argentinean Scientists in Germany (RCAA), Berlin, Germany
e-mail: c.lozano@fu-berlin.de

M.-T. Albert et al. (eds.), *50 Years World Heritage Convention: Shared Responsibility – Conflict & Reconciliation*, Heritage Studies,
https://doi.org/10.1007/978-3-031-05660-4_31

31.1 Introduction

The media and a growing body of detailed research from Latin American rural and cultural studies on environmental conflicts have placed natural assets, land and territorial management at the centre of public discussions on natural resource development and preservation (Svampa, 2019). This trend is also complemented by a growing body of natural science literature on biodiversity, the protection and recovery of native species, particularly native forests (Cabrol & Cáceres, 2017; Godoy et al., 2019; Grosfeld et al., 2019). An issue that has received less attention is the relationship between land conflicts and the forms of territorial distribution and the management of protected areas such as UNESCO World Heritage Sites.

This article aims to address conflicts over land use, and territorial and natural resource management in Latin America in relation to one of the protected areas in the Argentinean Patagonia, Los Alerces National Park (PNLA), inscribed by UNESCO as a World Heritage Site in 2017.[1] I argue that the changes driven by the commodity boom (2000–2014), i.e., the large-scale export of commodities such as grains, agro-industrial products and minerals, had and have a high environmental impact. This impact is especially visible in the accelerated clearing of native forests and the expansion of monocultures and hydrocarbon activities, which has unleashed the resistance of the local population affected by these processes. From the 2000s onwards, the continent has experienced a series of protests that made public the destructive impact of large-scale development projects (Gras & Hernández, 2013), indirectly including areas protected as part of humanity's cultural and natural heritage (Lozano, 2017, 205–211).

The article focuses primarily on conflicts caused by large- and small-scale extractive activities and the commodification of natural resources – water and forests – that affect Los Alerces Nacional Park and its buffer zones. Secondly, it addresses the parks management proposals aimed at protecting the integrity of the native forest and promoting a sustainable agro-silvo-pastoral culture, based on technical innovation and community participation.

[1] The nominated property comprises the totality of Los Alerces National Park. Located in the northwestern Andean region of the Province of Chubut. The denomination PNLA includes the National Park as well as the adjoining National Reserve. The National Park itself occupies an area of 188,379 ha. This area, which includes land and lakes, is completely free of human habitation and fragmentation caused by roads. The National Reserve includes 71,443 ha inhabited by rural settlers and management and control staff, as well as the main services and visitor facilities of the protected area. The denomination "Parque Nacional Los Alerces" or PNLA includes both the National Park and the National Reserve (Nomination of Los Alerces National Park as World Heritage Site, 2017: 4, 6; WHC/17/41.COM/18, 2017: 190).

31.1.1 Conflictive Projects and Traditions

If the conflicts associated with the expansion of the plantation and mining economy on a global scale were historically related to the claims of indigenous peoples, i.e., peasant communities and organisations asserting their territorial rights, this situation of particularism and exceptionality has changed.[2] Since the 1990s, demands related to limiting land grabbing, stopping open-pit mining and forest clearance and protecting native species have included other population groups, scientist, activists and entrepreneurs and their organisations. The expansion of the social and cognitive base and the plurality of values made it possible to promote and articulate legislative changes at provincial, national and transnational levels. These changes to the legislation refer, among other things, to territorial reorganisation policies, which expanded protected areas, national parks and nature reserves.[3] The Nomination document (State Party of Argentina & Ministry of Tourism, 2017, 80) highlights the existence of "conflicting projects" involving possible mining exploitation. Although the declaration of the region as part of the Andean-North Patagonian Biosphere Reserve since 2007 and a decree of the Argentinian Province of Chubut made it impossible to initiate open-pit mining megaprojects throughout the province, such projects periodically re-emerge on public agendas.[4]

This process took place throughout the continent and acquired peculiar characteristics in Argentina. Geographer Reboratti (2019,2) considers that certain characteristics of Argentina, such as the size of its territory, the richness and diversity of its natural resources and a relatively small and mainly urban population, explain the late emergence of socio-environmental conflicts. Similarly, the ecological awareness of local actors,[5] who until recently were excluded or self-excluded from negotiations and the distribution of responsibilities, costs and benefits of extractive

[2] In Argentinian Patagonia, the *"conquest of the desert"* (1879, 1881–1884) was a military campaign of extermination, whose implications were the end of the great chiefdoms, the dispossession of land and the cultural disintegration of the indigenous groups that inhabited Patagonia: Ranqueles, Vorogas, Araucanos or Mapuches and Tehuelches that inhabited Patagonia (Martinez Sarasola, 1992: 274–295; 355–368).

[3] In Argentina, most national parks, including those in the northern Patagonian region, were created after 1930. A series of scientific expeditions formed the basis of geographical knowledge that allowed the delimitation of the areas that are now part of the National Parks The central motif of the maps and the photographic records (the Encina and Moreno album) and watercolours (Meethfesel) are the great lakes. The records allowed the members of the scientific expeditions to transform mental images of the "wild" areas into "natural" areas to be preserved for scientific exploration and management by technically trained personnel (Penhos, 2017: 61).

[4] In the 6 May 2021 edition of the newspaper La Nación published a note on the treatment of the bill that enables exploitation in the central plateau region (Tronfi, 2021).

[5] According to the Census of Inhabitants carried out by the PNLA, in 2012, rural residents are spread across 38 localities, they are occupants of public lands with precarious occupation or grazing permits, private owners, engaged in livestock activities, forestry, tourism services and staff of public agencies in the category of temporary residents of Villa Futalaufquen (Plan de Gestión del Parque Nacional Los Alerces, 2019–2029, 2019: 78–83).

activities, is a recent development. This is why the large infrastructure projects of the 1960s with potential impacts on the environment remained unopposed and were hardly discussed at local and national levels until the 1990s.[6]

Argentina thus becomes a particularly relevant context in which to explore the interconnections between traditional and contemporary extractive activities, the inscription of a conservation area as a Natural World Heritage Site in 2017, in this case, the PNLA, and compliance with the protection and management requirements of the site and its buffer zones set by UNESCO. Compliance with the requirements is of fundamental importance because the appropriate technical management of agro-forestry and livestock activities on the edges of its buffer zones generate the basis for the integrity of the natural landscape, as well as for sustainable regional development in a region with a growing population (IUCN, 2017, 84; Godoy et al., 2019, 480).

Specifically, protected areas are conceptualised in the sense given to them by Mendoza (2018, 11), i.e., as a space under the strict control of National Parks, while family members living in the villages and private owners engaging in livestock, forestry, agriculture and tourism activities at different scales exercise subsidiary control over the park's tangible and intangible resources. The Park Service guards and monitors the territory and earns income from permit and entrance fees. The hydropower company, private landowners and tourism agents located in the buffer zones extract raw materials, transform them into products for human and animal consumption (pasture and firewood) and into material goods (energy, meat and timber) and immaterial goods (landscapes and enjoyment of nature) for the market.[7]

The following sections address the aforementioned conflicting projects, i.e., the construction of a hydroelectric power plant, traditional agropastoral activities and forest fires, in terms of threats to the integrity and potential fragmentation of the natural landscape. Subsequently, proposals for land use and sustainable forest management of the site's buffer zone are examined, highlighting its socio-economic potential as well as its technical and administrative constraints along two lines: First, the articulation between the demands of different social actors and the institutions and organisations involved in the containment of ecological and social risk situations. Secondly, territorial planning and the promotion of socio-productive

[6] In numerous articles and books, Argentine historiography has documented conflicts over boundaries and land distribution, ownership and use since the colonial period. The notion of environmental conflicts is relatively new. Today, there is an extensive and detailed literature on the subject. See Crespo, 2018; Hermosilla Rivera, 2019; Lobba Araujo, 2019.

[7] In Latin America (as in the rest of the countries of the West), a relationship of domination of nature, conceived as a source of resources, prevails. In this context, the natural sciences (biology, ecology) distinguish themselves from this model by considering human activity as a factor that disturbs the ecological balance and causes environmental damage (deforestation, pollution, invasive alien species). In the same vein, they argue that a degraded environment has effects on quality of life, health problems, natural disasters, food and water insecurity and cultural discontinuity (Roulier et al., 2020, 20).

strategies within the framework of the conservation and sustainability objectives of the protected area are also addressed.[8]

31.2 Integrity Protection

31.2.1 The Water

Los Alerces National Park is part of Argentina's National System of Protected Areas, which is under the jurisdiction of the National Parks Administration (APN). APN is a legally established autonomous body. The proposed site is a "National Protected Area" according to National Law 22.351 of 1980, established with legal objectives focused on "protection and conservation" as well as "scientific research, education and enjoyment of present and future generations" (IUCN, 2017, 79).

In 1971, the two protected areas of the PNLA and the Nature Reserve or buffer zone had their boundaries and areas redefined by Law No. 19,292 (see Fig. 31.1). According to the evaluation report, the site contains the largest, most-intact and least-degraded Valdivian Temperate Rainforest in Argentina. These conditions affirm its Outstanding Universal Value. The report also states that it is possible to expand the site to other areas of Argentina and Chile. This requires the technical elaboration of not only the factors that prevent the reduction of the forest mass, such as appropriate fire management and forestation on the edges of the buffer zone, but also changes in the economic, social and cultural matrix related to the forest and aquifer system (glaciers, lakes and rivers) as sources of natural resources and suppliers of raw materials and energy (IUCN, 2017, 79).

Despite its remoteness and the history of forest conservation that form the basis of its exceptional value, northern Patagonia was the scene of technical and economic progress and development linked to hydrocarbon extraction in the 1960s and the construction of three nuclear power stations and the Futaleufú hydroelectric power station inaugurated in 1978. The construction of the Futaleufú hydroelectric dam caused changes in the original lake system, altering the natural landscape of a considerable area of Los Alerces National Park. The construction of the dam, however, did not trigger lawsuits regarding the environmental impact assessment of the

[8]This article is based on the detailed study of nomination documents, reports, management plans and scientific articles available on the site (PNLA) and on an interview and email exchanges with them to clarify aspects of topics treated in those documents and articles, with Dr. Javier Grosfeld, Chair of the National Park Administration of the North-Patagonian Region and email exchanges, Prof. Dr. Guillermo Defossé, National Scientific and Technical Council CONICET, Andean Patagonian Forest Research and Extension Center, Forest Engineering University of Patagonia, Ruta 259 km 16, Esquel, 9200 Chubut, Argentina, and Dr. Veronica Rusch, Ing. Verónica Rusch INTA, EEA Bariloche (Instituto Nacional de Tecnología Agropecuaria) were conducted in April and Mai 2021.

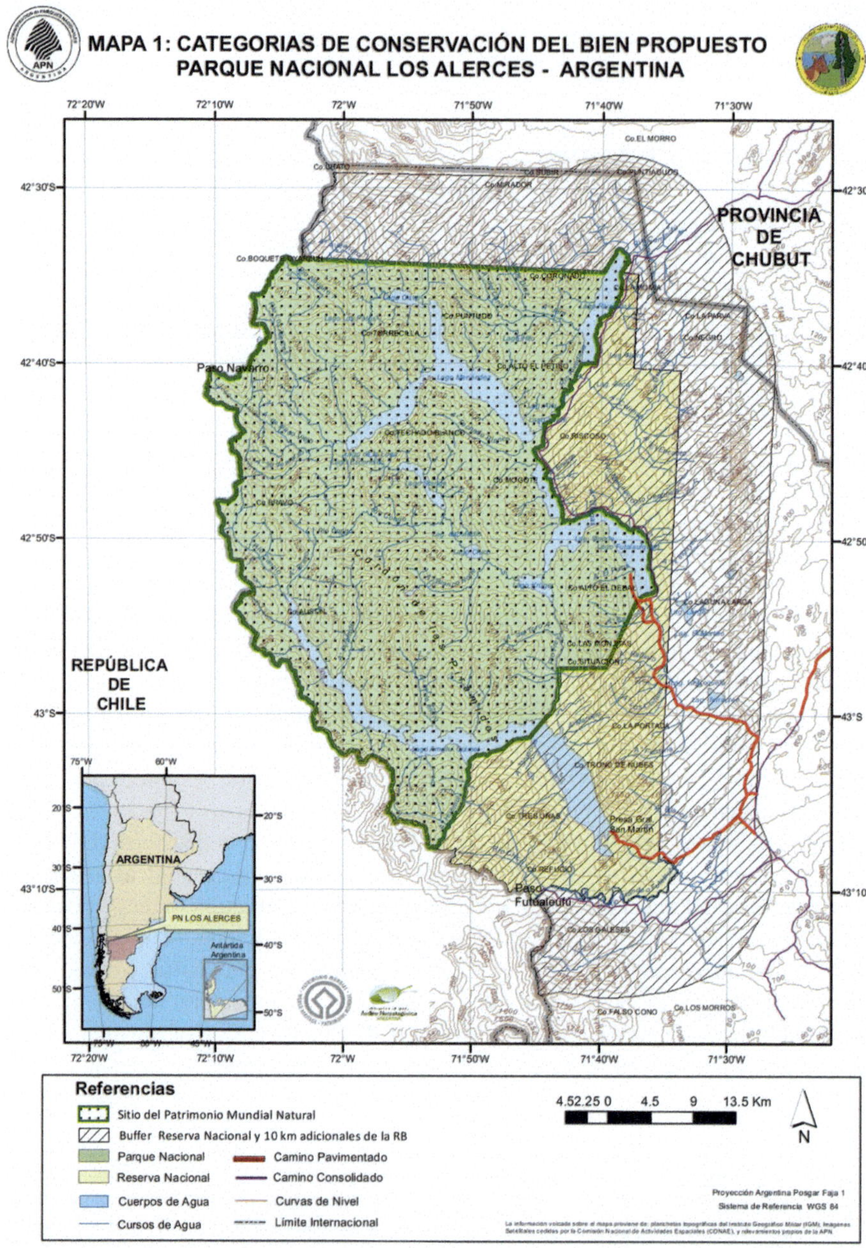

Fig. 31.1 Map: Administración de Parques Nacionales. (2017). Categorías de Conservación del bien propuesto Parque Nacional Los Alerces – Argentina. (*Note.* The information on the map comes from topographic planchettes of the Military Geographic Institute (IGM); Satellite Images provided by the Argentina National Space Activities Commission (CONAE), and National Park Administration (APN). Argentine Projection Posgar Faja 1 WGS 84 Reference System [Map] by IGM, CONAE, & APN. https://whc.unesco.org/en/list/1526/multiple=1&unique_number=2172)

work within a territorial space protected, among other reasons, for its "impressive scenic beauty" (IUCN, 2017, 84).

The indifference to the environmental impact and the use of available resources within protected areas took a radical turn from the 1990s onwards. In 1993, Argentina enacted Law 24.196 on the Promotion of Mining Activities, which granted advantages to companies engaged in prospecting and the development of mineral extraction. The law encouraged the expansion of such activities, and the resulting environmental impact gave rise to a series of conflicts. Reboratti (2012, 8, 9) highlights the particularity of the first of these, which originated in 2002 in the city of Esquel as a result of the establishment of a mining company and the environmental impact of its activity. The conflict highlighted the opposed visions of the company and the residents regarding water use, pollution, noise, employment and the distribution of economic benefits. The public presentation of the environmental impact study led local people to gather in public assemblies to demand that the authorities halt the project until a legal body regulating the activity was in place. As the region is particularly valued for its exceptional natural beauty and as a tourist attraction, the local population began to take any potential threat to nature seriously. The conflict and its resolution brought to the fore socially conscious and honest actors' doubts and mistrust of formal political organisations in relation to the uses of nature and the pursuit of environmental justice (Reboratti, 2012 8, 9).

The residents' demands gained the support of local indigenous people, the Mapuche, and national environmental organisations formed by a network of conservation activists who, thanks to the targeted use of social media, expanded a local conflict to a national scale.[9] The public dispute led to a call for a non-binding referendum by the Municipality for the resident population of Esquel. The result of the referendum showed that more than 80% of voters did not agree with the establishment of the mine.

The support network that integrated institutions dedicated to independent scientific research and technical assistance was just as important as the use of social networks and the resonance that the demands of the local assemblies found in environmental organisations. These contributed to the production of knowledge about the socio-ecological impact of profit-making projects, as well as conflicts over socio-ecological institutionality and the regulation of forest use in the site's buffer zones, to which we will return later.

In summary, the social processing of the confrontation unleashed by the conflict-generating mining projects led to the banning of open-pit mining in the province. In a similar vein, UNESCO's Technical Evaluation requires the State Party to strictly monitor activities linked to the Futaleufú dam, the reservoir and its associated infrastructure to avoid or mitigate adverse effects on the site's Outstanding Universal Value, and ensure that the maintenance and improvement plan is subject to prior environmental and social impact assessment (IUCN, 2017, 85).

---

[9] Indigenous communities were increasingly involved in environmental conflicts. Since the conflicts over the installation of gold mining in Esquel, the presence of Mapuche communities stands out (Wagner & Walter, 2020, 247).

31.2.2 The Forest

If mining projects have proved to be conflictive, the high rate of deforestation in Argentina also affects the native species that make up the Valdivian Temperate Rainforest. Illegal logging, forest fires and technically inadequate grazing management in the buffer and transition zones of the PNLA form part of a historical process of degradation of the native flora.

In 2007, National Congress passed Law 26.331 on Minimum Standards for the Environmental Protection of Native Forests. One problem with the application of this law is that the 1994 National Constitution grants limited competence to national authorities in environmental matters. The national authorities can establish a "minimum conservation threshold that the provinces must comply with, but it is up to the provinces to design and effectively implement environmental policies" (Langbehn, 2016, 141; IUCN, 2017, 79). These differences in competences prevent a homogeneous application of legislation across the national territory.[10] Official sources from the Directorate of Forestry in the province of Chubut, where the North Patagonian Andean Forest is located, state that its surface area has remained stable (Gómez Lende, 2018, 161).

Since 2015, the Argentine Association of Ecology has been conducting a debate on the possibility of achieving sustainable development of the Andean-North Patagonian Forest. Soler and Gowda (2019, 119) point out that the practical application of what the FAO defines as sustainable forest management, "the sustainable use and conservation of forests in order to maintain and enhance their multiple values through human intervention", is a difficult task. Forests support diverse activities, ranging from extensive cattle ranching, timber extraction, tourist recreation, conservation, real estate speculation and urban development. There are many examples of economic activities linked to forests, and the management of forests is deficient and does not tend to maintain and increase their value.

A considerable body of knowledge of the biology and ecology of the plant communities that make up forests and their response to different forms of management provides a solid basis for sustainable native forest management. However, the lack of communication and trust between the institutional academic sector and rural producers, who today depend on livestock as their main source of income, conditions the development of good management practices. Increased funding to generate an applied knowledge base and updating the 1984 Forestry Regulation according to Law 26.331 of 2007 are vital for promoting virtuous forest management among rural producers (Grosfeld et al., 2019, 158).

Another central problem is related to the distribution of territory within the park and the reserve. The authorities did not offer relocation plans from the buffer zones

[10] This law, known as the Forest Law, was a response to the accelerated deforestation of the previous decade. The clearing of native forests had triggered the concern and mobilisation of various actors in different parts of the country, particularly in the north of Argentina (Langbehn et al., 2020, 193).

to the transition areas of the Biosphere Reserve to all the inhabitants of the protected areas. To reverse this situation, the Management Plan 2020–2029 includes not only relocations but also the development of training projects in the management of a combination of productive activities, forestry, tourism, livestock and small-scale agriculture. The implementation of the projects will be carried out with the advice of regional technical assistance institutions. This is intended to contribute to the transition from traditional agroforestry management, based on outdated standards, to sustainable management.

As previously emphasised, the region has scientific research centres, technical assistance and a university in Esquel. These constitute a platform that promotes and sustains processes of productive, economic and social change based on the redesign and monitoring of productive activities associated with the forest. The configuration of a new production and commercialisation arrangement would allow scaling up through the appropriate technical management of the native forest in the buffer and transition zones surrounding the site.[11]

Soler and Gowda's (2019, 120) point about "ecological experiments" as a basis for new learning in the buffer and transition zones of the PNLA is relevant not only for ecologists but for all public and private actors involved in the conflicting interests. Its relevance arises from the contrast with the historical experience of the degrading impact of a deeply rooted traditional extractive economy, whose socio-ecological effects are outside the thresholds of contemporary sustainability. In the face of this, various research and technical assistance agencies are opening up a new perspective based on the healthy management of conflicts associated with common problems, the most hostile manifestation of which is anthropogenic forest fire ignitions.

31.2.3 Forest Fires and Conflict Management

One of the main motivations for the creation of Los Alerces National Park in 1937 was to eliminate anthropogenic fires. However, this intervention also led to the arrival of personnel from the National Parks Administration, as well as a public administration that has been expanding ever since. The growing social and cultural differentiation pits those in favour of the continuation of traditional extractive activities and families dedicated to cattle ranching in the buffer zones against those who, from a conservationist and recreational perspective, are against all extractive activities, including National Parks staff, eco-tourists and members of the Fishermen's Club. According to the literature on fire prevention, the existence of divergent value frameworks would explain the increase in forest fires (Defossé et al., 2015; Seijo et al., 2020).

[11] Interview with Dr. Javier Grosfeld, 03.05.2021.

In line with this explanation, forest fire prevention should be associated with the resolution of conflicts related to incompatible value frameworks, which are the basis of the increase in fires. Stakeholder participation in the design of preventive fuel management strategies through prescribed burning would mitigate the size and intensity of individual fire events, bringing the parties together towards a common goal.

The conflicts arising from human occupations and activities affecting the PNLA also call for research on the relationship between the National Parks Administration and the creole and indigenous, Mapuche, communities that reside, occupy and/or develop productive and commercial activities in the buffer zones. Such studies would make it possible to elaborate and define methods of awareness-raising, distribution of responsibilities and participation of local actors in the conservation of wildlife species of special value or cultural resources in accordance with the sustainability goals (8 and 15) set by UNESCO. These challenges are not only related to the ecological impacts of traditional natural resource management by communities and peoples today but also concern the technical and administrative management of the PNLA and its buffer zones. The proposal of a new agro-forest-pastoral culture whose objective goes beyond the extinction of individual fires becomes important for the future, attending to forms of joint management and administration that combine the transmission and reinvention of traditions in accordance with contemporary sustainability goals. This will connect the preservation of a unique forest ecosystem with the expansion of protected conservation areas (Nomination Document) to other places, e.g., the Patagonian steppe, in an ecological experiment based on technical and institutional innovation and community participation (Rusch et al., 2017).

The recommendations for bridging the gap between theory and practice identified from the detailed study of the available literature on the PNLA cannot simply be applied to the management of all conflicts associated with extractive activities in protected native forests in Latin America. The formulation of recommendations requires the development of detailed research projects related to the scientific and technical conditions available in each case and their particular way of conceptualising the problems arising from extractive practices.

31.3 Concluding Remarks

The starting point of this article was the problem of the destructiveness of regional development policies based on extractive activities whose input–output matrix does not consider criteria of landscape conservation and biodiversity protection or sustainable development. The article also showed that the extractivist basis of the productive activities that historically facilitated Latin America's incorporation into the world market, such as logging, large-scale mining and hydrocarbon projects, was recognised as a threat, in this case, to the Valdivian temperate forest (extractivism and anthropic ignitions), by natural scientists and conservationists in the late 19th and early 20th centuries. Studies highlighting the destructive aspects of productive development based on extractive activities justified the nation state's decision to create nature reserves and national parks.

Towards the end of the twentieth century and the beginning of the twenty-first century, the degradation of nature worsened. In response to this, environmental activist organisations and the local population directly affected by conflict-generating projects demanded that the state update the legislation regulating the use of the territory and the available tangible and intangible assets.

The updating of the legislation created the legal framework for the nomination of the PNLA. Institutions dedicated to scientific research and technical assistance contributed to the creation of databases, documentary archives, reports and analytical articles on the current state of the reserves, parks and sites in the Andean-North Patagonian region. They also generated an ongoing debate that supports the formulation of the Management Plan for the nominated site and its buffer zones. This is based on the reconciliation of conservation and nature protection criteria with the development of productive activities within contemporary sustainability thresholds.

Indeed, it is not a question of ignoring the traditional agro-silvo-pastoral activities that persist in both the buffer and transition zones of the site, but of redirecting them within the framework of a new input–output matrix based on concepts of native forest restoration and economic and social sustainability. The challenge, beyond the introduction of scientific research and technical innovation in forest management, is to create a culture based on responsibility. The strengthening of technical and cultural innovation justifies the expansion of the heritage area to protect the exceptional nature of the Valdivian temperate forest in Argentinean and Chilean Patagonia, as is clear from the PNLA Nomination document.

References

Administración de Parques Nacionales Argentina. (2019). *Plan de Gestión del Parque Nacional Los Alerces 2019–2029*. Administración de Parques Nacionales Argentina. https://sib.gob.ar/archivos/PG_Los_Alerces.pdf

Cabrol, D. A., & Cáceres, D. M. (2017). Las disputas por los bienes comunes y su impacto en la apropiación de servicios ecosistémicos. La Ley de Protección de Bosques Nativos, en la Provincia de Córdoba, Argentina. *Ecología Austral, 27*(1bis), 134–145.

Crespo, C. (2018). Promesas de desarrollo, despojos y daño moral. Experiencias Mapuche en torno a políticas forestales en Puerto Patriada, El Hoyo (Chubut, Argentina). *Antropologías del Sur, 5*(10), 113–129.

Defossé, G. E., Godoy, M. M., Bianchi, L. O., Lederer, N. S., & Kunst, C. (2015). Fire history fire ecology and management in argentine Patagonia: From ancient times to nowadays. In B. Leblon & M. E. Alexander (Eds.), *Current international perspectives on wildland fires, mankind and the environment* (pp. 177–210). Nova Publishers.

Godoy, M. M., Martinuzzi, S., Kramer, H. A., Defossé, G. E., Argañaraz, J., & Radeloff, V. C. (2019). Rapid WUI growth in a natural amenity-rich region in Central-Western Patagonia, Argentina. *International Journal of Wildland Fire, 28*, 473–484.

Gómez Lende, S. (2018). Destrucción de Bosques Nativos y Deforestación ilegal: El Caso Argentino (1998-2016). *Papeles de Geografía*, (64), 154–180. https://revistas.um.es/geografia/article/view/349551/256531

Gras, C., & Hernández, V. (2013). *El agro como negocio: Producción, Sociedad y Territorios en la Globalización*. Editorial Biblos.

Grosfeld, J., Chauchard, L., & Gowda, J. H. (2019). Debates: ¿Podemos manejar sustentablemente el bosque nativo de Patagonia Norte? *Ecología Austral, 29*, 156–163.

Hermosilla Rivera, C. (2019). Conflictividad territorial en Chubut: Una lectura en torno a los movimientos ambientalistas. *Geograficando, 15*(1). https://www.geograficando.fahce.unlp.edu.ar/article/download/GEOe050/10711

IUCN Evaluation Report. (2017, May). *IUCN technical Evaluation Los Alerces National Park (Argentina) – ID N° 1526*. UNESCO. https://whc.unesco.org/en/list/1526/documents/

Langbehn, L. (2016). Arenas públicas, modelos de desarrollo y políticas de protección del ambiente: la ley de bosques entre "conservación" y "producción". In G. Merlinsky (Ed.), *Cartografías del conflicto ambiental en Argentina* (Vol. 2, 1st ed., pp. 141–168). Ediciones CICCUS.

Langbehn, L., Schmidt, M., & Pereira, P. (2020). Las leyes ambientales en el ojo de la tormenta. Un análisis comparativo en torno a la legislación sobre glaciares, bosques y humedales en Argentina. In Berger, M., Merlinksky, G., (Coord.) *Cartografías del Conflicto Ambiental en al Argentina* (1 st ed., Vol 3., pp. 187–213). Ediciones CICCUS.

Lobba Araujo, J. (2019, October 9–11). *Conflicto en torno a los incendios forestales en el Noroeste del Chubut. El caso de la Horqueta, Cholila, 2015* [Paper Presentation]. VII Congreso Nacional de Geografía de Universidades Públicas y XXI Jornadas de Geografía de la Universidad Nacional de Plata. Facultad de Humanidades y Ciencias de la Educación. La Plata, Argentina. https://rid.unrn.edu.ar/bitstream/20.500.12049/4960/1/LobbaAraujo.pdf

Lozano, C. (2017). Feeling responsible for the good life on earth: The construction of social spaces and sustainability in the Andes. In M. Albert, F. Baldarin, A. Roders, & A. (Eds.), *Going beyond. Perceptions of sustainability in heritage studies* (Vol. Nr. 2, pp. 201–214). Springer.

Martinez Sarasola, C. (1992). *Nuestros Paisanos los Indios*. Emecé.

Mendoza, M. (2018). *La Sublime Patagónica: La Economía Verde y la Política Post-Neoliberal*. Rutgers University Press.

Penhos, M. (2017). Las fotografías del álbum de Encina, Moreno y Cía. (1883) y la construcción de la Patagonia como espacio geográfico y paisaje. In M. I. Rodríguez, J. Vezub, J. M. Raone, & A. Pozzo (Eds.), *Patrimonios Visuales Patagónicos. Territorios y Sociedades. Álbumes fotográficos de Antonio Pozzo y Encina, Moreno & Cia* (pp. 49–64). Instituto de Investigaciones Históricas.

Reboratti, C. (2012). Socio-environmental conflict in Argentina. *Journal of Latin American Geography, 11*(2), 3–20.

Reboratti, C. (2019). Tensiones Geográficas: controversias y conflictos ambientales en Argentina. *Investigaciones Geográficas. Instituto de Geografía, 100*, 1–9.

Roulier, C., Anderson, C. B., Ballari, & Nielsen, E. A. (2020). Estudios sociales y socioecológicos sobre restauración ecológica: Una revisión de la literatura a escala global e iberoamericana. *Ecología Austral, 30*(1), 19–32.

Rusch, V. E., López, D., Cavallero, L., Rusch, G. M., Garibaldi, L. A., Grosfeld, J., & Peri, J. (2017). Modelos de estados y transiciones de los ñirantales del NO de la Patagonia como herramienta para el uso silvopastoril sustentable. *Ecología Austral, 27*(2), 266–278.

Seijo, F., Godoy, M. M., Guglielmin, D., Ciampoli, C., Ebright, S., Picco, O., & Defossé, G. (2020). Conflicting frames about ownership and land use drive wildfire ignitions in a protected conservation area. *Environmental Management, 65*(4), 448–462.

Soler, R., & Gowda, J. (2019). ¿Podremos manejar en forma sostenible nuestros bosques nativos? *Ecología Austral, 29*(1), 120–121.

State Party of Argentina & Ministry of Tourism of Argentina. (2017). *Nomination of Los Alerces National Park as a World Heritage Site* (No. 1526–2172). UNESCO. https://whc.unesco.org/en/list/1526/documents/

Svampa, M., (2019). *Las Fronteras del neoextrativismo en América Latina. Conflictos socioambientales, giro ecoterritorial y nuevas dependencias*. Calas Maria Sibylla Merian Center and-Bielefeld University Pres.
Tronfi, A. (2021, May 6). Chubut: rechazaron la iniciativa popular que proponía prohibir la minería. La Nación. https://www.lanacion.com.ar/politica/chubut-rechazaron-la-iniciativa-popular-que-proponia-prohibir-la-mineria-nid06052021/
Wagner, L., & Walter, M. (2020). Cartografía de la conflictividad minera en Argentina (2003-2018). Un análisis desde el Atlas de Justicia Ambiental. In M. Berger & G. Merlinsky (Eds.), *Cartografías del conflicto ambiental en Argentina III* (pp. 247–278). Ediciones CICCUS.

Chapter 32
Shifting Scales in the Honghe Hani Rice Terraces: Traditional Knowledge, Commodification and Community Participation

Fabienne Wallenwein

Abstract In rapidly transforming Asian environments, traditional agricultural heritage systems struggle with increasing development pressure and out-migration. Drawing on the Chinese cultural landscape of the Honghe Hani Rice Terraces as a case study, the paper investigates how the concepts of scale and "politics of scale" can be fruitfully mobilised for critical heritage theory and provide practical solutions to overcome conservation–development tensions. In processes of ethnic tourism development and cultural commodification, government authorities pursue different scalar strategies to harness natural and cultural resources for heritage-led regeneration schemes. Such strongly tourism-oriented agendas, as prevailed in the initial stages of development, privilege natural and selected cultural values over social values, thereby contrasting with local inhabitants' aspirations to improve their living conditions. To encourage participation and sustainable cultural landscape management, the study suggests "upscaling" traditional knowledge and local interests.

Keywords Cultural heritage · Cultural landscape · China · Commodification · Community participation · Politics of scale

32.1 Introduction

At the beginning of the 1970s, a decade of professional debates, mobilisation and institutional rivalries crystallised in an unprecedented form of international cooperation to safeguard places and sites of outstanding significance across the globe – the World Heritage Convention (UNESCO, 1972). The successful adoption of

F. Wallenwein (✉)
Heidelberg Centre for Transcultural Studies, Heidelberg University, Heidelberg, Germany
e-mail: Fabienne.wallenwein@hcts.uni-heidelberg.de

M.-T. Albert et al. (eds.), *50 Years World Heritage Convention: Shared Responsibility – Conflict & Reconciliation*, Heritage Studies,
https://doi.org/10.1007/978-3-031-05660-4_32

shared conservation objectives and principles was not least a result of new anthropogenic threats to cultural and natural heritage, such as the often referenced construction of the Aswan Dam in Egypt and potential flooding of the Philae and Abu Simbel temples. Fifty years later, environmental degradation is becoming an ever more pressing risk to heritage on a global scale, exacerbated by unsustainable ways of production and consumption.

Based on conceptualisations put forward in the 1987 Brundtland Report, *sustainable* development mainly rests on three pillars: economic viability, environmental protection and social equity. Despite initial success in introducing culture into the sustainable development discourse, Agenda 21 for culture (UCLG, 2004) and the Hangzhou Declaration (UNESCO, 2013) in particular, its integration into practical planning and decision making has remained limited (Throsby, 2017, 141). However, the Sustainable Development Goals (SDGs) primarily acknowledge culture, cultural heritage and creativity as drivers and enablers for achieving these goals (Hosagrahar, 2017). This central role of culture as a fourth dimension becomes evident in agricultural landscapes where local ethnic communities live in an intimate relationship with their environment.

32.2 The Honghe Hani Rice Terraces

A remarkable and particularly noteworthy example of such a landscape is located between the Ailao Mountains 哀牢山 and the Hong River 红河 in China's southwestern Yunnan Province. The Honghe Hani Rice Terraces 红河哈尼梯田 that were inscribed as World Heritage cultural landscape in 2013 organically integrate four components: mountaintop forests, rural villages of the Hani ethnic community, rice terraces and a water system. Based on historical accounts in Chinese sources, notably Fan Chuo's 樊绰 Tang era (618–907) geographical report *Man Shu* 蛮书 ("Book of the Southern Tribes", Fan & Oey, 1961, 67), rice terrace agriculture in this area is estimated to date back at least 1000 years. In adaptation to the natural conditions of mountainous terrain and sub-tropical climate, the Hani developed a complex system of irrigated rice terraces on the mountain slopes, making use of shallow groundwater and water storage capacity of mountaintop forests (Jiao & Li, 2011, 33, see Fig. 32.1).

The maintenance of this agricultural land-use system relies on the rich traditional knowledge, also designated as "indigenous knowledge" or "traditional ecological knowledge", of its local communities. Such knowledge includes indigenous languages, water management systems, customary rules and technical skills. In the integrated farming system of the Hani Terraces, the breeding of ducks, fish and other livestock complements red rice cultivation. This mode of production is further supported by a distinctive social and spiritual system. Modifications of water sources, for example, have traditionally been regulated through customary rules based on consent. Moreover, cosmological conceptualisations of the environment where spirits reside in landscape components, such as streams, trees or caves, and

Fig. 32.1 Red rice cultivation in the Hani Rice Terraces, Yuanyang County, China. (Note: Photograph by Joana Wardenga, 2018)

circulate along topographically and hydrographically determined routes (Bouchery, 2011, 332) have contributed to a resource-friendly attitude towards nature.

With increasing globalisation, advantages of this close nature–culture relationship have come under serious threat. A major factor that negatively affects the Hani terrace system is climate change, which has caused severe landslides (Gao et al., 2020, 1916) and water shortages. Especially in Asian developing countries, cultural landscapes are increasingly disappearing as a result of industrialisation and rapid urbanisation (Takeuchi, 2010, 894; Lennon, 2012, 54). Due to demographic changes, not only external factors but also local livelihoods affect the integrity of the terraces: Traditionally, when a terrace system could no longer support a village's population, the human–land ratio was balanced again by splitting the village and creating another terrace (Jiao & Li, 2011, 36–38). With the introduction of the household responsibility system and strict land delimitation policies in the 1980s, this practice was prohibited, and human settlements became more and more concentrated. With continued population growth on the one hand and land scarcity on the other, cropland per capita dramatically decreased in the Hani villages. As a consequence, rural labourers turn to non-agricultural activities that provide higher income (Zhang et al., 2017) and move to larger urban centres such as the provincial capital Kunming, abandoning their homes and fields.

A popular strategy for poverty reduction in the southwestern Chinese provinces, which has been embraced by the central and local governments alike, is tourism development. The UN has acknowledged tourism's potential to create jobs and promote local culture and products in its eighth sustainable development goal (SDG 8.9). However, such strategies have proven to be limited in effectiveness if formulated as general political demands, ignoring relevant interests and power constellations (Albert & Ringbeck, 2015, 159).

32.3 Politics of Scale and Heritage Administration in China

A promising theoretical and methodological approach, which takes the political nature and complex power structures related to heritage into account, is the critical lens of "politics of scale" (Swyngedouw, 1997; Brenner, 2001). The concept of scale in its significance for the production of space was first brought into focus by French philosopher and social theorist Henri Lefebvre (1901–1991). One of its founders, Scottish geographer Neil Smith (1954–2012), conceptualised scales as platforms for social activity and spatial manifestations of power relations which form nested hierarchies (e.g. global, regional, national and local) in capitalist societies (Smith, 2000, 725). As a geographical concept, it triggered a productive debate in the Anglo-American *Radical Geography* at the beginning of the 1990s, a period of extensive worldwide spatial reconfigurations. Peter Taylor's (1982) seminal essay and Smith's (1984) *Uneven Development* still proceed from a three-tiered scale model (global, national and urban). In the following scale debate, the concept was more precisely defined as being socially constructed rather than natural or given (Herod, 1991; Marston, 2000).

In more recent discussions, the concept has been introduced to additional research fields, such as political ecology (McCarthy, 2005; Köhler, 2008) and heritage studies (Butland, 2012; Lähdesmäki et al., 2019). In the latter, an important strength of scale lies in its consideration of the tangled hierarchies and changing positionalities of cultural heritage production with other forms of sociospatial structuration. While every action is embedded in local contexts, a purely local–local interpretation would risk obscuring differences in access possibilities and interaction spaces of the agents involved (Schmitt, 2011, 88). In the context of Chinese cultural landscapes, relevant sociospatial processes include not only commodification but also the formation of national identity. The following are key questions for a better understanding of such processes: Does a commodification of heritage uphold or reshuffle existing hierarchies in heritage governance? How are social groups empowered/disempowered through an appropriation of traditional knowledge for income generation? How do reconfigurations of heritage management create forms of inclusion/exclusion?

In a globally highly interconnected world, supranational (and subnational) institutions gain importance as regulatory mechanisms in political, environmental, cultural and economic terms because "new forms of institutional organisation, political authority, and economic coordination" are generated "above and below the national scale of state power" (Brenner, 2004, 7). This trend can be well observed in the establishment of the Chinese heritage conservation and administration system. Since its ratification of the World Heritage Convention in 1985, China has become an active applicant for World Heritage inscription of its cultural and natural properties. Apart from a sense of national pride and local patriotism, this great enthusiasm or "world heritage craze" (Gu et al., 2012, 55; Yan, 2018) is nurtured by hopes for economic growth and development. The process of rendering heritage conservation economically feasible comes with significant risks for authenticity and integrity loss. Through the adoption of its own conservation principles in 2000, the *China Principles* (Zhongguo wenwu guji baohu zhunze 中国文物古迹保护准则), China primarily strives to "downscale" globally established concepts and practices to lower administrative levels (Zhu, 2019, 25), while, at the same time, actively "upscaling" its own key concerns and integrating them into the international heritage discourse.

Within the country, the state has undertaken a number of scale-making processes in the domain of heritage governance, ranging from jurisdictional frameworks to heritage inventories and new administrative mechanisms. In a strictly hierarchical structure of state regulation, the State Administration of Cultural Heritage 国家文物局 heads subordinate heritage departments on lower scales (provincial, prefectural, municipal, county). Equivalent to the World Heritage nomination process, Chinese cultural relic entities (*wenwu baohu danwei* 文物保护单位) must follow a hierarchical nomination process from one administrative scale to another. Based on this rigid administrative structure, sites can be upscaled or downscaled or their title revoked in designation rounds (Wallenwein, 2020, 259). It is precisely in this process of surveying, recording and conserving that heritage acquires symbolic value and becomes "amenable to commodification" (Kowalski, 2011, 87). The following section explores how scalar strategies can be and are already utilised by different agents to consolidate, shift or challenge existing power relations in the transformation and regulation of heritage.

32.4 National and Regional Scalar Strategies

The commodification of heritage and, notably, tourism development in the Hani Terraces is deeply entangled with policy objectives and personal aspirations on different scalar levels. In its long-term struggle to reduce poverty, the Chinese central government launched a number of campaigns such as "Open up the West" (Xibu da kaifa 西部大开发) at the turn of the millennium to narrow the gap between developed coastal regions and rural areas of the western interior. Besides fostering endogenous economic development and social stability, related policy objectives are to strengthen state capacity and nation-building (Goodman, 2004, 317). Ethnic tourism development enables the state to economically integrate minority groups while simultaneously implementing its cultural policies (Zhu, 2018, 20–21). The strictly hierarchical administrative system of heritage in China further allows central government agencies to "downscale" their value systems to heritage institutions on regional and local scales, enabling the cultivation of selected aspects of ethnic identity.

One aspect that may lead to conflicts with the local population is the question of how to deal with traditional residential buildings, some of which are dilapidated and in bad condition. In their attempt to develop cultural tourism, authorities aim to produce a homogenous landscape of traditional "mushroom houses" (local dwelling resembling a mushroom in shape) that reflect local cultural characteristics as they are widely known through touristic marketing. Homeowners, on the other hand, are striving to improve their living conditions, transform or even abandon their houses, as in the case of Azheke 阿者科 village, after having generated sufficient income (Zhang & Stewart, 2017, 43). Other such aspects of Hani culture include traditional folk songs and dances, as well as rituals and customs related to terrace cultivation. This intangible cultural heritage, as well as its transmitters, has been meticulously listed on different administrative levels.

At a regional scale, Honghe Prefectural Government strives to exert influence and accumulate resources via integrating the terrace landscape into institutional structures across different policy fields. Primarily, the decision to adopt a tourism-based development strategy was coupled with the promotion of the Hani Rice Terraces as a UNESCO World Heritage Site. For better protection of the terraces, which is also a requirement for World Heritage listing, the prefectural government has adopted provisional management measures and a master plan since 2001, set up the Hani Terrace Authority in 2007 and established regulatory agencies at the county level one year later (FAO, 2010). With successful World Heritage designation, such newly established heritage departments move up the institutional scale, acquiring further administrative functions and funding (Zhu, 2019, 31). Moreover, the Prefecture Government made efforts to "upscale" the Hani Rice Terraces in the closely related fields of forestry and agriculture. In 2007, the State Forestry Administration approved parts of the terraces as a National Wetland Park, and they were inscribed as a Globally Important Agricultural Heritage System (GIAHS) pilot site in 2010 (Gu et al., 2012, 55). This integration of the landscape into further national and global-scale environmental programmes and the marketing of corresponding titles foster the establishment of a strong tourism brand and a competitive advantage, considering the number of outstanding scenic areas in Yunnan Province.

32.5 Local Scalar Strategies

With China's introduction of market-oriented reforms and fiscal decentralisation at the beginning of the 1980s, state officials on local scales received new incentives to engage in profit-seeking economic activities and act as "state entrepreneurs" (Duckett, 2001). Following established practices to draw on heritage as a tool for "improvement" (Oakes, 2013), Yuanyang 元阳 County adopted a government-led approach for cultural tourism development. This decision and its accompanying process of cultural commodification created and still creates tensions between desired economic benefits and transformations of local culture and heritage.

Controversial measures as they appeared in the early stages of tourism development in the terraces included the erection of totem poles, which are not part of Hani culture, and regular ethnic cultural performances (Gu et al., 2012, 56). In his study on social spaces in tourist settings, MacCannell (1973) has explained similar phenomena with touristic attitudes and their "quest for authentic experiences", which makes them susceptible to settings of "staged authenticity". More recent studies of ethnic tourism have questioned ideas of touristic performances as mere reductions of culture to its exchange value and see marketed cultural enactments as social practices, bearing great potential for refiguring and reclaiming ethnic identity (Cohen, 1988, 380; Bruner, 2004, 7; Comaroff & Comaroff, 2009, 27).

Natural resources embodying cultural meaning can also become objects of dispute. In the Hani Terraces, the commodification of a forest with high spiritual significance triggered a conflict between the local government, Yuanyang County's tourism bureau and residents. Led by objectives to satisfy the tourist gaze (Urry, 1990), the tourism bureau built a footpath across this forest in the northwest of Qingkou 箐口 hamlet, one of the earliest and central touristic areas. To express their opposition, residents blocked the footpath that disturbed ancestors' resting places and violated local spiritual traditions (Gu et al., 2012, 57–58). The importance of social struggles for the production of scale has already been demonstrated for different contexts (e.g. US labour relations, see Herod, 1991) in the 1990s scale debate. By blocking the footpath, local inhabitants mobilised themselves and reached out to higher scales where conservation of intangible heritage is a priority policy goal. By expanding their "space of engagement", they strengthened their position as custodians of heritage in their "space of dependence" (Cox, 1998).

Most conflicts in the terraces, however, are related to water as a highly limited resource. As public tap water supply is insufficient to meet the demands of both local communities and rapidly expanding tourism facilities, natural water sources are fiercely contested. In 2016, local farmers of the Yi ethnic group's Shengcun 胜村 hamlet cut down water pipes to hotels, which drew water from the main water source for irrigating their terraces. Although the reasons for water scarcity are natural as well as anthropogenic, local communities hold tourism development responsible for the decline in water resources (Hua et al., 2018, 10).

Finally, on a horizontal scale, new agents such as migrant businesspeople participate in tourism-related activities. With strong profit-oriented objectives, this social group migrates to locations highly frequented by tourists for selling related products or services. As shown in previous studies, this phenomenon is not exclusive to the Hani Terraces but also evokes sharp local–migrant tensions at the World Heritage Town of Lijiang in the north-western part of Yunnan Province, where businesspeople skimmed off tourism-generated revenues to a significant extent (Su, 2013).

32.6 Upscaling

In this complex situation of competing agents and interests, one opportunity to avoid or at least curb exploitative forms of commodification is to initiate upscaling processes of traditional knowledge and local interests. In earlier preservation projects of inhabited Chinese landscapes, human agricultural activity was considered harmful to nature, and, not infrequently, inhabitants were resettled (Han, 2018, 71). At least partly, this one-sided protection of natural values was related to efforts for World Natural Heritage inscription and ecotourism development. In the case of the

Hani Rice Terraces cultural landscape, social and cultural values have been more strongly considered, and customary law has been included in the conservation scheme as one form of local traditional knowledge to be protected.

According to traditional management, the terraces are irrigated with spring water diverted from mountain streams and channelled to the pond fields. Water allocation is managed collectively, monitored by a village-elected specialist (*laoga-laepha*), and manipulations are punished with fines (Bouchery, 2011, 328ff). While some particular spiritual and cultural practices of the Hani ethnic minority have fallen victim to the Cultural Revolution, customary rules for resource management seem to have been more resilient (Xu et al., 1999, 130). By preventing overexploitation of forests and through forms of social collaboration such as labour exchange, the recognition of customary institutions for managing land and water resources contributes to sustainable use of ecosystem resources (SDGs 6, 15), protection of the world's cultural and natural heritage (SDG 11), and mitigation of climate change (SDG 13).

In order to improve local livelihoods and to tackle out-migration, landscape maintenance and red rice cultivation need to be economically feasible. Well-regulated and community-based tourism activities can create new income opportunities for younger generations but can only be part of an integrated set of measures. In Yuanyang County, traditional knowledge is further promoted to "upscale" food production and management capacity of farmers through a cooperative business model. This model is based on the cooperation of four partners: the state-owned Yuanyang County Grain Purchase and Marketing Co., Ltd. 元阳县粮食购销有限公司, the Hani Terraces Organic Red Rice Professional Cooperative of Yuanyang County 元阳县哈尼梯田有机红米专业合作社, red rice farmers and a number of e-commerce companies. To encourage red rice cultivation, seeds are subsidised by the local government, the agricultural cooperative handles supply, packaging and processing, and farmers receive an above-market price (Li et al., 2020). While tourists are just one target group, organic agricultural products are offered to a much wider range of customers through e-commerce sales platforms. Yuanyang County has already set up four e-commerce platforms, such as "Yuanyang Mall" 元阳商城, with more than 3600 red rice shops registered (Office of the Central Cyberspace Affairs Commission, 2020). While this model enables a significant value enhancement of local products (SDGs 1, 8) through public–private cooperation (SDG 17), conditions remain largely set by the authorities, and farmers' involvement seems to be limited to economic participation.

32.7 Conclusion

Taking a scalar approach to the processual commodification of natural resources and traditional knowledge in agricultural landscapes sheds light on associated sociospatial processes and their impact on sustainable cultural development. In China, the vulnerability of living heritage to economic exploitation and its detrimental effects on heritage value are related to strong development agendas on different administrative scales in a dynamic and globalising environment. Similarly, the recognition and transmission of traditional knowledge offer great potential to trigger community-based development and empower its custodians, albeit within the boundaries of state regulation.

In the Hani Rice Terraces, ethnic tourism development produced "politics of scale" where national heritage authorities are concerned with issues of poverty alleviation and national unity, regional governments aim at harnessing heritage for allocating resources, local governments pursue economic growth, businesspeople strive to make profits and local people try to improve their living conditions. While governments on higher scales pursue strategies of inventorying and conserving heritage to uphold existing power relations, the rescaling of traditional knowledge provides opportunities to expand local spaces of engagement. However, as domestic tourism in China is expected to strongly increase in the future, strict regulation and management will be essential to avoid adverse commodification effects, such as those that occurred at the initial stages of tourism development in the Hani Rice Terraces. Therefore, a scalar approach should not be merely considered as a descriptive and analytical tool but rather a tool to actively initiate upscaling processes of local knowledge and interests to foster positive effects of commodification and achieve a balance between stakeholder interests.

Acknowledgements The author would like to thank the anonymous reviewers for their valuable insights and constructive comments.

References

Albert, M.-T., & Ringbeck, B. (2015). *40 Jahre Welterbekonvention: Zur Popularisierung eines Schutzkonzeptes für Kultur- und Naturgüter (Heritage studies)* (Vol. 2). De Gruyter.

Bouchery, P. (2011). Terrace cultivation and mental landscapes in Southern Yunnan. In M. Lecomte-Tilouine (Ed.), *Nature, culture and religion at the crossroads of Asia* (pp. 318–343). Social Science Press.

Brenner, N. (2001). The limits to scale? Methodological reflections on scalar structuration. *Progress in Human Geography, 25*, 591–614.

Brenner, N. (2004). *New state spaces: Urban governance and the rescaling of statehood*. Oxford University Press.

Bruner, E. M. (2004). *Culture on tour: Ethnographies of travel*. University of Chicago Press.

Butland, R. (2012). Defining Angkor: The social, economic and political construction of scale in the management of cultural heritage areas. In K. Taylor & J. L. Lennon (Eds.), *Managing cultural landscapes (Key issues in cultural heritage)* (pp. 192–210). Routledge.

Cohen, E. (1988). Authenticity and commoditization in tourism. *Annals of Tourism Research, 15*(3), 371–386.

Comaroff, J. L., & Comaroff, J. (2009). *Ethnicity, Inc. (Chicago studies in practices of meaning)*. University of Chicago Press.

Cox, K. R. (1998). Spaces of dependence, spaces of engagement and the politics of scale, or: Looking for local politics. *Political Geography, 17*(1), 1–23.

Duckett, J. (2001). Bureaucrats in business, Chinese-style: The lessons of market reform and state entrepreneurialism in the People's Republic of China. *World Development, 29*(1), 23–37.

Fan, C., & Oey, G. P. (1961). *Man Shu (Book of the Southern Barbarians)* (G. H. Luce, Trans., G. P. Oey, Ed.). (Data paper no. 44). Southeast Asia Program. Cornell University.

FAO. (2010). *Hani Rice Terraces, China*. Globally Important Agricultural Heritage Systems Programme Proposal. Retrieved June 10, 2021, from http://www.fao.org/3/bp781e/bp781e.pdf

Gao, X., Roder, G., Jiao, Y., Ding, Y., Liu, Z., & Tarolli, P. (2020). Farmers' landslide risk perceptions and willingness for restoration and conservation of world heritage site of Honghe Hani Rice Terraces, China. *Landslides, 17*, 1915–1924.

Goodman, D. S. G. (2004). The Campaign to "Open Up the West": National, Provincial-level and Local Perspectives. *The China Quarterly (London), 178*(178), 317–334.

Gu, H., Jiao, Y., & Liang, L. (2012). Strengthening the socio-ecological resilience of forest-dependent communities: The case of the Hani Rice Terraces in Yunnan, China. *Forest Policy and Economics, 22*, 53–59.

Han, F. (2018). World Heritage Cultural Landscapes: An old or a new concept for China? *Built Heritage, 3*, 68–84.

Herod, A. (1991). The production of scale in United States labour relations. *Area, 23*(1), 82–88.

Hosagrahar, J. (2017). Culture: At the heart of SDGs. *UNESCO Courier*. Retrieved January 13, 2021, from https://en.unesco.org/ courier/april-june-2017/culture-heart-sdgs

Hua, H., Zhou, S., Ding, Z., & Pan, Y. (2018). The change mechanism of human-environment interactions from the perspective of contextualization: A case study of the Honghe Hani Rice Terraces as a World Cultural Heritage Site. *Sustainability, 10*(2230), 1–17.

Jiao, Y., & Li, X. (2011). Ethnic culture and nature: Interactions in the Hani Terrace Landscape. In S.-K. Hong, J. Wu, J.-E. Kim, & N. Nakagoshi (Eds.), *Landscape ecology in Asian cultures (Ecological research monographs)* (pp. 29–40). Springer.

Köhler, B. (2008). Die Materialität von Rescaling-Prozessen: Zum Verhältnis von Politics of Scale und Political Ecology. In M. Wissen, B. Röttger, & S. Heeg (Eds.), *Politics of Scale: Räume der Globalisierung und Perspektiven emanzipatorischer Politik* (pp. 208–223). Westfälisches Dampfboot.

Kowalski, A. (2011). When cultural capitalization became global practice: The 1972 World Heritage Convention. In N. Bandelj & F. Wherry (Eds.), *The cultural wealth of nations* (pp. 73–89). Stanford University Press.

Lähdesmäki, T., Thomas, S., & Zhu, Y. (Eds.). (2019). *Politics of scale: New directions in critical heritage studies*. Berghahn Books.

Lennon, J. L. (2012). Cultural landscape management: International influences. In K. Taylor & J. L. Lennon (Eds.), *Managing cultural landscapes (Key issues in cultural heritage)* (pp. 45–70). Routledge.

Li, S. 李树芬, Fang, T. 方婷, & Xu, S. 许生菲. (2020). *Honghe Zhou Yuanyang Xian: Titian hongmi rang rizi huo qi lai* 红河州元阳县:梯田红米让日子火起来 (Honghe Prefecture, Yuanyang County: Red rice from the terraces brings about better days). *Yunnan Daily* 云南日报. Retrieved March 9, 2021, from www.yn.xinhuanet.com/nets/2020-12/21/c_139606291.htm

MacCannell, D. (1973). Staged authenticity: Arrangements of social space in tourist settings. *The American Journal of Sociology, 79*(3), 589–603.

Marston, S. A. (2000). The social construction of scale. *Progress in Human Geography, 24*(2), 219–242.

McCarthy, J. (2005). Scale, sovereignty and strategy in environmental governance. *Antipode, 37*(4), 731–753.

Oakes, T. (2013). Heritage as improvement: Cultural display and contested governance in rural China. *Modern China, 39*(4), 380–407.

Office of the Central Cyberspace Affairs Commission 中共中央网络安全和信息化委员会办公室. (2020). *Yunnan Sheng Honghe Zhou Yuanyang Xian shenhua wangluo fupin, dazao 'liu ge pingtai' zhutui dianshang jingzhun fupin* 云南省红河州元阳县深化网络扶贫 打造"六个平台"助推电商精准扶贫 (Yuanyang County in Honghe Prefecture, Yunnan Province, intensifies poverty reduction by use of the internet, building "six platforms" to boost precision poverty alleviation through e-commerce), Cyberspace Administration of China. Retrieved June 13, 2021, from www.cac.gov.cn/2020-07/22/c_1596970848192771.htm

Schmitt, T. (2011). *Cultural Governance: Zur Kulturgeographie des UNESCO-Welterberegimes. Erdkundliches Wissen, 149*. Franz Steiner Verlag.

Smith, N. (1984). *Uneven development: Nature, capital and the production of space*. Basil Blackwell.

Smith, N. (2000). Scale. In R. Johnston, D. Gregory, G. Pratt, & M. Watts (Eds.), *The dictionary of human geography* (4th ed., pp. 724–727). Blackwell. (Original work published 1981).

Su, X. (2013). Tourism, migration and the politics of built heritage in Lijiang, China. In T. Blumenfield & H. Silverman (Eds.), *Cultural heritage politics in China* (pp. 101–114). Springer.

Swyngedouw, E. (1997). Neither global nor local: 'Glocalization' and the politics of scale. In K. R. Cox (Ed.), *Spaces of globalization* (pp. 137–166). Guilford Press.

Takeuchi, K. (2010). Rebuilding the relationship between people and nature: The Satoyama Initiative. *Ecological Research, 25*(5), 891–897.

Taylor, P. J. (1982). A materialist framework for political geography. *Transactions – Institute of British Geographers (1965), 7*(1), 15–34.

Throsby, D. (2017). Culturally sustainable development: Theoretical concept or practical policy instrument? *International Journal of Cultural Policy, 23*(2), 133–147.

UCLG. (2004). *Agenda 21 for culture. Approved on May 8, 2004, by the 4th Forum of Local Authorities for Social Inclusion of Porto Alegre in Barcelona*, Spain, and adopted as reference document by United Cities and Local Governments. Retrieved May 30, 2021, from https://www.agenda21culture.net/sites/default/files/files/documents/multi/ag21_en.pdf

UNESCO. (1972). *Convention concerning the protection of the world cultural and natural heritage*. UNESCO.

UNESCO. (2013). *The Hangzhou Declaration: Placing culture at the heart of sustainable development policies*. (CLT-2013/WS/16). Adopted on May 17, 2013, in Hangzhou, People's Republic of China. Retrieved May 25, 2021, from https://unesdoc.unesco.org/ark:/48223/pf0000221238

Urry, J. (1990). *The tourist gaze: Leisure and travel in contemporary societies*. Sage.

Wallenwein, F. (2020). Tackling urban monotony: Cultural heritage conservation in China's historically and culturally famous cities. *CrossAsia-eBooks*. https://doi.org/10.11588/xabooks.748

Xu, J., Fox, J., Lu, X., Podger, N., Leisz, S., & Ai, X. (1999). Effects of swidden cultivation, state policies, and customary institutions on land cover in a Hani village, Yunnan, China. *Mountain Research and Development, 19*(2), 123–132.

Yan, H. (2018). *World Heritage craze in China: Universal discourse, national culture, and local memory*. Berghahn Books.

Zhang, L., & Stewart, W. (2017). Sustainable tourism development of landscape heritage in a rural community: A case study of Azheke village at China Hani Rice Terraces. *Built Heritage, 4*, 37–51.

Zhang, Y., Min, Q., Zhang, C., He, L., Zhang, S., Yang, L., Tian, M., & Xiong, Y. (2017). Traditional culture as an important power for maintaining agricultural landscapes in cultural heritage sites: A case study of the Hani terraces. *Journal of Cultural Heritage, 25*, 170–179.

Zhu, Y. (2018). *Heritage and romantic consumption in China (Asian heritages)*. Amsterdam University Press.

Part III
The World Heritage Convention – The Day After Tomorrow

Chapter 33
Shared Responsibility – A Guiding Principle of the World Heritage Convention – Perception – Implementation – Future

Marie-Theres Albert

Abstract Responsibility is a central category required for the protection of human heritage. But what does responsibility mean for the protection of our heritage today? Who was and is responsible for which form of responsibility, and how is it communicated? These central questions are derived from the theoretical basis of Hans Jonas' approach to our ethical responsibility for the consequences of technological development and Max Weber's approach to our political responsibility, which arises from the role of the state as a legitimised system of rule. Last but not least, reference is made to Hannah Arendt, who argues for individual human responsibility based on human morality. For the learning of responsibility and its implementation, reference is made to education on the basis of international conventions.

Keywords Personal responsibility · Social responsibility · Political responsibility · UNESCO mandate · Education

33.1 Introduction[1]

"…since wars begin in the minds of men, it is in the minds of men that the defences of peace must be constructed" (UNESCO, 1945, Preamble). This is the most important message of the UNESCO Constitution, signed on 16 November 1945. It is the most fundamental declaration of the responsibility that the international community

[1]The text of this chapter is translated with Deepl. Translator (n.d.). Retrieved June 5, 2021 from https://www.deepl.com/translator#de/en/. Manual editing by the author.

M.-T. Albert (✉)
Institute Heritage Studies at the Internationale Akademie Berlin (INA) gGmbH, Berlin, Germany
e-mail: albert@inaberlin.org

M.-T. Albert et al. (eds.), *50 Years World Heritage Convention: Shared Responsibility – Conflict & Reconciliation*, Heritage Studies,
https://doi.org/10.1007/978-3-031-05660-4_33

assumes for achieving peace in the world. The status of the legal, organisational and financial independence of UNESCO as the representative organisation of the people of the world as well as its *tasks, competences and responsibilities* are defined in article 63 of the Charter of the United Nations (UN), which was adopted on 26 July 1945.

In its introductory statement, the following is stated:

> We the peoples of the United Nations determined to save succeeding generations from the scourge of war, which twice in our lifetime has brought untold sorrow to mankind, and to reaffirm faith in fundamental human rights, in the dignity and worth of the human person, in the equal rights of men and women and of nations large and small, and to establish conditions under which justice and respect for the obligations arising from treaties and other sources of international law can be maintained, and to promote social progress and better standards of life in larger freedom. (U.N. Charter, intro)

As formulated in the UN Charter, the attribution of responsibility to people for peace in the world goes hand in hand with the attribution of rights that they must sustainably preserve and protect. The attribution of rights and duties and the concomitant responsibility of the global community continue to be the basis of the "*Universal Declaration of Human Rights*" adopted in 1948. The focus of this Declaration is, therefore, an important progenitor for the attribution of responsibility that people must assume for the protection of their heritage.

Heritage is identity-forming and is thus a constituent part of people's rights. The destruction of heritage was and is, therefore, always based on a deliberate disregard for people's rights. The *Declaration of Human Rights* was accompanied not only by an internationally recognised understanding of rights but also duties and responsibilities. Moreover, the responsibility of societies for keeping peace in the world through protecting human beings and their heritage is also the message of many other treaties or conventions adopted by the United Nations and other organizations. Concerning heritage, the famous ones are here presented in chronological order: *The Hague Convention for the Protection of Cultural Property in the Event of Armed Conflict*, 1954; the *Convention on the Means of Prohibiting and Preventing the Illicit Import, Export and Transfer of Ownership of Cultural Property*, adopted in Paris on 14 November 1970; the *World Heritage Convention* adopted 1972 in Paris; *the Underwater Cultural Heritage Convention from* 2001a; the *Convention for the Safeguarding of the Intangible Cultural Heritage* from 2003; or the *Faro Convention of the Council of Europe*, 2005.

Last but not least, the responsibility we all have for the sustainable protection of our world – including our heritage – is part of the 17 SDGs of the 2030 Agenda, which is on the way to being implemented worldwide.

33.2 Perception

Constructing peace in the "minds of men" and thus in the world as a whole requires awareness of what responsibility means and how it is to be assumed internationally, nationally and also individually. Awareness of responsibility and the willingness to take responsibility both require knowledge about the identity of human beings and how that identity is shaped. Furthermore, it requires consciousness of the important components of human identity and how an awareness of oneself helps to achieve peace in the world.

Based on these perceptions, further agreements were adopted by the United Nations.[2] From the beginning, this included an understanding of cultural diversity, which states that people can only live together in peace if they recognise that the world is made up of many cultures encompassing their heritage, traditions, lifestyles and expressions. The cultures in their diversity, including their varied components, are equally valuable and must therefore be respected and sustainably protected.

Furthermore, with the implementation of UNESCO's mission to anchor peace in the world, global educational processes were initiated with the intention of promoting science and culture and different disciplinary and epistemological positions. They not only highlighted the message of UNESCO and other international organisations, but they precisely developed concepts of how peace in the world could be achieved. And peace could and still can be achieved through accepting the responsibility of the human being for the sustainable safeguarding of heritage.

I would like to highlight *Structural Functionalism* as an epistemological approach of Critical Sociology in the interpretation of Norbert Elias[3] and the version of Cultural Studies strongly associated with Stuart Hall.[4] The responsibility of humans and societies for their heritage is also the message of one of the most famous scientists in UNESCO's context, namely the French *Structural Ethnologist* Claude Lévi-Strauss.[5] It is not surprising that the responsibility to protect the heritage of humankind is, therefore, the guiding principle of the World Heritage Convention.

[2] E.g. *The International Covenant on Civil and Political Rights,* 1966a *and the International Covenant on Economic, Social and Cultural Rights*, 1966b; The *Mexico City Declaration on Cultural Policies* from 1982; the *Universal Declaration of Human Responsibilities*, adopted in Tokyo on 1 September 1997 by the *Inter Action Council*; the *UNESCO Universal Declaration on Cultural Diversity* 2001b; The *Convention on the Protection and Promotion of the Diversity of Cultural Expression*, adopted 2005.

[3] This reference to Norbert Elias is mainly based on his book: "Über den Prozess der Zivilisation" (*The Civilizing Process*) 1976, see Norbert Elias 1976.

[4] Stuart Hall, together with Richard Hoggart and Raimund Williams, was a founder of British Cultural Studies, which has influenced the theoretical and political discourse on culture and heritage to today (Stuart, 1980).

[5] A selection of articles concerning Claude Lévi Strauss' important scientific and political positions are published in: The UNESCO Courier, 2008, number 5/ www.unesco.org/en/courier/cls/

But what does responsibility mean, what does assumption of responsibility mean? On what legal, social or even cultural basis must people bear responsibility, or how is the assumption of responsibility attributed to people and societies?

33.3 Implementation

The theme of this paper is the assumption of responsibility by people and societies for World Heritage. I first reflect on theoretical approaches with which responsibility for the protection of heritage can be interpreted and implemented. To this end, I refer to Hans Jonas, Max Weber and Hannah Arendt. Their interpretations of responsibility were and are theoretically fundamental and therefore also applicable to the protection of heritage.

I first refer to the German American philosopher Hans Jonas. With his world-famous book The Imperative of Responsibility: In Search of an Ethics for the Technological Age,[6] he contributed to international reflections on the social, cultural or political impact of technological progress on people and their natural environment and our overall ethical responsibility. One of his most famous statements on this is, "Act in such a way that the effects of your action are compatible with the permanence of genuine human life on earth" (Jonas, 1984, 36).

Hans Jonas first dealt with questions of human development as they were reflected in *Existentialist Philosophy* and, on this basis, formulated his ethical positions on the relationship between humans and nature in the wake of the development of technology. His "Imperative of Responsibility" was written at the end of the 1970s. In it, he considered the ontological effects of technological change and assigned people and societies responsibility for the consequences of these developments.

On the one hand, he wrote that responsibility has something to do with current life, with the "here and now… It is those who are alive now… who have a claim on my behaviour" (Jonas, 1984, 23). For the future, however, a "new dimension" of responsibility is needed. For this, it is necessary to consider "modern technology". This has introduced "actions of such novel consequences that the framework of earlier ethics can no longer grasp them" (Jonas, 1984, 26).

This new dimension goes beyond previous knowledge. "That is, that the predictive knowledge lags behind the technical knowledge that gives power to our actions, itself takes on ethical significance" (Jonas, 1984, 28). The new knowledge must "take into account the global condition of human life, and the distant future…" but is not really prepared for this. This concerns, for example, "nature as a human responsibility". It concerns not least the "biosphere as a whole and in its parts,

[6] The reflection on Hans Jonas' position is based on the German version of this book: Hans Jonas, Das Prinzip Verantwortung, Versuch einer Ethik für die technologische Zivilisation, Suhrkamp Taschenbuch, 1984.

which is now subject to our power and has something like a moral claim on us" (Jonas, 1984, 29).

The problem outlined here also concerns our responsibility to protect our heritage. World Heritage represents the heritage of all humanity. It has been created, among other things, through technical developments over the centuries and, at the same time, is permanently exposed to "technological change". Responsibility for sustainable protection of heritage is, therefore, to be derived from the effects of technological change on World Heritage itself.

This attribution of responsibility is formulated in the preamble of the World Heritage Convention (UNESCO, 1972). It identifies the many dangers to which World Heritage has been, and still is, exposed over the years and which the international community is responsible for eliminating. For example, the first paragraph states that "… the cultural… and the natural heritage are increasingly threatened with destruction not only by the traditional causes of decay, but also by changing social and economic conditions…", e.g. based on technological developments (UNESCO, 1972, para. 1). As a consequence, the international community must take care to anticipate potential dangers of technological developments and prevent the negative consequences of such developments.

It remains to be seen whether the challenge of technological change for heritage conservation can be sustainably implemented with the assumption of responsibility demanded here. Much has been achieved in implementing the criteria of "Outstanding Universal Value" (OUV),[7] laid down in the "Operational Guidelines"[8] of the Convention, and in implementing the criteria of "Authenticity"[9] and "Integrity".[10] The sporadic transfer of a site to the so-called "List in Danger",[11] due to intermediate damage to the OUV, has also been able to correct threats in most cases.

In the face of real technological developments and growing threats to World Heritage, much remains to be done. To return to Hans Jonas, it is necessary to formulate and implement an ethic for technological civilisation to anticipate and avoid potential damages to our heritage. However, the world community is still far away from achieving those forward-looking assumptions in the implementation of the many criteria for the protection of World Heritage.

Hans Jonas was one of the theorists who recognised the positive and negative effects of technological developments on people relatively early and derived from this the duty to assume responsibility. In the case of World Heritage, the following are just a few examples: the destruction of landscapes by extractive industries; loss

---

[7] The Outstanding Universal Value (OUV) is determined by ten criteria for inscribing a Heritage Site as World Heritage on the World Heritage List.

[8] The Operational Guidelines are the basic principles for the implementation of the World Heritage Convention and for managing World Heritage Sites. The OPG are regularly updated, most recently in July 2019.

[9] Authenticity is one of the basic principles for meeting OUV, see Operational Guidelines § 79–86.

[10] Integrity is one of the basic principles for meeting OUV, see Operational Guidelines § 87–95.

[11] The List in Danger defines the circumstances under which the OUV is threatened and proposes measures to be taken. Operational Guidelines §177–191.

of biodiversity in natural heritage sites; ambivalence of renewable technologies with problematic impacts on habitats, humans and animals; displacement of local knowledge in agricultural regions by technological developments; the endangerment of natural catastrophes around the world; or the threatening of our heritage by climate change in general etc.

The relevance of his insight becomes clear not only by answering questions about who recognises the consequences of such developments and takes responsibility for them; his ethical reflection also motivates questions about how responsibility should be implemented. Does taking responsibility mean, for example, correcting technological developments or taking them back? Who is responsible for this? How should responsibility be implemented? Hans Jonas did not answer these questions. Nevertheless, they are part of his ethics and irreplaceable for the sustainable protection of World Heritage. After 50 years of the World Heritage Convention, it is high time to address them, to derive concepts for action from them and to implement them.

Complementary to the ethical responsibility of Hans Jonas, as a responsibility to anticipate the effects of technological development on our heritage, is the reflection of the German sociologist and national economist Max Weber. Max Weber became known in particular for his critical analyses of social structures at the beginning of the twentieth century. He presented his concept of political responsibility, summarised in his famous critiques of the "ethics of mind versus the ethics of responsibility" (Weber, 1992, 70f.),[12] which he presented in a student lecture in January 1919. The title of this lecture was later published with the title "Politics as Profession".

If one compares the political constellation of societies in 1918/19 with that of the founding period of the UN and UNESCO in 1945, then references to the experiences of people and societies in the years 1918/19 are easy to make. Both were epochs in which national competencies of nation states for peace in the world and for implementing responsibility were urgently needed and required international commitment through the founding of the League of Nations in 1920[13] and the United Nations in 1945.

In this lecture, he argued that any political responsibility derives essentially from the rules and norms enshrined in the constitution of the state, which attribute to it "the monopoly of legitimate physical violence" (Weber, 1992, 6). "The state is,… a relationship of domination of people over people based on the means of legitimate (…) violence" (Weber, 1992, 7f), regardless of the form of government. It concerns the individual state just as much as the state system of the United Nations with 193 member states or, via the latter, the monopoly of the adoption and enforcement of international agreements.

[12] The English texts are based on translations of the German edition, Reclam, Stuttgart 1992.

[13] The League of Nations was founded as an intergovernmental organisation as a result of the Paris Peace Conference on 10 January 1920. It can be described as the forerunner of the United Nations. It dissolved on 18 April 1946.

Relevant to the attribution of political responsibility to states for their populations is the abovementioned focus on an "ethics of responsibility", which he distinguishes from the "ethics of mind". He justifies this by saying that the responsibility of human action in the "ethics of mind" is not attributed to human beings themselves but to "God". This is precisely what makes people helpless and irresponsible. Therefore, it is man's "self-responsibility" to manage peace and justice as well as the protection of heritage whose implementation is formulated in the "ethics of responsibility".

Weber (1992, 70f) writes:

> …it is an abysmal contradiction whether one acts according to the ethical maxim – religiously speaking – 'the Christian does right and entrusts success to God', or according to the ethical maxim of responsibility: that one has to pay for the (foreseeable) consequences of one's actions.…
>
> "If the consequences of an action flowing from pure sentiment are bad, he [*the ethicist of sentiment*, emphasis added] does not hold the person acting responsible, but the world, the stupidity of other people", etc. "The ethicist of responsibility, on the other hand, will say: these consequences are attributed to my actions." (Weber, 1992, 70f.)

It is precisely this individual and societal ethic of responsibility that also defines the protection of humanity's heritage. With Max Weber's (1992, 8) position, this responsibility is based on the legitimate rule of the state as well as on the political system, which in modern societies is based on "domination by virtue of legality, by virtue of the belief in the validity of legal statutes and the factual 'competence' justified by rationally created rules". This includes "on the one hand, the attitude of human action" and "obedience" to the bearers of "legitimate power" and "on the other hand, by means of this obedience, the disposal of… material goods" (Weber, 1992, 8).

UNESCO, as a "specialized agency" within the United Nations (UN) community of states, can be interpreted on this basis as an institution comparable to a state system. With the "mandate of contributing to achieve peace in the world" (U.N. Charter, chap. X, art. 63), as a community of states, it bears the responsibility for peace defined by it in the same way as the states themselves. In Max Weber's political approach, this attribution of political power and rule is then delegated by the states to their political functionaries and implemented by them. This then also involves the people as representatives of a structure of rules based on legality (Weber, 1992, 8).

If one reflects on the responsibility for peace and justice, equality and dignity or the protection of heritage ascribed to the people of the United Nations and its member states – formulated in the charters, conventions or declarations already mentioned – then they are comparable in style. Heritage forms identity, and identity is necessary for the creation of peace, justice and equality. In order to achieve these goals, responsibility – to argue once again with Max Weber – is transferred to the "system of rule". In the case of responsibility for human heritage, this concerns the United Nations as a community of states. It concerns the member states and their protagonists, the people themselves.

This system can be exemplified by the Declaration of Human Rights. In the introductory statement, the following is said: "The General Assembly proclaims this

The Universal Declaration of Human Rights as a common standard of achievement for all peoples and all nations … to secure their universal and effective recognition and observance, both among the peoples of Member States themselves and among the peoples of territories under their jurisdiction" (United Nations, 1948, intro.).

Defining rights is one thing; implementing them is another. The implementation of rights and also of duties is rarely done on the basis of declarations alone. Rather, this requires, among other things, educational measures, which are also formulated in the introductory statement of this declaration. Thus, it is then stated that "to the end that every individual and every organ of society, keeping this declaration constantly in mind, shall strive by teaching and education to promote respect for these rights and freedoms and by progressive measures, national and international" (United Nations, 1948, Preamble).

In this respect, it can be said that the *Declaration of Human Rights* was accompanied by an internationally recognised understanding of the rights, duties and responsibilities of human beings to achieve and maintain peace in the world and to protect their heritage. It was furthermore accompanied by a variety of important scientific publications which provided theoretical backgrounds and scientific contexts as well as practical recommendations (Langfield et al., 2010; Levy & Sznaider, 2010).

If one looks at the international implementation of The *Declaration of Human Rights* and, in particular, at the non-observance and violations, grave doubts remain about the interest of the international community in its implementation. Have the states, the international community and the people failed? Was Max Weber wrong in his theoretical analysis and in his vision?

The conclusion is that the observance of human rights and obligations is not self-explanatory. Rather, it must be communicated, in general and within the signatory states of the Declaration. It must also be communicated to individuals. Only if the individual has sufficient knowledge about the rights and duties arising from the Declaration will they be able to interpret this Declaration for themselves. If, in addition, they have the corresponding moral values, they will also be prepared to bear responsibility. Only then can this Declaration be implemented successfully at national and international levels.

This brings me to Hannah Arendt's concept of responsibility, which seems important to me as another dimension of the responsibility outlined here – ethical, political, in person. On the basis of Hans Jonas' ethical principle and Max Weber's political principle, I will now reflect upon Hannah Arendt's remarks on individual responsibility. I will demonstrate the individual and personal responsibility that people must assume for the protection of heritage. As a Jewish German American scientist, Hannah Arendt became known internationally for proving that people were personally responsible for Hitler's dictatorship. To this day, she stands for a theoretical reflection on people's personal responsibility in and with the Third Reich that is so fundamental that it can be applied across societies, situations or systems.

The remarks I refer to here are set out in the lecture "What does personal responsibility mean in a dictatorship" (Arendt, 2020),[14] which she gave in the USA in the early 1960s.[15] In this lecture, she focuses on the individual and thus provides a theoretical reflection on the personal responsibility of people in and with the Third Reich that has not been repeated to this day. It is at the same time a critical reflection on the morality of people's right or wrong behaviour in and for the "maintenance (of laws that) we consider essential for the integrity of our human community" (Arendt, 2020, 15).

The basis of this lecture is a reflection on the guilt of the Germans for the Hitler dictatorship and on who bears what responsibility as a result. The central paradigms of Hannah Arendt already become clear in the introductory remarks of this lecture. There she deals with the "deceptive assumption of a collective guilt" of the Germans for the dictatorship.

Arendt (2020, 14) writes:

> The attribution of a 'fallacious assumption of collective guilt' (of the Germans)… is… a concept that was applied for the first time to the German people and their collective past, a conception that has proved in practice to be a highly effective cleansing…; where everyone is guilty, no one is guilty.

In spite of such a white washing interpretation of responsibility and guilt, there is the "morality of human behaviour", which is considered the basis of all humane values and from which the dimension of individual responsibility is derived. (Arendt, 2020, 15ff.) According to Hannah Arendt, morality is not based on laws. It is nevertheless her leitmotif as shown by the example of the international acceptance of the *Declaration of Human Rights*.

Her criticism of the attribution of collective guilt to the Germans for the Hitler dictatorship and the justification derived from it for the long-overdue assumption of individual responsibility for the system was, therefore, groundbreaking. It not only named individual responsibility for social processes. It also justified it with morality for human behaviour. In this respect, the responsibility of the individual and their responsibility for the heritage of humanity is also established. However, in order for individuals to be able to assume their responsibility, they must have the necessary knowledge to do so, because, in general, people are not prepared for such "moral questions" (Arendt, 2020, 16ff.) and do not have the necessary knowledge to implement their responsibility.

This is precisely why educational processes need to be initiated. These must impart both ethical-moral standards for the protection of World Heritage and the knowledge necessary for its sustainable protection, which brings me back to the declarations and conventions mentioned at the beginning of this paper. The call for responsible implementation of the Convention through education is formulated in §

[14] All English texts are based on the translation of this German edition.

[15] Since the lecture was not published during Hannah Arendt's lifetime, specific dates regarding the place and time of the lecture are not known. The first German-language publication was published in 2020 by Piper Verlag gGmbH.

27. Whereas in § 28, this task is delegated to those politically responsible in the sense of Max Weber.

Article 27(1) states:

> The States Parties to this Convention shall endeavour by all appropriate means, and in particular by educational and information programmes, to strengthen appreciation and respect by their peoples of the cultural and natural heritage defined in Articles 1 and 2 of the Convention. 2. They shall undertake to keep the public broadly informed of the dangers threatening this heritage and of the activities carried on in pursuance of this Convention. (UNESCO, 1972, art. 27(1))

Article 28 states:

> States Parties to this Convention which receive international assistance under the Convention shall take appropriate measures to make known the importance of the property for which assistance has been received and the role played by such assistance. (UNESCO, 1972, art. 28)

Responsibility for the protection of the heritage of humankind, therefore, entails an ethical responsibility in the sense of Hans Jonas and, in particular, a responsibility that looks to the future. This has arisen and continues to arise as a result of changing social, economic and technological developments in the world with their corresponding effects on the heritage of humankind. It refers to the mission and role of states and the community of states, which are responsible as "institutions of governance" in the sense of Max Weber. Last but not least, it refers to the individual responsibility and its assumption by individuals. Taking responsibility for the protection of heritage is, therefore, always a personal task. Only when the individual is willing and able to protect heritage can it be protected in the long term.

33.4 Future

I think that today, more than ever, more attention needs to be paid to communicating the importance of heritage for people and societies and for their identities. This happens through the initiation of educational processes and across institutions and multidimensionally. That is, from cognitive processes to analytical – from reflective processes to experimental – from empirical learning processes to processes that require abstractions. It is necessary to anchor the moral relevance of human rights or heritage protection in people's value scales and to enable them to implement them.

Only by communicating the importance of heritage for people can the responsibility and morality in the sense of Hannah Arendt become comprehensible, even though they are already named in many of the above mentioned conventions or charters. In this respect, the protection of our heritage can and must be anchored more strongly than before in formal and non-formal educational processes. There are many ways to implement this. Some of those possibilities were presented and discussed within the framework of the international conference that we conducted last year for the launch of this book, and others have been tested by the Institute Heritage Studies in several educational projects.

The project "Transboundary European World Heritage – a Topic for UNESCO Associated Schools", is an important example intended to encourage students and teachers to explore their joint European history and thus their heritage with the goal of learning responsibility for the future (World Heritage Education, https://worldheritage-education.eu/en). With a comparable goal and result, in the project "Our World Heritage – Mining Cultural Landscape Erzgebirge/Krušnohoří", we focussed on students and teachers in creating videos reflecting on historical and intercultural understandings (Institute Heritage Studies, 2021).

Nevertheless, it remains to be said that, against the background of the developments endangering World Heritage, the continuation of the discourse on responsibility is an ongoing challenge that societies, states and their communities and not least individuals must face.

References

Arendt, H. (2020). *Was heißt persönliche Verantwortung in einer Diktatur?* (What does personal responsibility mean in a dictatorship?). Piper Verlag.

Council of Europe. (2005). *Council of Europe framework convention on the value of cultural heritage for society (Faro Convention)*. Council of Europe.

DeepL Translator. (n.d.) *DeepL Translator*. Retrieved June 5, 2021, from https://www.deepl.com/translator#de/en/.%20Manual

Elias, N. (1976). *Über den Prozess der Zivilisation* (The civilizing process) (Vol. 1–2). Suhrkamp.

Hall, S. (1980). Race, articulation and societies structured in dominance. In UNESCO (Ed.), *Sociological theories: Race and colonialism* (pp. 305–345). UNESCO.

Institute Heritage Studies. (2021). *Category: Completed projects*. https://heritagestudies.eu/en/category/projects/completed-projects/

Inter Action Council. (1997, September 1). *Universal declaration of human responsibilities*. Inter Action Council.

Jonas, H. (1984). *Das Prinzip Verantwortung, Versuch einer Ethik für die technologische Zivilisation* (The imperative of responsibility: In search of an ethics for the technological age). Suhrkamp Taschenbuch.

Langfield, M., Logan, W., & Craith, M. N. (Eds.). (2010). *Cultural diversity, heritage and human rights: Intersections in theory and practice*. Routledge.

Levy, D., & Sznaider, N. (2010). *Human rights and memory*. Pennsylvania State University Press.

United Nations. (1945, June 26). *Charter of the United Nations, Introduction & Art. 63*.

United Nations. (1948). *Universal Declaration of Human Rights*. United Nations.

United Nations. (1966a). *The International Covenant on Civil and Political Rights*. United Nations.

United Nations. (1966b). *The International Covenant on Economic, Social and Cultural Rights*. United Nations.

UNESCO. (1945). *United Nations Educational, Scientific and Cultural Organization Constitution*. UNESCO.

UNESCO. (1954). *Hague convention for the protection of cultural property in the event of armed conflict*. UNESCO.

UNESCO. (1970). *Convention on the means of prohibiting and preventing the illicit import, export and transfer of ownership of cultural property*. UNESCO.

UNESCO. (1972). *Convention concerning the protection of the world cultural and natural heritage*. UNESCO.
UNESCO. (1982, July 26–August 6). *Part IV Mexico City Declaration on cultural policies, World Conference on Cultural Policies*. UNESCO.
UNESCO. (2001a). *Convention on the protection of the underwater cultural heritage*. UNESCO.
UNESCO. (2001b). *Universal declaration on cultural diversity*. UNESCO.
UNESCO. (2003). *Convention for the safeguarding of the intangible cultural heritage*. UNESCO.
UNESCO. (2005). *The Convention on the protection and promotion of the diversity of cultural expression*. UNESCO.
UNESCO. (2008). Claude Lévi-Strauss: The view from Afar. *The UNESCO Courier*, *2008*(5). https://en.unesco.org/courier/2008-5.
UNESCO World Heritage Centre. (2019). *Operational guidelines for the implementation of the World Heritage Convention*. UNESCO.
Weber, M. (1992). *Politik als Beruf* (Politics as profession). Reclam Verlag.
World Heritage Education. (n.d.). *Transboundary European World Heritage – A topic for UNESCO Associated Schools*. World Heritage Education. https://worldheritage-education.eu/en

Chapter 34
World Heritage and Reconciliation

Birgitta Ringbeck

Abstract The series of World Heritage sites that have been attacked demonstrates that the reconstruction of cultural properties after conflicts and crises is more than the rebuilding and restoration of material and substance; ideally, it is a recovery process regaining social cohesion and cultural identity, which leads to reconciliation in post-trauma societies. If this succeeds, reconstruction is a value and an attribute for authenticity. Thus, the World Heritage program contributes to the constitutional mandate of UNESCO and to reconciliation and peace as a central mission.

Keywords World Heritage · Reconstruction · Recovery · Reconciliation

> At this particular time in history, as the fabric of civilized human society seems increasingly under attack by forces that deny the very existence of a shared heritage, forces that strike at the very of our sense of community, I am convinced that the World Heritage holds out a contrary and positive vision of human society and human future. (Train, 2002, 3)

Saying this in the year after the blowing up of the Buddha statues in the Bamiyan Valley on 11 March 2001 and the terrorist attack on September 2001, Russel E. Train, founder of the WWF and one of the pathfinders of the Convention (Cameron & Rössler, 2013, 289), implored the reconciling power of World Heritage. At the conference celebrating its 30th anniversary in Venice on 16 November 2002, Train (2002, 3) underlined that the purpose of the World Heritage concept is more than simply helping to assure the protection and conservation of unique natural and cultural sites; just from the beginning, it aims to "instil in the world's peoples a new sense of our kinship with one another as part of a single, global community".

The World Heritage Programme thus contributes to the constitutional mandate of UNESCO and its central mission

B. Ringbeck (✉)
Federal Foreign Office, World Heritage Coordinating Body (retired November 2022), Berlin, Germany
e-mail: birgitta.ringbeck@freenet.de

M.-T. Albert et al. (eds.), *50 Years World Heritage Convention: Shared Responsibility – Conflict & Reconciliation*, Heritage Studies,
https://doi.org/10.1007/978-3-031-05660-4_34

> that a peace based exclusively upon the political and economic arrangements of governments would not be a peace which could secure the unanimous, lasting and sincere support of the peoples of the world, and that the peace must therefore be founded, if it is not to fail, upon the intellectual and moral solidarity of mankind. (UNESCO, 1945, preamble)

This task remained the key challenge in the first two decades of the twenty-first century demanding "UNESCO's response to the rise of violent extremism" (Bokova, 2021).

Despite the destruction of the giant Buddha statues by the Taliban, the Cultural Landscape and Archaeological Remains of the Bamiyan Valley were inscribed as a serial site on the World Heritage List with reference to all criteria relevant for cultural heritage in 2003; the justification of criteria (vi) refers to the intangible attributes highlighting that "... due to their symbolic values, the monuments have suffered at different times of their existence, including the deliberate destruction in 2001, which shook the whole world" (UNESCO, 2003a). In the decision document (UNESCO, 2003b), a reconstruction of the Buddha statues in the colossal niches in the Bamiyan Cliff is not envisaged, though an anastylosis, as indicated in Article 15 of the Venice Charter, is at least considered a reasonable way to protect the remaining fragments in the evaluation report (ICOMOS, 2003). In the following years, extensive safeguard measures were implemented with international support, keeping open both options (Petzet, 2009; Emmerling & Petzet, 2016).

Two years later, the World Heritage Committee had to decide upon the inscription of the Old Bridge in Mostar, which was completely destroyed on command of the Croatian defence council during the conflicts in the Balkans on 9 November 1993. Criteria (iv) and criteria (vi) were listed as relevant in the evaluation report, but the committee only recognised the latter stating that:

> With the "renaissance" of the Old Bridge and its surroundings, the symbolic power and meaning of the City of Mostar – as an exceptional and universal symbol of coexistence of communities from diverse cultural, ethnic and religious backgrounds – has been reinforced and strengthened, underlining the unlimited efforts of human solidarity for peace and powerful cooperation in the face of overwhelming catastrophes. (UNESCO, 2005a)

Accordingly, the statement of authenticity is based on the attribute that "...the reconstruction of the fabric of the bridge should be seen as the background to the restoration of the intangible dimensions of this property" (UNESCO, 2005a). *The Operational Guidelines* revised in the same year were supplemented by the provision that "In relation to authenticity, the reconstruction of archaeological remains or historic buildings or districts is justifiable only in exceptional circumstances. Reconstruction is acceptable only on the basis of complete and detailed documentation and to no extent on conjecture" (UNESCO, 2005b).

Thus, the committee followed up on the justification of the inscription of the Historic Centre of Warsaw. This early inscription documents that from the beginning reconstruction was not only a question of the "...verification of conservation doctrines and practices" but also about "...the inner strength and determination of the nation, which brought about the reconstruction of the heritage on a unique scale in the history of the world" (UNESCO, n.d.-a). The evaluation report of ICOMOS expressly points out that "... the criterion of authenticity may not be applied in its strict sense"; however, the conditions are met because the Historic Centre of Warsaw represents "…a finished concept of post-war reconstruction" realized from 1945 to 1965 (ICOMOS, 1980). Thus, even in the first phase of the implementation of the World Heritage Convention, post-war recovery was recognised as a value conveying the outstanding universal value of a site.

In 2012, the 40th anniversary of the World Heritage Convention was celebrated on the occasion of the 36th session of the committee held in Saint Petersburg, Russian Federation, from 24 June to 6 July 2012. The meeting was overshadowed by shocking news from Mali. In May and June 2012, members of the Islamic militant organization Ansar Dine had destroyed the Sidi Mahmud Ben Amar mausoleum in Timbuktu and announced further attacks on other mausoleums. Therefore, the committee included the site in the List of World Heritage in Danger on 28 June 2012. Ansar Dine reacted immediately by destroying the mausoleums of Sidi Mahmud, Sidi Moctar and Alpha Moya in mockery of UNESCO in the days that followed.

The World Heritage Committee condemned the destruction of World Heritage sites in Mali and decided on measures to help the country protect its heritage (UNESCO, 2012). However, this time, it was not just a stirring appeal, the international community reacted. Heritage protection became an integrated part of the peacekeeping mandate of the United Nations Multidimensional Integrated Stabilization Mission in Mali (MINUSMA, n.d.). Three years after their destruction by extremists, the Timbuktu mausoleums were nearly restored through the extraordinary work carried out by local craftsmen and with international support. On the sidelines of the 39th session of the World Heritage Committee meeting in Bonn (Germany 2015), on behalf of UNESCO's Director-General, Irina Bokova, a UNESCO medal was presented to Alassane Hasseye, head of the Timbuktu masons, in recognition of work carried out by his guild (UNESCO, 2015). One year later, the International Criminal Court (ICC) recognised the destruction in Timbuktu as a war crime and sentenced the rebel leader Ahmad Al Faqi Al Mahdi to nine years in prison. Al Mahdi pleaded guilty.

Other World Heritage sites have not been able to recover as fast from destruction resulting from armed conflicts in the last decade. Due to political reasons and the security situation, international assistance on site under the UNESCO umbrella could not be organized so far for the six Syrian World Heritage sites or the Old City of Sana'a and the Old City of Shibam in Yemen, which were inscribed on the List of World Heritage in Danger in 2013 and 2015, respectively. Moreover, the Islamic State of Iraq and the Levant (ISIL, also known as Daesh) extremists who publicly beheaded the Syrian archaeologist and guardian of Palmyra, Dr. Khaled al-Asaad, have not yet been brought to justice.

In the face of all this deliberate damage to cultural heritage, particularly in the Middle East, the Director-General of UNESCO, Irina Bokova, launched the Global Coalition, Unite for Heritage, during the 39th World Heritage Committee meeting in Germany. Well received by the international community, the #Unite4Heritage campaign has since become a widely expanding global movement devoted to culture's unifying force and the mobilization of the global audience, with young people as its core demographic (https://www.unite4heritage.org/). It expands on UNESCO's efforts to coordinate technical work among the different specialized agencies and institutions by facilitating experts meetings, for example, on the safeguarding of

Syria's heritage in Berlin 2016 (UNESCO, 2016) and flagship initiatives like "Revive the Spirit of Mosul" for the recovery of one of Iraq's iconic cities (UNESCO, n.d.-b). In addition, the campaign is designed to support endeavours of Member States, e.g. the resolution 69/281 of 28 May 2015 "Saving the cultural heritage of Iraq" jointly presented by Germany and Iraq to the Plenary Meeting of the UN General Assembly and cosponsored by a total of 91 Member States. The unanimously adopted resolution condemns the barbaric acts of destruction and looting of the cultural heritage of Iraq carried out by ISIL, deplores the rising incidence of intentional attacks against the cultural heritage of countries affected by armed conflicts and emphasizes that heritage has an important role to play in the efforts of national reconciliation and reconstruction.

The United Nations Security Council Resolutions 2199 (2015) and 2347(2017) condemning the destruction of cultural heritage and the United Nations General Assembly resolution 69/281 on Saving the Cultural Heritage of Iraq mark the mobilization of the international community. The protection of cultural heritage has become one of the priorities at the highest political level, confirmed again by the first meeting devoted to culture in the history of the G20 on July 29 and 30, 2021 (https://www.g20.org).

In their Rome Declaration, the G20 Ministers of Culture state the following:

> Convinced that cooperation and dialogue are vital in the fight against violent extremism we express our strongest condemnation of the deliberate destruction of tangible and intangible cultural heritage, wherever it takes place, as it irreversibly affects the identities of the communities, damages human rights and community identity, erasing legacies of the past and damaging social cohesion. We support initiatives taken to protect endangered cultural heritage and restore destroyed or damaged cultural heritage. (G20 Research Group, 2021)

Moreover, they express their conviction "…that multilateral efforts, with UNESCO at the core, are crucial…" (G20 Research Group, 2021).

Without the World Heritage Convention, which is primarily recognised through the World Heritage List, the protection and preservation of cultural heritage would not have achieved such international recognition. In the last 50 years, standards have been developed and set the course of the implementation of the Convention, which have shaped the doctrines and practice of monument preservation far beyond the preservation of World Heritage sites. The initially very Western approaches and concepts for the preservation of cultural properties have been revised. One reason for this is certainly that interdisciplinary discourses in the World Heritage community are never purely technical but always also political. In an exemplary way, the attitude to and evaluation of reconstructions demonstrates the shift from the Charter of Venice (1964), which actually only allows reconstruction if there are enough parts that can be reassembled, over the Nara Document on Authenticity (1994) underlining the significance of intangible values, to the Warsaw Recommendation on Recovery and Reconstruction of Cultural Heritage (2018) addressing post-trauma challenges.

The reconstruction of cultural properties after conflicts and crises is more than the rebuilding and restoration of material and substance; ideally, it is a recovery process regaining social cohesion and cultural identity, which leads to reconciliation in post-trauma societies. If this succeeds, reconstruction is a value and an attribute for authenticity, as recently confirmed by the decision of the World Heritage Committee to recognise the ShUM Sites of Speyer, Worms and Mainz. The property was listed as a "pioneering ensemble of Jewish diasporic community centres … from the High Middle Ages" bearing witness to "… profound developments in the formation phase of the continuing cultural tradition …" and "… cultural achievements of Ashkenazic Jews…" (UNESCO, 2021) in Central Europe north of the Alps in 2021. The history of the communities also includes that they have been repeatedly jeopardised by pogroms over the centuries. During the period of National Socialism between 1933 and 1945, Jewish life was almost exterminated in Germany. In November 1938, when all synagogues in the "Deutsche Reich" went up in flames, the one in Worms was also destroyed. The "post-trauma reconstructions" as noted with regard to authenticity in the site's statement of outstanding value have not just "… been carried out respectfully and …retained the heritage significance of the monuments" between 1957 and 1961; they mark the beginning of the long journey for peace and reconciliation after the Shoah.

References

Bokova, I. (2021). *UNESCO'S response to the rise of violent extremism: A decade of building international momentum in the struggle to protect cultural heritage.* J. Paul Getty Trust. https://www.getty.edu/publications/occasional-papers-5/

Cameron, C., & Rössler, M. (2013). *Many voices, one vision: The early years of the World Heritage Convention*. Routledge.

Emmerling, E., & Petzet, M. (Eds.). (2016). *The Giant Buddhas of Bamiyan II: Safeguarding the remains 2010–2015* (Monuments and Sites XXI). H. Bäßler.

ICOMOS. (1980). *Advisory Body Evaluation (ICOMOS).* https://whc.unesco.org/en/list/30/documents/30bis-ICOMOS-2068.pdf

ICOMOS. (2003). *Evaluations of cultural properties.* (WHC-03/27COM/INF.8A ADD). UNESCO. https://whc.unesco.org/archive/2003/whc03-27com-inf08a-adde.pdf

MINUSMA. (n.d.). *Cultural heritage.* MINUSMA. https://minusma.unmissions.org/en/cultural-heritage

Petzet, M. (Ed.). (2009). *The Giant Buddhas of Bamiyan. Safeguarding the remains*. (Monuments and Sites XIX). H. Bäßler.

G20 Research Group. (2021). *Rome Declaration of the G20 Ministers of Culture.* University of Toronto. http://www.g20.utoronto.ca/2021/210730-culture.html

Train, R. E. (2002, November 16). *Remarks of the Honorable Russell E. Train.* [Written Speech]. UNESCO. https://whc.unesco.org/archive/websites/venice2002/speeches/pdf/train.pdf

UNESCO. (1945). *United Nations Educational, Scientific and Cultural Organization Constitution*. UNESCO.

UNESCO. (2003a). *Cultural landscape and archaeological remains of the Bamiyan Valley.* UNESCO. https://whc.unesco.org/en/list/208

UNESCO. (2003b). *Decision: 27 COM 8C.43 cultural landscape and archaeological remains of the Bamiyan Valley (Afghanistan)*. UNESCO. https://whc.unesco.org/en/decisions/628

UNESCO. (2005a). *Old bridge area of the Old City of Mostar*. https://whc.unesco.org/en/list/946

UNESCO. (2005b). *Operational guidelines for the implementation of the World Heritage Convention.* (WHC.05/2). UNESCO. https://whc.unesco.org/archive/opguide05-en.pdf

UNESCO. (2012). *World Heritage Committee calls for end to destruction of Mali's heritage and adopts decision for its support*. UNESCO. https://whc.unesco.org/en/news/907/
UNESCO. (2015). *Reconstruction of Timbuktu mausoleums near completion*. UNESCO. https://whc.unesco.org/en/news/1307/
UNESCO. (2016). *Director General opens Berlin expert meeting on the safeguarding of Syria's heritage*. UNESCO. https://whc.unesco.org/en/news/1503
UNESCO. (2021). *Extended forty-fourth session of the World Heritage Committee: Item 8 of the provisional Agenda: Establishment of the World Heritage List and of the List of World Heritage in Danger*. (WHC/21/44.COM/8B). UNESCO. https://whc.unesco.org/archive/2021/whc21-44com-8B-en.pdf
UNESCO. (n.d.-a). *Historic Centre of Warsaw*. https://whc.unesco.org/en/list/30
UNESCO. (n.d.-b). *Revive the spirit of Mosul*. UNESCO. https://en.unesco.org/fieldoffice/baghdad/revivemosul
United Nations Security Council. (2015). *Resolution 2199 (2015) adopted by the Security Council at its 7379th meeting, on. 12 February 2015*. https://www.undocs.org/S/RES/2199%20(2015)
United Nations Security Council. (2017). *Resolution 2347 (2017) adopted by the Security Council at its 7907th meeting, on 24 March 2017*. https://www.undocs.org/S/RES/2347%20(2017)

Chapter 35
Sustainability – A Guiding Principle of the World Heritage Convention – What Has Been Achieved – What Is Missing – What Is the Future Perspective

Constanze Fuhrmann

Abstract The World Heritage Convention (WHC), as the premier charter for heritage protection, has long adopted sustainability as a core principle. With today's understanding of heritage management as an important driver of economic growth, social capital and environmental protection, its reconciliation with sustainable practices is inevitable. Fifty years on, the WHC faces new challenges concerning promoting and supporting sustainable development. With the broad adoption of SDGs, heritage actors working under the WHC framework frequently encounter conflicting objectives. While in theory, the social, ecological and economic dimensions of sustainable development can be reconciled, in practice, this often requires finding viable and balanced trade-offs. The growing awareness for climate change in the last ten years results in the need to reprioritise the different dimensions and can lead to hard choices that have been framed as dilemma situations. This paper takes a closer look at those dilemma situations and discusses how the WHC can help tackle these and make the right decisions in the face of complex sustainability choices.

Keywords World heritage convention · Sustainability · Sustainable development · Sustainable development goals · Dilemma · Conflicts

35.1 Introduction

In the face of climate change, the loss of biodiversity and a steadily growing world population, the world is facing major challenges. Only a fundamental transformation towards sustainability can overcome the resulting global problems. The Sustainable Development Goals (SDGs) as an important guidepost for this

C. Fuhrmann (✉)
German Federal Environmental Foundation (DBU), Osnabrück, Germany
e-mail: c.fuhrmann@dbu.de

M.-T. Albert et al. (eds.), *50 Years World Heritage Convention: Shared Responsibility – Conflict & Reconciliation*, Heritage Studies,
https://doi.org/10.1007/978-3-031-05660-4_35

transformation have found their way into political frameworks and policies at the local, national and global levels. However, the SDGs have turned out to be difficult to integrate into the everyday activities of businesses, administrations, NGOs and civil society at large as their complex interdependencies and inherent tensions make them hard to navigate. For many heritage actors, adopting the SDGs is accompanied by insecurity and a need for guidance and support.

The World Heritage Convention (WHC), as the premier charter for heritage protection and management, has long adopted sustainability as a core principle. With today's understanding of heritage management as an important driver of economic growth, social and cultural capital and environmental protection, its reconciliation with sustainable development is inevitable. Consequently, the 40th anniversary of the World Heritage Convention was, among others, dedicated to the four pillars of sustainable development, namely social, cultural, environmental and economic development (Albert et al., 2017, 18–45). Heritage management helps to "shape, delineate and drive the development of tomorrow's societies" (ICOMOS, 2011, 9), which connects it to sustainability in two fundamental ways: First, heritage preservation is a sustainable practice because it represents resources that are protected and passed on to future generations with their material and cultural value. Heritage's role in communities as an anchor of identity and cultural self-reflection strongly contributes to social cohesion and a more resilient society. Second, given that the heritage sector is a significant player in the broader social and economic eco-system in many regions, the sector has a responsibility to actively contribute to more sustainable ways of working, living and producing (Boccardi, 2007).

Fifty years on, the WHC faces new challenges concerning promoting and supporting sustainable development. With the broad adoption of the SDGs, heritage actors working under the WHC framework frequently encounter conflicting objectives. While, in theory, the social, ecological and economic dimensions of sustainable development can be reconciled, this often requires finding viable and balanced trade-offs in practice. The growing awareness of climate change (Council of Europe, 2018) in the last ten years means that different dimensions must be reprioritised, leading to hard choices that have been framed as dilemma situations.

This paper examines dilemma situations in heritage contexts and discusses how the WHC can help tackle these. The first section provides a short overview of sustainability and its inclusion in the WHC is provided. The second section discusses the role of dilemmas in the context of sustainable development. In the third section, typical dilemma situations are highlighted with examples from heritage sites around the world. The final section concludes with a summary of the role the WHC can play in supporting heritage actors to make the right decisions in the face of complex sustainability choices.

35.2 Sustainability and the World Heritage Convention

Sustainable development has long found its way into the "*Convention concerning the Protection of the World Cultural and Natural Heritage*", or World Heritage Convention (WHC), which the General Conference of UNESCO adopted in 1972. Since then, the WHC has contributed significantly to more sustainable development by improving conservation policies and capacity building at the national level. It also fostered the involvement and participation of stakeholders in discussion around heritage and created new opportunities through innovative financial mechanisms (Bandarin, 2015, 35).

According to its preamble, the WHC was intended to heighten awareness that "cultural and natural heritage is increasingly threatened with destruction not only by decay but also by changing social and economic conditions which aggravate the situation with even more formidable phenomena of damage or destruction" (UNESCO, 1972, 1). With the high ratification numbers and ensuing commitment among nations, the WHC is recognised as "one of the most effective international instruments and a milestone in the conservation world" (Bandarin, 2015, 35).

The WHC does not directly mention sustainable development. However, the concept of sustainability is reflected in articles 4 and 5 by calling on member states to ensure

> "the identification, protection, conservation, presentation and transmission to future generations of the cultural and natural heritage", as well as "to adopt a general policy which aims to give the cultural and natural heritage a function in the life of the community and to integrate the protection of that heritage into comprehensive planning programmes" (UNESCO, 1972, articles 4 and 5a, 3).

Contracting states must guarantee the preservation of their national cultural heritage and include development strategies in their policies. Important development goals of the WHC are tied to building strong communities and promoting sustainable tourism and other forms of economic activities.

Therefore, the WHC can be seen to have adopted sustainability as a core guiding principle for the conservation and preservation of heritage sites, recognising the role heritage plays for society and local and national economies. In line with this, the WHC Operational Guidelines were amended in 2011 to integrate the principles of sustainable development more firmly. In response to the United Nations Millennium Development Goals, these amendments (in paragraphs 6, 112, 119, 132, and in Annex 5, points 4.b and 5.e) aimed to ensure sustainable practices in the use and management of World Heritage properties and in maintaining the Outstanding Universal Value (UNESCO et al., 2013, 19).

With the adoption of the so-called 2015 policy,[1] sustainable development perspectives were finally integrated into the practice of the convention – coupled with

[1] In accordance with the Agenda 2030, the General Assembly of the States Parties to the World Heritage Convention adopted new guidelines for leveraging cultural heritage to achieve sustainable development goals.

the recognition of the necessity for a "wider change" (UNESCO, 2015, para 5, 2) without undermining the Outstanding Universal Value of the listed heritage in pursuit of sustainable development (Boccardi & Scott, 2018; Labadi, 2017, 49–51). Although a new set of requirements for all relevant dimensions of sustainable development was introduced in addition to the original text from 1972, the Operational Guidelines did not provide sufficient practical recommendations in this regard (Cave & Negussie, 2017, 30–31). Possible conflicting objectives (e.g. heritage conservation vs. Sustainable Development Goals) should not be regulated by the convention but remain the responsibility of the individual states.

35.3 New Goals, New Problems

With the growing importance of sustainable development in the WHC, problems and conflicts of interests that have been discussed in the wider context of sustainability have also been imported (Labadi, 2017). While the overarching idea of sustainability is straightforward and can be seen as a welcome extension of the WHC guidelines, the devil is in the details. It is easy to call for sustainability in broad terms. Yet, it has not always been clear how measures can be practically implemented, given inherent conflicts that can hinder protection efforts.

A key problem of sustainability in practice is conflicting objectives and the resulting dilemmas. A dilemma is defined as a situation in which "a difficult choice has to be made between two or more alternatives, especially ones that are equally undesirable" (Knowles, 2021) or, after Kirchner, between two (action) alternatives, but neither stands out or both have negative consequences (Regenbogen et al., 2013).

In most cases, the available alternatives both lead to suboptimal outcomes and sometimes even outright undesirable results. Often, dilemmas require a trade-off between two objectives. Regardless of which option is chosen, the outcome is far from perfect: The chosen alternative contributes to achieving one objective at the cost of other objectives. There must be a clear distinction between conflicts and dilemmas. According to Müller, conflicts arise from incompatible interests of actors and can be resolved if the interests of the situation are made consistent (Müller-Christ et al., 2017, 14–15.).

This situation becomes clear when looking at the United Nations 2030 Agenda for Sustainable Development and its Sustainable Development Goals (SDGs) set up in 2015 and building on the previous eight Millennium Development Goals. The 17 SDGs are each dedicated to a global challenge and set out 169 specific targets to achieve a range of objectives, e.g. ensuring human dignity, protecting the planet, securing peace, fostering prosperity and promoting global cooperation (UNHCR, 2017). However, the general approach to sustainability entails a fundamental difficulty. Considered one of the most significant global political agendas of this century and lauded for their commitment to overcoming sectorial and geographic siloes in favour of international collaboration, the SDGs are also criticised for setting competing objectives that impede broad adoption and swift implementation (Henkel et al., 2018, 147–149).

The competing goals are rooted in apparent difficulties simultaneously achieving economic growth, fostering social cohesion and stopping ecological degradation. SDG 8 calls for sustainable economic growth, and SDG 12 demands sustainable consumption and production, highlighting the inherent contradictions. Achieving ongoing economic expansion (with the desired minimum GDP growth rate of 7% for least-developed nations) is clearly at odds with avoiding further environmental degradation. Even with rapid advances in green/clean tech and more sustainable modes of production, a full reconciliation of these SDGs seems elusive, and the resulting tension between the objectives cannot be fully resolved. Conflicting situations also often emerge around sustainable development on a more practical level: While wind turbines can help accelerate the desired transition to renewable energy, at the local level, they can also negatively affect wildlife and thus lead to increased environmental damage.

As sustainability refers to heterogeneous objectives and different social groups claim sustainability for themselves, the concept loses contour. The concept of sustainability involves specific dilemmas resulting from incommensurable goals, criteria and interests, and the question arises, which criteria can be used to assess the sustainability of projects, topics or initiatives (Henkel et al., 2018, 147–149).

35.4 Dilemmas in Heritage Management

The SDGs play an important role in sustainable cultural and natural heritage management and result in conflicting objectives. Their adoption in the WHC creates a need to deal with dilemma situations constructively. Within cultural heritage, they come with unique characteristics and require awareness and appreciation to develop suitable solutions. Most of the conflicting situations within heritage protection and management result from the need to reconcile the desire to preserve unique sites and objects with the practical requirements dictated by the climate crisis and changes in usage resulting from the recognition as cultural heritage, e.g. with an official designation by UNESCO. Furthermore, the nomination of World Heritage sites is often driven by economic reasons and not by an understanding of sustainable development (Labadi, 2017, 47). In this context, the following three examples illustrate common dilemma situations usually found in practice, each with specific characteristics and challenges:

35.4.1 Dilemma 1: Protection vs. Authenticity

The core objective of any cultural heritage management is the preservation and protection of places and objects of significant cultural value, encapsulated in the concept of Outstanding Universal Value as a key criterion for designation by UNESCO (UNESCO, 1972). According to the World Heritage Operational Guidelines,

> "a property should meet the test of authenticity in design, materials, workmanship, and setting; authenticity does not limit consideration to original form and structure but includes all subsequent modifications and additions, over the course of time, which in themselves possess artistic or historical values" (UNESCO, 1978, 4).

Maintaining the authenticity and integrity of cultural and natural heritage sites is often a challenging goal.

Global human-induced climate change is increasingly impacting World Heritage sites. According to the International Union for Conservation of Nature, climate change has become the biggest threat to natural World Heritage, a third of which is endangered (UNESCO, 2020). UNESCO has long recognised this negative impact (von Schorlemer, 2020, 17–30), but the World Heritage Committee only recently updated its "Policy Document on the Impacts of Climate Change on World Heritage properties" to include broad consultation of all Word Heritage stakeholders (UNESCO, 2021).

Difficult trade-offs can result from the need to balance the preservation of a heritage site's authentic character with community wellbeing. The desire to deal with increasing heat in urban areas due to climate change is one example of this conflict. While installing shades or additional roofing to protect against the sun or planting trees may be highly desirable from a health standpoint, such alterations change the appearance of historic buildings and town centres and thus reduce their authenticity. Other examples of dilemma situations arise when necessary climate adaptation strategies come at the expense of the original substance and authenticity of heritage. To protect heritage sites from further degradation, they might require adaptation; for example, flood gates installed in parts of Hamburg are necessary to protect against potential disasters and rising sea levels but alter the appearance of the historic substance.

35.4.2 Dilemma 2: Protection vs. Economic Potential

Heritage sites offer significant economic benefits for their regions and local communities, especially once officially designated as World Heritage. In the last 50 years, the WHC has created many opportunities through official designations and the resulting increasing interest and attention. A designated heritage site attracts visitors that often become a vital source of income for local communities in the form of tourism and the various revenue streams (European Court of Auditors,

2020, 26, 38, 46). Harvesting the economic potential of heritage is frequently detrimental to sustainable management on two levels. On the first level, the increased number of visitors produces more wear and tear on a heritage site and is thus directly detrimental to its preservation for future generations. On the second level, the increased number of visitors leads to increased travel and thus an increased indirect carbon footprint attributable to the heritage site, aggravating ecological degradation and accelerating climate change (Albert & Ringbeck, 2015, 174–182).

The pyramids of Giza in Egypt are formidable examples of this dilemma. Having received World Heritage status in 1979, increased tourism led to a troubling physical transformation of the site, which is also the last remaining wonder of the ancient world. It has also led to a host of adjacent environmental problems due to a large number of visitors. Consequently, the site and its surroundings suffer from severe environmental problems caused by unsustainable tourism and increased waste, air and traffic pollution (Hemeda & Sonbol, 2020, 8). Another well-known example of this dilemma is the city of Venice, which was designated as a World Heritage site in 1987. The boom in tourism and the increase in boat traffic intensified the preservation problems of the historic city and its canals. Heavy swells not only threaten the structures of the historic buildings, but ship propellers also cause a higher oxygen level in the water, favouring the growth of putrefactive bacteria and increasing problems for the conservation of the facades and pillars. To accommodate the growth of boat traffic and ever-larger vessels, the port entries were deepened, leading to easier access for flood tides that threaten the historic substance, which are now mitigated by a system of locks with controversial ecological impacts.

35.4.3 Dilemma 3: Protection vs. Traditional Use

The WHC has always fostered the empowerment of local communities as an important objective of heritage management. Heritage sites are important anchors of identity for the local populace and contribute to its overall wellbeing and resilience. From this standpoint, it is essential to foster cultural heritage's role in the community and preserve traditional usage patterns and access to sites. The designation of heritage sites attracts many more visitors and often restricts local communities' access. While local communities might benefit economically through tourism and adjacent business models, the site often becomes detached from their everyday use. This leads to a dilemma situation: The very protection that preserves the role of a heritage site in the community breaks the traditional usage patterns and often restricts usage or access.

Illustrative examples of this dilemma are the old towns of Prague and Dubrovnik. While these used to be the centre of gravity for both cities with locals living and working in these culturally important districts, they now resemble theme parks primarily frequented by tourists. The ensuing rise in real estate prices in these areas force out local businesses that do not cater to tourists and prevent the local community from residing there. They are no longer vibrant parts of the community but special economic zones solely devoted to tourism.

35.5 How to Deal with Heritage Dilemmas

Dilemmas and the resulting conflicts of interest can be observed at many World Heritage sites. Despite the importance of this topic, it has been little discussed, let alone researched, in cultural heritage science. The heritage community needs to find constructive ways to deal with these challenging situations and achieve the best possible outcome for all stakeholders and within different disciplines. Preservation activities are not merely technical approaches regulated by standards established in a single discipline; they always represent a broader negotiation where compromises among various interests and expectations must be found.

Effective heritage management is about making the right decisions, which is difficult if the choice reflects a dilemma. Research on sustainability and dilemmas and strategies to deal with dilemma situations emerges almost entirely from economics, where coordination of independent actors in the face of conflicts of interest or objectives has been discussed extensively.[2] One upshot of this wider context has been a discussion on dilemma situations specific to sustainability and sustainable development with proposals for decision frameworks (Müller-Christ, 2007, 2010; Hahn et al., 2010, 17; Vilanova et al., 2009, 64f). In essence, the various proposed strategies could be subsumed under three main approaches: ignorance, prioritisation and ongoing balancing.

With a strategy of ignorance, inherent conflicts between opposing targets that lead to dilemma situations are often simply ignored. While, in practice, a conflict between opposing targets exists, the actors pretend that it does not. The most overt form of dealing with a dilemma by ignoring it is outright denial. This strategy is rhetorically often concealed by talk of a "win–win" situation, which claims that both opposing targets can be achieved simultaneously (Müller-Christ, 2007, 146–147). This optimism about achieving everything at once is often justified by vague references to technological progress or other outside forces. A more subtle form of ignorance is abstraction, which refers to hiding the existing conflict by lifting the conversation to a level of abstraction on which the opposing targets are not visible at first glance (Müller-Christ, 2007, 144–146).

Prioritisation represents a more intellectually honest form of dealing with a dilemma situation. The opposing objectives are acknowledged and made visible. One or both objectives are altered based on case-specific criteria, e.g. by lowering the targets for one of the opposing objectives to allow a viable trade-off. This is the

[2] See research project "Dilemmata der Nachhaltigkeit zwischen Evaluation und Reflexion" at Technical University of Braunschweig, Germany, which for the first time combines research methods from evaluation research, governance research, philosophy of science and social research. https://www.tu-braunschweig.de/philosophie/dilemmata (Retrieved August 16, 2021).

case when damage to a heritage site by a large number of visitors is accepted in return for the resulting economic benefits. Such a trade-off is often not stated explicitly but rather achieved by accepting that one objective is not fully reached. By making such choices and achieving a specific trade-off, conflicting objectives are prioritised. Ideally, such an alteration should be a sustainable compromise that does not lead to the complete abandonment of one of the objectives and all are still pursued, even if only to a lesser degree. Instead of lowering or reprioritising objectives, the balancing strategy is based on a reconciliation process in which the different options are continuously (re-)calibrated. This strategy entails moving from a one-time decision compromise to an ongoing series of decisions. Müller-Christ introduces the metaphor of the tightrope walker for actors that follow this strategy: Instead of permanently leaning to one side, there is a constant balancing act and understanding that the inherent conflict exists (Müller-Christ, 2007, 160–161). This marks a shift from a static to a dynamic perspective of handling dilemma situations. While arguably the most "mature" approach, this demands a great level of openness for debate and a high tolerance for ambiguity.

In real life, heritage management decisions are often a mixture of these archetypical approaches. There is no one-size-fits-all strategy for dealing with the inherent sustainability dilemmas of heritage. The best solution for a specific heritage site depends on its specifics and context.

35.6 The Role of the WHC in Dealing with the Dilemmas of Sustainability

As the main document for worldwide heritage protection, the WHC has adopted sustainable development as a guiding principle. It demands that the three dimensions of sustainable development (social, economic and ecological) are considered, and heritage protection activities contribute to all three. As discussed, the inherent conflicts often observed between these dimensions and respective targets create dilemma situations for stakeholders who adhere to the WHC.

Action is needed on three levels to support heritage stakeholders in finding suitable solutions and trade-offs: discourse, decision and nudging. The first step in constructively dealing with inherent tensions between Sustainable Development Goals is an open debate about the sometimes challenging compromises and trade-offs necessary in practice. Instead of just setting out objectives in all three sustainability categories and pretending that a harmonious equilibrium can be achieved, a discussion is needed around finding trade-offs that work. Guidelines are also needed on how national, regional and local actors can best engage in and moderate fruitful discussions about overcoming inherent tensions and reconciling conflicting goals.

Finding a balance between opposing targets does require decisions. Here, the WHC should provide frameworks and guidelines for such decision processes. These should be based on best practices, past findings and outcomes of reconciliation processes around heritage sites. Guidelines that help to begin and moderate a process of balancing sustainability targets need to be practical and specific to the challenges in the context of cultural or natural heritage sites. Commonly agreed practices and benchmarks can support decisions in difficult situations. Consistent frameworks support overcoming procedural complications and make it easier to define a sound compromise. Many dilemma problems are not new, but threats to heritage sites from tourism, development or conflict are considered site-specific without a unified approach to these common and recurring phenomena. In providing a forum for discussion and guidelines for dealing with dilemma situations and overcoming the lack of sustainability reports for World Heritage sites, the WHC would strengthen its position as a forward-thinking platform ready for the future. Such measures would also be easy to implement and immediately make a difference for heritage stakeholders worldwide.

In the face of the current climate crisis, some voices demand further action. Anthropogenic climate change presents a threat of such magnitude that it permanently shifts the priority of the different dimensions of sustainable development in favour of ecological questions. While social and economic concerns remain important, they can only be considered after ecological concerns are addressed. This thinking opposes the longstanding primacy of the social aspect of sustainability, as promoted by the WHC. Accordingly, the importance of local communities and their empowerment was highlighted, for example, during the 40th anniversary of the World Heritage Convention in 2012 (Albert & Ringbeck, 2015, 174–182). The Kyoto Vision document issued on this occasion focused on the role of the community in conservation management and stressed the need to strengthen the relationship between people and heritage. Ecological challenges were mostly addressed in the context of the relationship between World Heritage properties and community engagement. Reducing the risks of climate change has been linked to strengthening local communities and capacities among relevant actors who should be fully involved in management and conservation activities. The "Strategic Action Plan for the Implementation of the Convention, 2012–2022" also demanded to contribute to the sustainable development of the "world's communities and cultures" (UNESCO, 2011, para 1, 2). In the face of the global climate crisis, it is justifiable to question whether the WHC is leveraged enough to compel actors on all levels towards prioritising ecological factors in their decision-making. A reinforced stance on climate change is justified by its negative global impacts on heritage sites, causing irreversible damage with varying degrees of severity and frequency and greater consequences over time. Proponents of a more robust approach to climate change propose that UNESCO not only nudges heritage stakeholders to prioritise ecological concerns but also uses its power to designate heritage sites as World Heritage to influence actors on different levels directly. By withholding or withdrawing designations or placing heritage sites on the list of endangered sites, UNESCO can directly influence economic and social benefits for local actors. This can be used as a bargaining tool for more ecologically sustainable development.

The Great Barrier Reef in Australia is a current example that illustrates this approach. There is pressure from environmental initiatives to place the reef on the

List of World Heritage in Danger, using climate change as an argument for classification (Readfearn, 2021). The site has suffered significant degradation due to increased water temperatures and greenhouse gas emissions (Great Barrier Reef Marine Park Authority, 2019). However, the Australian government is accused of not addressing UNESCO's requests for adequate climate protection measures under the WHC and not supporting emission policies to keep global warming below 1.5 °C above pre-industrial levels. Inclusion on the endangered list might increase pressure on the Australian government to finally take the required protective measures.

While some would welcome such a robust stance from UNESCO, this could arguably represent mission creep, with UNESCO crossing a boundary towards an increasingly activist approach. Given the complexity of successfully mitigating greenhouse gas emissions on a national level, which involves hard choices regarding economic and energy policy, it is questionable how much such an approach by UNESCO would achieve in practice. Some activists place inflated hope in the power of UNESCO to enforce the WHC against resistance. Nevertheless, even these more drastic views of the appropriate role and level of enforcement of the WHC encourage a productive discussion around the self-understanding of UNESCO and the definition of its global role. In an ideal scenario, it can get national actors to achieve better internal coordination for heritage management.

In any case, UNESCO needs to engage in the ongoing debate around sustainable development and define a clear way forward to preserve and protect our World Heritage in times of significant change and accelerating risk. Outlining ambitious and often conflicting targets without providing explicit guidance on reconciliation and agreeing on necessary trade-offs will not be enough. Formulating suitable answers that heritage actors and stakeholders agree on requires time, ongoing exchange and additional interdisciplinary research. Recommended long-term management practices for heritage sites will not be achieved without critical interpretation, nor will the sustainability strategy be adhered to, which the EU Member States have recently committed (European Court of Auditors, 2020, 35, 8). This will especially be the case if culture and cultural heritage continue to be insufficiently included in countries' national sustainability strategies and political practice, and ongoing processes of negotiation and communication are still required (Merkel & Möller, 2017, 112–121).

References

Albert, M.-T., & Ringbeck, B. (2015). 7 prospects for the future. In M.-T. Albert & B. Ringbeck (Eds.), *40 years World Heritage Convention: Popularizing the protection of cultural and natural heritage* (pp. 174–182). De Gruyter.

Albert, M. T., Bandarin, F., & Pereira Roders, A. (Eds.). (2017). *Going beyond. Perceptions of sustainability in Heritage Studies* (Vol. No. 2). Springer.

Bandarin, F. (2015). Sustainability in the World Heritage Convention: The making of a policy framework. In M. Albert (Ed.), *Perceptions of sustainability in Heritage Studies* (pp. 35–46). De Gruyter.

Boccardi, G. (2007). *World Heritage and sustainability; Concern for social, economic and environmental aspects within the policies and processes of the World Heritage Convention*. Unpublished M.Sc. Thesis, University College London.

Boccardi, G., & Scott, L. (2018). A view from the inside. An account of the process leading to the adoption of the policy for the integration of a sustainable development perspective within the World Heritage Convention. In P. B. Larsen & W. S. Logan (Eds.), *World heritage and sustainable development: New directions in world heritage management* (pp. 21–36). Routledge.

Cave, C., & Negussie, E. (2017). World Heritage and sustainable development. In C. Cave & E. Negussie (Eds.), *World Heritage conservation* (pp. 15–44). Routledge.

Council of Europe. (2018). *Recommendation CM/Rec (2018)3 of the Committee of Ministers to Member States on cultural heritage facing climate change: increasing resilience and promoting adaptation* (CM/Rec (2018)3). Council of Europe. https://search.coe.int/cm/Pages/result_details.aspx?ObjectId=0900001680791160

European Court of Auditors. (2020). *Special report 08/2020: EU investments in cultural sites: topic that deserves more focus and coordination*. European Court of Auditors. https://op.europa.eu/webpub/eca/special-reports/cultural-investments-08-2020/en/

Great Barrier Reef Marine Park Authority. (2019). *Position statement: Climate change* (Document no. 100486). Great Barrier Reef Marine Park Authority. https://elibrary.gbrmpa.gov.au/jspui/handle/11017/3460

Hahn, T., Figge, F., Pinske, J., & Preuss, L. (2010). Trade-offs in corporate sustainability: You can't have your cake and eat it. *Business Strategy and the Environment, 19*, 217–229.

Hemeda, S., & Sonbol, A. (2020). Sustainability problems of the Giza pyramids. *Heritage Science., 8*, 8.

Henkel, A., Bergmann, M., Karafyllis, N., Siebenhüner, B., & Speck, K. (2018). Dilemmata der Nachhaltigkeit zwischen Evaluation und Reflexion. Begründete Kriterien und Leitlinien für Nachhaltigkeitswissen. In N. Lüdtke & A. Henkel (Eds.), *Das Wissen der Nachhaltigkeit. Herausforderungen zwischen Forschung und Beratung* (pp. 147–172). Oekom Verlag.

ICOMOS. (2011). 17th general assembly and scientific symposium 'Heritage, driver of development' 27 November – 2 December 2011. *ICOMOS News, 18*(1), 1–28.

Knowles, E. (2021). Dilemma. In *The Oxford dictionary of phrase and fable*. https://www.encyclopedia.com/humanities/dictionaries-thesauruses-pictures-and-press-releases/dilemma

Labadi, S. (2017). UNESCO, World Heritage, and sustainable development: International discourses and local impacts. In P. Gould & K. Pyburn (Eds.), *Collision or collaboration. One World Archaeology* (pp. 45–60). Springer.

Merkel, C., & Möller, L. (2017). Nachhaltigkeit und Kultur. Die Vielfalt kultureller Ressourcen für die Nachhaltigkeitsstrategie heben. In G. Michelsen (Ed.), *Die Deutsche Nachhaltigkeitsstrategie. Wegweiser für eine Politik der Nachhaltigkeit* (pp. 112–121). Hessische Landeszentrale für politische Bildung. https://hlz.hessen.de/wp-content/uploads/2020/01/Forum_04_Nachhaltigkeitsstrategie.pdf

Müller-Christ, G. (2007). Formen der Bewältigung von Widersprüchen. Die Rechtfertigung von Trade-offs als Kernproblem. In G. Müller-Christ, L. Arndt, & I. Ehnert (Eds.), *Nachhaltigkeit und Widersprüche: Eine Managementperspektive* (pp. 128–177). LIT Verlag.

Müller-Christ, G. (2010). *Nachhaltiges Management. Einführung in Ressourcenorientierung und widersprüchliche Managementrationalitäten*. Baden-Baden Nomos.

Müller-Christ, G., Giesenbauer Bror, M., & Tegeler, K. (2017). *Studie zur Umsetzung der SDG im deutschen Bildungssystem*. Rat für Nachhaltige Entwicklung.

Readfearn, G. (2021, June 4). UN body pushed to demand stronger climate action from Australia to save Great Barrier Reef. *The Guardian*. https://www.theguardian.com/environment/2021/jun/04/un-body-pushed-to-demand-stronger-climate-action-from-australia-to-save-great-barrier-reef

Regenbogen, A., Meyer, U., & Kirchner, F. (2013). *Wörterbuch der philosophischen Begriffe*. Meiner.

UN High Commissioner for Refugees (UNHCR). (2017). *The Sustainable Development Goals and addressing statelessness*. UNHCR. https://www.refworld.org/docid/58b6e3364.html

UNESCO. (1972). *Convention concerning the protection of the world cultural and natural heritage*. UNESCO.

UNESCO. (1978). *Operational guidelines for the implementation of the World Heritage Convention*. UNESCO. https://whc.unesco.org/archive/opguide78.pdf

UNESCO. (2011). *18th session of the general assembly of states parties to the convention concerning the protection of the world cultural and natural heritage. Future of the World Heritage Convention* (WHC-11/18.GA/11). UNESCO. https://whc.unesco.org/archive/2011/whc11-18ga-11-en.pdf

UNESCO. (2015). *Policy document for the integration of a sustainable development perspective into the processes of the World Heritage Convention* (WHC-15/20.GA/INF.13). UNESCO. https://whc.unesco.org/en/sustainabledevelopment/

UNESCO. (2020). *Climate change now top threat to Natural World Heritage – IUCN report*. UNESCO. https://whc.unesco.org/en/news/2219

UNESCO. (2021). *Convention concerning the protection of the world cultural and natural heritage. Draft updated policy document on the impacts of climate change on World Heritage properties* (WHC/21/44.COM/7C). UNESCO. https://whc.unesco.org/archive/2021/whc21-44com-7C-en.pdf

UNESCO, ICCROM, ICOMOS, & IUCN. (2013). *World Heritage Resource Manual – Managing Cultural World Heritage*. UNESCO.

Vilanova, M., Lozano, J., & Arenas, D. (2009). Exploring the nature of the relationship between CSR and competitiveness. *Journal of Business Ethics, 87*(1), 57–69.

Von Schorlemer, S. (2020). The sustainable development goals and UNESCO: Challenges for World Heritage. In S. von Schorlemer, S. Maus, & F. Schmermer (Eds.), *UNESCO World Heritage and the SDGs – Interdisciplinary perspectives. Special issue 1 of the papers of the UNESCO Chair in International Relations* (pp. 17–30). Technische Universität Dresden, UNESCO Chair in International Relations. https://tud.qucosa.de/api/qucosa%3A71540/attachment/ATT-0/

Chapter 36
World Heritage Education and the Next 50 Years of the Convention: Current Pitfalls and Future Potentials of World Heritage Education

Claudia Grünberg and Klaus-Christian Zehbe

Abstract Since its adoption in 1972, the World Heritage Convention has proven to be a remarkable global success story. Despite educational programmes being anchored from the outset in the World Heritage Convention (1972, Art. 27), UNESCO's own 1994 World Heritage Education Programme (WHEP) has not been as successful as the World Heritage Convention itself. WHEP's lack of grounding in educational theory, practical implementation and links to current educational debates cast serious doubts on the programme's relevance for the next 50 years of the World Heritage Convention, potentially even threatening the continued protection of World Heritage properties. This chapter examines the status of education in the implementation of Article 27 of the World Heritage Convention and offers some foundations in educational theory to tap the potential of World Heritage Education (WHE) in the wider framework of Education for Sustainable Development (ESD) and Global Citizenship Education (GCEd).

Keywords World Heritage Education · Education for Sustainable Development · World Heritage Paradoxes · Minimal Morality · Modular Approach

---

C. Grünberg (✉)
Institute Heritage Studies at the Internationale Akademie Berlin (INA) gGmbH, Berlin, Germany
e-mail: gruenberg@inaberlin.org

K.-C. Zehbe
Institut Allgemeine Erziehungswissenschaft und Berufspädagogik, TU-Dortmund, Dortmund, Germany
e-mail: klaus.zehbe@tu-dortmund.de

M.-T. Albert et al. (eds.), *50 Years World Heritage Convention: Shared Responsibility – Conflict & Reconciliation*, Heritage Studies,
https://doi.org/10.1007/978-3-031-05660-4_36

36.1 The "Social Turn" in World Heritage and the Need for Education

Since its adoption in 1972, the *Convention Concerning the Protection of the World Cultural and Natural Heritage* – henceforth World Heritage Convention – has proven a remarkable, global success story. At the time of its adoption, many societies faced the destruction of heritage on account of rapid modernization and fundamental economic and social change, thus technical processes of identification, protection and conservation of heritage put forth by the World Heritage Convention were – and are still – much needed. In the meantime, a total of 1154 properties around the world have been listed and protected by the 194 State Parties to the World Heritage Convention, ranging from State Parties inscribing up to 58 properties on the list – such as Italy – to 27 State Parties ratifying the World Heritage Convention, but not inscribing any properties so far. The World Heritage Committee's Global Strategy (1994) takes cue from these global and structural imbalances, showing that emphasis needs to shift away from "identification, protection, conservation, presentation" to social processes, as it is reflected in the 2002 Budapest Declaration (UNESCO, 1972, Art. 4; World Heritage Committee, 2002). The Budapest Declaration aims to strengthen the World Heritage Convention by introducing "Four Cs": increasing credibility of the World Heritage List as well as enhancing capacity building and communication, thereby effectively conserving World Heritage. This "social turn" in World Heritage was reinforced in 2007 by the addition of the "Fifth C" – communities – aiming to enhance the role of communities in the implementation of the World Heritage Convention (World Heritage Committee, 2007). It is becoming clear that such a "social turn" with the enhancement of community involvement needs educational programmes to provide the necessary knowledge and skills, as required by the World Heritage Convention (1972, Art. 27) itself:

> States Parties to this Convention shall endeavour by all appropriate means, and in particular by educational and information programmes, to strengthen appreciation and respect by their peoples of the cultural and natural heritage defined in Articles 1 and 2 of the Convention.

Although educational programmes have been set down in the World Heritage Convention, systematic implementation started relatively late with the founding of the World Heritage Education Programme (WHEP) in 1994 – in line with the World Heritage Committee's Global Strategy. William Logan (2012, 21) comments on this late initiative:

> The early lack of urgency is perhaps surprising. If the world's cultural heritage and diversity are to survive beyond the current generation of decision-makers and professionals, it would seem critically important to enable the next generation – today's young people – to appreciate the value of maintaining heritage in its various tangible and intangible forms and to bring them into the work of heritage protection and maintenance.

However, it was not until the turn of the millennium that World Heritage Education (WHE) received political and institutional attention (Dippon & Siegmund, 2010, 36). The idea of education was subsequently set down in different formal instruments and procedures of the World Heritage process: since 2005, the Operational Guidelines differentiate between awareness-raising and education and define the latter as the "development of educational materials, activities and programmes" (UNESCO, 2019, Art. 219). Although education is not an obligatory part of a World Heritage nomination, it is highly recommended to integrate it into the mandatory management plan (UNESCO, et al., 2013, 125), and State Parties to the World Heritage Convention are asked to provide information about their educational strategy and activities in Periodic Reporting (UNESCO, 2015a, 5). Parallel to this formal embedding of education in World Heritage procedures, didactical and practical actions for WHE were implemented in the framework of WHEP. WHE is a complex construct, working on different levels and involving various actors across the globe. In the following section, we focus on WHE as it is implemented by the World Heritage Centre (WHC) in the framework of WHEP since there is currently no readily available data on the practical operationalization of WHE across the world in schools and at World Heritage properties.

36.2 The World Heritage Education Programme and Its Current Pitfalls

The aim of the WHEP is "to encourage and enable tomorrow's decision-makers to participate in heritage conservation and to respond to the continuing threats facing our World Heritage" (UNESCO WHC, 2021). The WHEP consists of five different elements: (1) World Heritage Volunteers, (2) Youth Fora, (3) Media/Communication Training, (4) a Cartoon Series and (5) one Educational Resource Kit for Teachers (UNESCO WHC, 2021; Vuijcic-Lugassy, 2018, 38 f.). Table 36.1 shows the status of the respective elements' implementation:

As can be seen in Table 36.1, WHEP's main elements are (1) the camps of World Heritage Volunteers and (2) the Youth Fora at the World Heritage Committee Sessions. Both take place on an ongoing basis and reach a significant number of young participants from all over the world. The action camps are hands-on and conservation-oriented, while the Youth Fora are more political, giving young people opportunities to meet heritage professionals and to raise their voice in the context of the World Heritage Committee Sessions.

The other three elements (3–5) aim to enable students and teachers to transmit the World Heritage idea either through trainings or educational material. These latter three all lack recent updates and ongoing activities: Media/Communication Training only took place in 2013, 2014 and 2015; the Educational Resource Kit was last updated in 2002 and the Cartoon Series *Patrimonitos World Heritage Adventures*

Table 36.1 Implementation of UNESCO's World Heritage Education Programme

Element	Type of activity	Last update/ Activity	Number/Reach	Frequency
1: World Heritage Volunteers	International action camps at World Heritage properties	2020	350 actions camps in more than 60 countries	annually since 2008
2: Youth Fora	Workshops in the context of the World Heritage Committee Session	2019	44 Youth Fora	one to three times a year since 1995
3: Media/ Communication Training	Trainings to promote capacity building of youth for heritage protection	2015	three different countries	once in 2013, 2014 and 2015
4: Cartoon Series	*Patrimonitos World Heritage Adventures*: animated short films written by students	2012	14 episodes	irregular, from 1995 till 2012
5: Educational Resource Kit for Teachers	*World Heritage in Young Hands*: publication	2002	one educational kit, translated into 40 languages	one-time activity

Note. Table prepared by the authors

in 2012. During the last 10 years, no trainings have taken place, and no new teaching material has been published.

Among the five elements of WHE, the Educational Resource Kit is still considered "the most important tool" of WHE (Vuijcic-Lugassy, 2018, 38). The Educational Resource Kit for Teachers aims to "impart to students the desire to know, cherish and act in favor of World Heritage *conservation*" (UNESCO, 2002, 18, *original emphasis*). It focuses on three objectives:

> to encourage young people to become involved in heritage conservation on a local as well as on a global level; to promote awareness among young people of the importance of our common World Heritage and of the UNESCO, 1972 World Heritage Convention; and to develop effective educational approaches and materials [...] in order to introduce WHE into the schools [...] in all parts of the world. (UNESCO Bangkok, 2010, 39).

The Educational Resource Kit offers a good starting point for making teachers and students aware of the topic. However, it is mainly focused on the actual hands-on process of conserving properties for future generations. Current challenges to World Heritage, such as reconstruction in conflict areas, terrorism, illicit trafficking and climate change, are not covered. Neither does it reflect the progress in teaching methods, nor does it consider links to more recent educational discussions in the field, like Education for Sustainable Development (ESD) or Global Citizenship Education (GCEd). With the adoption of Agenda 2030 in 2015, the importance of World Heritage and also education in and for sustainable development was brought to the fore. On the one hand, the protection of World Heritage was anchored in aim 11.4 of Agenda 2030, and, on the other, a sustainable development perspective was integrated into the World Heritage Convention by the eponymous Policy Document (UNESCO, 2015b). Changes in education for operationalizing Agenda 2030 were set down in aim 4.7, and UNESCO took a leading role in developing and promoting ESD.

While it can be assumed that there is a certain time lag in new topics and policies being integrated into current programmes, we found that a discussion of educational themes and policies – which do not focus exclusively on conservation – is not taking place among the professional World Heritage community. Our review of UNESCO's publications *World Heritage Review*, the *Paper Series* and *Resource Manuals* showed that there is no online publication on the topic WHE as of July 2021. UNESCO's publications largely focus on nomination, conservation and management of properties. Topics related to education – such as interpretation, communication, capacity building, community involvement, tourism or sustainable development – are addressed in several UNESCO publications, but a comprehensive educational approach based on either Art. 27 of the World Heritage Convention or Agenda 2030 cannot be found in these publications.

The lack of guidance on WHE is reflected in the most recent results of Periodic Reporting from Latin America and the Caribbean (LAC) (2013) and Europe (2015). A total of 60% of the World Heritage properties in the LAC region and 40% of the sites in Europe reported to have no educational programme or work on an ad hoc basis (UNESCO, 2013, 103; UNESCO, 2015a, 58). In Europe, 20% of the cultural sites reported having an educational programme, and another 40% reported having an educational programme, which only partially worked and needed improvement (UNESCO, 2015a, 58). The actors of the European World Heritage properties identified education as a potentially serious management issue, which would require attention by the World Heritage Committee (UNESCO, 2015a, 63).

Having examined the reports of the WHC on the implementation of the World Heritage Committee's decisions, we found that there is no operational action plan or strategy on WHE as of August 2021. The latest report prepared by the WHC for the 44th World Heritage Committee Session merely summarizes the different actions of the WHEP and mentions that WHE shall be given the highest priority for the 2022–2025 period (UNESCO, 2021, 18). Apart from this report, no decision of the World Heritage Committee on WHE exists. There are no references to Art. 27 in the preparatory documents for the World Heritage Committee whatsoever.

Concluding our analysis on WHE and WHEP, we see a fundamental lack of theoretical and practical guidance regarding new developments, which may even threaten the success of the World Heritage project in the coming years. Apart from the Youth Fora and the hands-on workshops of the World Heritage Volunteers, the overall concept of WHEP – as reflected in the Educational Resource Kit – seems to be frozen in time and stands isolated from recent developments in both fields of heritage *and* education. Although the potentials of merging WHE and ESD have been recognized, no conceptual basis has been developed to implement them in practice. In short, UNESCO has neglected the implementation of Art. 27 of the World Heritage Convention and WHE remains an educational niche concept, which seems to have no significance for the professional community. Nor does WHE play any significant role in ESD, UNESCO's most prominent educational approach. Realizing this, one must ask whether WHE is a priority for UNESCO after all.

36.3 The Two Paradoxes of World Heritage Education and Their Potential

Despite our critique of the implementation of WHEP, we consider WHE an important concept, which is grounded in the educational potential of the World Heritage properties themselves. It is frequently proposed by UNESCO, heritage professionals and academics that World Heritage properties are places of learning (Dornbusch et al., 2018; Ströter-Bender, 2010, 72; UNESCO WHC, 2021). The properties' Outstanding Universal Value (OUV) as well as their authenticity offer visitors "the opportunity [...] to engage in experiences that make them learn" (Van Lakerfeld & Gussen, 2011, 15) and "history to touch" (Schefers & Vieregg, 2010, 19). By personally exploring the properties, visitors can learn about history and values in the context of the property. Notwithstanding, there is a scarcity of scholarly literature dealing with the theoretical foundations of such learning processes and heritage education (Van Boxtel et al., 2016 2; Doorsselaere, 2021, 2). This is reflected in a corresponding lack of concepts in WHE. This lack of concepts has a profound influence on the implementation of WHE in practice: heritage education carried out in schools is most often limited to national or local history (Doorsselaere, 2021, 1) or arts education (Gesche-Koning, 2018, 9 f.), with the concomitant risk of limiting and instrumentalizing heritage in hegemonic narratives of cultural supremacy. The potential of heritage – and especially World Heritage – in and for education is thus not fully realized.

For WHE, this situation is aggravated by the paradoxical fact that World Heritage properties are always local properties, which are situated in territories of State Parties. Even though the World Heritage Committee recognized the current 1154 cultural, natural and mixed World Heritage properties (as of July 2021) for their outstanding value to all humanity, the global dimension cannot be fully assessed from a local perspective because it belongs to an "ideal" of human achievement, which is introduced by the World Heritage Committee in the nomination process. This paradoxical confusion of the local and the global, as well as the particular and the universal in World Heritage properties, throws light on the second paradox, the status of the relationship between the universal and the particular. This second paradox cannot be easily reconciled through normative intervention as Raymond Williams (1961/2011, 61) fittingly observes in the framework of analyzing cultural expressions:

> There is [...] the 'ideal', in which culture is a state in the process of human perfection, in terms of certain absolute or universal values. The analysis of culture, if such a definition is accepted, is essentially the discovery and description, in lives and works, of those values which can be seen to compose a timeless order, or to have permanent reference to the universal human condition.

In this "ideal" sense, the totality of World Heritage properties can be understood as a canon of unique properties, offering exemplary opportunities to discover the potential of human development. Such an "ideal" of universal value – while politically and practically desirable – is theoretically unsatisfactory, not least because it can be instrumentalized as a goal in political projects and hegemonic narratives. Without theoretical grounding, such "ideals" of the World Heritage Convention run the risk of arbitrariness and being subject to the political power plays and geopolitical ambitions of State Parties, seriously jeopardizing World Heritage as a project of the world community.

Consequently, from a general educational perspective, the notion of an "ideal" must be shifted away from hierarchical and normative goal orientations of human development towards the qualities of the development process itself and the historic nature and openness of this process (Benner, 1987/2015). Kant (1803/2007, 437 & 439), in his lectures on pedagogy (1803), already emphasized that human qualities can only come into existence through education, thus emphasizing the need for educational activities, which would be able to introduce desirable qualities into human development. Such educational activities – due to their necessarily historic and cultural nature as "social reactions to the fact that human beings are developing" – have to differ from place to place and in respect to the social norms and values of local communities (Bernfeld, 2012, 51). These paradoxes of WHE are shared to a bigger or lesser extent by other educational approaches aiming at a global context, such as Education for Sustainable Development (ESD) or Global Citizenship Education (GCEd).

In this paradoxical situation for WHE, we suggest that an approach of "minimal morals" may clarify the status of the relationship between the universal and the particular, the global community and the State Party and society and the individual, considering the diversity of human and biotic development processes (de Waal, 2013). According to de Waal (2013, 178), such morality may be based on the two principles of "not hurting" and "helping", which should govern both one-on-one relationships and group relations. While such minimal morals can be indefinitely extended into the future, they do not readily offer answers to questions on the quality of future developments. However, we hold that such answers can be derived from this minimal morality. From an educational perspective, this would have to include the right of the younger generation – and generations to come – to live their lives unhurt by the present, older generation. At the same time, it would require the present, older generation to help the younger generation not to fall behind what has been achieved by previous generations. Failure to do so would hurt the younger and coming generations without their knowing. Here education and heritage – and especially World Heritage – can play a crucial role because heritage properties manifest what may be possible to achieve in the future by showing what has been achieved in the past. This includes a non-positivistic, non-hierarchical, contra-factual dimension, opening up perspectives for transforming human thought and action without being governed by present conditions and needs. Consequently, a minimal morality would be able to reconcile the paradoxes of WHE, serving as a starting point to tap into the future-oriented potential of WHE.

There is significant future potential in WHE's connection to diversity and identity work. Heritage is actively used by people to construct their identities by transforming, adapting and re-using heritage for their own purposes, as well as transforming themselves in the very process (Van Boxtel et al., 2016, 2). This process is described by Laurajane Smith (2006, 1) as "heritage work". As a result, heritage becomes a meaningful element in the mosaics of people's identity and/or culture. The particular potential of WHE in the context of heritage work is to use World Heritage properties to create a local sense of belonging with a transnational or global outlook. Thus, WHE can serve as a starting point to learn about oneself, relate to others and gain an understanding of the diversity of identities and cultures. Consequently, the potential of WHE for young people lies not only in showcasing what former generations have achieved but also in enabling them to create a sense of belonging for themselves, to understand the present and to "consciously write their future" (Penna, 2018, 4).

In summary, the future potential of WHE does not lie in acquiring factual knowledge about World Heritage properties, their conservation or management, but in understanding how the properties came to be, their meaning for local and global communities and what opportunities they may offer for shaping the future. In the context of minimal morality, the potential of WHE can be mapped in terms of ESD. ESD aims to equip people with the knowledge and skills to sustainably shape their futures within their respective environments. In this sense, the aim of WHE is not to create conservation experts but to empower people to be "knowledgeable enough to make sound decisions about the preservation of their environment" (Penna, 2018, 7). Implementing a sustainable development perspective into WHE can thus provide occasions for acquiring and sharing key competencies and qualifications (Ströter-Bender, 2010, 13; Van Lakerfeld & Gussen, 2011, 9). Applying these considerations from the context of ESD to WHE opens World Heritage properties for educational processes, global learning and sustainability, by translating the slogan "act local, think global" into concrete educational experiences.

36.4 Perspectives on Improving WHE

This contribution focused on emphasizing the relevance and importance of WHE for sustainably safeguarding World Heritage today and in the future, on showing its current pitfalls and on tapping its potential for education. This last section provides an outlook on how WHE can be improved to better realize the potential of Word Heritage for education and vice versa. We see such improvements in mainly two interrelated dimensions: the organizational strengthening of WHE within the UNESCO system and increasing theoretical and practical efforts to conceptualize and implement WHE.

To close the current gap between the potential and the actual implementation of WHE, UNESCO and the WHC should actively prioritize WHE and the implementation of Art. 27 of the World Heritage Convention. Common decisions by the World

Heritage Committee, implementation strategies and action plans, as well as the allocation of staff and funds, are essential to this end. Making WHE a priority from 2022–2025, as set out in the report of the WHC at the 44th World Heritage Committee Session, can be a starting point to strengthen the role of WHE within the World Heritage Programme.

As has been shown in this contribution, the lack of grounding in educational theory and educational approaches is currently one of the main pitfalls of WHE. Unsolved paradoxes between the global and the local – as well as the universal and the particular – prevent WHE's wider application and the building of synergies between WHE, ESD and GCEd. Therefore, priority should be given to the development of theoretical foundations for WHE. With our minimal morality approach, we offer one such possibility. From our considerations on the cultural and historic nature of education processes, we recommend a modular approach to WHE, which is grounded in local cultures. Rather than updating the existing Educational Resource Kit for current needs, we hold that different modules are better suited to reflect upon the diverse educational settings and different actors in WHE across the globe. The different modules would be able to present a wide range of topics and approaches in WHE, while being flexible enough to integrate recent developments and debates in education and heritage. For example, such a modular approach could adapt the aims, methods and content of other educational approaches, such as ESD, to WHE.

Reflecting our analysis of paradoxes of World Heritage and practical experiences gathered through self-conducted intercultural education projects at European transnational World Heritage sites, the following perspectives seem essential to tap the future potential of World Heritage in education and vice versa (Institute Heritage Studies, 2021a, b). These perspectives build on topics and approaches to WHE, which the WHEP has not focused on so far. At the same time, they bring specific questions into focus which can be addressed by educational activities.

Perspective on the Global Story The global meaning of World Heritage sites manifests itself in the OUV; however, the OUV is presented in the technical language of the WHC. To be used in educational contexts, this language needs to be translated into narratives which connect the local and the global. The following questions are in the focus of this perspective: What is the global meaning of World Heritage sites? How are they connected across the globe? How do World Heritage sites inform about universal values and human and biotic development on this planet?

Perspective on Sustainable Development Goals (SDGs) World Heritage can contribute to the SDGs by showing sustainable development in the past and how World Heritage is presently threatened by unsustainable development. This perspective can also reveal conflicting goals – such as heritage conservation versus necessary heritage adaptation in the context of climate change. The following questions are in the focus of this perspective: How do World Heritage sites show sustainable livelihoods on this planet? What sustainable solutions to everyday problems are represented in their materials and processes? What kind of social processes contributed to the continued – and therefore presumably sustainable – safeguarding over the span of several hundreds of years?

Perspective on Transnational Educational Approaches The 42 inscribed transboundary World Heritage sites are nominated by at least two different State Parties and offer perspectives which transcend national borders, potentially opening up a global perspective. The following questions are in the focus of this perspective: What does this shared World Heritage mean for people of the different involved State Parties? What are differences and similarities in interpretation? How can the property be jointly safeguarded?

Perspective on Evaluation and Impact Assessment Evaluation and impact assessment of WHE can help to better understand the effects of educational activities, like volunteer camps, resource kits, workshops, media campaigns etc., on young people. The following questions are in the focus of this perspective: How did people's attitude towards World Heritage change through participation in WHE activities? Which measures were successful and which were not? What are indicators of successful WHE, and how can the impact be measured?

References

Benner, D. (2015). *Allgemeine Pädagogik: Eine systematisch-problemgeschichtliche Einführung in die Grundstruktur pädagogischen Denkens und Handelns* [General education: A systematic and historically-informed introduction into the structure of pedagogic thought and action] (8th ed.). Beltz Juventa (Original work published 1987).

Bernfeld, S. (2012). *Sisyphos oder die Grenzen der Erziehung* [Sisyphos or the limits of education]. Suhrkamp (Original work published 1925).

de Waal, F. (2013). *The Bonobo and the atheist: In search of humanism among the primates*. W. W. Norton and Company.

Dippon, P., & Siegmund, A. (2010). Der Bildungsanspruch von UNESCO-Welterbestätten – eine aktuelle Bestandsaufnahme im Spannungsfeld zwischen Welterbekonvention und lokaler Praxis. In J. Ströter-Bender (Ed.), *World Heritage Education. Positionen und Diskurse zur Vermittlung des UNESCO-Welterbes*. Tectum.

Doorsselaere, J. (2021). Connecting sustainable development and heritage education? An analysis of the curriculum reform in Flemish public secondary schools. *Sustainability, 13*(4), 1857. https://doi.org/10.3390/su13041857

Dornbusch, R., Hansell, F., & Manz, K. (2018). Communicating World Heritage. A task of the World Heritage Convention. In R. Dornbusch, F. Hansell, & K. Manz (Eds.), *Communicating World Heritage As task of the World Heritage Convention* (Series: Industrie Archäologie) (Vol. 19, pp. 24–27). Sächsisches Industriemuseum/IWTG TU Bergakademie Freiberg. https://tu-freiberg.de/sites/default/files/media/technikgeschichte-und-industriearchaeologie-3412/Sonstiges/welterbe_vermitteln_ein_unesco-auftrag_ia_vol_19_k_2018.pdf

Gesche-Koning, N. (2018). Research for CULT Committee – Education in cultural heritage. European Parliament. PE 617.486. https://www.europarl.europa.eu/RegData/etudes/STUD/2018/617486/IPOL_STU(2018)617486_EN.pdf.

Institute Heritage Studies. (2021a). *Our World Heritage – Mining Cultural Landscape Erzgebirge/ Krušnohoří.* https://heritagestudies.eu/en/world-heritage-mining-cultural-landscape-erzgebirge-krusnohori/

Institute Heritage Studies. (2021b). *Transboundary European World Heritage – A Topic for UNESCO Project Schools.* https://worldheritage-education.eu/en#start

Kant, I. (2007). Lectures on pedagogy. In G. Zöller & R. B. Louden (Eds.), *Anthropology, History, and Education* (pp. 434–485). Cambridge University Press. (Original work published 1803).

Logan, W. (2012). Cultural diversity, cultural heritage and human rights: Towards heritage management as human rights-based cultural practice. *International Journal of Heritage Studies, 18*(3), 231–244. https://doi.org/10.1080/13527258.2011.637573

Penna, K. N. (2018 June). Cultural heritage as an educational base for the traditional pillars of sustainable development [Paper Presentation]. In *6th International Conference in Heritage and Sustainable Development, Granada, Spain,* https://www.researchgate.net/publication/330881714_Cultural_heritage_as_an_educational_base_for_the_traditional_pillars_of_sustainable_development

Schefers, H., & Vieregg, H. K. (2010). World Heritage Education. Positionen und Diskurse zur Vermittlung des UNESCO-Welterbes: World Heritage Education. Studie zur Annäherung an eine Definition. In J. Ströter-Bender (Ed.), *World Heritage Education. Positionen und Diskurse zur Vermittlung des UNESCO-Welterbes.* Tectum.

Smith, L. (2006). *Uses of heritage.* Routledge.

Ströter-Bender, J. (Ed.). (2010). *World Heritage Education. Positionen und Diskurse zur Vermittlung des UNESCO-Welterbes.* Tectum.

UNESCO. (1972). *Convention concerning the protection of the world cultural and natural heritage.* UNESCO. https://whc.unesco.org/en/conventiontext/

UNESCO. (2002). *World Heritage in young hands. To know, cherish and act. An educational resource kit for teachers.* UNESCO. https://whc.unesco.org/uploads/activities/documents/activity-54-19.pdf

UNESCO. (2013). *WHC-13/37.COM/10A Final report on the results of the second cycle of the periodic reporting exercise for Latin American and the Caribbean.* UNESCO. http://whc.unesco.org/archive/2013/whc13-37com-10A-en.pdf

UNESCO. (2015a). *WHC-15/39.COM/10A Final reports on the results of the second cycle of the periodic reporting exercise for the European region and Action plan.* UNESCO. http://whc.unesco.org/archive/2015/whc15-39COM-10A-en.pdf

UNESCO. (2015b). *Policy on the integration of a sustainable development perspective into the processes of the World Heritage Convention.* UNESCO. https://whc.unesco.org/en/sustainabledevelopment/

UNESCO. (2019). *Operational guidelines for the implementation of the World Heritage Convention.* UNESCO. https://whc.unesco.org/en/guidelines/

UNESCO. (2021). *Report of the World Heritage Centre on its activities and the implementation of the World Heritages Committee's Decisions* (WHC21/44.COM/5A). UNESCO. https://whc.unesco.org/archive/2018/whc18-42com-5A-en.pdf

UNESCO Bangkok, Asia and Pacific Regional Bureau for Education. (2010). *A teacher's guide: Incorporating education for sustainable development into World Heritage education.* UNESCO Bangkok. https://unesdoc.unesco.org/ark:/48223/pf0000190006

UNESCO, ICCROM, ICOMOS, & IUCN. (2013). *Managing cultural World Heritage resource manual.* UNESCO.

UNESCO World Heritage Center. (2021). *World Heritage Education Programme.* https://whc.unesco.org/en/wheducation/

Van Boxtel, C., Gever, M., & Klein, S. (Eds.). (2016). *Sensitive pasts. Questioning heritage in education. Making sense of History* (Vol. 27). Berghahn Books.

Van Lakerfeld, J., & Gussen, I. (2011). *Acquiring key competences through heritage education.* Landkommanderij Alden-Biesen.

Vuijcic-Lugassy, V. (2018). Communicating World Heritage: The World Heritage Convention and World Heritage Education Programme. In R. Dornbusch, F. Hansell, & K. Manz (Eds.), *Communicating World Heritage as task of the World Heritage Convention* (Series: Industrie Archäologie) (Vol. 19, pp. 32–41). Sächsisches Industriemuseum/IWTG TU Bergakademie Freiberg. https://tu-freiberg.de/sites/default/files/media/technikgeschichte-und-industriearchaeologie-3412/Sonstiges/welterbe_vermitteln_ein_unesco-auftrag_ia_vol_19_k_2018.pdf

Williams, R. (2011). *The long revolution*. Parthian.

World Heritage Committee. (2002). *Decision: CONF 202 9 Budapest Declaration on World Heritage*. UNESCO. https://whc.unesco.org/en/decisions/1217/

World Heritage Committee. (2007). *Decision: 31 COM 13B The "fifth C" for "Communities"*. UNESCO. https://whc.unesco.org/en/decisions/5197/

Chapter 37
Young Professionals' Perspectives on World Heritage – Transformation from an Expert-Dominated Concept to a Project for the People it is Made for

Roland Bernecker, Juan Carlos Barrientos García, Elisabeth Korinth, Isabelle Rupp, Giulia Tomasi, and Klaus-Christian Zehbe

Abstract In this roundtable discussion, emerging heritage experts address their personal involvement with the World Heritage Convention, their assessment of achievements and failures of the 50 years of its history, and perspectives of future developments as seen by the younger generation. The discussion reveals a strong emphasis on more convincingly participatory procedures, community involvement, global equity and sustainable development. Heritage is what we take from the past to shape our future. From this conceptual stance, the emerging experts develop their claim of a far more substantial involvement of the younger not only in conceptual perspectivizations of heritage, but also in decision-making bodies. This would allow

R. Bernecker (✉)
Chair of Cultural Management, Brandenburg University of Technology Cottbus-Senftenberg, Cottbus, Germany
e-mail: bernecker@outlook.de

J. C. Barrientos García
European Heritage Volunteers – World Heritage Volunteers, Weimar, Germany
e-mail: office@heritagevolunteers.eu

E. Korinth
Blue Shield Germany, Berlin, Germany
e-mail: korinth@blue-shield.de

I. Rupp
World Heritage Studies, Brandenburg University of Technology Cottbus-Senftenberg, Cottbus, Germany
e-mail: ruppisab@b-tu.de

G. Tomasi
Italian Youth Association for UNESCO, Rome, Italy
e-mail: info@aiguofficial.it

K.-C. Zehbe
International Association of World Heritage Professionals (IAWHP e. V.), Cottbus, Germany
e-mail: info@iawhp.com

M.-T. Albert et al. (eds.), *50 Years World Heritage Convention: Shared Responsibility – Conflict & Reconciliation*, Heritage Studies,
https://doi.org/10.1007/978-3-031-05660-4_37

them to transform and shape the institutions they are working with to be sustainable, diverse, inclusive and future-oriented.

Keywords Heritage · Youth involvement · Governance · Sustainability

37.1 A Round Table with Young Heritage Experts Moderated by Roland Bernecker on June 22nd, 2021

Roland Bernecker After the conversations we had in preparing this meeting, I was impressed by your commitment, your projects and ideas, your energy. Let me come straight to the point and start with a simple question: Approaching the 50th anniversary of the World Heritage Convention, what is, for you personally, its most meaningful success, and in what respect do you feel frustration or criticism?

Juan Carlos Barrientos García I think the Convention, in itself, was one of the most important landmarks in the history of humanity, as it managed to bring together most nations to concur in the idea that cultural heritage and natural heritage do not belong to a single state or nation, but they are the shared heritage of humankind. I think this novel notion, which inspired and has remained the essence of the Convention, effectively erases the political borders between humans. It creates a sense of a global community where the Mayan pyramids in Honduras are also the heritage of the people in Nepal, or how the Great Barrier Reef in Australia is also my coral reef, not just a natural resource belonging to Australians. It belongs to all of us. It's a wonderful idea.

However, the Convention was designed to stand on the pillars of education, conservation and tourism. Tourism has become one of the main motivators nowadays, and, sadly, education is not playing the important role it should play. There is a competition to inscribe sites, and most of the efforts are put into these inscriptions. But once an inscription is achieved, instead of considering it as the beginning of the work, it is often where it ends. Education must be the main force driving the World Heritage project.

Elisabeth Korinth I think there are a lot of achievements that are worth celebrating, like building an international community around heritage protection, subscribing to a set of shared principles and values—and having over a thousand World Heritage sites protected. But what I think is maybe a more subtle achievement, for me as a heritage professional, is that the Convention itself has fostered a global discourse on heritage. It has also enabled us to build a network of heritage professionals for exchange. I think this is something that we are all part of, and that is something very valuable and very precious to keep.

At the same time, we can identify a range of failures with the Convention, such as the politicized decision-making machinery, an unbalanced World Heritage List, as well as the lack of sufficient funding. I think we should not forget that the Convention itself was originally based on a technocratic conservation agenda that promotes heritage as something static, material. In the past decades, there have been numerous attempts to change this. The concept of heritage has developed. But at the core, the Convention itself remains focused on physical substance. I think this still causes problems that can harm heritage as well as communities.

Isabelle Rupp One of the greatest achievements of the World Heritage programme is the positive public image it has fostered for heritage protection. My own motivation for studying heritage at BTU was the focus on World Heritage. I was inspired by UNESCO's work and the purpose of the World Heritage programme. UNESCO's achievements with World Heritage contribute to reinforcing what I would consider the programme's most glaring weaknesses. The ideas it espouses and its modalities of operation have become somewhat reified and sealed to change. The way heritage is defined has become institutionalized to the point that dynamic reform is less discernible than it could, and probably should, be.

Giulia Tomasi The Convention itself is fantastic; it created a sense for the concept of a world community with a responsibility to take care of cultural and natural heritage of outstanding universal value. It really gives this sense of belonging by overcoming cultural, ethnic and geographical boundaries. It does, however, exclude the younger generations. We seem to be kept out of the active process of managing, of preserving heritage sites. In Italy, the complex national and local governmental structures behind heritage management tend to exclude young people. We protect heritage for the world and for future generations. But young people are not perceived as actors in contemporary processes of cultural and heritage site management. We want to be involved in the responsibilities expressed in the World Heritage Convention and contribute to the decisions that need to be taken for the future. World Heritage projects need to open up to the creativity and the ideas of the young.

Klaus-Christian Zehbe I also think one of the biggest achievements of the Convention is offering a political arena across nations, which allows the mediation of conflicts on a global scale. Having said that, the Convention adapted to current challenges, which is also a major achievement. It is still with us after 50 years and has not been taken away. However, it also carries the legacy of being conceived in the Global North, and maybe one of its biggest shortfalls is a very Eurocentric division of nature and culture, as well as its failure to give voice to Indigenous Communities and their ways of looking at the world. As the Convention has been with us for 50 years, the institutions dealing with it have developed their own lives. I think this process has contributed to political power games over which sites might be inscribed and which not, and what kind of trade-offs are being made. This is one major threat. In terms of youth involvement, we have to consider that the young generation has its own right to devise its future. Rarely, though, do we have an idea of what the future can be. Therefore, I think heritage is an important element for the young generation to understand what might be possible to achieve. There is no true measure of saying where humanity is going. Heritage is a very important touchstone for the young to define themselves in relation to the future. Where do the older generations stand, and where are we going?

Roland Bernecker I would like now to come to the question of youth involvement, a point you all have mentioned. Giulia, you have adopted a "youth manifesto" with the Italian Youth Forum in Parma in March this year. I have seen it, and I think it is an impressive document. Let me come back to what you said about your involvement as a younger heritage expert in cultural policies: Is your generation addressed? Is there an interest to capture the perspective of your generation, and what would be your main request in this regard?

Giulia Tomasi As a Youth Association for UNESCO, we really believe that education and spreading knowledge of UNESCO's values is an essential aspect. So, for us, this is the real starting point and plan of action, especially for creating a new active citizenship. We have an educational program that reflects the importance of global agendas and topics but also provides a sense of empowerment. We asked students to create something for their peers, for their community. Through this programme, we aim to make sure that young people discover how their voices can be heard in a multi-stakeholder system, a complex system such as culture and World Heritage. We have over 300 members, and we are organized throughout Italy in regional teams. We create programmes and concepts developed across regions in Italy and try to reach as many students as possible. I firmly believe that education is the starting point. We actually go into schools and make young people acknowledge that they have the power and potential to contribute to important causes. By doing this, society can benefit as a whole. Education, therefore, is the major asset of the Italian Youth Association for UNESCO.

However, as previously said, the job market always seems a little behind. So, this year we started another approach, which is organizing technical roundtables composed of young experts, members of the association. Together we drafted strategic proposals to be submitted to the government. This year we worked specifically on four technical tables: school, culture, food and cities. We proposed some operative approaches for the upcoming post-pandemic recovery fund, which were well received by the Italian government. This is an example of how we have tried to achieve youth participation and community outreach, simply by empowering them. We focus on education and on giving space to young members of the association to reach out to the government through us, with plans of actions regarding culture, school and education, and other important topics. Obviously, there should be more national and international frameworks for official youth participation at decision-making tables. This should be a methodology to be applied generally.

Roland Bernecker Elisabeth, how serious is the commitment which is asked of young people? You are working on a concrete project with the Syrian community in Germany. This is a particular challenge in the current context. What are your experiences with the involvement of these exiled communities, and what can you do to help them maintain links to their own heritage?

Elisabeth Korinth I would like to tackle this question first from a conceptual point of view and then look at it again from the perspective of Blue Shield Germany, because if we want to find out how to improve community involvement, we need to better understand who our community actually is. The World Heritage system has increasingly addressed the importance of community involvement, but it has not given a clear definition of community, and it somehow assumes that there is a community that can be clearly defined and that it can be included in a management system. But if we really go into the technicalities of the question and ask, "who belongs to a community?" which necessarily also raises the questions of, "who doesn't belong to a community, and who has the right to decide that?", then we are faced with very difficult questions that confront us with the controversial and exclusive notion of community they can be entangled with. In order to improve participation, we need to look at criteria of inclusion and diversity, which we are, of course, doing in the Syrian Heritage Initiatives of the Museum for Islamic Art. We do not need to look at the people we are already including, but the ones we are excluding.

I would also like to shortly look at the notion of community as a fixed and harmonic geographical group, which is problematic and becomes apparent when you look at community involvement at World Heritage sites in conflict. Conflict is characterized by a division within a community or between communities. Long-term armed conflicts, in particular, are characterized by movement and by a very dramatic change of population as people are being displaced. There is a fluctuation of local communities, disrupting the social life that was there before the conflict. From a heritage perspective, we see a loss of expertise on the ground that needs to be responded to. So, if we want to improve community involvement at World Heritage sites in conflict zones, I think we need to ask ourselves: "who belongs to the community that is not directly at the site?"

I want to stress this issue of displaced communities here because it is no longer a temporary phenomenon. In 2020, UNHCR estimated the number of forcibly displaced people to be 82 million, among them many children and young adults. These people have the right to belong to a community, but they are usually not recognized by participatory programmes and community involvement. They are important carriers of knowledge and expertise needed to protect World Heritage sites. They are important advocates for their heritage worldwide, and they can play a vital role in safeguarding measures. In the case of Syria, we have seen numerous initiatives and grassroots community projects that have been initiated by Syrian heritage professionals from abroad, which have played a key role in documentation, in raising awareness about heritage under threat, and also in supporting local communities on the ground. If the World Heritage system can strengthen these ties and networks and can include these groups of people, especially youth, in a way that they are not competing but complement each other, we can move away from a safeguarding effort focused on the physical matter to a holistic and sustainable approach of community involvement, which includes and values the most vulnerable parts of the community that are, of course, scattered around the world.

Roland Bernecker Klaus, you have studied World Heritage at BTU and are now focusing on educational sciences. Is World Heritage a political framework for global citizenship education? What is the role of heritage in our modern concepts of education? Can it play a role in educating global citizens?

Klaus-Christian Zehbe Actually, it should. From an educational perspective, we have a double responsibility. Youth, in one respect, has the responsibility not to fall back behind what has been achieved by previous generations. The older generation, on the other hand, has the responsibility to give access to these stores of knowledge. Heritage can play an important part in opening that store of knowledge and providing a benchmark. There is maybe a third responsibility. It is sometimes framed as a responsibility towards future generations, which should have the right to enjoy living on this planet in the same way as the current younger and older generations. Heritage plays an important part in all of this because it intersects the younger and the older generation and can also provide valuable insights into the processes which have been relevant in the past and provide a perspective for how things might develop. World Heritage is crucial for these responsibilities, in terms of sustainability and reconciliation between groups but also reconciliation with the mistakes and errors of the past. This might not be present in all the sites, but because the sites have been preserved by the older generation they are important enough to provide some kind of touchstone—where the different generations can convene, transmit and discuss their different perspectives. As Giulia said, the young generation has to have the right to inscribe their own meanings and stories in this heritage because they are going to carry it into the future.

Roland Bernecker How do students at universities, how does academia contribute to making better use of heritage? Do you think that university gives you sufficient tools to understand these complex issues and to get involved?

Isabelle Rupp A lot of studying is theoretical; we don't get as many practical tools as we probably should. We should learn more through doing or having a job in the field. I think we're exposed to a lot of ideas, and so we can pick up on things that, further down the line, we will then have the tools to actually do something about or engage with in a way that makes sense professionally. But it is something that still needs to be improved.

Roland Bernecker Juan Carlos, you do not speak on behalf of *World Heritage Volunteers*, but you speak as a manager involved in the volunteer projects. What is your experience with the involvement of young people in these activities?

Juan Carlos Barrientos García This is a good follow-up question to what Isabelle said. It has become the role of organizations like the one I am working with to provide these opportunities for young heritage professionals to explore the field. The World Heritage Volunteers initiative, for example, is a worldwide campaign started by UNESCO. It seeks to foster youth commitment towards World Heritage and to provide projects that include concrete awareness raising and hands-on activities at the sites and the opportunity to interact with people who come from the same background, to make practical experiences and nourish enthusiasm for heritage. The campaign is organized and implemented on the ground by different organizations. The campaign creates a link between them. I have been involved with the World Heritage Volunteers campaign for several years now through the organization European Heritage Volunteers. European Heritage Volunteers is the leading organization implementing educational heritage volunteering initiatives in Europe, uniting a wide network of international partners, local heritage activists, heritage institutions and community stakeholders at heritage sites; all coming together for organizing and implementing volunteer projects at World Heritage sites in Germany and around Europe. There is an entire network of partner organizations supporting European Heritage Volunteers with volunteering activities all over Europe, not just at World Heritage sites.

I was first a participant and then became a group coordinator for projects linked to the World Heritage Volunteers campaign, organized by European Heritage Volunteers. Today, I am a concept developer and coordinator for the projects of European Heritage Volunteers. European Heritage Volunteers' programme is focused on providing opportunities for hands-on activities, addressing young heritage professionals and heritage enthusiasts through volunteering projects and training courses. They create a space for real hands-on experiences in the field that is perhaps lacking in universities. The programme develops a well-structured educational experience and provides a platform to connect communities and local stakeholders with their own heritage through their collaboration in these volunteer projects, both at World Heritage sites as well as other heritage sites.

The volunteering projects we create are always centred on different themes and heritage topics, including the impact of climate change on heritage sites, reconstruction of elements, interpretation of heritage, traditional skills, documentation of heritage, site management, conservation work on neglected structures, and other themes. The participants get real immersion in the heritage field. They widen their perspectives of heritage through an encounter with heritage in a well-organized cultural and educational experience.

These seminars and training courses contribute to the dissemination of the values of heritage and the World Heritage Convention. I am very proud to be a part of this organization. Through the work of this organization, the ideas that inspire the World Heritage Convention are put into practice. To continue to support a wonderful initiative like the World Heritage Volunteer initiative and also to support dynamic regional organizations like European Heritage Volunteers is actually a practical way to involve young people and communities with World Heritage.

Roland Bernecker What you do is a very good example of the practical involvement of young people, with their passion and their interest in the real thing. The point that still intrigues me is the political dimension of involvement, which is very ambitious. What needs to be done to get a serious involvement of young people in the policy side of things, an involvement that goes beyond rhetoric declarations?

Giulia Tomasi It's a very complex issue. The political territory is intricate and complex to begin with. As previously said, having a framework—an international framework—would be a start, a framework in which young people are invited to technical tables, not because they organize them on their own and then present what is done to the government, but officially, as a side arm of the government itself. Officially recognizing this process when it comes to culture and sustainability is important. For the moment, young people should be provided with the tools to develop something on their own and present it. Make them understand that they can do this and have the power and the knowledge to do it and be part of creating something that is better for the future. For the moment, we need to start on our own and actively promote and network with each other, especially among youth associations, and then propose something until we will have a real space on the stage for decision-making.

Roland Bernecker There are a few questions from the audience. I'll address the first question to Klaus. It's more a comment: The Eurocentric silos (nature, culture; tangible, intangible) are not simply in conflict with Indigenous communities. They are constructs that do not fit well with most of the Global South and non-urban regions of the Global North. Would you like to comment?

Klaus-Christian Zehbe Unfortunately, it seems that these attempts to address the dichotomy between culture and nature, to value non-Western concepts, are again led by Western organizations which operationalize them. These silos were meant to address the problems. However, they seem to be aggravating them now because these institutions all work in different ways. The problem is, more or less, trying to find local ways of involving people and broadening horizons to include voices that are largely unheard. I think that might be a way of strengthening civil society and forming coalitions to develop political pressure in order to have an impact on state-level institutions, such as ministries and intergovernmental agencies.

Roland Bernecker A question to Juan Carlos: Today's level of tourism was not anticipated in the 1972 Convention. It did not exist at that time; it arrived later and commodification likewise. This is linked to your criticism of the focus on tourism and commodification.

Juan Carlos Barrientos García The Convention was not originally conceived to be focused on tourism. It was built with a focus on education and on awareness raising for the protection of World Heritage. This evolved on the side and has become one of the main motors right now of the World Heritage brand. In its essence, the Convention was conceived to promote education about protection and the necessity of protecting heritage. We should return to this focus. It is why and what World Heritage was created for.

Roland Bernecker The next comment is for Elisabeth. The notion of "homogenous" is top down, artificial and time dependent, fluid in both space and time. This is a remark to your intervention.

Elisabeth Korinth I cannot agree more. That is what I was trying to point out by emphasizing that it is a problematic notion. It is a concept; communities aren't a thing. Communities are complex. They are constantly changing, and they are deeply intertwined. Sometimes members of a community may not even like each other, so we need to better understand what we mean precisely when we speak of community involvement. I agree totally with the point made in the comment.

Roland Bernecker A last question to Isabelle and Giulia. How should young people be educated, when, where and by whom, to be capable of receiving and using the messages coming from various types of heritage?

Isabelle Rupp If there was a stronger emphasis on heritage interpretation, we would need less of a formal education process. People would be engaged. These places would matter to them, and they could be involved in the interpretation. If we do not look at it as fixing something, a deficit, but rather as if these places were theirs from the beginning, then the education process wouldn't necessarily be formal.

Giulia Tomasi There are a lot of associations like the Italian Youth Association. Juan Carlos was talking about how education has been prioritized by youth organizations or other independent organizations. I think there are exceptional initiatives at universities. So, a lot can be improved, but a lot is already being done. Specifically, there could be more attention to what civil society is. Culture in a broader sense and some aspects of the Convention could be studied in schools as part of creating a global community. This would be essential to learn to care for each other beyond geographical or ethnic boundaries. There is too much dispersion in channelling these competencies. When people specialize and learn about heritage, where can they apply it? This often is a waste of energy.

Roland Bernecker A 50th anniversary is an occasion to reflect on a perspective for the 50 years to come. What would be, very briefly, your main expectation, your hopes or requests for the future of the World Heritage Convention?

Klaus-Christian Zehbe To put it like the Lakota: We should not think of just 50 years, but for the coming seven generations. That would make some 200 years to live peacefully and sustainably on this planet.

Isabelle Rupp It is time to reconsider, or at least reconfigure, some of the core concepts that were introduced and defined in the Convention and through its implementation but have since become problematic or ineffective. The foundational concepts of the Convention need to be updated to reflect contemporary understandings of heritage and also the paradigm shifts happening in the broader heritage sector.

Juan Carlos Barrientos García The message I would like to leave people with is that there are organizations like European Heritage Volunteers, providing and trying to create platforms for youth to get educated and expand their views on heritage, which should receive stronger support from UNESCO. More support is needed to help these organizations grow and continue to provide platforms for youth and young heritage professionals.

Elisabeth Korinth I think we all agree that there is a lot of opportunity for growth, and I hope we will find solutions for the many challenges that World Heritage is facing today, including climate change and rising conflicts. As Birgitta Ringbeck has pointed out earlier, the World Heritage Convention gives us a framework, and the system addresses all these issues. It's up to us as heritage professionals to create political awareness and also to create awareness among the youth about the importance of this valuable Convention. My hope and request for the next 50 years as a heritage professional would be, to bring it to a personal level, that we learn how we can be part of the solution. We should learn to enable communities to be involved so that not only my heritage is protected, but other peoples' heritage is protected, too. That we do our very best to transform and shape the institutions we are working in to be sustainable, to be diverse, to be future-oriented and inclusive.

Giulia Tomasi This has been a wonderful meeting. I hope we can find ways to collaborate further, which would be in the interest of everyone. My hope for the global scale is that the Convention will continue to promote this incredible sense of belonging and caring for World Heritage. On a more local scale, international protocols and methodologies need to be better enforced by promoting collaboration and alignment with the Convention. We need to shorten the distance between what is internationally perceived as World Heritage and what is actually happening on the local level and its effect on the people. Although we have a role in this, and we actively want to improve the world in which we live, we need support from the political level and from institutions like UNESCO to be able to work successfully in this sense.

Roland Bernecker Thank you, Giulia, Elisabeth, Juan Carlos, Isabelle and Klaus, for sharing your thoughts and also for the work you are doing in the organizations you are engaged with. And, finally, thanks for your dedication to the cause of World Heritage.

Chapter 38
Outlook

Marie-Theres Albert, Roland Bernecker, Claire Cave, Anca Claudia Prodan, and Matthias Ripp

Abstract The chapter "Outlook" brings together the main messages of this book; formulates concluding comments and reflects on the way forward. Out of the many conflicts affecting World Heritage, the chapter highlights some, which appear as obstacles that must be overcome for a sustainable protection. The unequal geographical distribution of World Heritage properties and of the decision-making bodies; and the difficulty to reconcile economic interests with conservation and development needs are two examples. Further examples refer to the discrepancies in the interpretation of the meaning of World Heritage between experts and the civil society; and the climate and biodiversity crises, which require full participatory and inclusive approaches that integrate culture and nature protection. In light of these examples, the chapter concludes by underlining that the future of the World Heritage Convention can only be envisioned if such challenges were confronted and resolved.

Keywords Future perspectives · Development · Sustainability · Equality · Justice · Participation

M.-T. Albert (✉) · A. C. Prodan
Institute Heritage Studies at the Internationale Akademie Berlin (INA) gGmbH, Berlin, Germany
e-mail: albert@inaberlin.org; prodan@inaberlin.org

R. Bernecker
Chair of Cultural Management, Brandenburg University of Technology Cottbus-Senftenberg, Cottbus, Germany
e-mail: bernecker@outlook.de

C. Cave
School of Archaeology, University College Dublin, Dublin, Ireland
e-mail: Claire.cave@ucd.ie

M. Ripp
Organisation of World Heritage Cities, Regensburg, Germany
e-mail: ripp.matthias@regensburg.de

M.-T. Albert et al. (eds.), *50 Years World Heritage Convention: Shared Responsibility – Conflict & Reconciliation*, Heritage Studies,
https://doi.org/10.1007/978-3-031-05660-4_38

Heritage creates identity and the destruction of heritage destroys identity. Therefore, individuals and societies are responsible for the sustainable safeguarding of their heritage. It is this message with which this book began and with which it ends. It is a message that relates to all of our heritage and to World Heritage in particular. And it is the message that legitimises the criteria for designating heritage as World Heritage, namely the so-called "Outstanding Universal Value". It is also the message on the basis of which the standard measures for protection and use of World Heritage have been established. And last but not least, it is the message that can be used to verify whether and in what way the social, political, ecological, cultural and economic goals of the World Heritage Convention (UNESCO, 1972) have been met in the 50 years of practical application.

Fifty years of implementation of the World Heritage Convention – together with other heritage conventions, which have emerged in the meantime – have ensured that its fundamental message has been received worldwide. In practice, implementation is expressed by the list of 1154 World Heritage sites currently inscribed and by 1719 sites on the so-called Tentative Lists of the 194 signatory states to the World Heritage Convention (all figures are as of December 2021). However, it is also expressed in a growing list of heritage sites in danger, currently 52 sites, and in the fact that criticism of the inscription criteria and their implementation practices is becoming louder and more public (see Alexander, 2021; also Schwiering, 2021).

The above-mentioned numbers of signatory states to the Convention, of inscribed sites as well as sites on the national Tentative Lists prove the quantitative success in the 50-year practice of applying the Convention. In this respect, success can be clearly stated. This is true even if, as was also formulated in the introduction to this book, the worldwide geographical distribution of the sites shows a high degree of Eurocentrism, or as the geographical participation in the political decisions of the General Assembly of the World Heritage Convention shows a Eurocentrism and Asiacentrism (see Chap. 2 by Birgitta Ringbeck in this publication).

The message of this book was developed on the basis of its editors' reflections on the mandate of such a Convention in view of its 50th anniversary and UNESCO's commitment to implement the 2030 Agenda. It also emerged from a critical reflection on its successes and failures. On the occasion of the Convention's 50th anniversary, we wanted to know whether and how the global community has aligned the changing social framework with the Convention's goals.

Considering the current living, working, recreational spaces, which have developed through political, social, cultural, ecological and economic processes worldwide and have affected our heritage we have identified 6 areas of conflict of and for World Heritage. These areas have been discussed and reflected by a total of 61 authors from 28 countries. They form the core of this book. The authors have formulated a variety of approaches to dealing with conflicts. Additionally, central themes that are important for anchoring the ideas and goals of the Convention, especially in a changing world, were presented in part III. There are categories such as responsibility, reconciliation, sustainability, education and diversity, which are important for a further sustainable implementation of the Convention and are conveyed in particular through and with the voices of "young professionals" in our book.

A few aspects remain to be pointed out in the concluding comments. On the one hand, there is the need to address the unequal geographical distribution of inscribed World Heritage Sites, which has been demanded in a multitude of resolutions since the 1990s and has been tackled with diverse proposals. It is the geographically, culturally, professionally and gender-equitable composition of the various decision-making bodies of UNESCO, which has also been demanded for years. Adaptation in the implementation of the Convention, demanded by means of the principles of equality, justice and sustainability, would, for example, also allow the repeated criticisms of the increasing quantitative orientation of the Convention to be put into perspective.

Quantity alone is not the problem. The emphasis on inscription and the annual addition of more and more sites on the list are often based on economic interests. A focus on economic values without reference to the Convention's wider mandate contradicts its spirit. For example, nominations of World Heritage sites are increasingly justified by tourism and the associated potential for economic development. In many cases, the resulting damage to World Heritage is ignored or accepted. The fact that economic interests only emerged with the establishment of the World Heritage Convention and that economic development policies have become as important as encouraging people to identify with their heritage can be seen as a product of the Convention's 50 years of implementation. This should be cause for a critical examination of this approach.

However, not every economic development of sites, communities, cultural landscapes or regions that are made possible by World Heritage designations should be assessed negatively per se. On the contrary, the identification of people with their sites is fundamental to the sites' preservation and thus fundamental to the preservation of heritage, especially in these times of massive upheaval and rapid change. People identify with their locations and with their sites if they offer them quality of life. This includes an economic quality. And that is exactly what tourism can contribute to.

This is the crucial challenge, namely developing models that combine economic development with sustainability and responsibility. For tourism, as for many other uses of World Heritage sites, this means developing alternative models. Guidelines and experience are available, for example, in Günter Faltin's entrepreneurship concept, which he developed and successfully implemented in the 1990s as part of an economically oriented but developmentally just and relevant tea campaign. With an application of approaches based on entrepreneurship, economic use of World Heritage is brought together with local/regional or cultural/nature or object-related development and thus also fulfils the idealistic goals of the Convention (Faltin, 2013).

The quantification of heritage in an economic interest and its contradicting developments does not only concern tourism, but also the growing rate of urbanisation on a global scale. It is posing threats on the one hand. On the other hand, heritage sites in an urban setting represent a great potential and resource to stimulate sustainable development and improve the quality of life for local communities. Especially the focus on the "Historic Urban Landscape Recommendation" (HUL), which was adopted by UNESCO`s General Conference already in November 2011 was a starting point for a holistic interpretation of the meaning of heritage for people (UNESCO, 2011). The application of HUL initiated a new understanding of how the different elements and entities of heritage are interconnected and relevant concerning current challenges like for example climate change and urban resilience. The topics addressed in this book will therefore serve as a reference and inspiration for World Heritage Cities to contribute to a new urban reality that the citizens are thriving for. World Heritage Cities are organised within the Organisation of World Heritage Cities (OWHC) (https://www.ovpm.org), which was also a cooperation partner in the development of this publication.

Another aspect that is relevant for the outlook is the increasing difference in the interpretation of the meaning of World Heritage between experts on the one hand and the civilian population on the other. This is due to the fact that communicating the significance of heritage has not been the focus of its representatives. However, in view of the dangers to which World Heritage is increasingly exposed, its communication is becoming more and more important. The significance of World Heritage for people and their societies must therefore be communicated in a holistic and sustainable manner. This concerns formal as well as non-formal, official as well as unofficial processes, which we have discussed earlier in this publication.

In light of climate change and the biodiversity crisis, a full participatory and inclusive approach is also emphasised in conservation strategies for nature friendly, sustainable, and climate-resilient development. Nature-based solutions (NbS) has emerged as an approach which can help society to move away from the destructive global economic model centred around GDP and infinite growth, to one where social and ecological well-being are the decisive measures that identify a healthy economy. As defined by IUCN "*Nature-based solutions are actions to protect, sustainably manage and restore natural and modified ecosystems in ways that address societal challenges effectively and adaptively, to provide both human well-being and biodiversity benefits*" (IUCN, 2020, 1). The IUCN Global Standard for NbS is available for all interested parties including policy makers, heritage practitioners and local communities to have a common understanding and interpretation of the NbS concept (IUCN, 2020). Also recognising the urgency of the need to address society's largely dysfunctional relationship with biodiversity, a Working Group has been established under the Convention on Biological Diversity to develop the Post-2020 Global Biodiversity Framework. This aims to realise a hugely ambitious plan "to ensure that, by 2050, the shared vision of living in harmony with nature is fulfilled" (Working Group, 2021, Annex A1). The intention is for this Framework to be used under all the biodiversity-related conventions, including the World Heritage Convention (Working Group, 2021, Para. 5a). In communicating the role that World Heritage can play in valuing, conserving, restoring and equitably and sustainably using biodiversity, the World Heritage Convention can support the effort needed to enable governments and society to implement changes for a sustainable future.

In this outlook, we would like to refer to an international initiative that is explicitly future-oriented as a constituent part of the communication of heritage. This is the initiative "Our World Heritage", which was launched by experts from the broad context of the World Heritage Convention in 2020. Organised as a digital and global discussion forum, its message is: "Join us – citizens and professionals – as we mobilize to renew and reinforce heritage protection for the next 50 years. Without action, the legacy of the past will not be here for tomorrow's generation" and thus it contributes to the sustainable anchoring of the Convention. (www.ourworldheritage.org).

In summary, the success of the World Heritage Convention, which has been implemented for 50 years, can be considered unique and positive. Nevertheless – as has been formulated in a number of contributions – contradictions to the intended goals can also be observed in its implementation. When it comes to formulating goals for the future, it is long past time that such contradictions were confronted and resolved. To do this, one must first identify causes. With this publication we have begun to do so. We assume that this discourse will continue.

References

Alexander, M. (2021, August 2). Weltkulturerbe: Nur Italien vor uns. [World Cultural Heritage: Only Italy before us]. *Frankfurter Allgemeine Zeitung*. https://www.faz.net/aktuell/feuilleton/debatten/matthias-alexander-ueber-das-deutsche-weltkulturerbe-17464448.html

Faltin, G. (2013). *Brains versus capital*. Stiftung Entrepreneurship.

IUCN. (2020). *Guidance for using the IUCN global standard for nature-based solutions* (1st ed.). https://portals.iucn.org/library/sites/library/files/documents/2020-021-En.pdf

Schwiering, K. (Director). (2021). *Welterbe in Gefahr*. [World Heritage in Danger]. [Film]. Kobalt for ZDF/3sat. https://www.3sat.de/kultur/kulturdoku/welterbe-in-gefahr-100.html

UNESCO. (1972). *Convention concerning the protection of the world cultural and natural heritage*. UNESCO.

UNESCO. (2011). *Recommendation on the historic urban landscape, including a glossary of definitions*. UNESCO.

Working Group. (2021). *First draft of the post-2020 global biodiversity framework, convention on biological diversity*, Open Ended Working Group on The Post-2020 Global Biodiversity Framework, Third meeting, 23 August – 3 September 2021, CBD/WG2020/3/3. https://www.cbd.int/doc/c/634e/15be/78d817a6d4ef4520408ed501/wg2020-03-1-01-en.pdf

Index

M.-T. Albert et al. (eds.), *50 Years World Heritage Convention: Shared Responsibility – Conflict & Reconciliation*, Heritage Studies,
https://doi.org/10.1007/978-3-031-05660-4

F

G

H

I

N

O

The manufacturer's authorised representative in the EU is Springer Nature Customer Service Centre GmbH, Europaplatz 3, 69115 Heidelberg, Germany. If you have any concerns regarding our products, please contact ProductSafety@springernature.com

Printed and bound by CPI Group (UK) Ltd, Croydon, CR0 4YY
30/09/2026
02225317-0016